W9-BFC-708

THE TOTAL SYNTHESIS
OF NATURAL PRODUCTS

The Total Synthesis
of Natural Products

VOLUME 4

Edited by

John ApSimon

Department of Chemistry
Carleton University, Ottawa

A WILEY-INTERSCIENCE PUBLICATION

JOHN WILEY & SONS,　New York　•　Chichester　•　Brisbane　•　Toronto

Library of Congress Cataloging in Publication Data:

ApSimon, John.
   The total synthesis of natural products.

   Includes bibliographical references and index.
   1. Chemistry, Organic–Synthesis.   I. Title.
QD262.A68   547'.2   72-4075
ISBN 0-471-05460-7 (v. 4)

Printed in the United States of America

10 9 8 7 6 5 4 3 2 1

Contributors
to Volume 4

Yvonne Bessière, University of Lausanne, Lausanne, Switzerland
Jasjit S. Bindra, Pfizer Central Research, Groton, Connecticut
Kenji Mori, University of Tokyo, Bunkyo-ku, Tokyo, Japan
Raj K. Razdan, SISA Incorporated, Cambridge, Massachusetts
Alan F. Thomas, Firmenich SA, Geneva, Switzerland
Wendell Wierenga, The Upjohn Company, Kalamazoo, Michigan

Preface

Throughout the history of organic chemistry, we find that the study of natural products frequently has provided the impetus for great advances. This is certainly true in total synthesis, where the desire to construct intricate and complex molecules has led to the demonstration of the organic chemist's utmost ingenuity in the design of routes using established reactions or in the production of new methods in order to achieve a specific transformation.

These volumes draw together the reported total syntheses of various groups of natural products and commentary on the strategy involved with particular emphasis on any stereochemical control. No such compilation exists at present, and we hope that these books will act as a definitive source book of the successful synthetic approaches reported to date. As such, it will find use not only with the synthetic organic chemist but also perhaps with the organic chemist in general and the biochemist in his specific area of interest.

One of the most promising areas for the future development of organic chemistry is synthesis. The lessons learned from the synthetic challenges presented by various natural products can serve as a basis for this ever-developing area. It is hoped that these books will act as an inspiration for future challenges and outline the development of thought and concept in the area of organic synthesis.

The project started modestly with an experiment in literature searching by a group of graduate students about thirteen years ago. Each student prepared a summary in equation form of the reported total syntheses of various groups of natural products. It was my intention to collate this material and possibly publish it. During a sabbatical leave in Strasbourg in 1968-69, I attempted to prepare a manuscript, but it soon became apparent that the task would take many years and I wanted to enjoy some of the other benefits of a sabbatical leave. Several colleagues suggested that the value of such a collection would be enhanced by commentary. The only way to encompass the amount of data collected and the inclusion of some words was to persuade experts in the various areas to contribute.

Volume 1 presented six chapters describing the total synthesis of a wide variety of natural products. The subject matter of Volume 2 was somewhat more related, being a description of some terpenoid and steroid syntheses. Volume 3 concentrated on alkaloid synthesis and appeared in 1977. The present volume contains three chapters on new areas of synthetic endeavor and two more encompassing the progress in synthetic work in the areas of monoterpenes and prostaglandins since the appearance of Volume 1.

It is intended that Volume 5 of this series will contain predominantly up-dating chapters in order that this series may continue to be of timely use to those with interests in synthetic chemistry.

John ApSimon

Ottawa, Canada
March 1981

Contents

THE TOTAL SYNTHESIS
OF NATURAL PRODUCTS

The Synthesis
of Insect Pheromones

KENJI MORI

Department of Agricultural Chemistry,
The University of Tokyo,
Bunkyo-ku, Tokyo, Japan

1. INTRODUCTION

Man has wondered at spectacular scenes of metamorphosis, aggregation, and mating of insects for many years. During the past two decades it has gradually become clear that these biological phenomena are regulated by chemical substances known as insect hormones and pheromones. Insect chemistry, the study of natural products of insect origin, is now regarded as an established branch of natural products chemistry.

After the discovery of bombykol 1, the first insect pheromone, by Butenandt and his associates,[1] the term "pheromone" was defined by Karlson and Lüscher.[2] The name is derived from the Greek *phrein*, to transfer, and *hormon*, to excite. Pheromones are substances that are secreted to the outside by an individual and received by a second individual of the same species, in which they release a specific reaction, for example, a definite behavior or a developmental process.

From the beginning the synthetic approach was very important in pheromone researches because of the limited availability of natural pheromones from insects (usually less than several milligrams). Synthetic work in insect pheromones may be classified into three categories: (1) synthesis as the final proof of the proposed structure, including olefin geometry and relative as well as absolute stereochemistry; (2) synthesis that provides sufficient material for biological study, such as field tests; and (3) synthesis of a number of isomers and analogs to clarify the structure-pheromone activity relationship. Synthesis thus ensures ample supplies of otherwise inaccessible pheromone and facilitates the practical uses of pheromones in agriculture and forestry.

Pheromone structures are scattered among various types of volatile compounds ranging from alkanes to nitrogen heterocycles. Recent studies on structure-activity relationships reveal the importance of stereochemistry in pheromone perception by insects. Three types of isomerism, structural, geometrical, and optical, are all shown to effect the biological activity, as described below.

*Bombykol, the Pheromone of the Silkworm Moth (*Bombyx mori*), and its Geometrical Isomers*

Butenandt et al.[3,4] and Truscheit and Eiter[5] synthesized all of the four possible geometrical isomers of bombykol 1 and compared their attractancy to the male silkworm moth. The results are shown in Table 1. The biological activity, as well as physical properties, of $(10E, 12Z)$-10,12-hexadecadien-1-ol was almost identical with that of the natural bombykol. The geometry of the diene system in bombykol was thus established as $10E, 12Z$ by this synthetic work. It should be noted that the other three geometrical isomers possess only moderate or weak biological activities. A highly stereoselective synthesis of the most active isomer is therefore of paramount importance both scientifically and economically.

Table 1.   Biological Activity of Natural Bombykol and Synthetic Geometrical
Isomers of 10,12-Hexadecadien-1-ol

	Activity ($\mu g/ml$)[a]		
	Butenandt[3]	Butenandt[4]	Eiter[5]
$10Z, 12Z$	1	1	—
$10Z, 12E$	10^{-3}	10^{-2}	10^{-5}
$10E, 12Z$	10^{-12}	10^{-12}	10^{-12}
$10E, 12E$	10	100	10
Natural bombykol	10^{-10}	10^{-10}	10^{-10}

[a]The activity is expressed by *die Lockstoffeinheit* (LE). This is the lower limit of
the pheromone concentration ($\mu g/ml$) to which 50% of the test insects show
reaction.

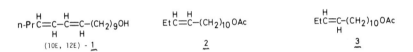

The Pheromone of Red-banded Leaf Roller (Argyrotaenia velutinana) and Its Geometrical Isomer

Roelofs et al. identified (Z)-11-tetradecenyl acetate **2** as the sex pheromone of
red-banded leaf roller moths.[6] They then demonstrated that a large amount of
the (E)-isomer **3** is inhibitory to pheromone action.[7] Here again stereochemistry
was shown to be important. Roelofs' argument on this subject was based on his
bioassay results with many pheromone analogs, some of which work in an inhibitory manner, while others act synergistically.[7] Subsequently Klun et al.[8] and

Beroza et al.[9] reported the very interesting observation that a small amount of opposite geometrical isomer was critical to pheromone attraction. Klun found that a geometrically pure preparation of **2** was very weakly attractive to the moth and that the presence of 7% of (*E*)-isomer **3** was necessary for maximum activity.[8] Previous syntheses of **2** employed either the Wittig reaction or the Lindlar semihydrogenation, and neither of them was 100% stereoselective. It is therefore obvious that a highly pure geometrical isomer is required to study this kind of very subtle biological phenomena. Beroza's relevant work was on the pheromone of the oriental fruit moth *Grapholitha molesta*. The biological activity of the synthetic pheromone (*Z*)-8-dodecenyl acetate increased 25 times by the addition of a small amount of the (*E*)-isomer.[9]

*Gossyplure, the Pheromone of Pink Bollworm Moth (*Pectinophora gossypiella*)*

In the case of gossyplure the pheromone consists of a mixture of two geometrical isomers in an equal amount: (7*Z*, 11*Z*)-7,11-hexadecadienyl acetate **4** and its 11*E*-isomer **5**.[10] Neither is biologically active alone. This suggests the existence of two different receptor sites on the pheromone receptor of the pink bollworm moth.

The Pheromone of Dendroctonus *Bark Beetles*

Two stereoisomers of 7-ethyl-5-methyl-6,8-dioxabicyclo[3.2.1]octane were isolated from the frass of the western pine beetle (*Dendroctonus brevicomis*).[11] Only one of them, *exo*-brevicomin **6**, is biologically active as a component of the aggregation pheromone of the western pine bettle. The other isomer, *endo*-brevicomin **7**, is inactive to the western pine beetle and even inhibits the olfactory response of flying male and female southern pine beetles (*Dendroctonus frontalis*) to the female-produced pheromone, frontalin **8**.[12] In this case the *endo-exo* stereoisomerism is of utmost importance for biological activity. This necessitated the stereoselective synthesis of these pheromones.

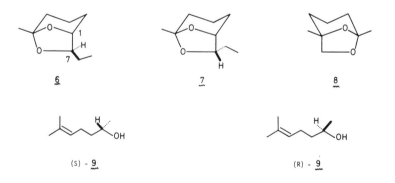

6

7

8

(S) - 9

(R) - 9

Biological Activities of the Optical Isomers of Pheromones

exo-Brevicomin **6** and frontalin **8** are chiral molecules. They therefore can exist in two enantiomeric forms. Both enantiomers of these pheromones were synthesized, ensuring biological evaluation of the isomers.[13,14] The biologically active isomers were (1*R*, 5*S*, 7*R*)-(+)-*exo*-brevicomin **6** and (1*S*, 5*R*)-(−)-frontalin **8**.[15] In these cases only one enantiomer of the two optical isomers possesses pheromone activity.

Sulcatol is the aggregation pheromone produced by males of *Gnathotrichus sulcatus*.[16] Both (+)-sulcatol[(*S*)-**9**] and (−)-isomer [(*R*)-**9**] were synthesized.[17] Surprisingly, neither of them was biologically active. However, when combined to give a racemic mixture the synthetic sulcatol was more active than the natural pheromone, which was a mixture of 65% of (*S*)-**9** and 35% of (*R*)-**9**.[18] This situation is somewhat similar to that encountered in the case of gossyplure and suggests the presence of enantiomer-specific active sites on receptor proteins in the same or different cells of *Gnathotrichus sulcatus*. These examples illustrate the importance of stereochemistry in pheromone researches.

The aim of this chapter is to provide a compilation of synthetic works on pheromones. As one of the major synthetic problems in this field is the stereoselective construction of olefinic linkages, Section 2 deals mainly with preparative methods for disubstituted olefins. Then synthesis of individual pheromones is detailed according to a classification based on the type of compound and functional groups present. As the trend in modern organic synthesis is to develop new methods for providing chiral molecules in a stereocontrolled manner, synthesis of chiral pheromones are treated comprehensively. It is hoped that this chapter will be useful not only to synthetic chemists but also to entomologists who wish to prepare pheromones of particular interest to them.

There are a number of monographs and reviews on pheromones. Especially noteworthy is the recent chapter on Insect Chemistry in *Annual Reports on the Progress of Chemistry*, which is a thorough survey on pheromone chemistry.[19,20] Four reviews on pheromone synthesis are available: Katzenellenbogen's review focuses on the methodological point of view[21]; Henrick discusses selected pheromones (*Lepidoptera, Coleoptera,* and *Diptera*) in depth[22]; and Rossi reviews the synthesis of both achiral[23] and chiral[24] pheromones. Aspects of pheromone chemistry is reviewed in Refs. 25-29. For those who are interested in pheromone biology and its application a plethora of monographs and reviews is available: e.g., Refs. 30-32 (general treatises) and Refs. 33-35 (insect behavior and the practical application of pheromones). Bark beetle pheromones are reviewed in Refs. 36-40. References 41 and 42 are concerned with the terpenoid pheromones, and Ref. 43 is a readable review on the pheromone receptor of moths.

2. GENERAL METHODS

Before the advent of pheromone and juvenile hormone chemistry the stereo-selective construction of di- and trisubstituted olefins was of only limited interest to oil and terpene chemists. During the past decade the situation has changed, and we now have many ingenious new methods as well as modifications of older methods for olefin synthesis. References 44 and 45 are excellent reviews on the stereoselective synthesis of olefins. In this section reactions that have been used or may be useful in pheromone synthesis are presented. Synthetic methods for trisubstituted olefins are omitted, since they are the theme of another review on juvenile hormone synthesis.[46]

A. Synthesis of (E)-Alkenes

Metal-Ammonia Reduction of Alkynes

The reduction of alkynes with sodium in liquid ammonia is the standard method (Equation 1).[47] Warthen and Jacobson recommend the use of a large volume of liquid ammonia to minimize the recovery of the starting alkynes.[48]

$$RC{\equiv}C-(CH_2)_nOTHP \xrightarrow{\text{Na / NH}_3} \underset{H}{\overset{H}{R-C{=}C-(CH_2)_nOTHP}} \qquad (1)$$

Lithium Aluminum Hydride Reduction of Alkynes

The reduction of 2-alkyn-1-ols to 2-alken-1-ols with lithium aluminum hydride in ether usually proceeds in an excellent yield (Equation 2).[49] Other alkynols such as 3-alkyn-1-ol, 7-alkyn-1-ol, and 8-alkyn-1-ol can also be reduced to the corresponding alkenols by reacting them at 140° for 48-55 h, under nitrogen, with a large excess of lithium aluminum hydride in a mixture of diglyme and tetrahydrofuran (Equation 3).[50]

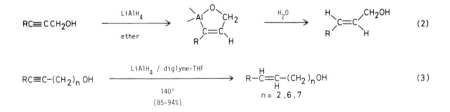

$$RC{\equiv}C-(CH_2)_n OH \xrightarrow[\substack{140° \\ (85-94\%)}]{\text{LiAlH}_4 \text{ / diglyme-THF}} \underset{H}{\overset{H}{R-C{=}C-(CH_2)_n OH}} \qquad n = 2,6,7 \qquad (3)$$

Reductive Elimination of Allylic Substituents

3-Alken-1-ols can be prepared from alkynes by the route shown in Equation 4. The key stereoselective step (97% E) is the reductive elimination of the allylic *t*-butoxy group.[51] A highly stereoselective synthesis of an (*E*)-alkene employs the reduction of a phosphonate ester as the key step (Equation 5).[52] The yield is moderate to excellent.

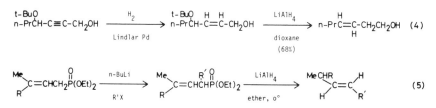

Rearrangement of Allylic Dithiocarbamates

The rearrangement of allylic dithiocarbamates is applicable to the synthesis of various alkenol pheromones (Equation 6).[53] The yield is good to excellent.

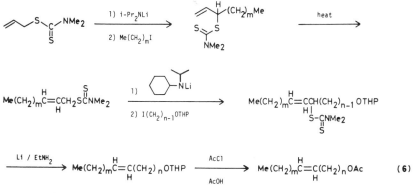

The Wittig Reaction

The Schlosser modification of the Wittig reaction as shown in Equation 7 gives an (*E*)-alkene in 60-72% yield with 90-96% stereoselectivity.[54-56] For the stereochemistry of this reaction see Ref. 22, p. 1875.

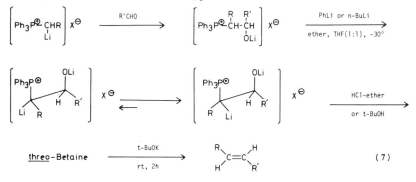

Utilization of Organoaluminum Compounds

Disubstituted (E)-alkenes can be prepared by the reaction of (E)-alkenyl trialkyl-aluminates with alkyl halides and sulfonates (Equation 8).[57] The yield is good (44-79%) for allylic halides and moderate (41-44%) for primary halides. Secondary and tertiary halides gave poor results. (E)-Vinyl iodides are obtainable in 94% stereoselectivity, as shown in Equation 9.[58] Reaction with lithium di-alkylcuprate (R_2CuLi) converts the iodide to (E)-alkene. A vinylalane is converted to (E)-homoallylic alcohol in the yield of 81-88% (Equation 10).[59]

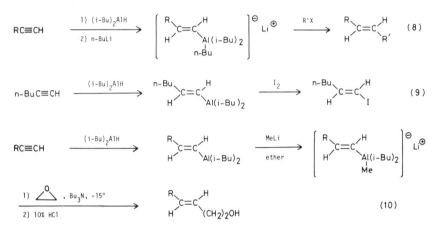

Utilization of Organotin Compounds

(E)-Allylic alcohols can be prepared from propargylic alcohol via an organotin compound (Equation 11).[60]

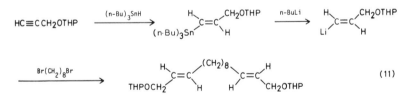

Utilization of Organoboranes

(E)-Alkenes are prepared by the reaction of boranes with palladium acetate (Equation 12).[61] (E, E)-Conjugated dienes are obtainable via hydroboration (Equation 13).[62]

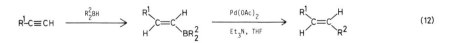

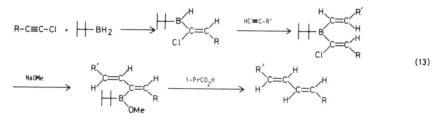

(13)

Utilization of Organozirconium Compounds

(*E, E*)-Conjugated dienes are synthesized by the palladium-catalyzed reaction of
(*E*)-1-alkenylzirconium derivatives with alkenyl halides (Equation 14).[63]

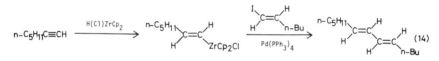

(14)

Reduction of β-Acetoxysulfones

(*E*)-Alkenes are obtained by reduction of β-acetoxysulfones with sodium amalgam (Equation 15).[64]

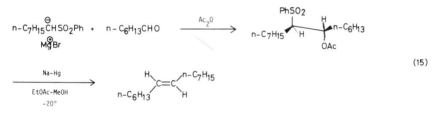

(15)

B. Synthesis of (Z)-Alkenes

Hydrogenation of Alkynes

Hydrogenation of alkynes is the standard method for the preparation of (*Z*)-
alkenes (Equation 16). Lindlar's palladium catalyst is widely used for this purpose.[65, 66] This catalyst produces alkenes containing only small amounts (~5%)
of (*E*)-isomer. Frequently at low catalyst ratios no (*E*)-alkene is detectable at the
point of disappearance of the substrate alkyne.[67] Two alternative catalysts are
5% palladium on barium sulfate in methanol containing a small amount of
quinoline[68] and P-2 nickel in ethanol containing a small amount of ethylenediamine.[69] Although hydrogenation with P-2 nickel is said to be highly stereoselective (*E* : *Z* = 1 : 200), the best method is low-temperature hydrogenation over
Lindlar catalyst. Thus at −10 to −30° in pentane, hexane, or hexane-THF the
(*Z*)-alkenes obtained contain ~0.5% of the (*E*)-isomer.[70]

$$R-C{\equiv}C-R' \xrightarrow{\text{H}_2 \text{ / Lindlar-Pd or Pd-BaSO}_4 \text{ or P-2 Ni}} \underset{\text{H}}{R-C}{=}\underset{\text{H}}{C}-R' \qquad (16)$$

The Wittig Reaction

The Wittig olefination carried out in DMSO is known to give a (Z)-disubstituted alkene as the major product (Equation 17).[71, 72] The use of potassium in HMPA as the base favors the (Z)-olefination (Equation 18).[73, 74] The use of salt-free ylid solution is recommended for the preparation of (Z)-alkenes (Equation 19).[75] The stereochemistry of this reaction is discussed by Schlosser.[56]

$$RCH{=}PPh_3 \quad + \quad R'CHO \xrightarrow{\text{DMSO}} \underset{\text{H}}{R-C}{=}\underset{\text{H}}{C}-R' \ (\sim 90\% \ Z) \qquad (17)$$

$$\left[RCH_2\overset{\oplus}{P}Ph_3\right]X^{\ominus} \xrightarrow[\text{HMPA}]{\text{K}} RCH{=}PPh_3 \xrightarrow{\text{R'CHO}} \underset{\text{H}}{R-C}{=}\underset{\text{H}}{C}-R' \cdot \qquad (18)$$

$$\left[RCH_2\overset{\oplus}{P}Ph_3\right]X^{\ominus} \xrightarrow[\text{THF, } -78°]{\text{NaN(SiMe}_3)_2} RCH{=}PPh_3 \ \text{(salt free)} \qquad (19)$$

$$\xrightarrow[-78°]{\text{R'CHO}} \underset{\text{H}}{R-C}{=}\underset{\text{H}}{C}-R' \quad (Z:E = 98:2)$$

Utilization of Organoboranes

(Z)-Alkenes (98-99% Z) can be prepared from disubstituted alkynes by hydroboration-protonolysis (Equation 20).[76] The protonolysis is also achieved under neutral conditions by treatment with catalytic amounts of palladium acetate.[77] (Z)-Vinyl iodides are obtainable from iodoacetylenes by the same process (Equation 21).[78] They are convertible to (Z)-alkenes by treatment with organocopper reagents.

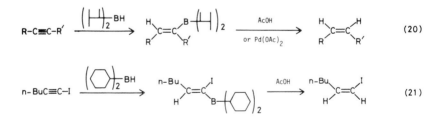

C. Carbon-Carbon Bond Formation

Alkylation of Alkynes

Alkylation of 1-alkynes gives disubstituted acetylenes, which are the starting materials for both (E)- and (Z)-alkenes. The traditional procedures are listed in Ref. 79. Recently a convenient and efficient procedure for the alkylation was proposed independently by two groups (Equation 22).[47,80] The procedure is particularly convenient for small-scale preparations. The yield is excellent if the alkylating agent is a primary and unbranched halide (see also Ref. 22; p. 1847.) An interesting 1,3-disubstitution reaction of 1,3-dilithiopropyne may be useful in pheromone synthesis (Equation 23).[81]

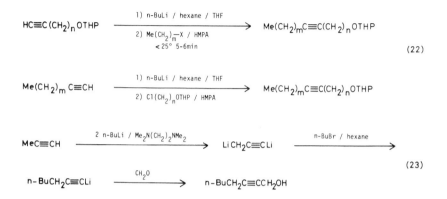

(22)

(23)

Coupling Reactions with Organometallic Reagents

An acetoxy group occupying the allylic position or a tosyloxy group can be re-placed by the hydrocarbon moiety of a Grignard reagent. The replacement is regio- and stereoselective (Equation 24).[82] The coupling of an (E)-terminal vinyllithium with an alkyl halide gives an (E)-alkene stereoselectively (Equation 25).[83]

A new regio- and stereoselective olefin synthesis applicable to pheromone syn-thesis is direct substitution of hydroxyl groups of allyl alcohols with alkyl groups by the reaction of lithium allyloxyalkylcuprates with N,N-methylphenyl-aminotriphenylphosphonium iodide (Equation 26).[84] By this method both (E)- and (Z)-alkenes are obtainable.

(Z)-Alkenols can be prepared from (Z)-vinylic organocopper reagent in a moderate yield (Equation 27).[85]

The palladium-catalyzed cross-coupling reactions of vinylic iodides with a variety of Grignard reagents occurs with retention of configuration (~97%) (Equation 28).[86]

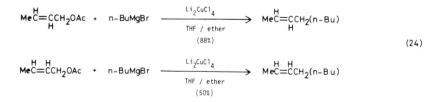

(24)

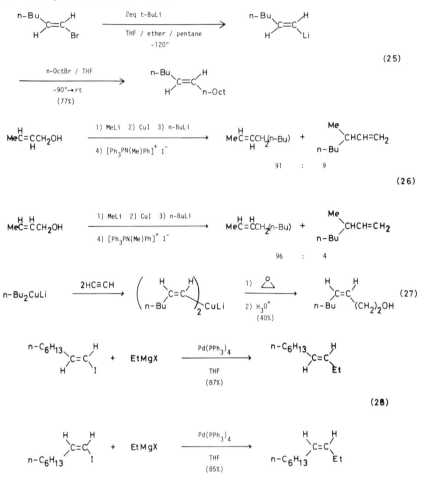

$$ (25) $$

$$ (26) $$

$$ (27) $$

$$ (28) $$

D. Other Useful Methods

Multipositional Isomerization of Alkynes and Alkynols

Potassium 3-aminopropylamide (**KAPA**) is an exceptionally active catalyst for multipositional isomerization of alkynes and alkynols to terminal acetylenes (Equation 29).[87-89] This is a convenient method to obtain 1-alkynes, popular starting materials for the *Lepidoptera* sex pheromones.

$$ (29) $$

Conversion of Alkyl Chlorides to Bromides

For the alkylation of alkynes (Equations 22 and 23), alkyl chlorides are generally not so reactive. Therefore an efficient method is desirable for the conversion of alkyl chlorides to bromides. A new procedure employs ethyl bromide as the source of bromine (Equation 30).[90] The high volatility of ethyl chloride, the by-product, is the driving force for the completion of the reaction.

$$\text{Cl(CH}_2)_6\text{OTHP} \quad \xrightarrow[\substack{\text{(solvent)} \\ 60\text{-}70^\circ, \ 5\text{days, (84\%)}}]{\text{1eq NaBr, 30eq EtBr}} \quad \text{Br(CH}_2)_6\text{OTHP} \qquad (30)$$

Conversion of Tetrahydropyranyl (THP) Ethers to Acetates

Many pheromones are acetates of long-chain alcohols. The THP group is the most commonly employed protecting group in the course of pheromone synthesis. The direct conversion of THP ethers into acetates can be achieved in 85-90% yield (Equation 31).[47]

$$\text{ROTHP} \quad \xrightarrow[\substack{35\text{-}40^\circ, \ \text{overnight} \\ (85\text{-}90\%)}]{\text{AcCl-AcOH (1:10)}} \quad \text{ROAc} \qquad (31)$$

Conversion of THP Ethers to Bromides

It is possible to carry out the direct conversion of THP ethers to halides (Equation 32).[91]

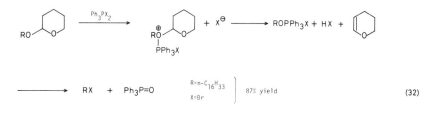

$$\xrightarrow{} \quad \text{RX} \quad + \quad \text{Ph}_3\text{P=O} \qquad \left. \substack{\text{R=n-C}_{16}\text{H}_{33} \\ \text{X=Br}} \right\} \quad 87\% \ \text{yield} \qquad (32)$$

Solid-phase Synthesis

The application of solid-phase synthesis in the pheromone field has been reported by Leznoff (Equation 33).[92-95] Solid-phase synthesis gives comparable or better overall yields than previous methods, uses inexpensive symmetrical diols as starting materials, and has the potential for being adapted to an automated procedure. It is questionable, however, whether this method is suitable for the large-scale preparation of pheromones required for field tests.

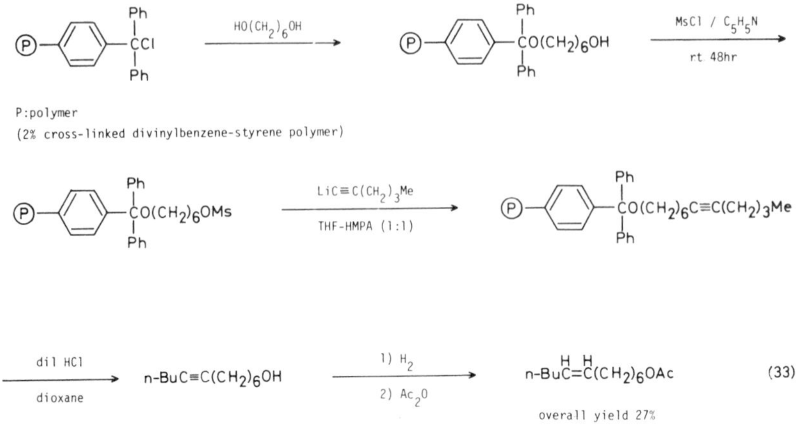

P:polymer
(2% cross-linked divinylbenzene-styrene polymer)

(33)

overall yield 27%

Separation of Geometrical Isomers by Formation of Urea Complexes

Mixtures of geometrical isomers can be conveniently separated on a large scale by the relatively inexpensive method of urea inclusion complex formation.[482] The recovery of both isomers from the separation procedure is almost quantitative. Urea inclusion complexes are formed preferentially with (E)-isomers. By applying this method of separation at a convenient stage, several insect phero-

Table 2.   Separation of Geometrical Isomers of $RR_1C{=}CH(CH_2)_nX$ by Urea Complex Formation[482]

R	R_1	n	X	Separation[a]
n-Pr	Me	9	CO_2Me	+
n-Pr	Me	9	CHO	−
n-Bu	H	9	CO_2Me	−
n-Bu	H	9	CN	+
n-Bu	H	9	CH_2OH	−
n-Bu	H	9	CHO	+
n-Bu	H	7	CO_2Me	−
n-Bu	H	7	CH_2OH	+
n-Bu	H	2	CO_2Me	+
n-PrCH=CH[b]	H	8	CH_2OH	+
n-PrC≡C[b]	H	8	CO_2Et	+
n-PrC≡C[b]	H	8	CH_2OH	+

[a] The plus sign indicates that at least one isomer could be obtained in essentially pure form; the minus sign indicates that no separation was achieved.
[b] From Butenandt et al.[4]

mones were prepared without the necessity of a stereoselective step. The separations that were achieved by means of the urea complex technique are summarized in Table 2.

Annual surveys are available for methods in olefin synthesis and olefin inversion.[96],[97]

3. ALKANES AS PHEROMONES

15-Methyltritriacontane 10 and 15,19-Dimethyltritriacontane 11

Examination of the cuticular hydrocarbon content of the stable fly, *Stomoxys calcitrans* L., indicated that the male was aroused to mating behavior by the externally borne saturated hydrocarbons of the female fly. These were methyl-branched and 1,5-dimethyl-branched alkanes.[98] 15-Methyl- and 15,19-dimethyl-

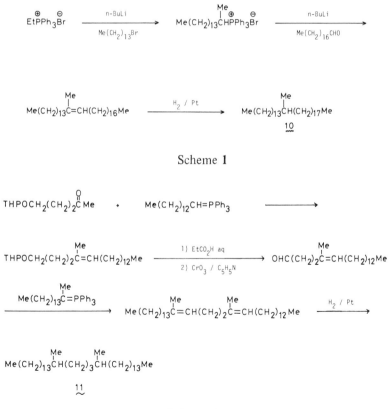

Scheme 1

Scheme 2

tritriacontanes showed the highest biological activity. The synthesis of 15-methyltritriacontane **10** and 15,19-dimethyltritriacontane **11** were carried out by the Wittig synthesis (Schemes 1 and 2).[99,100] A general synthetic route to insect hydrocarbons is shown in Scheme 3, which employs thiophenes as intermediates.[101] Optical isomers of these hydrocarbons are yet to be synthesized.

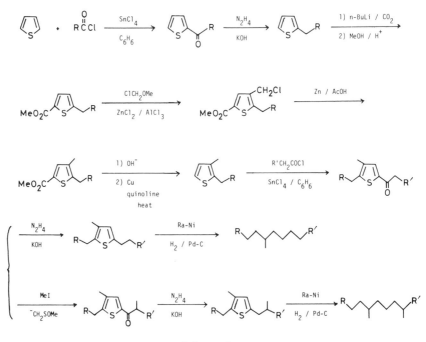

Scheme **3**

15,19-Dimethylheptatriacontane **12**, *17,21-Dimethylheptatriacontane* **13**, *and 15,19,23-Trimethylheptatriacontane* **14**

These title hydrocarbons are the sex pheromones isolated from the cuticle of the female tsetse fly (*Glossina morsitans morsitans*), which is the major vector of Rhodesian sleeping sickness.[102] The alkanes release mating behavior in the male fly at ultrashort range or on contact with baited decoys. The synthesis of 15,19-dimethylheptatriacontane **12** was carried out as shown in Scheme 4[100] or by a similar route as shown in Scheme 2. The synthesis of 17,21-dimethylheptatri-acontane **13** is shown in Scheme 5.[102] The Julia cyclopropane cleavage reaction was used here as well as in the case of the synthesis of 15,19,23-trimethyl-heptatriacontane **14** (Scheme 6).[102] The trimethylalkane **14** released responses four times more often than **12** and 14 times more than **13**. No synthesis of the optically active forms of these hydrocarbons has been reported.

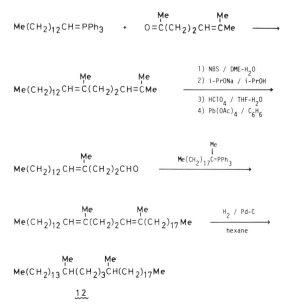

Me(CH$_2$)$_{12}$CH=PPh$_3$ +   O=C(CH$_2$)$_2$CH=CMe
(with Me groups above)

Me(CH$_2$)$_{12}$CH=C(CH$_2$)$_2$CH=CMe
(with Me groups above)
→
1) NBS / DME-H$_2$O
2) i-PrONa / i-PrOH
3) HClO$_4$ / THF-H$_2$O
4) Pb(OAc)$_4$ / C$_6$H$_6$

Me(CH$_2$)$_{12}$CH=C(CH$_2$)$_2$CHO
(with Me group above)
→
Me(CH$_2$)$_{17}$C=PPh$_3$
(with Me group above)

Me(CH$_2$)$_{12}$CH=C(CH$_2$)$_2$CH=C(CH$_2$)$_{17}$Me
(with Me groups above)
→
H$_2$ / Pd-C
hexane

Me(CH$_2$)$_{13}$CH(CH$_2$)$_3$CH(CH$_2$)$_{17}$Me
(with Me groups above)

<u>12</u>

Scheme **4**

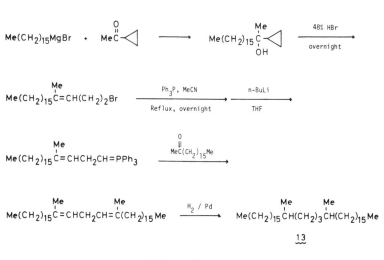

Me(CH$_2$)$_{15}$MgBr   +   MeC(=O)◁   →   Me(CH$_2$)$_{15}$C(Me)(OH)◁
→
48% HBr
overnight

Me(CH$_2$)$_{15}$C=CH(CH$_2$)$_2$Br
(with Me group above)
→
Ph$_3$P, MeCN
Reflux, overnight
→
n-BuLi
THF

Me(CH$_2$)$_{15}$C=CHCH$_2$CH=PPh$_3$
(with Me group above)
→
MeC(=O)(CH$_2$)$_{15}$Me

Me(CH$_2$)$_{15}$C=CHCH$_2$CH=C(CH$_2$)$_{15}$Me
(with Me groups above)
→
H$_2$ / Pd
→
Me(CH$_2$)$_{15}$CH(CH$_2$)$_3$CH(CH$_2$)$_{15}$Me
(with Me groups above)

<u>13</u>

Scheme **5**

17

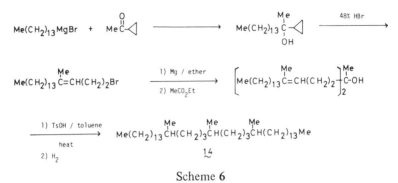

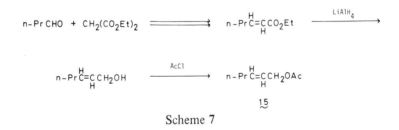

Scheme 6

4. PHEROMONES WITH AN *E*–DOUBLE BOND

(E)-2-Hexen-1-yl Acetate 15

This is the sex pheromone of Indian water bug (*Lethocerus indicus*) and serves to excite the female immediately before or during mating. Its synthesis is shown in Scheme 7.[103] Pattenden, however, states that **15** is neither sex- nor species-specific.[104]

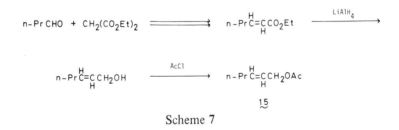

Scheme 7

(E)-7-Dodecen-1-yl Acetate 16

This is the sex pheromone of false codling moth (*Cryptophlebia leucotreta*).[105] Two similar syntheses were reported.[105, 106] Henderson's synthesis is shown in Scheme 8.[106]

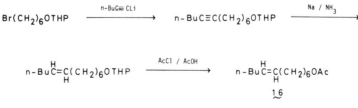

Scheme 8

(E)-11-Tetradecenal **17**

The eastern spruce budworm (*Choristoneura fumiferana*) uses this compound as its sex pheromone.[107] The original synthesis was based on acetylene chemistry (Scheme 9).[107] The second one employed the highly stereoselective phosphonate method (Scheme 10, see Equation 5).[52] The third one is an application of the solid-phase synthesis (Scheme 11, see Equation 33).[95]

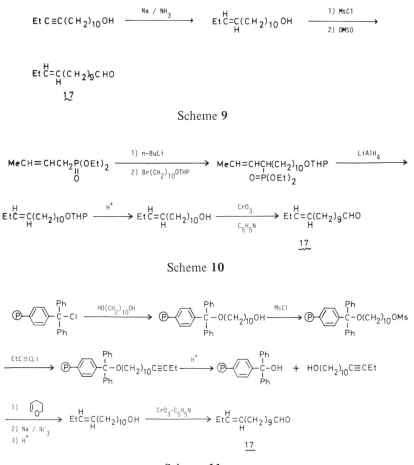

Scheme **9**

Scheme **10**

Scheme **11**

(E)-11-Hexadecen-1-yl Acetate **18**

This is the sex pheromone produced by female sweet potato leaf folder moth, *Brachmia macroscopa.*[108] This alkene was synthesized by the inversion of olefin geometry of its (*Z*)-isomer obtainable by the conventional Wittig reaction

(Scheme 12).[108] The inversion was carried out by treating the (Z)-isomer with thiols and diphenylphosphine in the presence of azoisobutyronitrile.[109] The equilibrium mixture obtained by this method usually contains 75-80% of (E)-alkenes. The current methods of olefin inversion employ epoxides as intermediates.[110, 111]

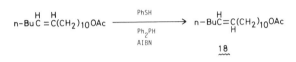

Scheme 12

5. PHEROMONES WITH A Z–DOUBLE BOND

Muscalure, (Z)-9-Tricosene 19

This is a sex pheromone isolated from the cuticle and feces of the female house-fly (Musca domestica) and attracts the male fly.[112] The first synthesis was accomplished via a Wittig route (Scheme 13).[112] This Wittig synthesis was later modified, by using potassium in HMPA as the base, to give 94% of the (Z)-alkene plus 6% of its (E)-isomer (Scheme 14).[73] The second type of synthesis employs the Lindlar semihydrogenation of the alkyne (Scheme 15).[113] The use of the naturally occurring erucic acid as the starting material yielded the pheromone only in two steps (Scheme 16), although the acid was rather expensive.[114] Reaction of the methanesulfonate of erucyl alcohol with methylmagnesium

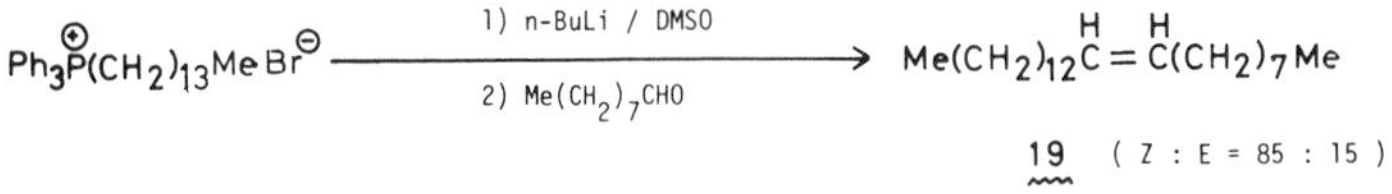

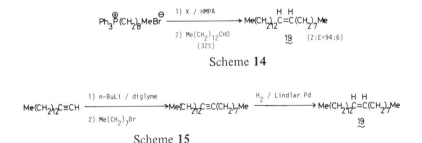

Scheme 13

Scheme 14

Scheme 15

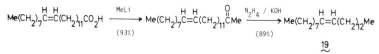

Me(CH₂)₇C=C(CH₂)₁₁CO₂H →(MeLi)(93%)→ Me(CH₂)₇C=C(CH₂)₁₁ĊMe →(N₂H₄ / KOH)(89%)→ Me(CH₂)₇C=C(CH₂)₁₂Me

19

Scheme 16

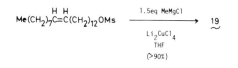

Me(CH₂)₇C=C(CH₂)₁₂OMs →(1.5eq MeMgCl)(Li₂CuCl₄)(THF)(>90%)→ **19**

Scheme 17

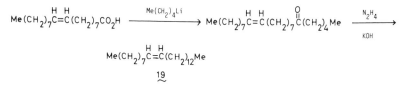

Me(CH₂)₇C=C(CH₂)₇CO₂H →(Me(CH₂)₄Li)→ Me(CH₂)₇C=C(CH₂)₇Ċ(CH₂)₄Me →(N₂H₄)(KOH)→

Me(CH₂)₇C=C(CH₂)₁₂Me

19

Scheme 18

Me(CH₂)₇C=C(CH₂)₇CO₂H + Me(CH₂)₅CO₂H →(Kolbe electrolysis)(14%)→

Me(CH₂)₇C=C(CH₂)₁₂Me

19

Scheme 19

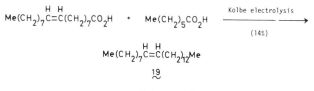

Me(CH₂)₇C=C(CH₂)₇COCl + HO-⟨quinoline⟩ →(NaOH)(THF)(88%)→ Me(CH₂)₇C=C(CH₂)₇CO-O-⟨quinoline⟩

→(Me(CH₂)₄MgCl)(ether / THF)(-90°, (75%))→ Me(CH₂)₇C=C(CH₂)₇Ċ(CH₂)₄Me →(N₂H₄)(KOH)(84%)→ **19**

Scheme 20

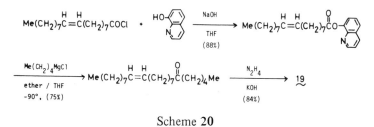

Me(CH₂)₇C=C(CH₂)₈Br + Me(CH₂)₄MgBr →(Cu⁺)(THF)(0-5°, 2.5hr)(quant.)→ **19**

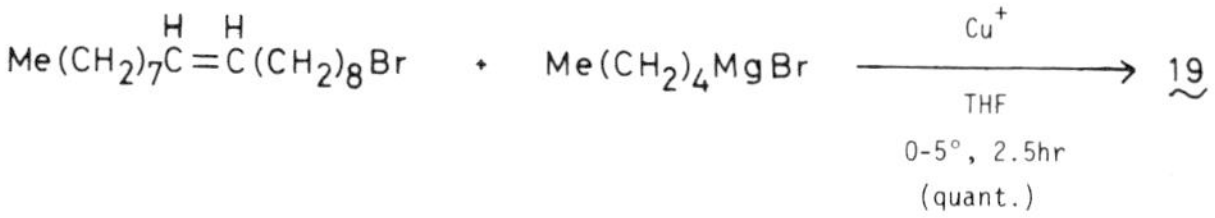

Scheme 21

chloride in the presence of lithium tetrachlorocuprate gave muscalure in >90% yield (Scheme 17).[115] Cheaper oleic acid was also converted to muscalure in several ways. A two-step synthesis similar to Scheme 16 is shown in Scheme 18.[116] A mixed Kolbe electrolysis of oleic and n-heptanoic acids gave muscalure in 14% yield (Scheme 19)[117] in a single step. A three-step synthesis similar to Scheme 18, but less direct, was recently reported and is shown in Scheme 20.[118] The best procedure used to prepare 150-kg batches of muscalure by Zoecon Corporation is the coupling of oleyl bromide with 1.15 eq n-amylmagnesium bromide in THF in the presence of 0.03 eq lithium chlorocyanocuprate. The yield is reported to be nearly quantitative (Scheme 21).[115]

(Z)-14-Nonacosene 20

This is the most active component of the sex pheromones of the female face fly (*Musca autumnalis*) and effective for mating stimulation at short range.[119] The synthesis was carried out either by the Wittig reaction in THF-HMPA (70% yield of **20** containing ~94-96% of the (Z)-isomer)[119] or by the acetylene route (Scheme 22).[120]

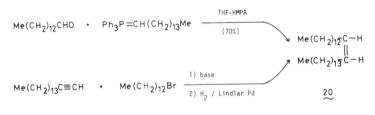

Scheme 22

Disparlure, (Z)-7,8-Epoxy-2-methyloctadecane 21

The gypsy moth (*Porthetria dispar*) is a serious despoiler of forests. The sex pheromone was extracted from 78,000 tips of the last two abdominal segments of female moths and shown to be an epoxide **21**.[121] The olefinic precursor, (Z)-2-methyl-7-octadecene, is present in the female sex pheromone gland of the gypsy moth[121] and inhibits male attraction to disparlure.[122] As little as 2 pg of **21** was active in laboratory bioassay. The first synthesis employed the Wittig reaction (Scheme 23).[121] The stereoselectivity of the reaction, however, was unsatisfactory, and the final product had to be purified by chromatography over silica gel-silver nitrate. A more stereoselective version of this Wittig synthesis was reported by Bestmann (Scheme 24).[123] Bestmann further improved the stereoselectivity to >98% by employing sodium bis(trimethylsilyl)amide as the base (Scheme 25).[124] Two syntheses were reported by Chan employing organosilicon intermediates. The earlier synthesis gave a 1 : 1 mixture of the *cis*- and *trans*-isomers of **21** (Scheme 26).[125] The later synthesis was based on the reac-

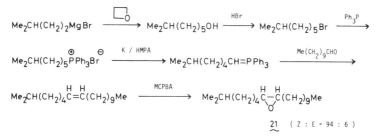

Scheme 23

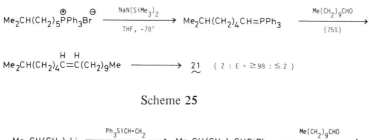

Scheme 24

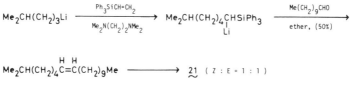

Scheme 25

Scheme 26

tion of carbonyl compounds with carbanions α to silicon and proceeded in a more stereoselective manner (Scheme 27).[126] A synthesis via an acetylenic inter-mediate was reported by two groups (Scheme 28).[127, 128] 1,5-Cyclooctadiene served as the common starting material in two syntheses of **21**. Klünenberg and

Schäfer employed Kolbe electrolysis as the key step (Scheme 29).[129] Tolstikow et al. used Schlosser's coupling method in their synthesis (Scheme 30).[130,131] The synthesis of optically active disparlure will be discussed later (Schemes 264-267).

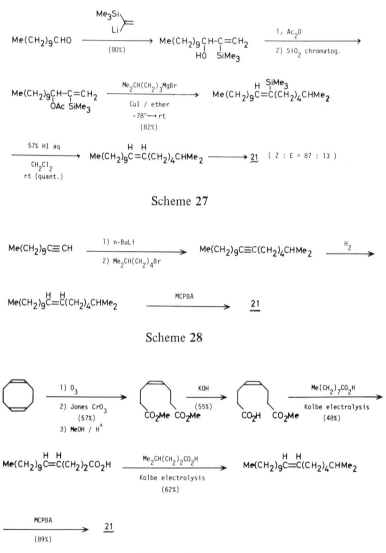

Scheme 27

Scheme 28

Scheme 29

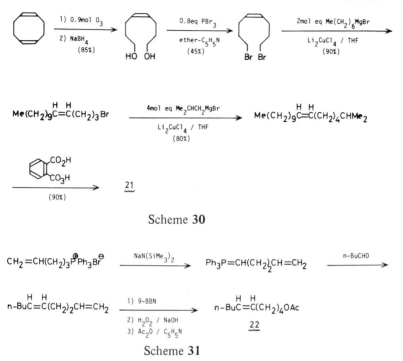

Scheme **30**

CH$_2$=CH(CH$_2$)$_3$$\overset{\oplus}{P}Ph_3Br^{\ominus}$ $\xrightarrow{\text{NaN(SiMe}_3)_2}$ Ph$_3$P=CH(CH$_2$)$_2$CH=CH$_2$ $\xrightarrow{\text{n-BuCHO}}$

n-BuC=C(CH$_2$)$_2$CH=CH$_2$ (H H) $\xrightarrow[\substack{2) \text{ H}_2\text{O}_2 \text{ / NaOH} \\ 3) \text{ Ac}_2\text{O / C}_5\text{H}_5\text{N}}]{1) \text{ 9-BBN}}$ n-BuC=C(CH$_2$)$_4$OAc (H H)

22

Scheme **31**

(Z)-5-Dodecen-1-yl Acetate 22

This is a sex attractant for the male turnip moth *Agrotis segetum*, a ubiquitous grain pest.[132] A synthesis was carried out by a Wittig route (Scheme 31).[132]

(Z)-7-Dodecen-1-yl Acetate 23

This is the pheromone of the cabbage looper (*Trichoplusia ni*).[133] The first synthesis is straight-forward via acetylenic intermediates (Scheme 32).[133] A synthesis by Schäfer et al. employed Kolbe electrolysis in the key step (Scheme 33).[134] Schäfer used ^{1}H- and ^{13}C-NMR to determine the Z/E ratio of the key intermediate, 4-nonenoic acid. Useful information was obtained by the inspection of the ^{1}H-NMR spectrum of the acid in the presence of Eu(fod)$_3$. Its ^{13}C-NMR, however, was more informative. The chemical shift value of C-6 was 27.13 for Z and 32.43 for E and that of C-3 was 22.55 for Z and 27.84 for E. The $Z : E$ ratio was shown to be 4 : 1.[134]

(Z)-8-Dodecen-1-yl Acetate 24

The oriental fruit moth, *Grapholitha molesta*, uses this compound as the sex pheromone.[135] The structure was confirmed by synthesis via both the Wittig and acetylenic routes.[135] A new version of the acetylenic route is shown in Scheme 34.[136] Another synthesis by an acetylenic route was recently reported by a Chinese group, who used palladium-calcium carbonate as the catalyst for partial hydrogenation.[137] A recent synthesis employed cyclohexane-1,3-dione as a C_6 synthon (Scheme 35).[138]

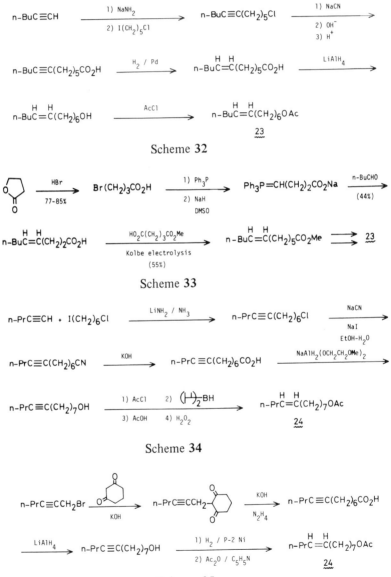

Scheme 32

Scheme 33

Scheme 34

Scheme 35

(Z)-9-Dodecen-1-yl Acetate 25

This is the pheromone of the female grape berry moth (*Paralobesia viteana*)[139] and *Eupoecillia ambiguella*.[140] Its synthesis by the acetylenic route is shown in Scheme 36.[141] A new synthesis used cheap hexamethylene glycol as a starting material instead of more expensive octamethylene glycol, but proceeded in somewhat lower overall yield (Scheme 37).[142]

(Z)-7-Tetradecen-1-yl Acetate 26

This is the pheromone of *Amathes c-nigrum* found both in Japan[143] and in Germany.[144] Its synthesis was carried out by the Wittig route (Scheme 38).[144]

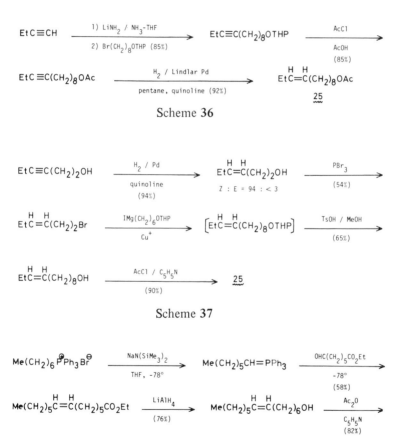

Scheme 36

Scheme 37

Scheme 38

(Z)-9-Tetradecen-1-yl Acetate 27

This is the pheromone of the fall armyworm (*Spodoptera frugiperda*),[145] the smaller tea tortrix (*Adoxophyes fasciata*),[146] and some other insects. The first synthesis used methyl myristolate as the starting material (Scheme 39).[145] The second synthesis was based on conventional acetylene chemistry (Scheme 40),[147] and the third one was a partial synthesis from **23** (Scheme 41).[148] A Wittig synthesis was also reported (Scheme 42).[149]

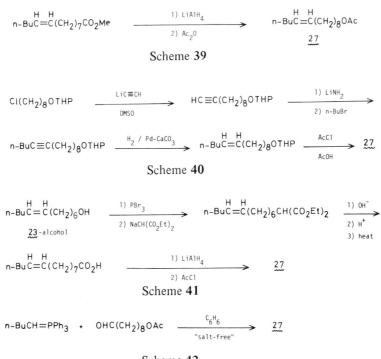

Scheme 39

Scheme 40

Scheme 41

Scheme 42

(Z)-11-Tetradecen-1-yl Acetate 2

The red-banded leaf roller (*Argyrotaenia velutinana*),[6] the smaller tea tortrix (*Adoxophyes fasciata*),[146] and several other insects use this compound as a pheromone. The synthesis was carried out either through the acetylenic route[6] or by employing the Wittig reaction (Scheme 43).[6, 146]

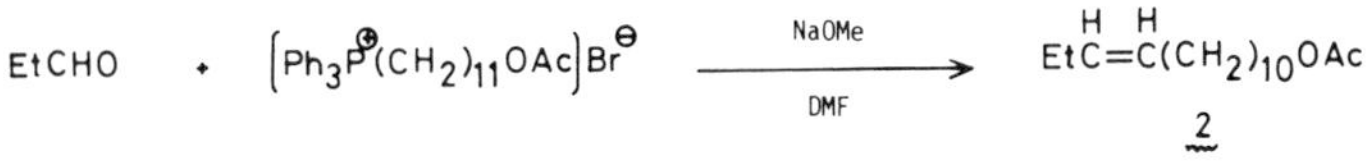

Scheme 43

(Z)-11-Hexadecen-1-yl Acetate 28

This is the pheromone of the purple stem borer (*Sesamia inferens*), a noctuid moth whose larvae attack a wide range of graminaceous crops.[150] This is also the pheromone of *Mamestra brassicae*.[151] The synthesis was carried out through the acetylenic route (Scheme 44).[150, 151]

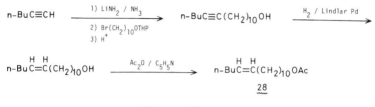

Scheme **44**

(Z)-11-Hexadecenal 29 and (Z)-13-Octadecenal 30

The striped rice borer, *Chilo suppressalis*, is a serious pest of rice in Asian countries. Its female sex pheromone is a 5 : 1 mixture of **29** and **30**.[152] They were synthesized by the conventional acetylenic route (Scheme 45).[152]

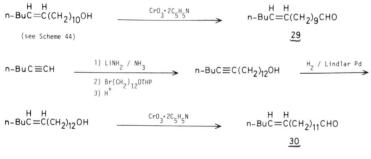

Scheme **45**

(Z)-3-Decenoic Acid 31

The sex pheromone of the furniture carpet beetle (*Anthrenus flavipes*) was identified as **31** (Scheme 46).[153]

(Z)-14-Methyl-8-hexadecen-1-ol 32 and Methyl (Z)-14-methyl-8-hexadecenoate 33

These were isolated as the sex pheromones of the female dermestid beetle (*Trogoderma inclusum*).[154] Later these were shown to be artifacts in the course of isolation of the genuine pheromone, (Z)-14-methyl-8-hexadecenal.[155] The earlier two syntheses employed the Wittig reaction, as shown in Schemes 47[154] and 48.[156] The third synthesis utilized an interesting Cope rearrangement reac-

$\text{Me(CH}_2)_5\text{C}\equiv\text{C(CH}_2)_2\text{OH} \xrightarrow{\text{CrO}_3} \text{Me(CH}_2)_5\text{C}\equiv\text{CCH}_2\text{CO}_2\text{H}$

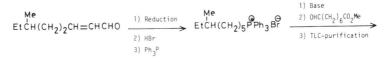

$\xrightarrow[\text{C}_5\text{H}_5\text{N-EtOH}]{\text{H}_2 / \text{Pd-C}} \text{Me(CH}_2)_5\overset{\text{H}}{\text{C}}=\overset{\text{H}}{\text{C}}\text{CH}_2\text{CO}_2\text{H}$

31

Scheme **46**

$\text{MeCH}=\text{CHMe} \xrightarrow[\text{2) CH}_2=\text{CHCHO}]{\text{1) B}_2\text{H}_6} \text{Et}\overset{\text{Me}}{\text{CH}}(\text{CH}_2)_2\text{CHO} \xrightarrow{\text{Ph}_3\text{P}=\text{CHCHO}}$

$\text{Et}\overset{\text{Me}}{\text{CH}}(\text{CH}_2)_2\text{CH}=\text{CHCHO} \xrightarrow[\substack{\text{2) HBr} \\ \text{3) Ph}_3\text{P}}]{\text{1) Reduction}} \text{Et}\overset{\text{Me}}{\text{CH}}(\text{CH}_2)_5\overset{\oplus}{\text{P}}\text{Ph}_3\overset{\ominus}{\text{Br}} \xrightarrow[\substack{\text{2) OHC(CH}_2)_6\text{CO}_2\text{Me} \\ \text{3) TLC-purification}}]{\text{1) Base}}$

$\text{Et}\overset{\text{Me}}{\text{CH}}(\text{CH}_2)_4\overset{\text{H}}{\text{C}}=\overset{\text{H}}{\text{C}}(\text{CH}_2)_6\text{CO}_2\text{Me} \xrightarrow{\text{LiAlH}_4} \text{Et}\overset{\text{Me}}{\text{CH}}(\text{CH}_2)_4\overset{\text{H}}{\text{C}}=\overset{\text{H}}{\text{C}}(\text{CH}_2)_7\text{OH}$

32

Scheme **47**

$\text{Et}\overset{\text{Me}}{\text{CH}}\text{C}\equiv\text{CBr} \xrightarrow[\text{O}_2, \text{ EtNH}_2, \text{ CuCl}]{\text{HC}\equiv\text{CCH}_2\text{OH}} \text{Et}\overset{\text{Me}}{\text{CH}}\text{C}\equiv\text{C}-\text{C}\equiv\text{CCH}_2\text{OH} \xrightarrow{\text{H}_2 / \text{Pt}}$

$\text{Et}\overset{\text{Me}}{\text{CH}}(\text{CH}_2)_4\text{CH}_2\text{OH} \xrightarrow{\text{CrO}_3\cdot2\text{C}_5\text{H}_5\text{N}} \text{Et}\overset{\text{Me}}{\text{CH}}(\text{CH}_2)_4\text{CHO} \xrightarrow[\text{NaOMe / DMF}]{\text{Ph}_3\text{P}^+(\text{CH}_2)_7\text{CO}_2\text{MeBr}^-}$

$\text{Et}\overset{\text{Me}}{\text{CH}}(\text{CH}_2)_4\overset{\text{H}}{\text{C}}=\overset{\text{H}}{\text{C}}(\text{CH}_2)_6\text{CO}_2\text{Me} \xrightarrow{\text{LiAlH}_4} \text{Et}\overset{\text{Me}}{\text{CH}}(\text{CH}_2)_4\overset{\text{H}}{\text{C}}=\overset{\text{H}}{\text{C}}(\text{CH}_2)_7\text{OH}$

33   **32**

Scheme **48**

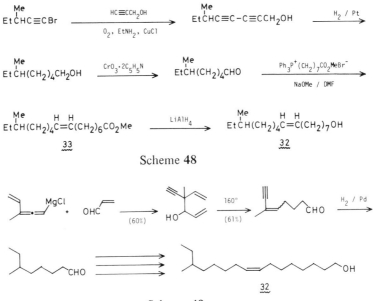

32

Scheme **49**

tion (Scheme 49).[157] The synthesis of optically active pheromones will be described later (Schemes 221 and 222).

6. PHEROMONES WITH A CONJUGATED DIENE SYSTEM

(7E, 9Z)-7,9-Dodecadien-1-yl Acetate 34

The sex pheromone of the grape vine moth (*Lobesia botrana*) was identified as **34**.[158] The first synthesis by Roelofs is shown in Scheme 50.[158] A synthesis by Descoins involves a new method of stereoselective synthesis of a conjugated diene with *E*, *Z*-geometry. Dicobalt octacarbonyl was used as the protecting group for the triple bond.[159] The bulkiness of this protected acetylene makes the Julia cleavage of the cyclopropane ring stereoselective (Scheme 51).[160]

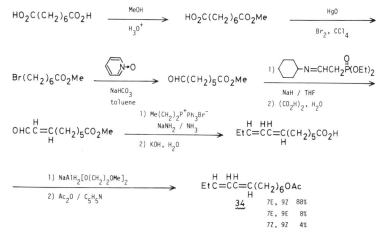

Scheme 50

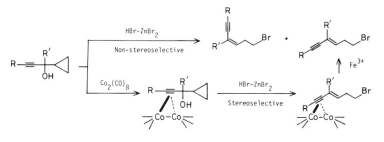

Scheme 51

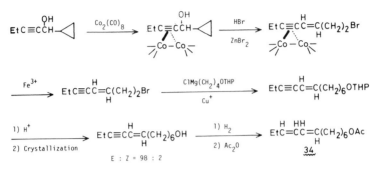

Scheme 52

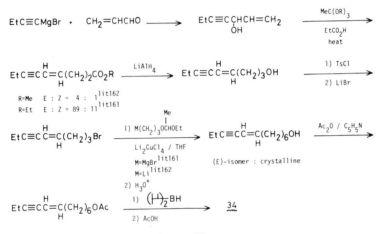

Scheme 53

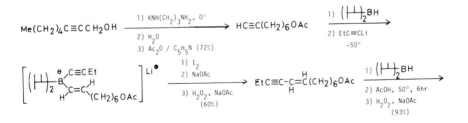

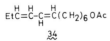

Scheme 54

32

The synthesis based on this idea is shown in Scheme 52.[161] A synthesis employing the orthoester Claisen rearrangement as the key step was reported by two groups (Scheme 53).[161,162] Organoborane chemistry was also applied in the synthesis of this pheromone (Scheme 54).[163] The acetylene "zipper" reaction was used in the first step to prepare the 1-alkyne (cf. Equation 29).[163] Recently a simple synthesis of an internal conjugated (Z)-enyne was reported, which might be useful in preparing the starting material for **34** (Scheme 55).[164] In this case the undesired (E, E)-isomer was easily removed by the formation of its Diels-Alder adduct with tetracyanoethylene in THF.

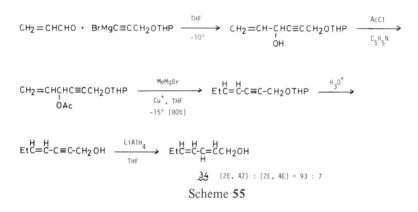

Scheme 55

(8E, 10E)-8,10-Dodecadien-1-ol 35

Roelofs et al. isolated this compound as the pheromone of the codling moth (*Laspeyresia pomonella*), a notorious pest of apple orchards.[165] The first synthesis used the Wittig reaction (Scheme 56),[165] and the later syntheses were all carried out by the Grignard coupling reaction. The first Zoecon synthesis was more stereoselective than Roelofs's synthesis, but employed rather inaccessible C_7 and C_5 synthons (Scheme 57).[166] However, the Zoecon group found that pure **35** is crystalline. This facilitated later work when it became possible to isolate pure **35** by recrystallization from pentane at low temperature. Then Mori used two readily accessible C_6 synthons in his synthesis to obtain crystalline

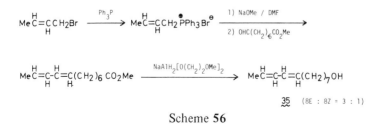

Scheme 56

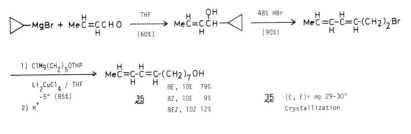

Scheme **57**

35. (Scheme 58).[167] Sorbic acid and hexamethylene glycol were the starting materials. The second Zoecon synthesis is shown in Scheme 59.[168] Recent syntheses, including the third Zoecon synthesis,[169] used sorbyl acetate as the starting material and were more efficient than any of the previous syntheses (Scheme 60).[169-171] The synthesis of **35** is discussed in detail by Henrick.[22]

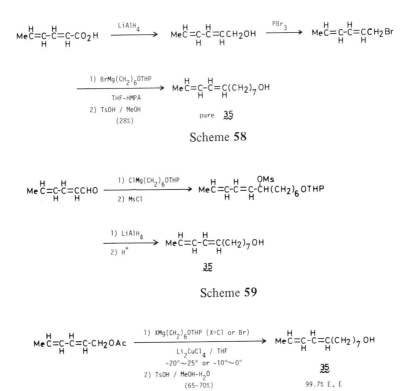

Scheme **58**

Scheme **59**

Scheme **60**

(E)-9,11-Dodecadien-1-yl Acetate 36

The red-bollworm moth (*Diparopsis castanea*) is a major cotton pest in south-eastern Africa. The most potent of the sex pheromones produced by the virgin female was shown to be (*E*)-9,11-dodecadien-1-yl acetate 36 by Nesbitt et al.[172] The components of the pheromones are 9,11-dodecadien-1-yl acetate (*E* : *Z* = 80 : 20), (*E*)-9-dodecen-1-yl acetate, 11-dodecen-1-yl acetate, and dodecan-1-yl acetate. The first synthesis by Nesbitt et al. was accomplished by the Wittig route (Scheme 61).[173] The stereoselectivity of the olefination step was low (*E* : *Z* = 3 : 2). However, the desired (*E*)-isomer was converted to a sulfolene by treatment with liquid sulfur dioxide and separated as such from the unreacted

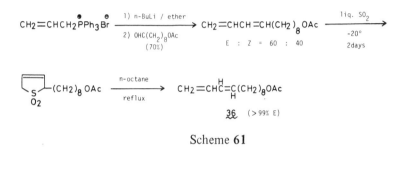

Scheme 61

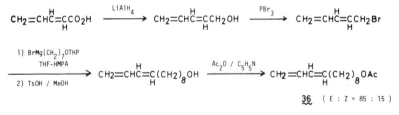

Scheme 62

(*Z*)-isomer by chromatography over Florisil. The sulfolene gave back pure 36 on thermolysis. Mori's synthesis was based on the Grignard coupling reaction (Scheme 62).[167] Since the C_{12}-alcohol was an oil at room temperature, its purification was not so successful as in the case of the codling moth pheromone 35 and the yield as well as purity of 36 was not very high in this synthesis. The third synthesis by Yamamoto et al. was based on their new general method for the preparation of 1,3-dienes (Scheme 63).[174] The fourth synthesis by Babler employed the Wittig reaction, as shown in Scheme 64.[175] Recently Tsuji synthesized a mixture of 36 and its (*Z*)-isomer from butadiene, employing a palladium-catalyzed reaction (Scheme 65).[176]

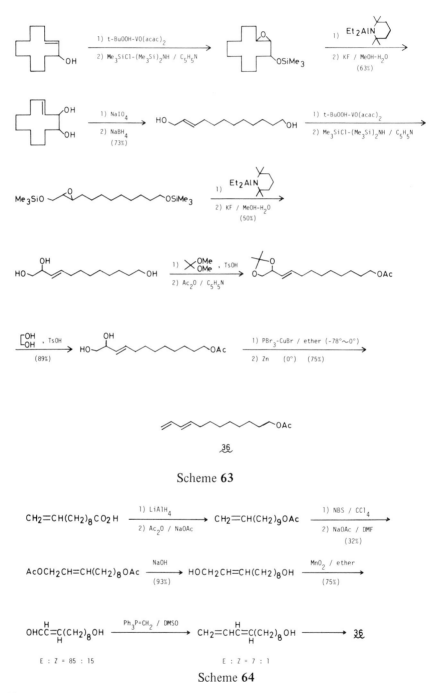

Scheme **63**

CH₂=CH(CH₂)₈CO₂H $\xrightarrow[\text{2) Ac}_2\text{O / NaOAc}]{\text{1) LiAlH}_4}$ CH₂=CH(CH₂)₉OAc $\xrightarrow[\substack{\text{2) NaOAc / DMF}\\(32\%)}]{\text{1) NBS / CCl}_4}$

AcOCH₂CH=CH(CH₂)₈OAc $\xrightarrow[\substack{(93\%)}]{\text{NaOH}}$ HOCH₂CH=CH(CH₂)₈OH $\xrightarrow[\substack{(75\%)}]{\text{MnO}_2 \text{ / ether}}$

OHCC=C(CH₂)₈OH $\xrightarrow{\text{Ph}_3\text{P=CH}_2 \text{ / DMSO}}$ CH₂=CHC=C(CH₂)₈OH ⟶ **36**

E : Z = 85 : 15   E : Z = 7 : 1

Scheme **64**

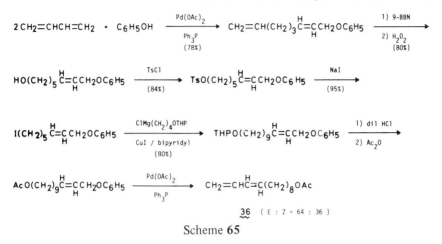

Scheme 65

(3Z, 5E)-3,5-Tetradecadien-1-yl Acetate 37

This is the sex attractant for the female carpenterworm moth (*Prionoxystus robiniae*).[177] The synthesis outlined in Scheme 66 yielded a 1 : 2 mixture of (3Z, 5E)- and (3E, 5E)-isomers, which were separated by GLC, HPLC, or spinning band distillation.[177] The Wittig reaction with (3-hydroxypropyl)triphenyl-phosphonium iodide was unsuccessful when 2 eq sodium hydride in DMSO was employed to generate the phosphorane. Use of 1 eq sodium hydride and 1 eq *n*-butyllithium was essential for the success.

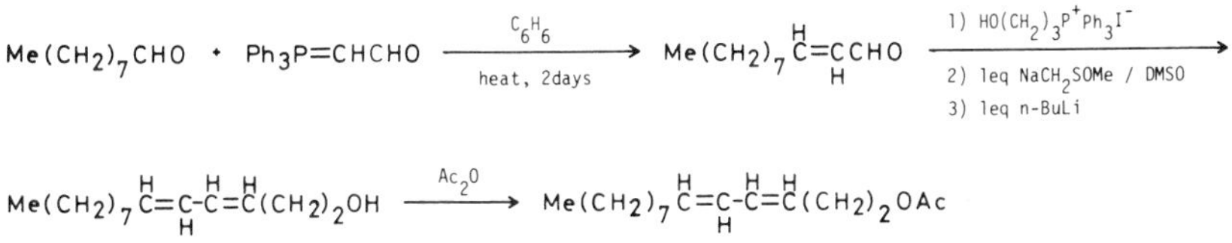

$$Me(CH_2)_7CHO \; + \; Ph_3P{=}CHCHO \xrightarrow[\text{heat, 2days}]{C_6H_6} Me(CH_2)_7\overset{H}{\underset{H}{C}}{=}CCHO \xrightarrow{\substack{1) \; HO(CH_2)_3P^+Ph_3I^- \\ 2) \; 1eq \; NaCH_2SOMe \; / \; DMSO \\ 3) \; 1eq \; n\text{-BuLi}}}$$

$$Me(CH_2)_7\overset{H}{\underset{H}{C}}{=}\overset{H}{C}{-}\overset{H}{C}{=}\overset{H}{C}(CH_2)_2OH \xrightarrow{Ac_2O} Me(CH_2)_7\overset{H}{\underset{H}{C}}{=}\overset{H}{C}{-}\overset{H}{C}{=}\overset{H}{C}(CH_2)_2OAc$$

37 (3Z, 5E) : (3E, 5E) = 1 : 2

Scheme **66**

(9Z, 11E)-9, 11-Tetradecadien-1-yl Acetate 38

This compound is the major component of the sex pheromone of *Spodoptera litura*, a serious pest of vegetable crops in Japan.[178,179] It was also isolated from the Egyptian cotton leafworm (*Spodoptera littoralis*).[180] The first synthesis was nonstereoselective (Scheme 67).[178] The second synthesis was 80-90% stereoselective and carried out independently by two groups (Scheme 68).[71,72] The Wittig olefination in DMSO was used to generate the (Z)-olefin and the undesired (E, E)-diene was removed as a Diels-Alder adduct with tetracyanoethylene.

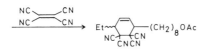

Scheme 67

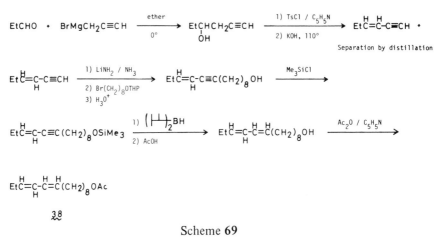

(Removable by chromatography)

Scheme 68

Scheme 69

Another synthesis was accomplished by the acetylenic route (Scheme 69).[181] Spinning band distillation of an isomeric mixture of hex-3-en-1-yne yielded pure (*E*)-enyne, which was converted to **38** with 98.7% purity containing 0.5% of the (*E*, *E*)-isomer.

(3E, 5Z)-3,5-Tetradecadienoic Acid 39

This is the pheromone of black carpet beetle (*Attagenus megatoma*), and was synthesized as shown in Scheme 70.[182]

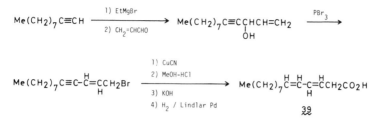

Scheme **70**

Bombykol, (10E, 12Z)-10,12-Hexadecadien-1-ol 1

The isolation[183] and structure elucidation[184] of bombykol were fully described in 1961. An earlier nonstereoselective synthesis was achieved by Butenandt et al.[3,4] and by Truscheit and Eiter.[5] The separations of the intermediates as shown in Schemes 71[4], 72[4], 73[5], 74[5], and 75[5] enabled them to firmly establish

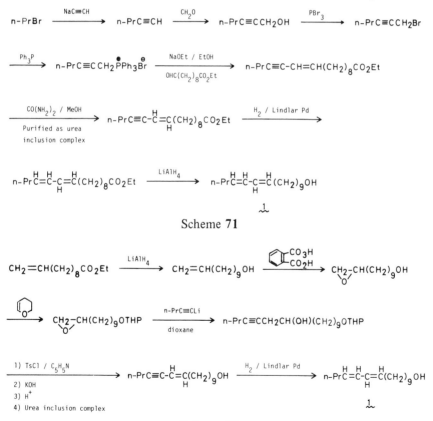

Scheme **71**

Scheme **72**

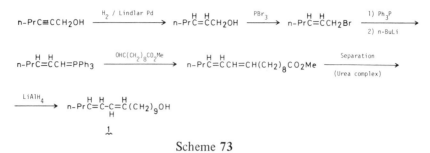

Scheme 73

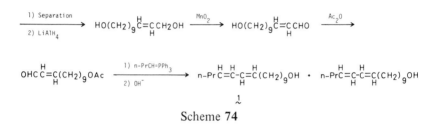

Scheme 74

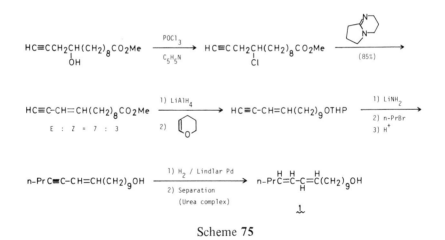

Scheme 75

the structure of bombykol as **1** by comparing the pheromone activity of the geometrical isomers. Bestmann recently modified Eiter's Wittig route (Scheme 74) to achieve higher stereoselectivity (Scheme 76).[185] Organoborane chemistry was successfully applied in a recently reported stereoselective synthesis (Scheme

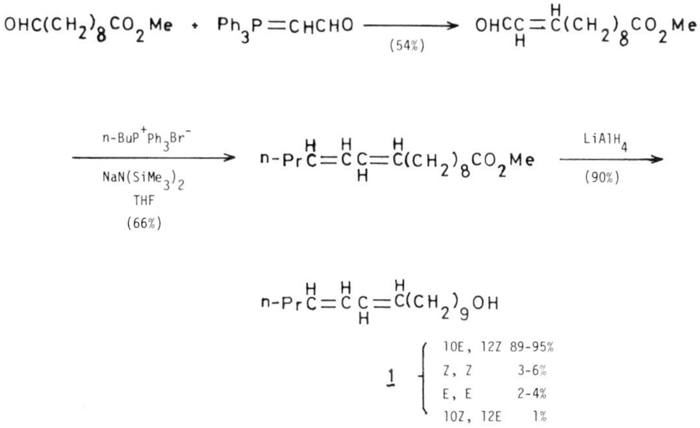

$$OHC(CH_2)_8CO_2Me \ + \ Ph_3P=CHCHO \xrightarrow[(54\%)]{} OHC\overset{H}{C}=\overset{H}{C}(CH_2)_8CO_2Me$$

$$\xrightarrow[\substack{NaN(SiMe_3)_2 \\ THF \\ (66\%)}]{n\text{-}BuP^+Ph_3Br^-} \ n\text{-}Pr\overset{H}{C}=\overset{H}{C}\overset{H}{C}=\overset{H}{C}(CH_2)_8CO_2Me \xrightarrow[(90\%)]{LiAlH_4}$$

$$n\text{-}Pr\overset{H}{C}=\overset{H}{C}\overset{H}{C}=\overset{H}{C}(CH_2)_9OH$$

$$\underline{1} \quad \begin{cases} 10E, \ 12Z & 89\text{-}95\% \\ Z, \ Z & 3\text{-}6\% \\ E, \ E & 2\text{-}4\% \\ 10Z, \ 12E & 1\% \end{cases}$$

Scheme 76

$$HC\equiv C(CH_2)_9OSiMe_3 \xrightarrow[THF]{(\boxed{})_2\text{-}BH} \left(\boxed{}\right)_2 \overset{B}{\underset{H}{\diagdown}}\overset{H}{\underset{(CH_2)_9OSiMe_3}{C=C}} \xrightarrow[\substack{THF\text{-}hexane \\ -50°}]{n\text{-}PrC\equiv CLi}$$

$$\left[\left(\boxed{}\right)_2 \overset{\ominus}{B}\overset{C\equiv C(n\text{-}Pr)}{\underset{\overset{H}{C=C}\underset{(CH_2)_9OSiMe_3}{}}{}} \right] \overset{\ominus}{Li} \xrightarrow[\substack{2) \ 3N\text{-}NaOH \\ (63\%)}]{1) \ I_2 \ / \ THF, \ -78\sim25°}$$

$$n\text{-}PrC\equiv C\overset{H}{C}=\overset{H}{C}(CH_2)_9OSiMe_3 \xrightarrow[\substack{2) \ i\text{-}PrCO_2H \ / \ THF \\ reflux}]{1) \ (\boxed{})_2\text{-}BH,0°, \ 1hr \quad LiAlH_4} \ n\text{-}Pr\overset{H}{C}=\overset{H}{C}\overset{H}{C}=\overset{H}{C}(CH_2)_9OH$$

$$\underline{1}$$

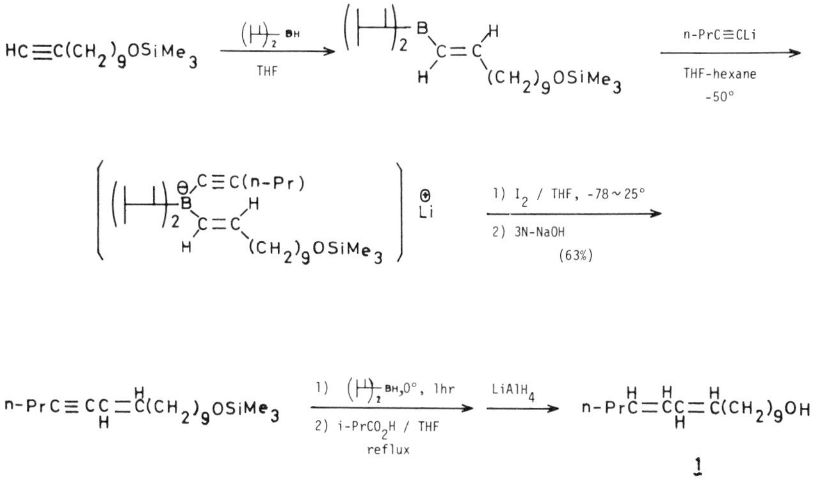

Scheme 77

77).[186] Another stereoselective synthesis was based on organocopper and related carbanion chemistry (Scheme 78).[187] Various syntheses of bombykol are a reflection of the synthetic methodology available at the time of achievement. The most recent two syntheses are remarkable in their high stereoselectivities. In 1978 another pheromone of the silkworm moth was isolated and identified as bombykal (10E, 12Z)-10,12-hexadecadienal.[188] The composition of the silkworm moth pheromone was 92.8% bombykol and 7.2% bombykal.[188a] Bombykal acts as an inhibitor against the action of bombykol.[188b]

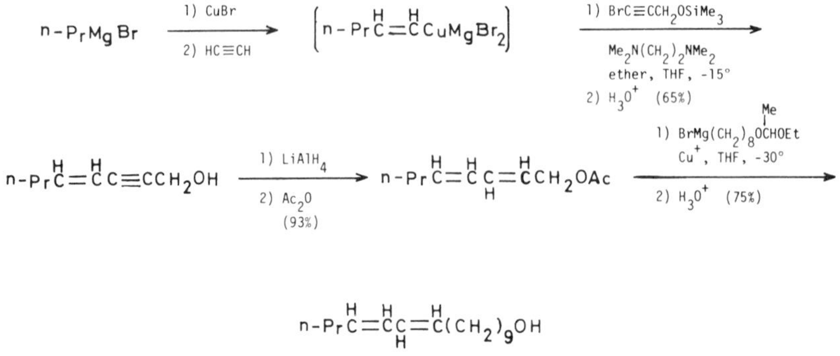

$$n-P_rMgBr \xrightarrow[\text{2) HC}\equiv\text{CH}]{\text{1) CuBr}} \left[n-PrC\overset{H}{=}\overset{H}{C}CuMgBr_2 \right] \xrightarrow[\substack{Me_2N(CH_2)_2NMe_2 \\ \text{ether, THF, } -15^\circ \\ \text{2) } H_3O^+ \ (65\%)}]{\text{1) BrC}\equiv\text{CCH}_2OSiMe_3}$$

$$n-PrC\overset{H}{=}\overset{H}{C}C\equiv CCH_2OH \xrightarrow[\substack{\text{2) Ac}_2O \\ (93\%)}]{\text{1) LiAlH}_4} n-PrC\overset{H}{=}\overset{H}{C}C\overset{H}{=}\overset{H}{C}CH_2OAc \xrightarrow[\substack{Cu^+, \text{ THF, } -30^\circ \\ \text{2) } H_3O^+ \ (75\%)}]{\substack{\text{1) BrMg(CH}_2)_8O\overset{Me}{C}HOEt}}$$

$$n-PrC\overset{H}{=}\overset{H}{C}C\overset{H}{=}\overset{H}{C}(CH_2)_9OH$$

$$\underline{1}$$

Scheme 78

7. PHEROMONES WITH A NONCONJUGATED DIENE SYSTEM

(4E., 7Z)-4,7-Tridecadien-1-yl Acetate 40

This is a component of the sex pheromone of the potato tuberworm moth (*Phthorimaea operculella*) and unusual as an odd-carbon-chain compound.[189]

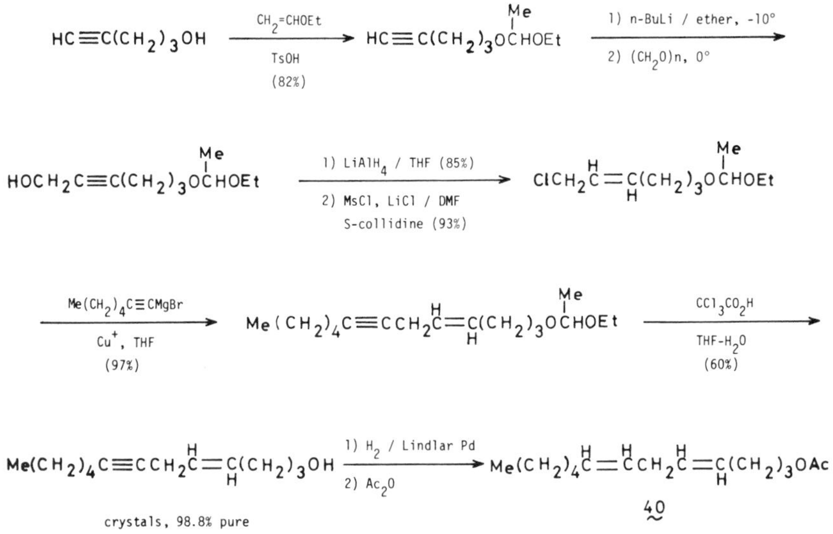

$$HC\equiv C(CH_2)_3OH \xrightarrow[\substack{TsOH \\ (82\%)}]{CH_2=CHOEt} HC\equiv C(CH_2)_3O\overset{Me}{C}HOEt \xrightarrow[\text{2) } (CH_2O)n, \ 0^\circ]{\text{1) n-BuLi / ether, } -10^\circ}$$

$$HOCH_2C\equiv C(CH_2)_3O\overset{Me}{C}HOEt \xrightarrow[\substack{\text{2) MsCl, LiCl / DMF} \\ \text{S-collidine (93\%)}}]{\text{1) LiAlH}_4 \text{ / THF (85\%)}} ClCH_2\overset{H}{C}=\underset{H}{C}(CH_2)_3O\overset{Me}{C}HOEt$$

$$\xrightarrow[\substack{Cu^+, \text{ THF} \\ (97\%)}]{Me(CH_2)_4C\equiv CMgBr} Me(CH_2)_4C\equiv CCH_2\overset{H}{C}=\underset{H}{C}(CH_2)_3O\overset{Me}{C}HOEt \xrightarrow[\substack{THF-H_2O \\ (60\%)}]{CCl_3CO_2H}$$

$$Me(CH_2)_4C\equiv CCH_2\overset{H}{C}=\underset{H}{C}(CH_2)_3OH \xrightarrow[\text{2) Ac}_2O]{\text{1) H}_2 \text{ / Lindlar Pd}} Me(CH_2)_4\overset{H}{C}=\overset{H}{C}CH_2\overset{H}{C}=\underset{H}{C}(CH_2)_3OAc$$

crystals, 98.8% pure

$$\underset{\sim}{40}$$

Scheme 79

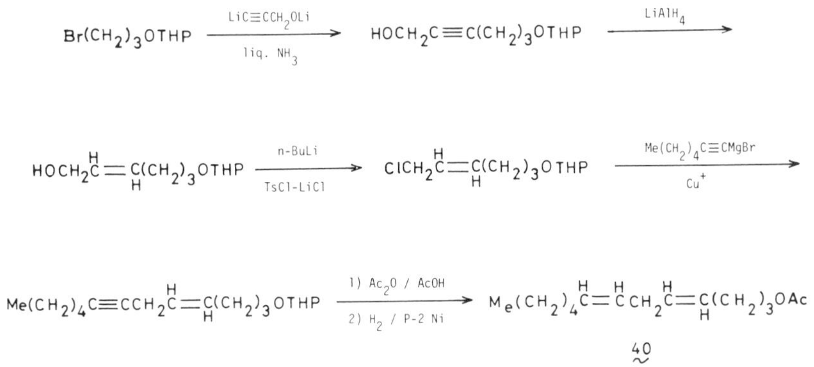

Br(CH$_2$)$_3$OTHP $\xrightarrow[\text{liq. NH}_3]{\text{LiC}\equiv\text{CCH}_2\text{OLi}}$ HOCH$_2$C≡C(CH$_2$)$_3$OTHP $\xrightarrow{\text{LiAlH}_4}$

HOCH$_2$C=C(CH$_2$)$_3$OTHP $\xrightarrow[\text{TsCl-LiCl}]{\text{n-BuLi}}$ ClCH$_2$C=C(CH$_2$)$_3$OTHP $\xrightarrow[\text{Cu}^+]{\text{Me(CH}_2)_4\text{C}\equiv\text{CMgBr}}$

Me(CH$_2$)$_4$C≡CCH$_2$C=C(CH$_2$)$_3$OTHP $\xrightarrow[\text{2) H}_2\text{ / P-2 Ni}]{\text{1) Ac}_2\text{O / AcOH}}$ Me(CH$_2$)$_4$C=CCH$_2$C=C(CH$_2$)$_3$OAc

40

(overall yield 14%)

Scheme **80**

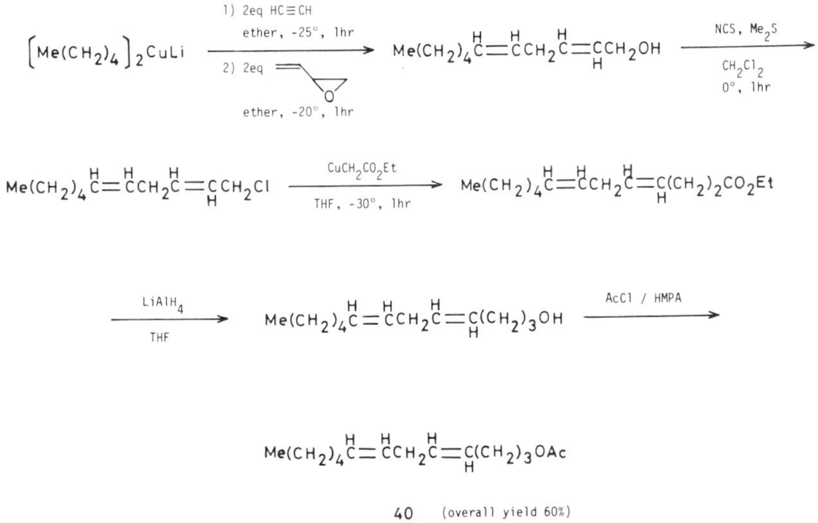

[Me(CH$_2$)$_4$]$_2$CuLi $\xrightarrow[\text{2) 2eq}\,\triangle\,\text{ether, -20°, 1hr}]{\text{1) 2eq HC≡CH ether, -25°, 1hr}}$ Me(CH$_2$)$_4$C=CCH$_2$C=CCH$_2$OH $\xrightarrow[\text{CH}_2\text{Cl}_2\ 0°, 1hr]{\text{NCS, Me}_2\text{S}}$

Me(CH$_2$)$_4$C=CCH$_2$C=CCH$_2$Cl $\xrightarrow[\text{THF, -30°, 1hr}]{\text{CuCH}_2\text{CO}_2\text{Et}}$ Me(CH$_2$)$_4$C=CCH$_2$C=C(CH$_2$)$_2$CO$_2$Et

$\xrightarrow[\text{THF}]{\text{LiAlH}_4}$ Me(CH$_2$)$_4$C=CCH$_2$C=C(CH$_2$)$_3$OH $\xrightarrow{\text{AcCl / HMPA}}$

Me(CH$_2$)$_4$C=CCH$_2$C=C(CH$_2$)$_3$OAc

40   (overall yield 60%)

Scheme **81**

The synthesis is shown in Scheme 79.[189] An alternative but similar synthesis was also achieved by the acetylenic route (Scheme 80).[190] The third synthesis by Normant et al. was based on organocopper chemistry (Scheme 81).[191] Another component of the pheromone is (4E, 7E, 10Z)-4,7,10-tridecatrien-1-yl acetate whose synthesis is discussed later (Scheme 100).

(9Z, 12E)-9,12-Tetradecadien-1-yl Acetate **41**

This was first isolated and synthesized by Jacobson et al. as a component of the pheromone of the southern armyworm moth (*Spodoptera eridiana*).[192] Another

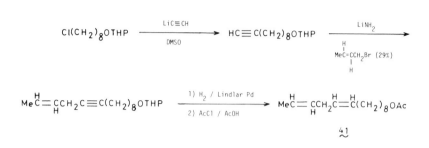

Scheme **82**

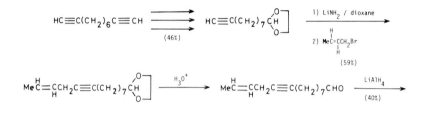

Scheme **83**

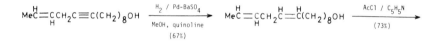

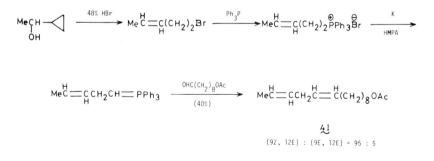

Scheme **84**

component of the pheromone of this insect was (*Z*)-9-tetradecen-1-yl acetate **27**. Later **41** was shown to be the pheromone of the almond moth (*Cadra cautella*), the Indian meal moth (*Plodia interpunctella*),[193, 194] and several other insects. Jacobson's synthesis is shown in Scheme 82.[192] Brady's synthesis started from 1,9-decadiyne as shown in Scheme 83.[195] Bestmann's synthesis was a combination of the Julia cyclopropane cleavage and the Wittig reaction (Scheme 84).[196] Bestmann recently synthesized the (9*E*, 12*E*)-isomer of **41** by Kondo's phosphonate method (Equation 5) as shown in Scheme 85.[197]

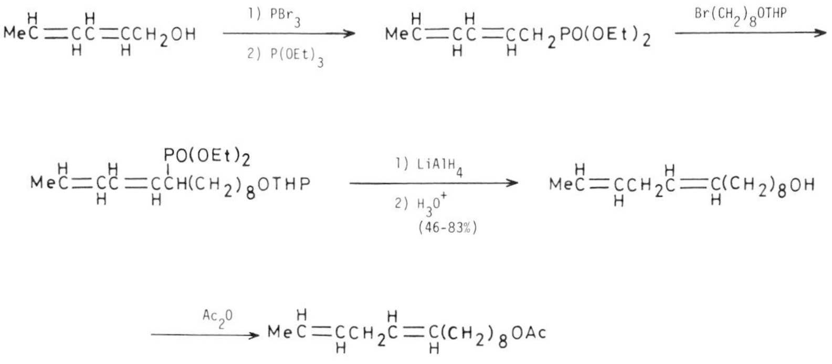

$$\text{MeC}\!\!=\!\!\text{CC}\!\!=\!\!\text{CCH}_2\text{OH} \xrightarrow[\text{2) P(OEt)}_3]{\text{1) PBr}_3} \text{MeC}\!\!=\!\!\text{CC}\!\!=\!\!\text{CCH}_2\text{PO(OEt)}_2 \xrightarrow{\text{Br(CH}_2)_8\text{OTHP}}$$

$$\text{MeC}\!\!=\!\!\text{CC}\!\!=\!\!\text{CCH(CH}_2)_8\text{OTHP} \xrightarrow[\text{2) H}_3\text{O}^+]{\text{1) LiAlH}_4} \text{MeC}\!\!=\!\!\text{CCH}_2\text{C}\!\!=\!\!\text{C(CH}_2)_8\text{OH}$$
$$\overset{\text{PO(OEt)}_2}{} \qquad (46\text{-}83\%)$$

$$\xrightarrow{\text{Ac}_2\text{O}} \text{MeC}\!\!=\!\!\text{CCH}_2\text{C}\!\!=\!\!\text{C(CH}_2)_8\text{OAc}$$

41

Scheme 85

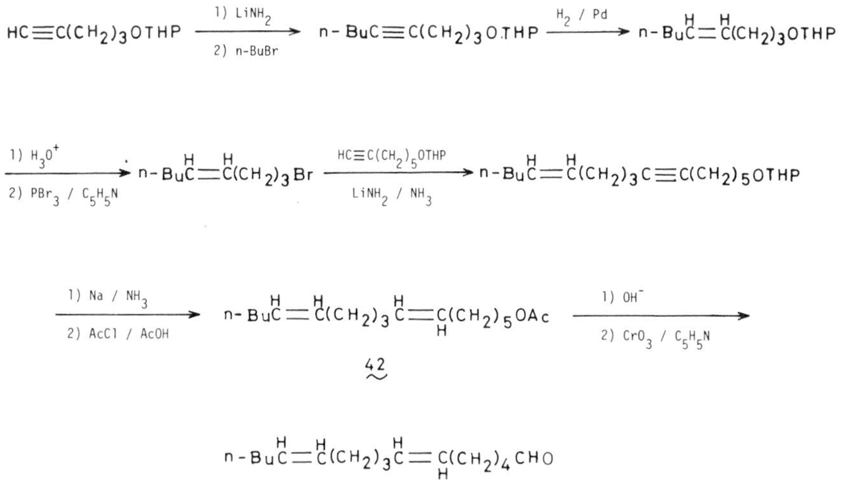

$$\text{HC}\!\!\equiv\!\!\text{C(CH}_2)_3\text{OTHP} \xrightarrow[\text{2) n-BuBr}]{\text{1) LiNH}_2} \text{n-BuC}\!\!\equiv\!\!\text{C(CH}_2)_3\text{OTHP} \xrightarrow{\text{H}_2\text{ / Pd}} \text{n-BuC}\!\!=\!\!\text{C(CH}_2)_3\text{OTHP}$$

$$\xrightarrow[\text{2) PBr}_3\text{ / C}_5\text{H}_5\text{N}]{\text{1) H}_3\text{O}^+} \text{n-BuC}\!\!=\!\!\text{C(CH}_2)_3\text{Br} \xrightarrow[\text{LiNH}_2\text{ / NH}_3]{\text{HC}\!\equiv\!\text{C(CH}_2)_5\text{OTHP}} \text{n-BuC}\!\!=\!\!\text{C(CH}_2)_3\text{C}\!\!\equiv\!\!\text{C(CH}_2)_5\text{OTHP}$$

$$\xrightarrow[\text{2) AcCl / AcOH}]{\text{1) Na / NH}_3} \text{n-BuC}\!\!=\!\!\text{C(CH}_2)_3\text{C}\!\!=\!\!\text{C(CH}_2)_5\text{OAc} \xrightarrow[\text{2) CrO}_3\text{ / C}_5\text{H}_5\text{N}]{\text{1) OH}^-}$$

42

$$\text{n-BuC}\!\!=\!\!\text{C(CH}_2)_3\text{C}\!\!=\!\!\text{C(CH}_2)_4\text{CHO}$$

43   (overall yield ~1%)

Scheme 86

(6E, 11Z)-6,11-Hexadecadien-1-yl Acetate 42 *and (6E, 11Z)-6, 11-Hexadecadienal* 43

These were isolated and synthesized by Roelofs et al. as the sex pheromone of the wild silkmoth (*Antheraea polyphemus*).[198] The pheromone is a 90 : 10 mixture of 42 and 43. The synthesis is shown in Scheme 86.[198] This synthesis proceeded in extremely low overall yield (~1%). A modified synthetic procedure as shown in Scheme 87 gave 42 in 21% overall yield.[199]

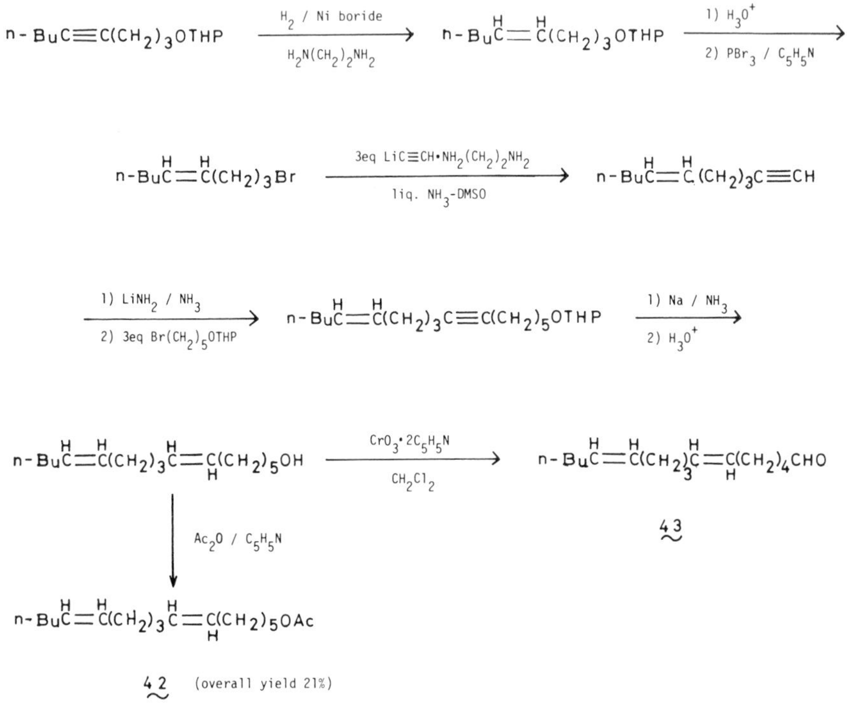

$$n\text{-}Bu\text{C}\equiv\text{C}(CH_2)_3OTHP \xrightarrow[\text{H}_2N(CH_2)_2NH_2]{\text{H}_2 \text{ / Ni boride}} n\text{-}Bu\overset{H}{\text{C}}=\overset{H}{\text{C}}(CH_2)_3OTHP \xrightarrow[\text{2) PBr}_3 \text{ / } C_5H_5N]{\text{1) H}_3O^+}$$

$$n\text{-}Bu\overset{H}{\text{C}}=\overset{H}{\text{C}}(CH_2)_3Br \xrightarrow[\text{liq. NH}_3\text{-DMSO}]{\text{3eq LiC}\equiv\text{CH}\cdot\text{NH}_2(CH_2)_2NH_2} n\text{-}Bu\overset{H}{\text{C}}=\overset{H}{\text{C}}(CH_2)_3C\equiv CH$$

$$\xrightarrow[\text{2) 3eq Br(CH}_2)_5OTHP]{\text{1) LiNH}_2 \text{ / NH}_3} n\text{-}Bu\overset{H}{\text{C}}=\overset{H}{\text{C}}(CH_2)_3C\equiv C(CH_2)_5OTHP \xrightarrow[\text{2) H}_3O^+]{\text{1) Na / NH}_3}$$

$$n\text{-}Bu\overset{H}{\text{C}}=\overset{H}{\text{C}}(CH_2)_3\overset{H}{\text{C}}=\overset{}{\text{C}}(CH_2)_5OH \xrightarrow[\text{CH}_2\text{Cl}_2]{\text{CrO}_3\cdot 2C_5H_5N} n\text{-}Bu\overset{H}{\text{C}}=\overset{H}{\text{C}}(CH_2)_3\overset{H}{\text{C}}=\overset{}{\text{C}}(CH_2)_4CHO$$

43

$$\Big\downarrow \text{Ac}_2O \text{ / } C_5H_5N$$

$$n\text{-}Bu\overset{H}{\text{C}}=\overset{H}{\text{C}}(CH_2)_3\overset{H}{\text{C}}=\overset{}{\text{C}}(CH_2)_5OAc$$

42   (overall yield 21%)

Scheme 87

Gossyplure, a Mixture of (7Z, 11Z)-7,11-Hexadecadien-1-yl Acetate 4 *and Its (7Z, 11E)-Isomer* 5

The pheromone produced by the female pink bollworm moth (*Pectinophora gossypiella*), a severe cotton pest in the United States, was studied for many years. It was erroneously assigned the structure (*E*)-10-*n*-propyl-5,9-tridecadien-1-yl acetate (propylure),[200] and a number of syntheses were accomplished.[21] However, in 1973 Hummel et al. identified it as a 1 : 1 mixture of 4 and 5 and named it gossyplure.[10] A synthesis reported by Bierl et al. was rather complicated and based on the coupling of two allylic halides to construct the carbon

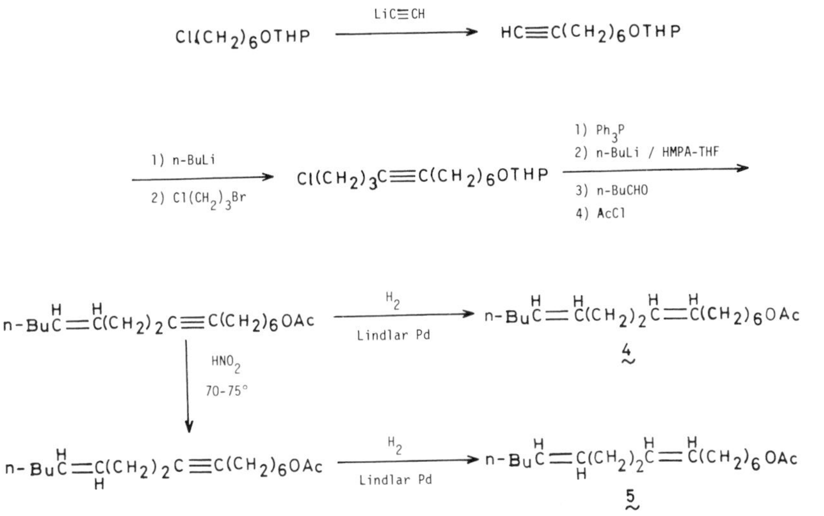

$Cl(CH_2)_6OTHP$ $\xrightarrow{LiC\equiv CH}$ $HC\equiv C(CH_2)_6OTHP$

$\xrightarrow[\text{2) } Cl(CH_2)_3Br]{\text{1) } n\text{-BuLi}}$ $Cl(CH_2)_3C\equiv C(CH_2)_6OTHP$ $\xrightarrow[\substack{\text{3) } n\text{-BuCHO} \\ \text{4) } AcCl}]{\substack{\text{1) } Ph_3P \\ \text{2) } n\text{-BuLi / HMPA-THF}}}$

$n\text{-Bu}\overset{H}{C}=\overset{H}{C}(CH_2)_2C\equiv C(CH_2)_6OAc$ $\xrightarrow[\text{Lindlar Pd}]{H_2}$ $n\text{-Bu}\overset{H}{C}=\overset{H}{C}(CH_2)_2\overset{H}{C}=\overset{H}{C}(CH_2)_6OAc$ **4**

$\downarrow$ $\substack{HNO_2 \\ 70\text{-}75°}$

$n\text{-Bu}\overset{H}{\underset{H}{C}}=C(CH_2)_2C\equiv C(CH_2)_6OAc$ $\xrightarrow[\text{Lindlar Pd}]{H_2}$ $n\text{-Bu}\overset{H}{C}=\underset{H}{C}(CH_2)_2\overset{}{C}=\overset{H}{C}(CH_2)_6OAc$ **5**

Scheme **88**

$n\text{-Bu}C\equiv CLi$ $\xrightarrow[\text{HMPA}]{\triangle O}$ $n\text{-Bu}C\equiv C(CH_2)_2OH$ $\xrightarrow[\substack{\text{2) } Ph_3P\cdot Br_2 \\ \text{dioxane-}C_5H_5N}]{\text{1) } H_2}$ $n\text{-Bu}\overset{H}{C}=\overset{H}{C}(CH_2)_2Br$

$\xrightarrow{LiC\equiv C(CH_2)_6OTHP}$ $n\text{-Bu}\overset{H}{C}=\overset{H}{C}(CH_2)_2C\equiv C(CH_2)_6OTHP$ $\xrightarrow[\text{2) } AcCl / AcOH]{\text{1) } H_2}$

$n\text{-Bu}\overset{H}{C}=\overset{H}{C}(CH_2)_2\overset{H}{C}=\overset{H}{C}(CH_2)_6OAc$ **4**

$n\text{-Bu}C\equiv C(CH_2)_2OH$ $\xrightarrow[\substack{\text{2) } Ph_3P\cdot Br_2 \\ \text{dioxane-}C_5H_5N}]{\text{1) } Na / NH_3}$ $n\text{-Bu}\overset{H}{\underset{H}{C}}=C(CH_2)_2Br$ $\xrightarrow{LiC\equiv C(CH_2)_6OTHP}$

$n\text{-Bu}\overset{H}{\underset{H}{C}}=C(CH_2)_2C\equiv C(CH_2)_6OTHP$ $\xrightarrow[\text{2) } AcCl / AcOH]{\text{1) } H_2}$ $n\text{-Bu}\overset{H}{\underset{H}{C}}=C(CH_2)_2\overset{H}{\underset{H}{C}}=C(CH_2)_6OAc$ **5**

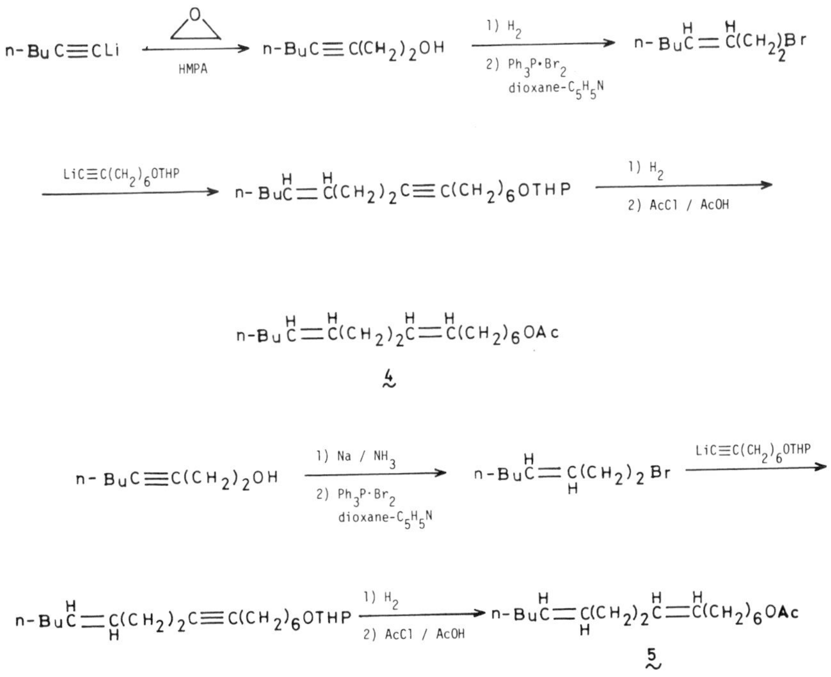

Scheme **89**

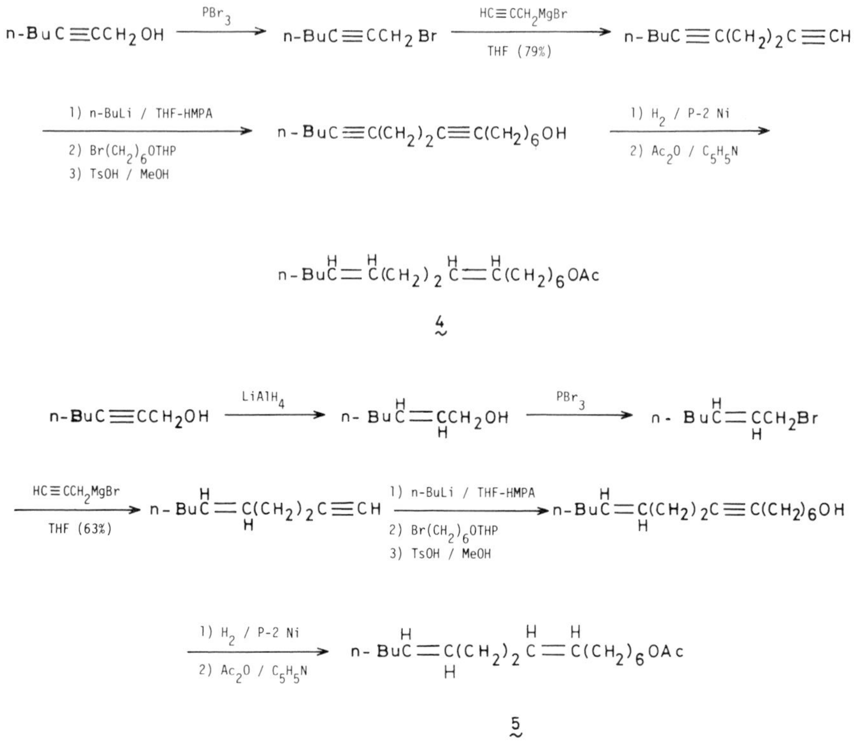

n-BuC≡CCH$_2$OH $\xrightarrow{\text{PBr}_3}$ n-BuC≡CCH$_2$Br $\xrightarrow[\text{THF (79%)}]{\text{HC≡CCH}_2\text{MgBr}}$ n-BuC≡C(CH$_2$)$_2$C≡CH

$\xrightarrow[\begin{array}{l}\text{2) Br(CH}_2)_6\text{OTHP}\\ \text{3) TsOH / MeOH}\end{array}]{\text{1) n-BuLi / THF-HMPA}}$ n-BuC≡C(CH$_2$)$_2$C≡C(CH$_2$)$_6$OH $\xrightarrow[\text{2) Ac}_2\text{O / C}_5\text{H}_5\text{N}]{\text{1) H}_2\text{ / P-2 Ni}}$

$$\text{n-Bu}\overset{\text{H}}{\text{C}}=\overset{\text{H}}{\text{C}}(\text{CH}_2)_2\overset{\text{H}}{\text{C}}=\overset{\text{H}}{\text{C}}(\text{CH}_2)_6\text{OAc}$$

<u>4</u>

n-BuC≡CCH$_2$OH $\xrightarrow{\text{LiAlH}_4}$ n-Bu$\overset{\text{H}}{\underset{\text{H}}{\text{C}}}$=CCH$_2$OH $\xrightarrow{\text{PBr}_3}$ n-Bu$\overset{\text{H}}{\underset{\text{H}}{\text{C}}}$=CCH$_2$Br

$\xrightarrow[\text{THF (63%)}]{\text{HC≡CCH}_2\text{MgBr}}$ n-Bu$\overset{\text{H}}{\underset{\text{H}}{\text{C}}}$=C(CH$_2$)$_2$C≡CH $\xrightarrow[\begin{array}{l}\text{2) Br(CH}_2)_6\text{OTHP}\\ \text{3) TsOH / MeOH}\end{array}]{\text{1) n-BuLi / THF-HMPA}}$ n-Bu$\overset{\text{H}}{\underset{\text{H}}{\text{C}}}$=C(CH$_2$)$_2$C≡C(CH$_2$)$_6$OH

$\xrightarrow[\text{2) Ac}_2\text{O / C}_5\text{H}_5\text{N}]{\text{1) H}_2\text{ / P-2 Ni}}$ n-Bu$\overset{\text{H}}{\underset{\text{H}}{\text{C}}}$=C(CH$_2$)$_2$$\overset{\text{H}}{\text{C}}$=$\overset{\text{H}}{\text{C}}$(CH$_2$)$_6$OAc

<u>5</u>

Scheme **90**

EtO$_2$C(CH$_2$)$_3\overset{\oplus}{\text{P}}Ph_3\overset{\ominus}{\text{Br}}$ $\xrightarrow[\text{2) n-BuCHO}]{\begin{array}{l}\text{1) NaN(SiMe}_3)_2\\ \text{THF, -78°}\end{array}}$ n-Bu$\overset{\text{H}}{\text{C}}$=$\overset{\text{H}}{\text{C}}$(CH$_2$)$_2CO_2$Et $\xrightarrow{\text{LiAlH}_4}$

n-Bu$\overset{\text{H}}{\text{C}}$=$\overset{\text{H}}{\text{C}}$(CH$_2$)$_3$OH $\xrightarrow{\text{PBr}_3}$ n-Bu$\overset{\text{H}}{\text{C}}$=$\overset{\text{H}}{\text{C}}$(CH$_2$)$_3$Br $\xrightarrow[\begin{array}{l}\text{2) NaN(SiMe}_3)_2\\ \text{3) OHC(CH}_2)_5\text{CO}_2\text{Et}\end{array}]{\text{1) Ph}_3\text{P}}$

n-Bu$\overset{\text{H}}{\text{C}}$=$\overset{\text{H}}{\text{C}}$(CH$_2$)$_2$$\overset{\text{H}}{\text{C}}$=$\overset{\text{H}}{\text{C}}$(CH$_2$)$_5CO_2$Et $\xrightarrow[\text{2) Ac}_2\text{O}]{\text{1) LiAlH}_4}$ n-Bu$\overset{\text{H}}{\text{C}}$=$\overset{\text{H}}{\text{C}}$(CH$_2$)$_2$$\overset{\text{H}}{\text{C}}$=$\overset{\text{H}}{\text{C}}$(CH$_2$)$_6$OAc

<u>4</u>

n-Bu$\overset{\text{H}}{\underset{\text{H}}{\text{C}}}$=C(CH$_2$)$_3\overset{\oplus}{\text{P}}Ph_3\overset{\ominus}{\text{Br}}$ $\Longrightarrow$ n-Bu$\overset{\text{H}}{\underset{\text{H}}{\text{C}}}$=C(CH$_2$)$_2$$\overset{\text{H}}{\text{C}}$=$\overset{\text{H}}{\text{C}}$(CH$_2$)$_6$OAc

<u>5</u>

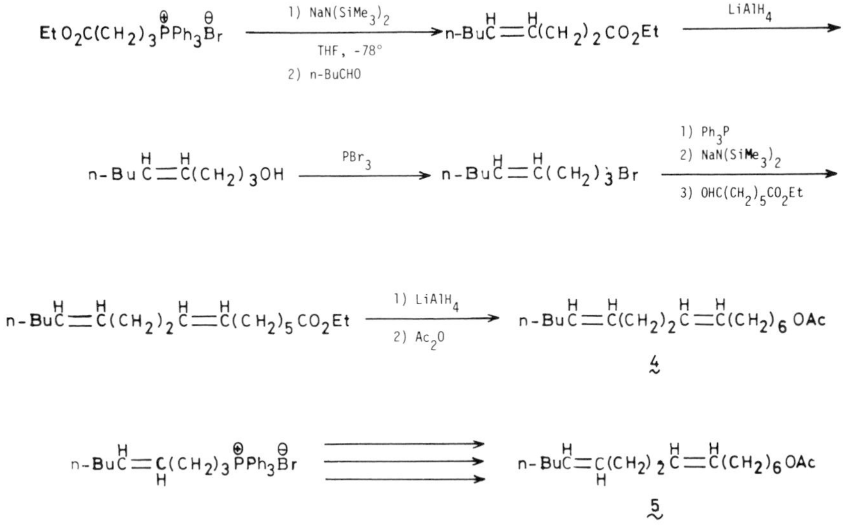

Scheme **91**

48

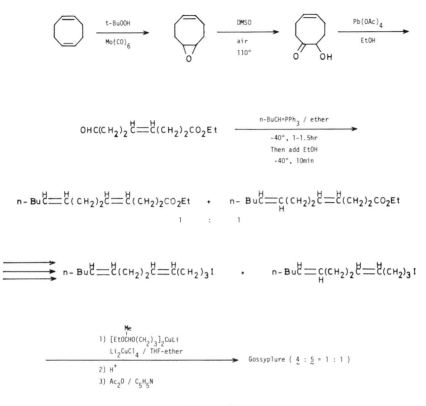

Scheme 92

chain, which proved to be unsatisfactory.[201a] Their paper, however, contains a lot of biological data. The second synthesis by Sonnet was stereoselective in synthesizing 4 but not so in preparing 5, as shown in Scheme 88.[201b] An acetylenic route to 4 and 5 was devised by Su and Mahany[202] and later modified by Eiter et al. to give products with 95% purity (Scheme 89).[203] ^{13}C-NMR of 4,5 and other isomers of 7,11-hexadecadien-1-yl acetate was recorded by Eiter et al.[203] Mori's synthesis was based on the Grignard coupling reaction and is highly stereoselective in preparing the both isomers (Scheme 90).[204] The next two syntheses illustrate the two different strategies in employing the Wittig reaction. Bestmann et al. synthesized the two isomers separately by the stereoselective Z-olefination method developed by them (Scheme 91).[205] Anderson and Henrick, however, directly obtained gossyplure by controlling the reaction conditions to give a 1 : 1 ratio of the two isomers 4 and 5 (Scheme 92).[206] This is convenient for practical purposes. It is also interesting to note their utilization of 1,5-cyclo-octadiene as a precursor of a Z-olefinic aldo ester. The synthesis of gossyplure

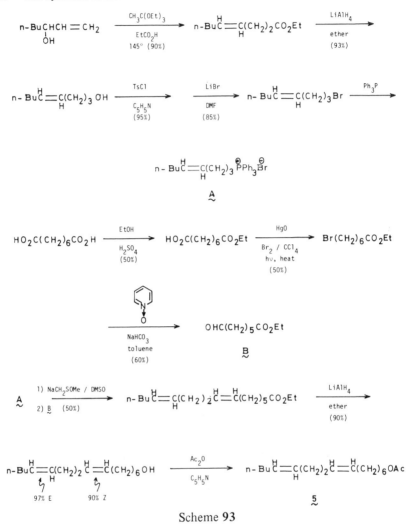

Scheme **93**

was discussed in detail by Henrick.[207] (7Z, 11E)-Hexadecadien-1-yl acetate **5** was also isolated as the pheromone of the angoumois grain moth (*Sitotraga cerealella*) and synthesized by Hammoud and Descoins (Schemes 93 and 94).[208] The stereoselectivity of the synthesis was ca. 90%.

(3Z, 13Z)-3,13-Octadecadien-1-yl Acetate 44 and *(3E, 13Z)-3,13-Octadecadien-1-yl acetate* 45

The female peachtree borer (*Sanninoidea exitiosa*) uses **44** and the female lesser peachtree borer (*Synanthedon pictipes*) uses **45** as sex pheromones.[209] These

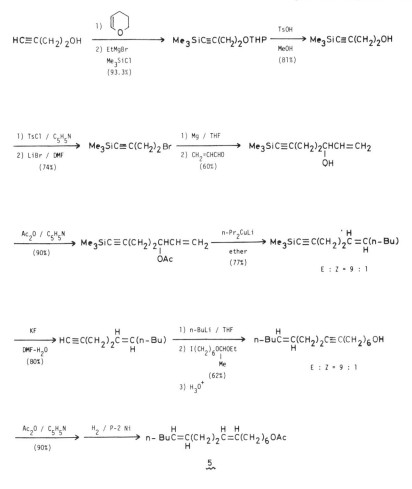

Scheme 94

compounds are the largest molecules isolated thus far as pheromones of lepidopterous species. *S. exitiosa* males did not respond to the (3*E*, 13*Z*)-isomer and low concentrations of it in the (3*Z*, 13*Z*)-isomer did not interfere with their response to the (3*Z*, 13*Z*)-isomer. In contrast, even very low concentrations of the (3*Z*, 13*Z*)-isomer (1%) in the (3*E*, 13*Z*)-isomer significantly inhibited the response of *S. pictipes* males. Tumlinson's synthesis was carried out through the acetylenic route (Scheme 95).[209] A modification of this synthesis was published recently (Scheme 96).[210] A 1 : 1 mixture of 44 and 45 were shown to be an effective trap bait for the male cherry tree borer (*Synanthedon hector*), an economic pest in Japanese peach orchards.[210,211] Mori devised a convenient synthesis of the 1 :

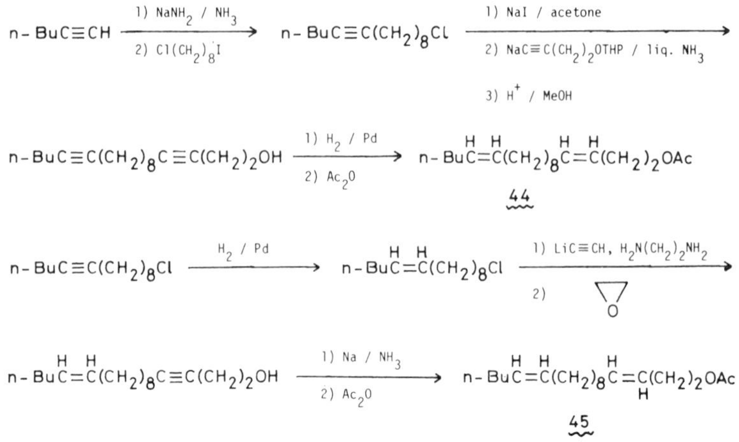

$n-\text{BuC} \equiv \text{CH}$ $\xrightarrow[\text{2) Cl(CH}_2)_8\text{I}]{\text{1) NaNH}_2 \text{ / NH}_3}$ $n-\text{BuC} \equiv \text{C(CH}_2)_8\text{Cl}$ $\xrightarrow[\substack{\text{2) NaC} \equiv \text{C(CH}_2)_2\text{OTHP / liq. NH}_3 \\ \text{3) H}^+ \text{ / MeOH}}]{\text{1) NaI / acetone}}$

$n-\text{BuC} \equiv \text{C(CH}_2)_8\text{C} \equiv \text{C(CH}_2)_2\text{OH}$ $\xrightarrow[\text{2) Ac}_2\text{O}]{\text{1) H}_2 \text{ / Pd}}$ $n-\text{BuC} \overset{\text{H H}}{=} \text{C(CH}_2)_8\text{C} \overset{\text{H H}}{=} \text{C(CH}_2)_2\text{OAc}$

$$\underset{\underset{\sim}{44}}{}$$

$n-\text{BuC} \equiv \text{C(CH}_2)_8\text{Cl}$ $\xrightarrow{\text{H}_2 \text{ / Pd}}$ $n-\text{BuC} \overset{\text{H H}}{=} \text{C(CH}_2)_8\text{Cl}$ $\xrightarrow[\text{2)}]{\text{1) LiC} \equiv \text{CH, H}_2\text{N(CH}_2)_2\text{NH}_2}$

$n-\text{BuC} \overset{\text{H H}}{=} \text{C(CH}_2)_8\text{C} \equiv \text{C(CH}_2)_2\text{OH}$ $\xrightarrow[\text{2) Ac}_2\text{O}]{\text{1) Na / NH}_3}$ $n-\text{BuC} \overset{\text{H H}}{=} \text{C(CH}_2)_8\overset{\text{H}}{\underset{\text{H}}{\text{C}}} = \text{C(CH}_2)_2\text{OAc}$

$$\underset{\underset{\sim}{45}}{}$$

$$\text{Scheme 95}$$

$\text{HC} \equiv \text{C(CH}_2)_8\text{OTHP}$ $\xrightarrow[\substack{\text{n-BuBr} \\ \text{2) TsOH / MeOH (88\%)}}]{\text{1) n-BuLi / THF-HMPA}}$ $n-\text{BuC} \equiv \text{C(CH}_2)_8\text{OH}$ $\xrightarrow[\substack{\text{2) LiI / acetone} \\ \text{(77\%)}}]{\text{1) TsCl / C}_5\text{H}_5\text{N}}$

$n-\text{BuC} \equiv \text{C(CH}_2)_8\text{I}$ $\xrightarrow[\substack{\text{n-BuLi / THF-HMPA} \\ \text{2) TsOH / MeOH} \\ \text{(43\%)}}]{\text{1) HC} \equiv \text{C(CH}_2)_2\text{OTHP}}$ $n-\text{BuC} \equiv \text{C(CH}_2)_8\text{C} \equiv \text{C(CH}_2)_2\text{OH}$ $\xrightarrow[\substack{\text{Pd-BaSO}_4 \\ \text{MeOH} \\ \text{quinoline} \\ \text{(82\%)}}]{\text{H}_2}$

$n-\text{BuC} \overset{\text{H H}}{=} \text{C(CH}_2)_8\text{C} \overset{\text{H H}}{=} \text{C(CH}_2)_2\text{OH}$ $\xrightarrow[\substack{\text{C}_5\text{H}_5\text{N} \\ \text{(82\%)}}]{\text{Ac}_2\text{O}}$ $n-\text{BuC} \overset{\text{H H}}{=} \text{C(CH}_2)_8\text{C} \overset{\text{H H}}{=} \text{C(CH}_2)_2\text{OAc}$

$$\underset{\underset{\sim}{44}}{}$$

$n-\text{BuC} \equiv \text{C(CH}_2)_8\text{OH}$ $\xrightarrow[\substack{\text{MeOH-quinoline} \\ \text{(89\%)}}]{\text{H}_2 \text{ / Pd-BaSO}_4}$ $n-\text{BuC} \overset{\text{H H}}{=} \text{C(CH}_2)_8\text{OH}$ $\xrightarrow[\substack{\text{2) NaI / acetone} \\ \text{(95\%)}}]{\text{1) TsCl / C}_5\text{H}_5\text{N}}$

$n-\text{BuC} \overset{\text{H H}}{=} \text{C(CH}_2)_8\text{I}$ $\xrightarrow[\substack{\text{n-BuLi / THF-HMPA} \\ \text{2) TsOH / MeOH} \\ \text{(35\%)}}]{\text{1) HC} \equiv \text{C(CH}_2)_2\text{OTHP}}$ $n-\text{BuC} \overset{\text{H H}}{=} \text{C(CH}_2)_8\text{C} \equiv \text{C(CH}_2)_2\text{OH}$ $\xrightarrow[\substack{\text{2) Na / NH}_3 \\ \text{3) TsOH / MeOH}}]{\text{1)}}$

$n-\text{BuC} \overset{\text{H H}}{=} \text{C(CH}_2)_8\overset{\text{H}}{\underset{\text{H}}{\text{C}}} = \text{C(CH}_2)_2\text{OH}$ $\xrightarrow[\substack{\text{C}_5\text{H}_5\text{N} \\ \text{(quant.)}}]{\text{Ac}_2\text{O}}$ $n-\text{BuC} \overset{\text{H H}}{=} \text{C(CH}_2)_8\overset{\text{H}}{\underset{\text{H}}{\text{C}}} = \text{C(CH}_2)_2\text{OAc}$

$$\underset{\underset{\sim}{45}}{}$$

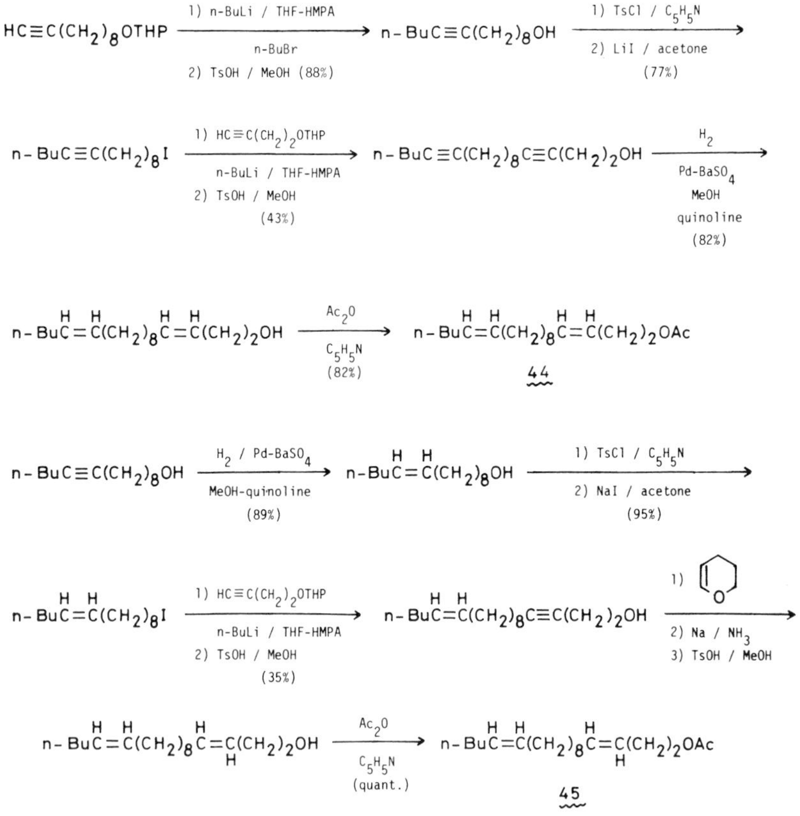

$$\text{Scheme 96}$$

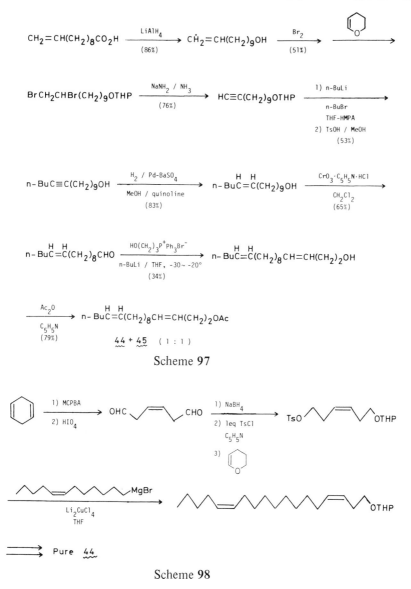

Scheme 97

Scheme 98

1 mixture employing the Wittig reaction whose stereoselectivity was adjusted to give the products in the required ratio (Scheme 97).[212] A stereoselective synthesis of (3Z, 13Z)-3,13-octadecadien-1-yl acetate **44** was recently achieved to secure >99.5% purity at C-3 geometry. The strategy was to employ (Z)-3-hexene-1,6-diol as the source of pure (Z)-olefinic linkage as shown in Scheme

98.[213] Thus obtained, highly pure **44** is attractive for the male smaller clear wing moth (*Synanthedon tenuis*), a pest in Japanese persimmon orchards. Several other insects are found to use **44, 45** or their corresponding alcohols as pheromones.

8. PHEROMONES WITH A TRIENE SYSTEM

(3Z, 6Z, 8E)-3,6,8-Dodecatrien-1-ol 46

This is the trail-following pheromone of a southern subterranean termite (*Reticulitermes virginicus*).[214] The termite trail-following substance is secreted by various species of termite workers to mark the source of suitable wood to other workers of the same species. The synthesis was nonstereoselective and carried out as shown in Scheme 99.[215] The minimum amount of the pheromone required to stimulate the worker termites to follow was 0.01 pg.

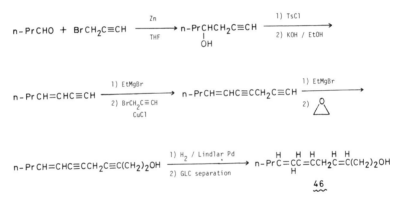

Scheme **99**

Scheme **100**

(4E, 7Z, 10Z)-4,7,10-Tridecatrien-1-yl Acetate 47

This is a component of the pheromone of the potato tuberworm moth (*Phthorimaea operculella*)[190, 216] and was synthesized by an acetylenic route (Scheme 100).[190] For the synthesis of another component see Schemes 79 to 81.

Methyl (2E)-(−)-2,4,5-tetradecatrienoate 48

Horler isolated this (−)-allenic ester from the male dried bean beetle (*Acanthoscelides obtectus*) as a sex pheromone.[217] The unique structure 48 attracted attention of many synthetic chemists and several syntheses were reported. The key problem of the synthesis was, of course, the formation of the allenic system. The first synthesis was achieved by Landor employing a reductive elimination reaction to generate the allenic linkage (Scheme 101).[218] The second synthesis used the S_N2' substitution reaction at C-1 of a 3-acetoxy-1-alkyne with an organocopper reagent, as the key step (Scheme 102).[219] Gore's synthesis employed the coupling of two acetylenic units with concomitant elimination of a

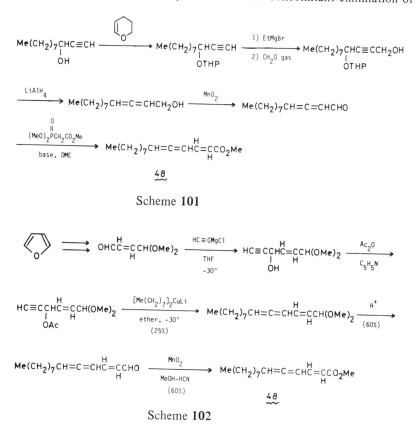

Scheme 101

Scheme 102

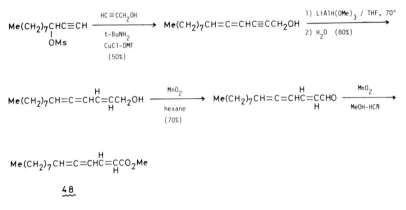

Scheme **103**

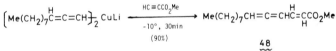

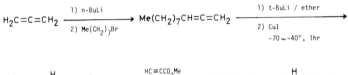

Scheme **104**

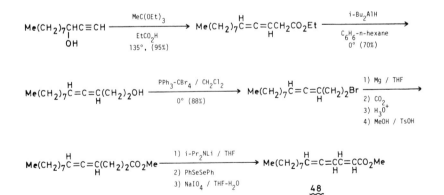

Scheme **105**

mesyloxy group (Scheme 103).[220] Linstrumelle employed organocopper chemistry (lithium diallenylcuprate) to achieve a simple synthesis of **48** (Scheme 104).[221] Orthoester Claisen rearrangement was also used to generate allene as shown in Scheme 105.[222] The synthesis of optically active forms of **48** are described in Schemes 262 and 263.

9. PHEROMONES WITH A CARBONYL GROUP (ALDEHYDES AND KETONES)

Undecanal 49

Although there are some aldehydes among insect pheromones, they are commercially available or easy to prepare. A synthesis of undecanal **49** was shown in Scheme 106.[223] It is the pheromone produced by male greater wax moth (*Galleria mellonella*).[224]

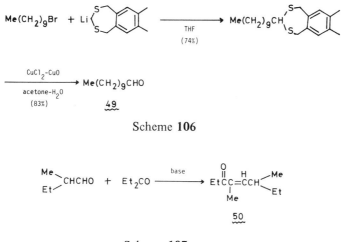

Scheme **106**

Scheme **107**

Manicone, (E)-4,6-Dimethyl-4-octen-3-one 50

The alarm pheromones present in the mandibular glands of *Manica mutica* and *Manica bradleyi* are dominated by manicone **50** with two minor ketones, 4-methyl-3-hexanone and 3-decanone.[225] The first synthesis of manicone was achieved by the conventional aldol condensation (Scheme 107).[225] The second one was based on dianion and organocopper chemistry as shown in Scheme 108.[226] In this synthesis the stereoselectivity of the olefin formation step was satisfactory. Another synthesis by Kocienski yielded a 45 : 55 mixture of **50** and its (Z)-isomer (Scheme 109).[227] They were separated by preparative GLC.

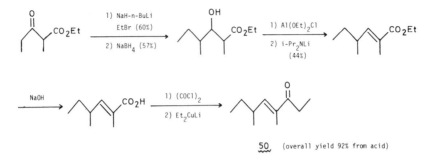

Scheme **108**

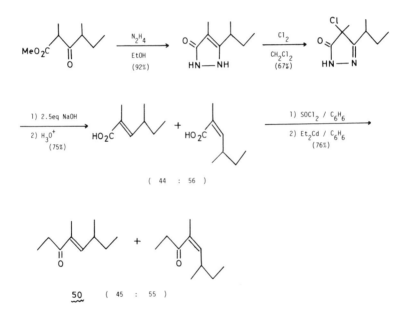

Scheme **109**

A recent synthesis by Nakai et al. employed the highly stereoselective [3,3] sig-matropic rearrangement of allylic thiocarbamates as the key step (Scheme 110).[228]

(Z)-7-Nonadecen-11-one 51 and (Z)-7-Eicosen-11-one 52

Tamaki et al. isolated and synthesized **51** and **52** as active components of the female sex pheromone of the peach fruit moth (*Carposina niponensis*), which

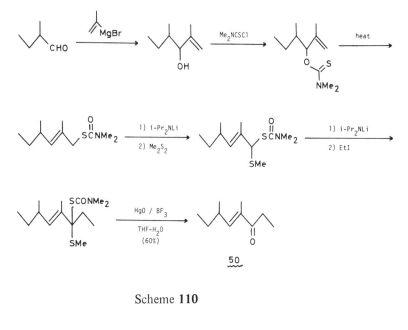

Scheme **110**

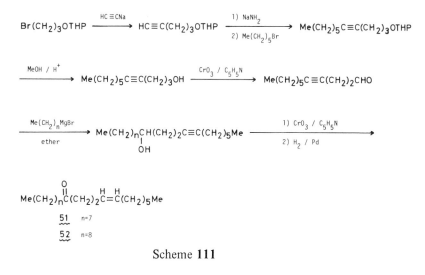

Scheme **111**

is a major economic pest of apple, peach, and other fruits in Japan.[229] The synthesis is shown in Scheme 111.[229] Another synthesis by Mori et al. employed commercially available 3-decyn-1-ol as the starting material (Scheme 112).[230]

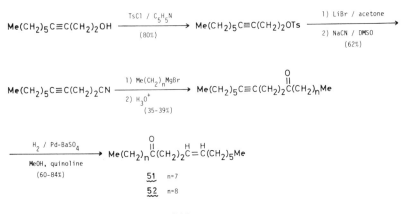

Scheme **112**

(Z)-6-Heneicosen-11-one **53** *and (Z)-1,6-Heneicosadien-11-one* **54**

These are the principal (**53**) and minor (**54**) sex pheromone components of the Douglas fir tussock moth (*Orgyia pseudotsugata*), which is a severe defoliator of fir forests of western North America.[231,232] Many syntheses of **53** were reported. The synthesis of **53** by Smith et al. was based on the alkylation of a 1,3-dithiane (Scheme 113).[233] Two independent syntheses of **53** used the Eschenmoser ring cleavage of an epoxy ketone as the key reaction (Schemes 114[234] and 115[138]). The Zoecon synthesis was achieved by an acetylenic route (Scheme 116).[235] The alcoholic intermediates in this synthesis were both purified by crystallization from pentane at $-35°$. A synthesis by the Wittig route is shown in Scheme 117.[236] The Raman spectrum (double bond stretching at 1652 cm^{-1}; cf. E = 1674, Z = 1658) and ^{13}C-NMR spectrum (allylic carbons at 26.76 and 27.03 ppm) of **53** are discussed in Ref. 236. An interesting synthesis of **53** and its E-isomer was reported using the -ene reaction (Scheme 118).[237] The steric

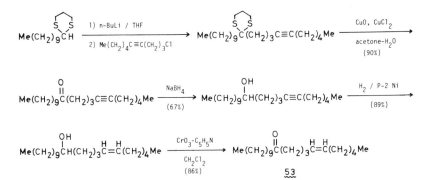

Scheme **113**

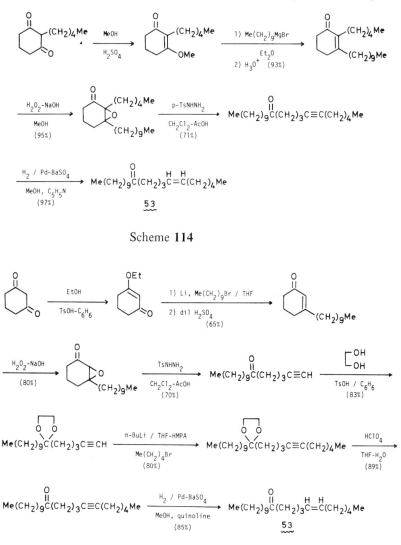

Scheme **114**

Scheme **115**

course of the -ene reaction between methyl acrylate and 1-octene, however, favored the formation of the undesired (E)-isomer (E : Z = 86 : 14). Therefore the olefin inversion reaction was necessary to obtain **53**. In a recent synthesis undecanoyl chloride was reacted with a Grignard reagent prepared in a novel manner from an organoborane (Scheme 119).[238] It should be mentioned that the pheromone of the southern armyworm moth (**27**) was also prepared by this new

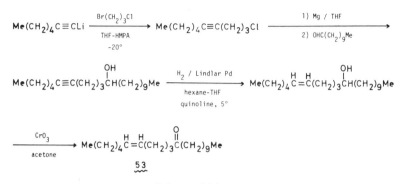

$$\text{Me(CH}_2)_4\text{C}\equiv\text{CLi} \xrightarrow[\substack{\text{THF-HMPA} \\ -20°}]{\text{Br(CH}_2)_3\text{Cl}} \text{Me(CH}_2)_4\text{C}\equiv\text{C(CH}_2)_3\text{Cl} \xrightarrow[\substack{2) \text{ OHC(CH}_2)_9\text{Me}}]{1) \text{ Mg / THF}}$$

$$\text{Me(CH}_2)_4\text{C}\equiv\text{C(CH}_2)_3\overset{\text{OH}}{\underset{|}{\text{CH}}}\text{(CH}_2)_9\text{Me} \xrightarrow[\substack{\text{hexane-THF} \\ \text{quinoline, 5°}}]{\text{H}_2 \text{ / Lindlar Pd}} \text{Me(CH}_2)_4\overset{\text{H}}{\text{C}}=\overset{\text{H}}{\text{C}}(\text{CH}_2)_3\overset{\text{OH}}{\underset{|}{\text{CH}}}(\text{CH}_2)_9\text{Me}$$

$$\xrightarrow[\text{acetone}]{\text{CrO}_3} \text{Me(CH}_2)_4\overset{\text{H}}{\text{C}}=\overset{\text{H}}{\text{C}}(\text{CH}_2)_3\overset{\text{O}}{\overset{\|}{\text{C}}}(\text{CH}_2)_9\text{Me}$$

$$\underset{\underset{\sim}{53}}{}$$

Scheme **116**

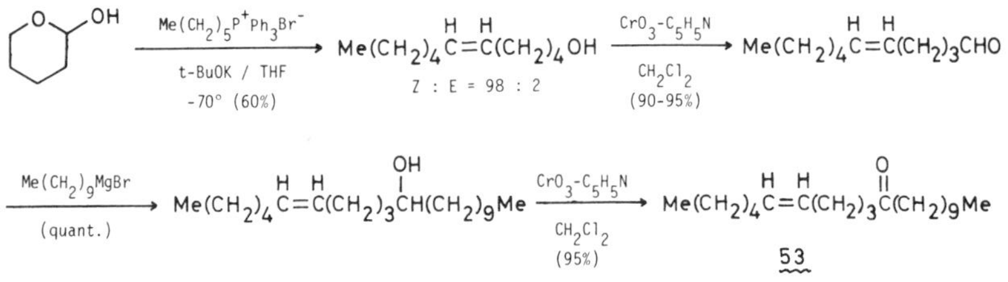

$$\xrightarrow[\substack{\text{t-BuOK / THF} \\ -70° \ (60\%)}]{\text{Me(CH}_2)_5\text{P}^+\text{Ph}_3\text{Br}^-} \text{Me(CH}_2)_4\overset{\text{H}}{\text{C}}=\overset{\text{H}}{\text{C}}(\text{CH}_2)_4\text{OH} \xrightarrow[\substack{\text{CH}_2\text{Cl}_2 \\ (90\text{-}95\%)}]{\text{CrO}_3\text{-C}_5\text{H}_5\text{N}} \text{Me(CH}_2)_4\overset{\text{H}}{\text{C}}=\overset{\text{H}}{\text{C}}(\text{CH}_2)_3\text{CHO}$$

$$\text{Z : E = 98 : 2}$$

$$\xrightarrow[\text{(quant.)}]{\text{Me(CH}_2)_9\text{MgBr}} \text{Me(CH}_2)_4\overset{\text{H}}{\text{C}}=\overset{\text{H}}{\text{C}}(\text{CH}_2)_3\overset{\text{OH}}{\underset{|}{\text{CH}}}(\text{CH}_2)_9\text{Me} \xrightarrow[\substack{\text{CH}_2\text{Cl}_2 \\ (95\%)}]{\text{CrO}_3\text{-C}_5\text{H}_5\text{N}} \text{Me(CH}_2)_4\overset{\text{H}}{\text{C}}=\overset{\text{H}}{\text{C}}(\text{CH}_2)_3\overset{\text{O}}{\overset{\|}{\text{C}}}(\text{CH}_2)_9\text{Me}$$

$$\underset{\underset{\sim}{53}}{}$$

Scheme **117**

$$\text{CH}_2=\text{CHCO}_2\text{Me} + \text{CH}_2=\text{CH(CH}_2)_5\text{Me} \xrightarrow[\substack{100°, 16\text{hr} \\ (40\%)}]{\text{AlCl}_3\text{-NaCl-KCl}}$$

$$\text{Me(CH}_2)_4\overset{\text{H}}{\underset{\text{H}}{\text{C}}}=\text{C(CH}_2)_3\text{CO}_2\text{Me} + \text{Me(CH}_2)_4\overset{\text{H}}{\text{C}}=\overset{\text{H}}{\text{C}}(\text{CH}_2)_3\text{CO}_2\text{Me} \xrightarrow[\substack{\text{CH}_2\text{Cl}_2 \\ (92\%)}]{\text{NBS-HCl}}$$

$$\text{86} \quad : \quad \text{14}$$

$$\text{Me(CH}_2)_4\overset{\text{Br}}{\underset{|}{\text{CHCH}}}\underset{|}{\overset{}{}}(\text{CH}_2)_3\text{CO}_2\text{Me} \xrightarrow[\substack{\text{DMF} \\ (99\%)}]{\text{NaI}} \text{Me(CH}_2)_4\overset{\text{H}}{\text{C}}=\overset{\text{H}}{\text{C}}(\text{CH}_2)_3\text{CO}_2\text{Me} \xrightarrow[\substack{\text{EtOH-H}_2\text{O} \\ (95\%)}]{\text{KOH}}$$

$$\underset{\text{Cl}}{} \qquad \text{Z : E = 80 : 20}$$

$$\xrightarrow[\text{(62\%)}]{\text{Me(CH}_2)_9\text{Li}} \text{Me(CH}_2)_4\overset{\text{H}}{\text{C}}=\overset{\text{H}}{\text{C}}(\text{CH}_2)_3\overset{\text{O}}{\overset{\|}{\text{C}}}(\text{CH}_2)_9\text{Me}$$

$$\underset{\underset{\sim}{53}}{} \quad (\text{ Z : E = 80 : 20 })$$

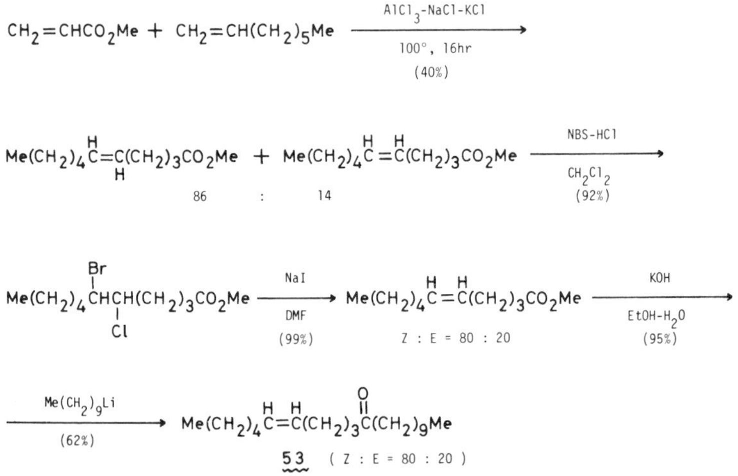

Scheme **118**

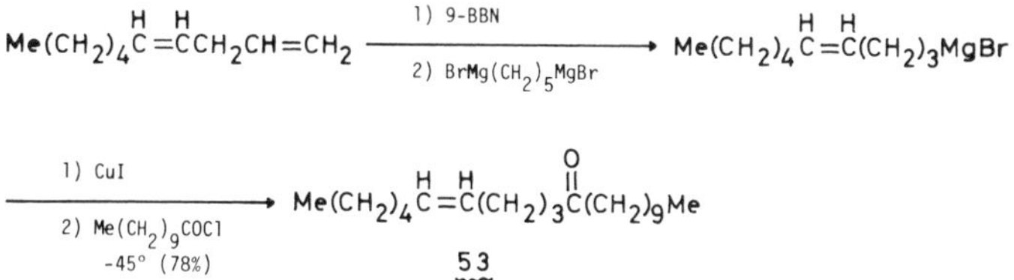

$$Me(CH_2)_4\overset{H}{C}=\overset{H}{C}CH_2CH=CH_2 \xrightarrow[\text{2) BrMg(CH}_2\text{)}_5\text{MgBr}]{\text{1) 9-BBN}} Me(CH_2)_4\overset{H}{C}=\overset{H}{C}(CH_2)_3MgBr$$

$$\xrightarrow[\underset{-45°\ (78\%)}{\text{2) Me(CH}_2\text{)}_9\text{COCl}}]{\text{1) CuI}} Me(CH_2)_4\overset{H}{C}=\overset{H}{C}(CH_2)_3\overset{\overset{O}{\|}}{C}(CH_2)_9Me$$

53

Scheme 119

$$HC\equiv C(CH_2)_4OH \xrightarrow[\text{2) CH}_2\text{=CH(CH}_2\text{)}_3\text{Br}]{\text{1) 2eq n-BuLi / THF-HMPA}} CH_2=CH(CH_2)_3C\equiv C(CH_2)_4OH \xrightarrow{\text{H}_2 \text{ / P-2 Ni}}$$

$$CH_2=CH(CH_2)_3\overset{H}{C}=\overset{H}{C}(CH_2)_4OH \xrightarrow[\text{CH}_2\text{Cl}_2]{\text{CrO}_3 \text{ C}_5\text{H}_5\text{N HCl}} CH_2=CH(CH_2)_3\overset{H}{C}=\overset{H}{C}(CH_2)_3CHO \xrightarrow{\text{Me(CH}_2\text{)}_9\text{MgBr}}$$

$$CH_2=CH(CH_2)_3\overset{H}{C}=\overset{H}{C}(CH_2)_3\overset{\overset{OH}{|}}{C}H(CH_2)_9Me \xrightarrow[\text{CH}_2\text{Cl}_2]{\text{CrO}_3 \text{ C}_5\text{H}_5\text{N HCl}} CH_2=CH(CH_2)_3\overset{H}{C}=\overset{H}{C}(CH_2)_3\overset{\overset{O}{\|}}{C}(CH_2)_9Me$$

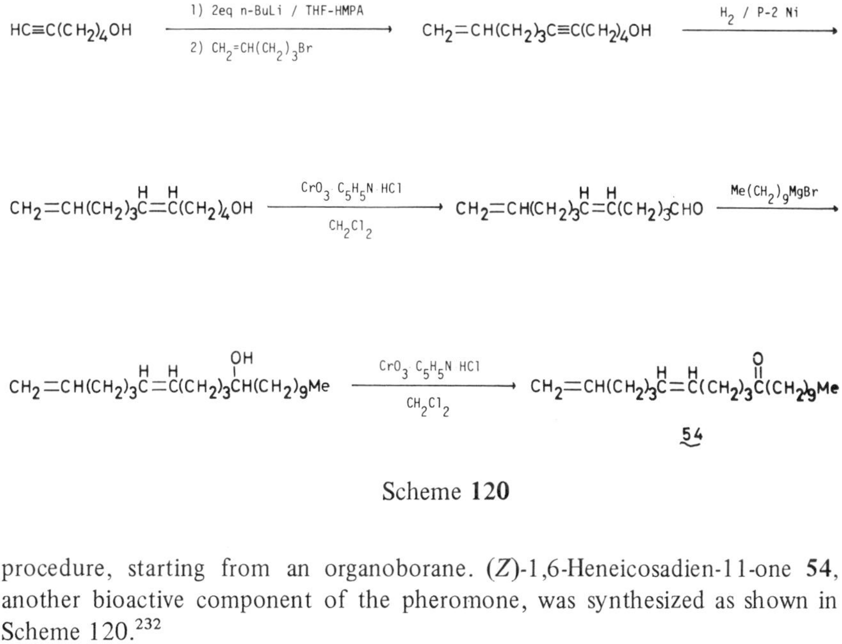

54

Scheme 120

procedure, starting from an organoborane. (Z)-1,6-Heneicosadien-11-one **54**, another bioactive component of the pheromone, was synthesized as shown in Scheme 120.[232]

3,11-Dimethylnonacosan-2-one 55 and 29-Hydroxy-3,11-dimethyl-nonacosan-2-one 56

From the cuticular wax of sexually mature females of the German cockroach (*Blattella germanica*) Fukami et al. isolated **55** and **56** as the sex pheromone, which, on contact with antennae, elicited wing-raising and direction-turning response from the male adults at the first stage of their sequential courtship behavior.[239, 240] Both **55** and **56** were synthesized in several ways as a mixture of all of the possible stereoisomers. The first synthesis of **55** by Fukami et al. is shown in Scheme 121.[239, 241] The synthetic **55** melted at 29-31°, while the m.p. of the natural **55** was 44.0-45.5° with $[\alpha]_D^{22} + 5.1°$ (*n*-hexane). Scheme 122 illustrates the synthesis by Schwarz et al.,[242] whose clever use of a stabilized ylid and

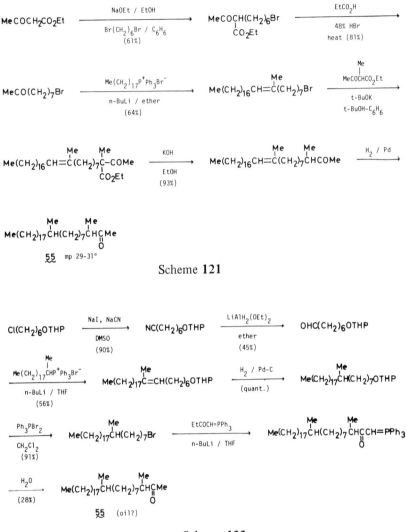

Scheme **121**

Scheme **122**

transformation of a THP ether to a bromide are noteworthy. The third synthesis by Burgstahler et al. is summarized in Scheme 123.[243] Organocopper chemistry was applied to the synthesis of **55** by the Zoecon group (Scheme 124).[244] A recent unique synthesis of **55** by Gore was based on the reaction of a vinylallenic Grignard reagent and the Cope rearrangement, as shown in Scheme 125.[245] A

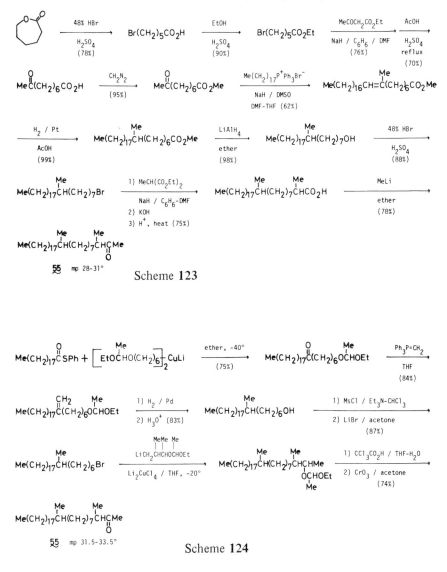

Scheme 123

Scheme 124

synthesis of a stereoisomeric mixture of **56** was achieved by Fukami et al. in a similar manner as that for **55** (Scheme 126).[246] The synthetic **56** melted at 40-41°, while the m.p. of the natural **56** was 42-43°. Another synthesis of **56** by Burgstahler et al. is shown in Scheme 127.[247] Synthesis of optically active **55** and **56** will be described later (Schemes 249-252).

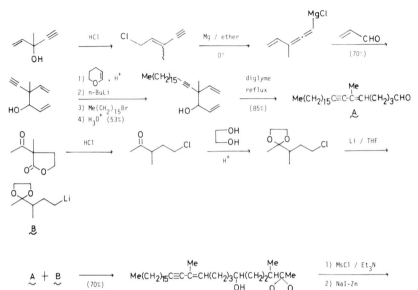

Scheme **125**

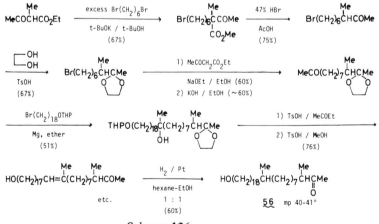

Scheme **126**

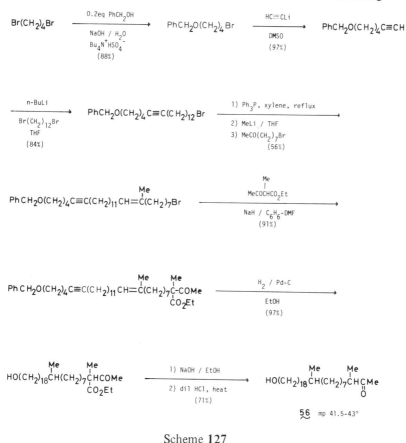

Scheme 127

10. PHEROMONES WITH AN INTRAMOLECULAR KETAL LINKAGE

These ketal pheromones are dissymmetric. In this section the argument is restricted to the synthesis of racemates. For the synthesis of optically active compounds, see Section 15. A review is available on the chemistry of the 6,8-dioxabicyclo[3.2.1]octane series.[248]

Chalcogran, 2-Ethyl-1,6-dioxaspiro[4.4]nonane 57

This is the pheromone produced by male "Kupferstecher" (*Pityogenes chalcographus*).[249] A stereoisomeric mixture of 57 was synthesized as shown in Scheme 128.[249] The stereochemistry of the natural 57 is unknown.

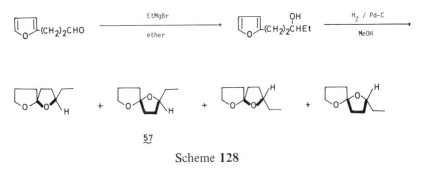

Scheme **128**

Frontalin, 1,5-Dimethyl-6,8-dioxabicyclo[3.2.1]octane 8

Extraction of about 6500 hindguts of the male western pine beetle (*Dendroc-tonus brevicomis*) led to the isolation of frontalin 8.[250] The first synthesis by Kinzer et al. was based on the Diels-Alder reaction (Scheme 129).[250] Mundy's synthesis also used the Diels-Alder reaction (Scheme 130).[251] D'Silva employed a similar approach (Scheme 131).[252] The fourth synthesis was a linear and

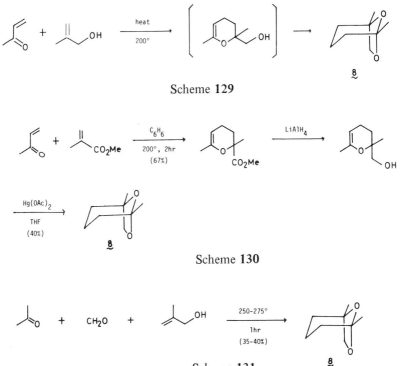

Scheme **129**

Scheme **130**

Scheme **131**

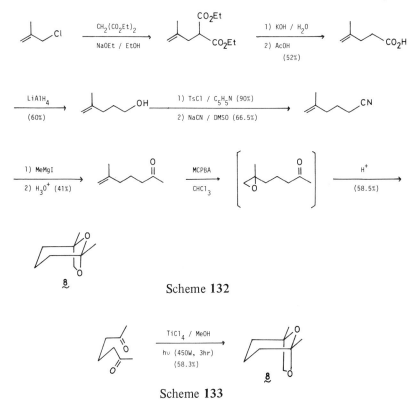

Scheme 132

Scheme 133

building-block-type approach (Scheme 132).[253] Recently a unique photochemical synthesis of (±)-frontalin was reported by Sato et al. (Scheme 133).[254] This is based on their findings that saturated ketones give 1,2-diols, accompanied in some cases by ketals as minor products, when irradiated in methanol with quartz-filtered light in the presence of titanium(IV) chloride (cf. Ref. 255).

exo-*Brevicomin*, exo-*7-Ethyl-3-methyl-6,8-dioxabicyclo[3.2.1]octane* 6

Silverstein et al. isolated exo-brevicomin **6** as the principal aggregation pheromone in the frass of the female western pine beetle (*Dendroctonus brevicomis*).[11] They achieved two different syntheses of its racemate (Schemes 134 and 135).[256] One of them (Scheme 135) was stereoselective. Wasserman's synthesis employed a thermal rearrangement of an epoxy ketone as the key step (Scheme 136).[257] A synthesis by Mundy et al. used the Diels-Alder reaction (Scheme 137).[251] A mixed Kolbé electrolysis reaction was used to prepare **6** in moderate yield (Scheme 138).[258] In this synthesis, (*E*)-6-nonen-2-one, the key intermediate, was obtained pure only after GLC separation. Kocienski's synthesis was

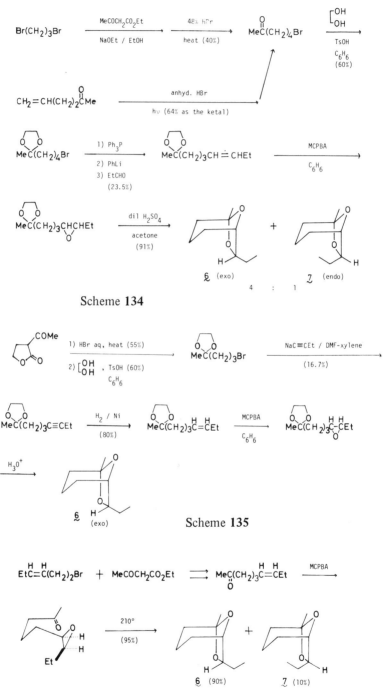

Br(CH$_2$)$_3$Br
$\xrightarrow[\text{NaOEt / EtOH}]{\text{MeCOCH}_2\text{CO}_2\text{Et}}$
$\xrightarrow[\text{heat (40\%)}]{\text{48\% HBr}}$
MeC(CH$_2$)$_4$Br
$\xrightarrow[\substack{\text{TsOH}\\ \text{C}_6\text{H}_6 \\ (60\%)}]{\text{OH OH}}$

CH$_2$=CH(CH$_2$)$_2$CMe
$\xrightarrow[\text{h}\nu \text{ (64\% as the ketal)}]{\text{anhyd. HBr}}$

MeC(CH$_2$)$_4$Br
$\xrightarrow[\substack{\text{2) PhLi}\\ \text{3) EtCHO}\\ (23.5\%)}]{\text{1) Ph}_3\text{P}}$
MeC(CH$_2$)$_3$CH=CHEt
$\xrightarrow[\text{C}_6\text{H}_6]{\text{MCPBA}}$

MeC(CH$_2$)$_3$CHCHEt
$\xrightarrow[\substack{\text{acetone}\\ (91\%)}]{\text{dil H}_2\text{SO}_4}$
6 (exo)   +   7 (endo)

4   :   1

Scheme **134**

$\xrightarrow[\substack{\text{2) OH OH , TsOH (60\%)}\\ \text{C}_6\text{H}_6}]{\text{1) HBr aq, heat (55\%)}}$
MeC(CH$_2$)$_3$Br
$\xrightarrow[(16.7\%)]{\text{NaC}\equiv\text{CEt / DMF-xylene}}$

MeC(CH$_2$)$_3$C≡CEt
$\xrightarrow[(80\%)]{\text{H}_2 \text{ / Ni}}$
MeC(CH$_2$)$_3$C=CEt
$\xrightarrow[\text{C}_6\text{H}_6]{\text{MCPBA}}$
MeC(CH$_2$)$_3$C-CEt

$\xrightarrow{\text{H}_3\text{O}^+}$
6 (exo)

Scheme **135**

EtC=C(CH$_2$)$_2$Br + MeCOCH$_2$CO$_2$Et ⟶ MeC(CH$_2$)$_3$C=CEt
$\xrightarrow{\text{MCPBA}}$

$\xrightarrow[(95\%)]{210°}$
6 (90%)   +   7 (10%)

Scheme **136**

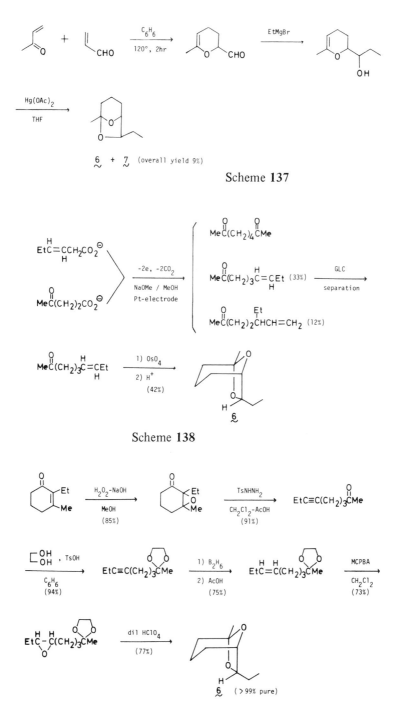

Scheme 137

Scheme 138

Scheme 139

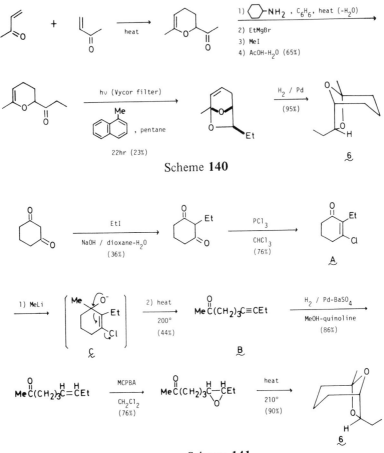

Scheme **140**

Scheme **141**

stereoselective and used the Eschenmoser fragmentation of an epoxy ketone as the key step (Scheme 139).[259] Kossanyi devised an easy access to **6** by employing the Diels-Alder reaction and a photochemical reaction as shown in Scheme 140.[260] A new fragmentation reaction of 3-chloro-2-cyclohexen-1-one was devised and used in a synthesis of **6** (Scheme 141).[261] The fragmentation (**A → B**) involves the thermal cleavage of the intermediate alkoxide **C**.

endo-Brevicomin, endo-7-Ethyl-3-methyl-6,8-dioxabicyclo[3.2.1]-octane 7

This was also isolated by Silverstein et al. as a biologically inactive component of the frass of the western pine beetle.[11] Later work showed it to be a pheromone inhibitor to the southern pine beetle (*Dendroctonus frontalis*).[12] This *endo*-isomer was a by-product of a nonstereoselective syntheses of *exo*-brevico-min.[251,256] Wasserman's thermal rearrangement of a *trans*-epoxy ketone was about 90% stereoselective to give **7** (Scheme 142).[257] The recognition of this substance as a pheromone inhibitor, which may be useful as a pesticide, prompted synthetic work, and four stereoselective syntheses were reported. All of them employ the epoxidation of a pure (*E*)-olefinic ketone as the key step. Kocienski and Ostrow generated the (*E*)-olefin by the Birch reduction (Scheme 143).[259] Look's synthesis employed a preformed (*E*)-olefinic alcohol as the starting material (Scheme 144).[262] An interesting synthesis based on organopalla-

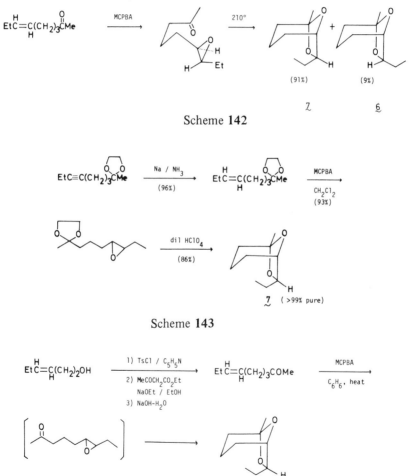

Scheme **142**

Scheme **143**

Scheme **144**

dium chemistry was reported by Byrom et al., in which an (*E*)-olefin was generated by the dimerization of butadiene (Scheme 145).[263] Kondo's phosphonate method (Equation 5) was applied to obtain 7 (Scheme 146).[264]

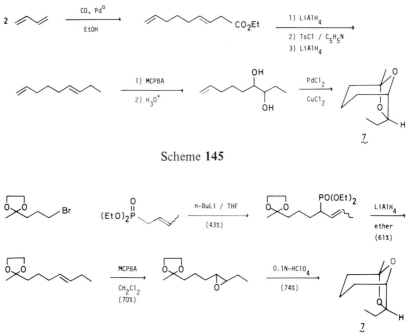

Scheme **145**

Scheme **146**

α-Multistriatin, 2, 4-Dimethyl-5-ethyl-6,8-dioxabi-cyclo[3.2.1]octane 58

Silverstein et al. found that the attractant for the smaller European elm bark beetle (*Scolytus multistriatus*) is a mixture of (−)-4-methylheptan-3-ol, (−)-α-multistriatin **58**, and a sesquiterpene, (−)-α-cubebene.[265] All three compounds were isolated from the volatile components collected by aerating elm bolts infested with virgin female beetles. Individually each compound was inactive in the laboratory bioassay, but a mixture of all three showed activity nearly equivalent to that of the original extract. Curiously the German populations of *S. multistriatus* did not aggregate in response to the mixture containing (−)-α-multistriatin. Instead, (−)-δ-multistriatin proved attractive when combined with 4-methylheptan-3-ol and cubeb oil.[266] Several syntheses of **58** were reported. Silverstein's earlier syntheses are shown in Schemes 147[265] and 148.[267] The latter Diels-Alder approach coupled with a detailed NMR analysis[268] enabled

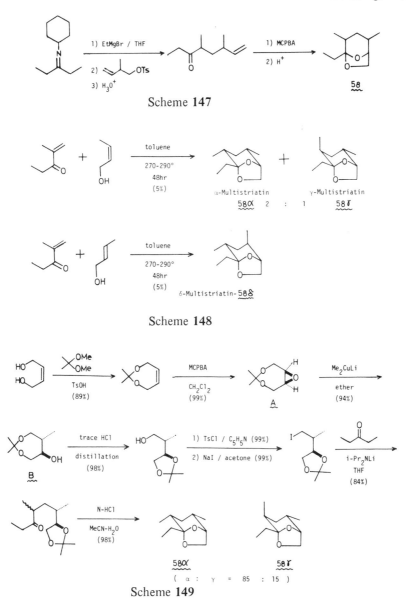

Scheme 147

Scheme 148

Scheme 149

Silverstein et al. to assign **58α** to the biologically active α-isomer. Elliot and Fried reported a stereoselective synthesis of (±)-α-multistriatin **58α** as shown in Scheme 149.[269] The key step was the stereoselective *trans*-opening of an epoxide with lithium dimethylcuprate (**A → B**).

11. MONOTERPENOID PHEROMONES

Ipsenol, 2-Methyl-6-methylene-7-octen-4-ol 59

Ipsenol **59**, ipsdienol **60**, and *cis*-verbenol **82** are the monoterpene aggregation
pheromones isolated from the frass produced by male California five-spined ips
(*Ips paraconfusus*).[270] The first synthesis was carried out by the dithiane alkylation reaction as shown in Scheme 150.[271] Vig adopted a building block-type
approach (Scheme 151).[272] Katzenellenbogen and Lenox devised a simple synthesis, as shown in Scheme 152.[273] Thermolysis of a cyclobutene generated the
characteristic conjugated diene system of **59** (Scheme 153).[274] In another
synthesis a Claisen rearrangement was used to construct the diene moiety
(Scheme 154).[275] Kondo et al. reported a simple synthesis of **59** employing a
Grignard reagent prepared from chloroprene (Scheme 155).[276] A *retro*-Diels-
Alder reaction was also used in an ipsenol synthesis (Scheme 156).[277] A modification of Scheme 154 was reported recently (Scheme 157).[278] Instead of the
classical Claisen rearrangement, an orthoester Claisen rearrangement was used in
this synthesis. Sakurai et al. recently published an interesting application of
organosilane chemistry in ipsenol synthesis (Scheme 158).[279] The synthesis of
optically active ipsenol is described in Schemes 224 and 225.

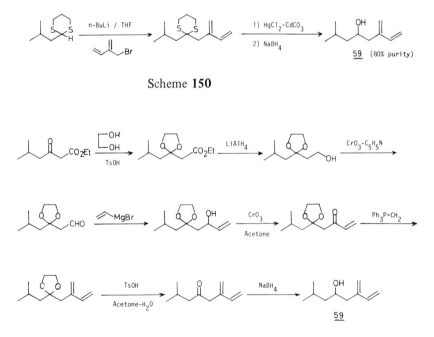

Scheme **150**

Scheme **151**

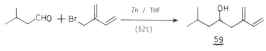

Scheme **152**

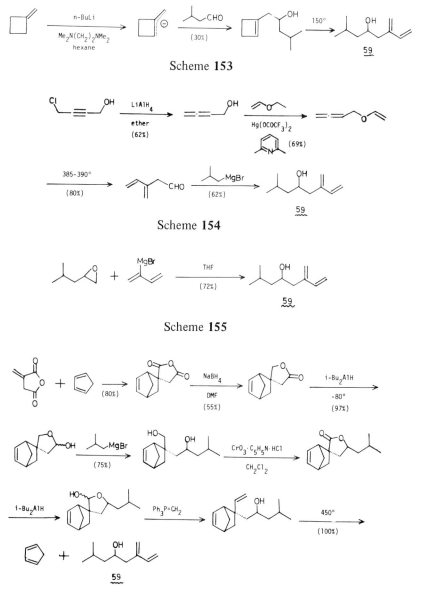

Scheme **153**

Scheme **154**

Scheme **155**

Scheme **156**

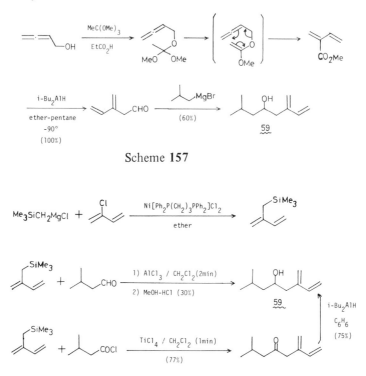

Scheme **157**

Scheme **158**

Ipsdienol, 2-Methyl-6-methylene-2,7-octadien-4-ol 60

Silverstein synthesized this trienol by dithiane alkylation (Scheme 159).[271] Extension of Katzenellenbogen's ipsenol synthesis to ipsdienol was particularly successful (Scheme 160).[280] Mori developed a synthesis of ipsdienol from myrcene, a readily available monoterpene hydrocarbon (Scheme 161).[281] In another synthesis the diene moiety was generated by the pyrolysis of an acetate (Scheme 162).[282] 3-Methylene-4-pentenal, the key intermediate in ipsenol synthesis, also served as such in ipsdienol synthesis (Scheme 163).[275, 278] 2-Trimethylsilylmethyl-1,3-butadiene was useful in ipsdienol synthesis as a new reagent for isoprenylation (Scheme 164).[279] The synthesis of optically active ipsdienol will be described later in Schemes 226-228.

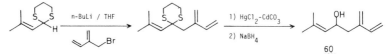

Scheme **159**

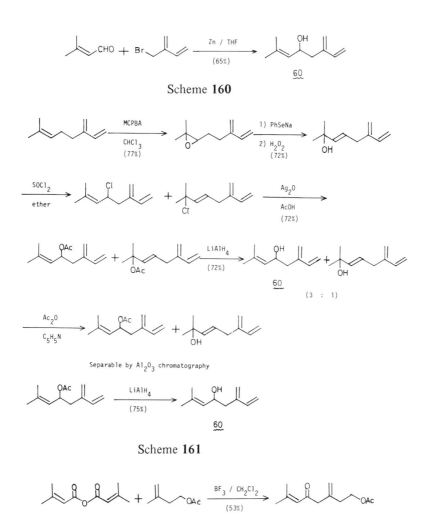

Scheme **160**

Scheme **161**

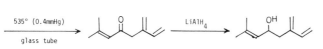

Scheme **162**

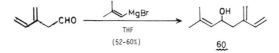

Scheme **163**

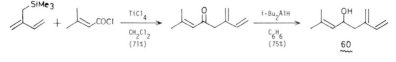

Scheme **164**

Grandisol, 1-Methyl-1-(2-hydroxy)ethyl-2-isopropenyl-cyclobutane 61

This is one of the pheromones elicited by male boll weevils (*Anthonomus grandis*). The volatile components of the pheromone complex were obtained by steam distillation of the crude extracts of 4,500,000 weevils and 54.7 kg of weevil feces.[283] This cyclobutane compound has attracted the attention of many synthetic chemists and a number of syntheses have appeared.[21] Photocycloaddition has often been used to construct the cyclobutane ring, as exemplified by

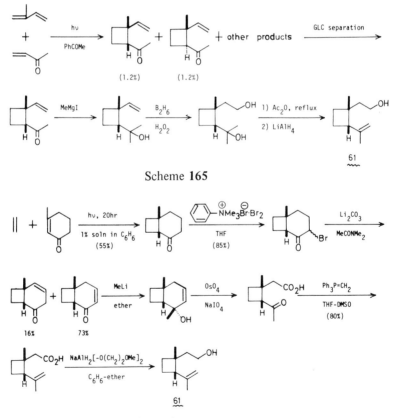

Scheme **165**

Scheme **166**

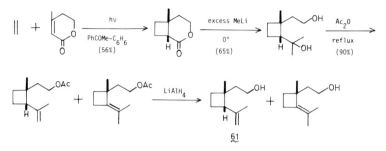

Scheme **167**

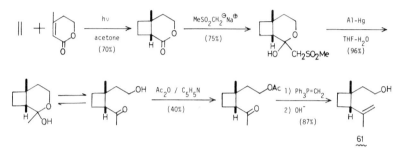

Scheme **168**

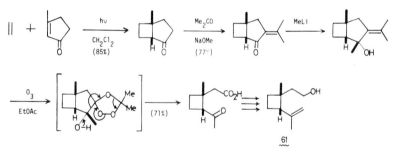

Scheme **169**

the first synthesis (Scheme 165).[283,284] The relative stereochemistry of grandisol was established as *cis* by this synthesis. The second and stereoselective synthesis was carried out by the Zoecon chemists as shown in Scheme 166.[285] Gueldner et al. then published a shorter synthesis (Scheme 167).[286] The final dehydration step, however, was not regioselective. A similar approach was reported by Uda et al. (Scheme 168).[287] This synthesis was both regio- and stereoselective. Another photochemical synthesis by Cargill and Wright involved a new fragmentation of an ozonide (Scheme 169).[288]

The first and most efficient nonphotochemical synthesis was the two-step preparation of Billups et al. (Scheme 170).[289] Catalytic dimerization of isoprene to yield a *cis*-1,2-divinylcyclobutane was the key step of this synthesis. The desired cyclobutane could be separated by fractional distillation at $0°$ after removal of the unreacted isoprene. Above room temperature the *cis*-divinyl-cyclobutane undergoes a Cope rearrangement to give 1,5-dimethylcycloocta-diene. Ayer's transformation of carvone into (±)-grandisol employed an intra-molecular photocyclization for the formation of the cyclobutane ring (Scheme 171).[290] Stork's epoxynitrile cyclization was successful in generating the correct stereochemistry of grandisol (Scheme 172).[291] An intramolecular cyclization of a chloro ester as shown in Scheme 173 yielded a mixture of the desired *cis* and undesired *trans* compounds in a ratio of 65 : 35.[292] The Trost synthesis is a unique application of his secoalkylation reaction based on organosulfur chemis-try (Scheme 174).[293, 294] The introduction of an isopropenyl group by pyrolysis

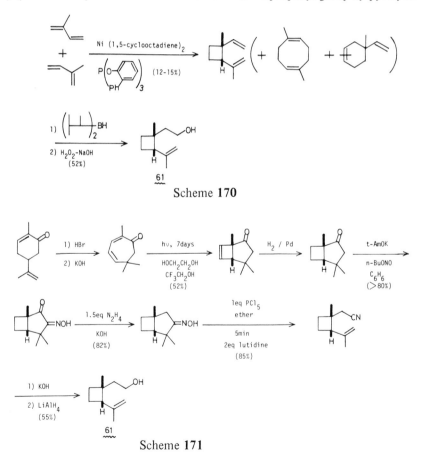

Scheme 170

Scheme 171

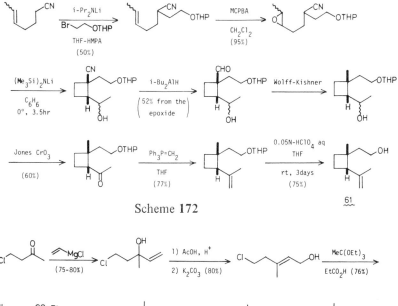

Scheme 172

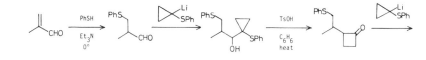

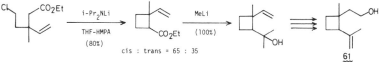

Scheme 173

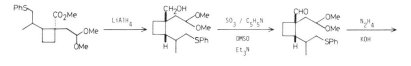

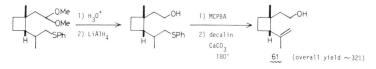

Scheme 174

of a sulfoxide yielded only the desired product without producing any double-bond isomer. However, his synthetic grandisol **61** contained about 20% of its *trans* isomer, known as fragranol.[295] Recently Wenkert reported a synthesis of grandisol (Scheme 175a).[296] He employed a regiospecific cyclopropanation and the hydrolytic rearrangement of the resulting methoxycyclopropane to cyclobutanone as the key steps. The final transformation of the bicyclo[3.2.0]hepta-

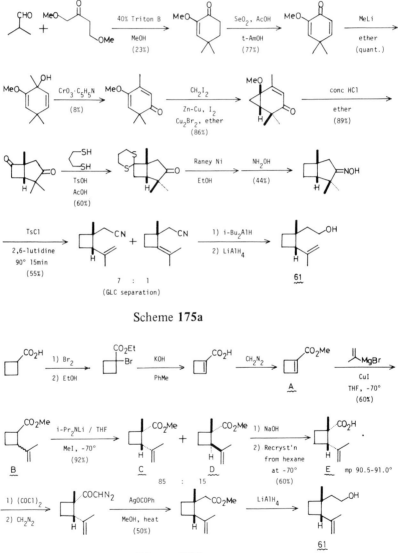

Scheme **175a**

Scheme **175b**

none system to **61** was carried out in a similar manner as that reported by Ayer (Scheme 171). Another synthesis of (±)-grandisol was recently reported by Clark (Scheme 175b).[478] The key reaction was the addition of an organocuprate to methyl 1-cyclobutene-1-carboxylate **A** to give **B** as a mixture of stereoisomers. This was methylated to give a mixture of *cis*-ester **C** and its isomer **D**. Separation was readily achieved by recrystallizing the corresponding acid **E**.

Finally a synthesis of the grandisol skeleton from the geraniol skeleton should be mentioned (Scheme 176).[297] In this synthesis, displacement of an epoxide O-atom by a stabilized allyl anion gave cyclobutane compounds **A** and **B** together with **C**. Unfortunately, the compound **A** (fragranol skeleton) was the predominant product. The compound **B**, obtained in only 29% yield, was a grandisol derivative.

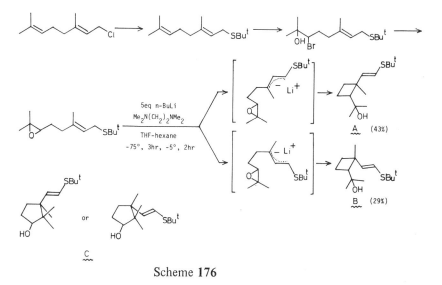

Scheme **176**

Lineatin, 3,3,7-Trimethyl-2,9-dioxatricyclo[3.3.1.0[4,7]]nonane **62**

Lineatin is an attractant compound isolated from the frass produced by female beetles of *Trypodendron lineatum* boring in Douglas fir.[298] Originally its structure was proposed to be one of the two isomeric tricyclic acetals (**62** or **62'**, Scheme 177) without assignment of the absolute configuration.[298] A synthesis shown in Scheme 177 was attempted to confirm one or the other of the postulated structures.[298] However, separation of the individual components of the isomeric mixture was unsuccessful and the structure could not be determined by this synthesis. Mori and Sasaki recently reported two syntheses of **62**. The earlier nonselective synthesis is shown in Scheme 178.[299] The two ketones **A** and **B** were separated by chromatography and converted to **62'** and **62**, re-

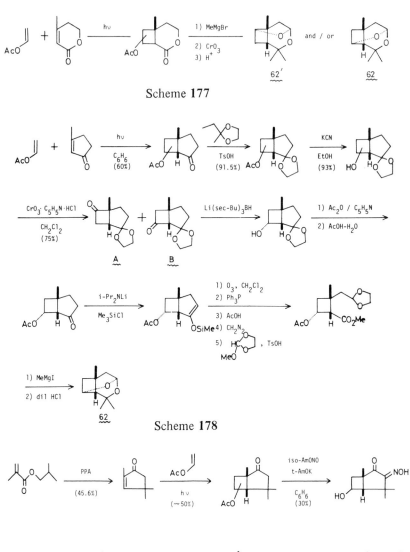

Scheme **177**

Scheme **178**

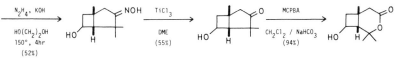

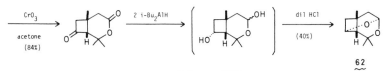

Scheme **179**

spectively. Comparison of the spectral data revealed **62** to be the racemate of lineatin. A selective synthesis of lineatin also employed photocycloaddition to generate the cyclobutane ring (Scheme 179).[300] 2,4,4-Trimethyl-2-cyclopenten-1-one, the starting material, was prepared from isobutyl methacrylate by treatment with polyphosphoric acid as described recently.[301] Although a mixture of acetoxy ketones was obtained as a photoadduct, subsequent oximation gave the desired keto-oxime as the only crystalline product. The undesired acetoxy ketone presumably resinified via a retroaldol cleavage of the cyclobutane ring of the corresponding β-hydroxy ketone.

(E)-3,3-Dimethyl-Δ$^{1,\alpha}$-cyclohexaneacetaldehyde 63, (Z)-3,3-Dimethyl-Δ$^{1,\alpha}$-cyclohexaneacetaldehyde 64, and (Z)-3,3-Dimethyl-Δ$^{1,\beta}$-cyclohexane ethanol 65

These monoterpenes are also the components of the pheromone produced by the male boll weevil.[283, 284] Tumlinson et al. synthesized them as shown in Scheme 180.[284] All other syntheses also used cyclohexanones as starting material

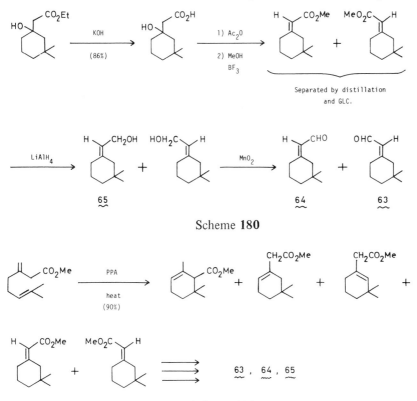

Scheme 180

Scheme 181

except one reported by Wolinsky.[302] Bedoukian and Wolinsky found that acid-catalyzed cyclization of methyl γ-geranate affords a mixture composed primarily of methyl 3,3-dimethylcyclohexenylacetates as shown in Scheme 181.[302] These were separated by preparative GLC and converted into **63,64** and **65** in the usual manner. Babler employed the Horner condensation to attach the side chain (Scheme 182).[303] For that purpose Pelletier used acetylene (Scheme 183a)[304a] and deSouza used ethyl vinyl ether (Scheme 183b).[304b] The Vilsmeier

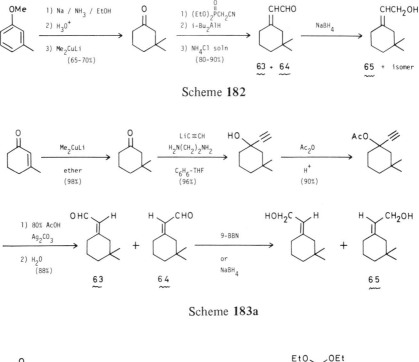

Scheme **182**

Scheme **183a**

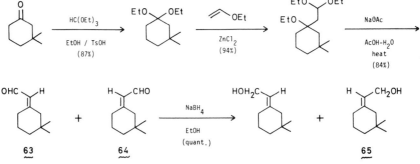

Scheme **183b**

reaction was employed as the key-step of a two-step synthesis of a mixture of **63** and **64** from isophorone (Scheme 184).[305] Nakai synthesized a mixture of **63** and **64** using the [3,3] sigmatropic rearrangement of an allylic thionocarbamate as the key-reaction (Scheme 185).[306] (For another application of this reaction see Scheme 110.) (Z)-2-Ethoxyvinyllithium was used as a nucleophilic acetaldehyde equivalent in the synthesis recently reported by Wollenberg (Scheme 186).[307] (Using this reagent Wollenberg also synthesized the red bollworm moth pheromone **36** in 36% yield from 9-decen-1-ol.[307])

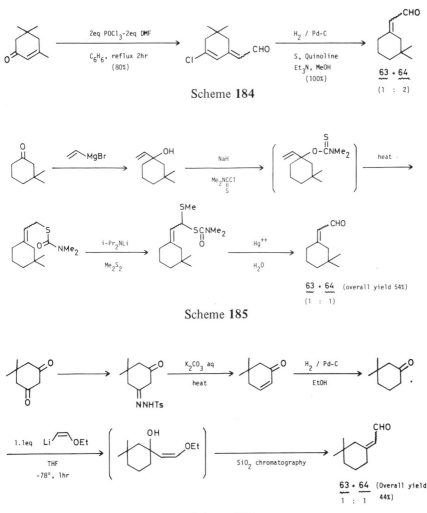

Scheme 184

Scheme 185

Scheme 186

12. PHEROMONES BELONGING TO HOMOMONOTERPENOID, SESQUITERPENOIDS, DEGRADED SESQUITERPENOIDS, AND DITERPENOIDS

(2Z, 6Z)-7-Methyl-3-propyl-2,6-decadien-1-ol 66

This was once claimed to be the sex pheromone of the codling moth.[308] The codling moth pheromone is now known to be **35**. Apart from this claim, the interesting tetrahomoterpene structure **66** attracted attention of chemists and several syntheses were recorded. The synthetic strategies devised for this compound were similar to those used for juvenile hormones. Cooke's synthesis is shown in Scheme 187.[309] The final stage of this synthesis was the application of Vedejs's method for the inversion of olefin geometry.[110a] Another synthesis was reported simultaneously by Katzenellenbogen (Scheme 188).[310, 311] Conjugate addition of lithium di-*n*-propylcuprate to an acetylenic ester effectively generated the (Z)-olefinic ester when carried out at −75°. The third synthesis used a Norrish Type I photofragmentation of 2-methyl-2-propylcyclopentanone as the

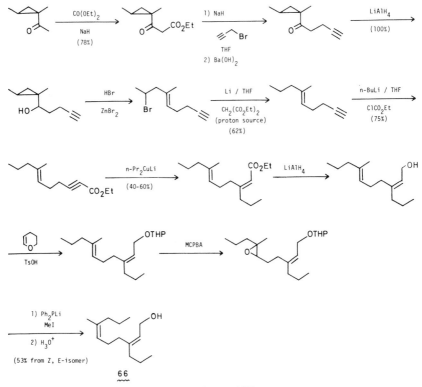

Scheme **187**

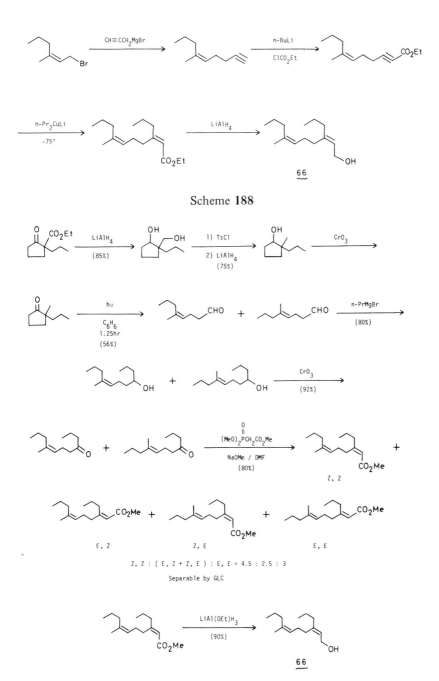

Scheme 188

Scheme 189

key step (Scheme 189).[312] The fragmentation lacked stereoselectivity and the Horner condensation was also nonstereoselective. The desired (Z, Z)-isomer therefore had to be separated by preparative GLC. A stereoselective synthesis as shown in Scheme 190 using a substitution reaction by an organocopper reagent and the alkylation of a dianion was reported by Ouannès and Langlois.[313] These authors synthesized all of the possible isomers stereoselectively. Another stereoselective synthesis was based on magnesium iodide induced rearrangement of α,β-epoxysilanes to β-ketosilanes (Scheme 191a).[314] Further treatment of the β-ketosilane with methyllithium followed by sodium acetate in acetic acid[315] gave a trisubstituted olefin in a stereoselective manner. This was converted to the acetylenic ester which had been transformed to 66 previously.[309] Another syn-

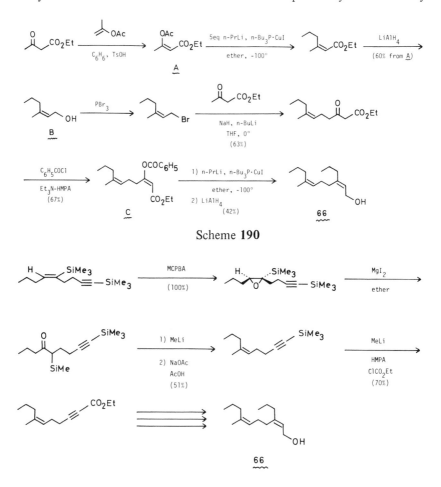

Scheme 190

Scheme 191a

thesis of **66** was recently published by Helquist et al. as shown in Scheme 191b.[479] The iterative construction of trisubstituted olefin units used in this synthesis seems to be of wide applicability. The overall yield was satisfactory.

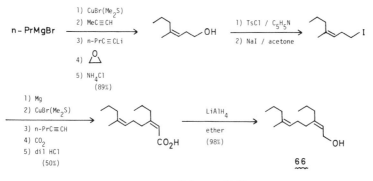

Scheme **191b**

3-Methyl-6-isopropenyl-9-decen-1-yl Acetate 67 and
(Z)-3-Methyl-6-isopropenyl-3,9-decadien-1-yl Acetate 68

These are pheromone components of female California red scale (*Aonidiella aurantii*).[316,317] Inspection of the structure **67** suggests synthetic pathways involving alkylation of citronellol. A synthesis of (±)-**67** along this line was reported by Snider and Rodini as shown in Scheme 192.[318] The key step of this synthesis was the aluminum chloride-catalyzed ene reaction of methyl propiolate with citronellyl acetate. The ene reaction introduced the isopropenyl group regiospecifically and provided appropriate functionality for conversion to (±)-

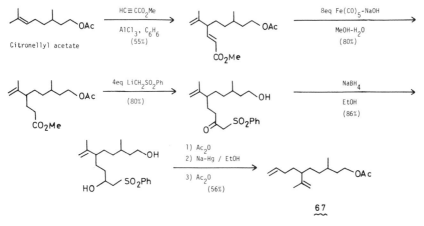

Scheme **192**

67. The final product was obtained as a mixture of all four stereoisomers. The synthesis was repeated with (R)-(+)- and (S)-(-)-citronellyl acetate and led to the pheromone with controlled stereochemistry at C-3. The (3S) -(-)-67 was more active than the (3R) -(+)-67. The natural pheromone therefore seems to have the (3S) -configuration.

An ingenious synthesis of (±)-68 was reported by Still and Mitra (Scheme 193).[319] Their success was based on a convenient and efficient [2,3]-sigmatropic Wittig rearrangement, which gave Z-trisubstituted homoallylic alcohols in high yields and with >95% stereoselectivity (A → B → C → D). It is remarkable that in the transition state C, a pseudoaxial butyl substitutent is strongly preferred. Using this reaction, they synthesized (±)-68 stereoselectively starting from the ethyl ester of β,β-dimethylacrylic acid. The synthesis of optically active 68 will be discussed later (Scheme 242).

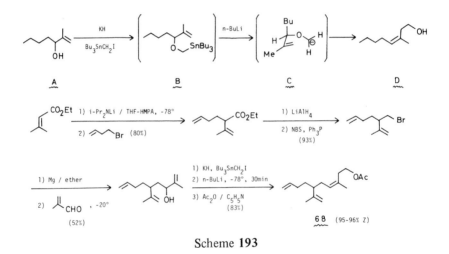

Scheme 193

(E)-3,9-Dimethyl-6-isopropyl-5,8-decadien-1-yl Acetate 69

This is the pheromone of the yellow scale (*Aonidiella citrina*), a pest of citrus crops.[320] Anderson and Henrick synthesized a racemic mixture of (E) and (Z)-isomers (4 : 1 ratio).[321] Their synthetic route is unavailable at present. A synthesis of (±)-69 and its (Z)-isomer was recently reported as shown in Scheme 194.[322] In this synthesis the (Z)-isomer of the pheromone was first synthesized starting from A and B and found to be different from the (E)-isomer by NMR and GLC. Then the intermediate C was submitted to Vedejs's olefin inversion reaction to obtain the (E)-isomer. Subsequent transformation to (±)-69 was accomplished as shown in the scheme.

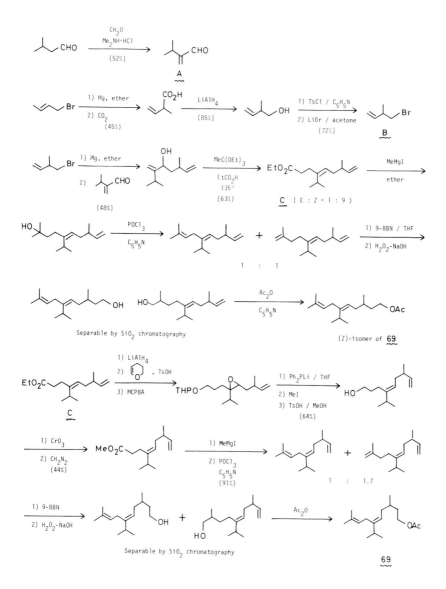

Scheme **194**

(2E, 6E)-10-Hydroxy-3,7-dimethyl-2,6-decadienoic acid 70

Meinwald isolated this compound as a major component in the "hairpencil" secretion of the male monarch butterfly (*Danaus plexippus*).[323] Its methyl ester acetate **70'** was prepared by degradation of the (±)-Juvenile Hormone III (Scheme 195).[323] Dianion alkylation and a new substitution reaction of enol phosphates with organocopper reagents were employed in a stereoselective synthesis of **70** by Sum and Weiler (Scheme 196).[324]

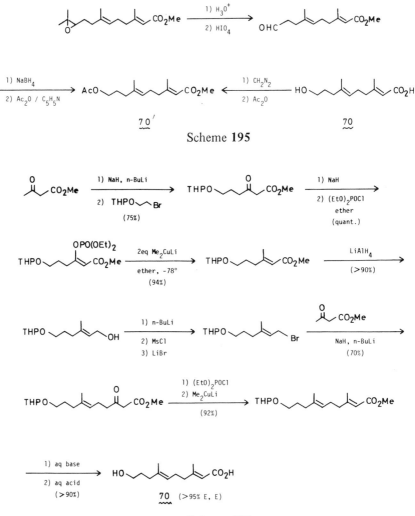

Scheme 195

Scheme 196

(2E, 6E)-3,7-Dimethyl-2,6-decadiene-1,10-diol 71

This is one of the components of the hairpencil secretion of the male queen butterfly (*Danaus gilippus berenice*) and was prepared from farnesol as shown in Scheme 197.[325]

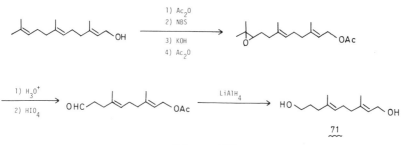

Reagents above arrows:
1) Ac$_2$O 2) NBS 3) KOH 4) Ac$_2$O

1) H$_3$O$^+$ 2) HIO$_4$

LiAlH$_4$

71

Scheme 197

(E)-3,7-Dimethyl-2-octene-1,8-diol 72

Meinwald isolated this as one of the components of the hairpencil secretion of the African monarch butterfly (*Danaus chrysippus*).[326] Its racemate was synthesized from geranyl acetate (Scheme 198).[326] Another synthesis of (±)-**72** used the Norrish Type I photofragmentation of ethyl-1-methylcyclopentan-2-one-1-carboxylate as the key step (Scheme 199).[327, 328]

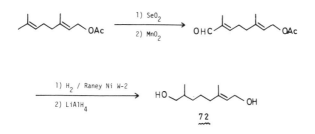

1) SeO$_2$ 2) MnO$_2$

1) H$_2$ / Raney Ni W-2 2) LiAlH$_4$

72

Scheme 198

(2E, 6E)-3,7-Dimethyl-2,6-decadiene-1,10-dioic acid 73

Extraction of hairpencils of about 6500 male monarch butterflies (*Danaus plexippus*) gave 11.8 mg of a solid identified as **73**.[329] A synthesis was accomplished starting from farnesol (Scheme 200).[329] Johnson and Meinwald devised a short and stereoselective synthesis of these degraded terpenes utilizing the

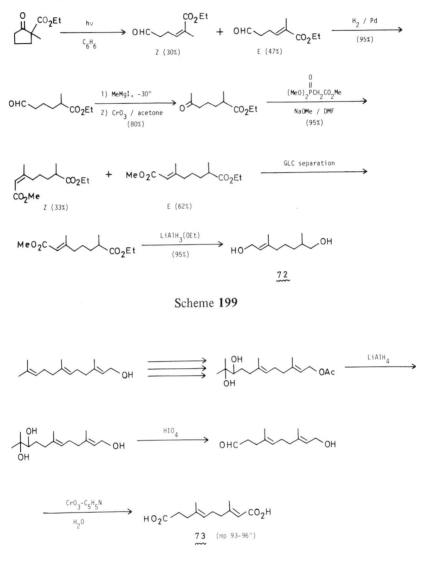

Scheme 199

Scheme 200

Claisen orthoester rearrangement (Scheme 201).[330] Trost recently reported a synthesis of the dimethyl ester of **73** by means of organopalladium chemistry (Scheme 202).[331] Trost's other synthesis of the dimethyl ester of **73** was based on his oxasecoalkylation reaction via a cyclobutanone intermediate (Scheme 203).[332]

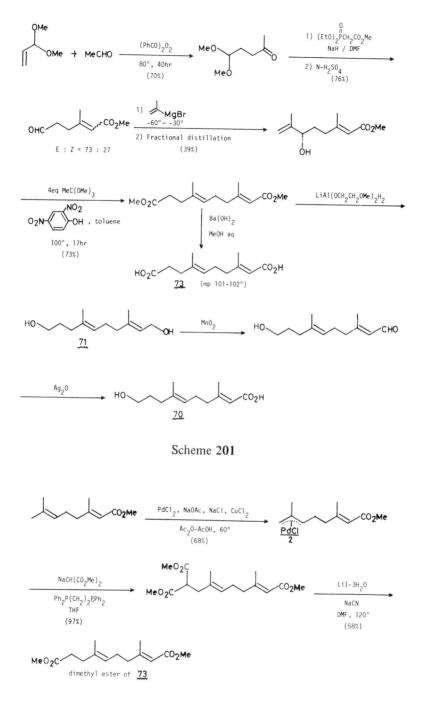

Scheme **201**

Scheme **202**

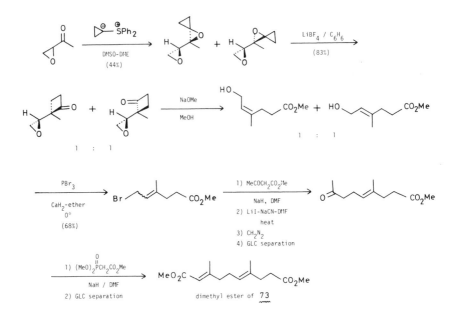

Scheme **203**

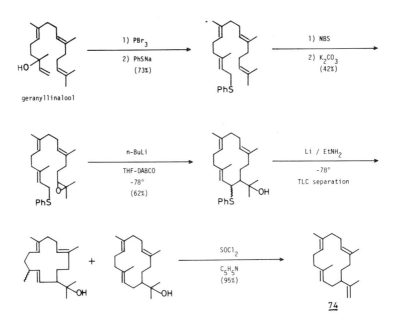

Scheme **204**

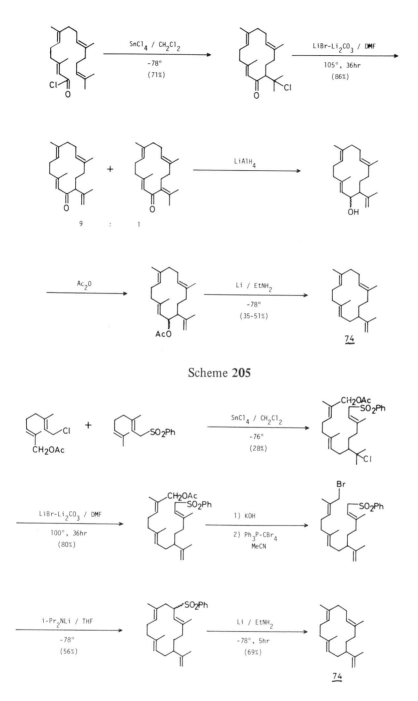

Scheme **205**

Scheme **206**

Neocembrene A **74**

Birch et al. isolated this macrocyclic diterpene hydrocarbon as a termite trail pheromone of *Nasutitermes exitosus, N. walkeri,* and *N. graveolus.*[333] The stereochemistry was elucidated by Sukh Dev et al.[334] Syntheses were achieved by Itô et al.[335] and by Kitahara et al.[336] utilizing different modes of polyene cyclization (Schemes 204[335] and 205[336]). An alternative synthetic route to the cembrane skeleton was reported recently by Kato et al. (Scheme 206).[337] For other syntheses of the cembrane skeleton see Ref. 337 and references cited therein.

13. QUEEN SUBSTANCE OF HONEYBEE

Butler et al. isolated (*E*)-9-oxo-2-decenoic acid **75** as the "Queen substance" of the honeybee (*Apis mellifera*).[338, 339] Queen honeybees secrete material, which, distributed through the colony, affects the bees in two ways; it inhibits the development of ovaries in workers and influences their behavior by inhibiting Queen rearing (that is, Queen-cell construction). Butler's synthesis is shown in Scheme 207.[339] Barbier et al. employed cycloheptanone as the starting material (Scheme 208).[340] Bestmann's synthesis was based on a unique application of ylid chemistry, as shown in Scheme 209.[341] A photochemical fragmentation reaction was used by Doolittle et al. to synthesize the Queen substance (Scheme 210).[342]

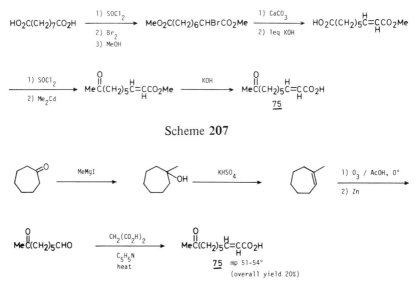

Scheme **207**

Scheme **208**

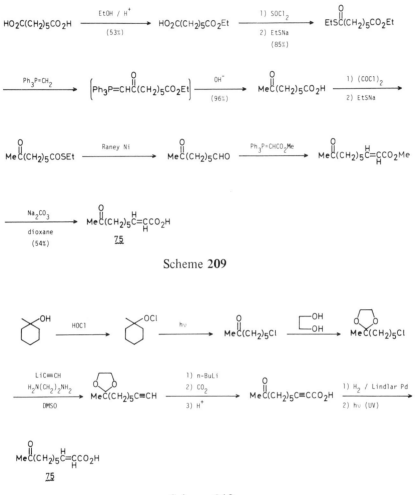

Scheme **209**

Scheme **210**

Trost devised an efficient modification of the Butler synthesis by the application of new organosulfur chemistry (Scheme 211).[343, 344] In this synthesis α-sulfenylation of an ester followed by oxidation and thermal elimination of methylsulfenic acid gave the Queen substance stereoselectively in high yield. Another interesting synthesis of Trost and Hiroi involves an oxidative cleavage as the key step, which is accompanied by rearrangement of 2,2-dithiocyclononan-1-ol, as shown in Scheme 212.[345] They called this process "oxidative seco rearrangement." Tsuji et al. developed a new synthesis from the butadiene telomer (Scheme 213).[346] Both organopalladium and organoselenium chemistry were employed

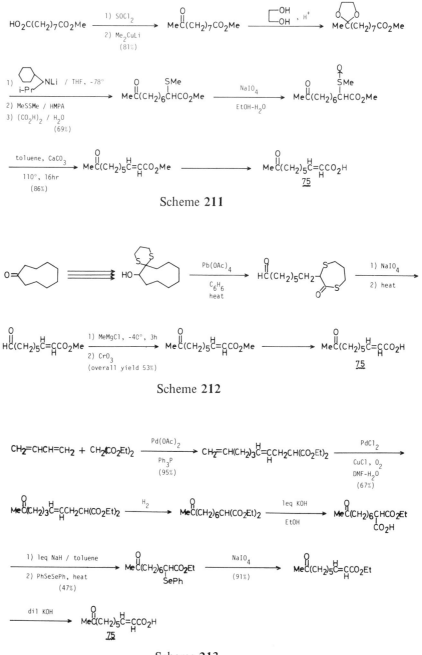

Scheme **211**

Scheme **212**

Scheme **213**

successfully in this synthesis. In a synthesis reported by Tamaru et al. the thiophene ring was used as the source of four methylene units (Scheme 214).[347, 348] For the similar use of a thiophene ring in pheromone synthesis, see Scheme 3. Subramaniam et al. recently reported a synthesis in which no exotic reagent was used (Scheme 215).[349] Manganese(III) acetate successfully generated the $MeCOCH_2$ group. The salient feature of the synthesis is this terminal olefin-acetone reaction with a three-carbon homologation. Recently Hase and McCoy

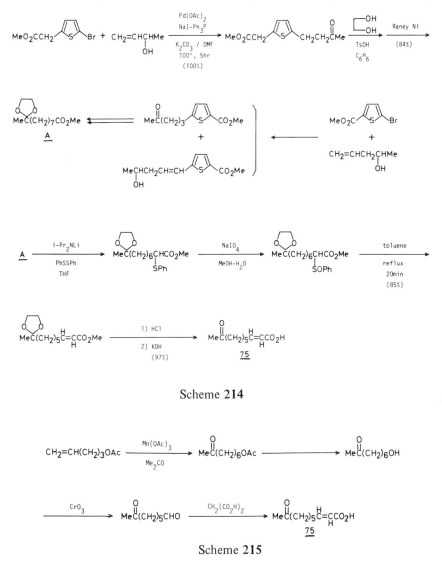

Scheme **214**

Scheme **215**

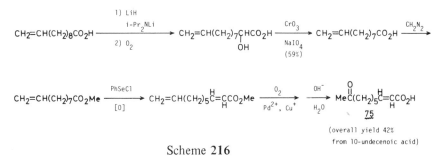

Scheme 216

converted 10-undecenoic acid to the Queen substance as shown in Scheme 216.[350] This synthesis utilized a new method for the one-carbon chain shortening of carboxylic acid.

14. HETEROCYCLES AS PHEROMONES

2,3-Dihydro-7-methyl-1H-pyrrolizin-1-one 76

This heterocyclic ketone was isolated by the Meinwalds as a component of the hairpencil secretion of the male Trinidad butterfly (*Lycorea ceres ceres*).[351] Later this was also isolated from the queen butterfly (*Danaus gilippus berenice*)[325] and from the African monarch (*Danaus chrysippus*).[326] In the case of *D. gilippus berenice*, electrophysiological and behavioral studies indicate that this heterocyclic ketone 76 serves as a pheromone. The role of other compounds such as 70-73 remains unknown. The synthesis is shown in Scheme 217.[351]

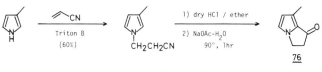

Scheme 217

1-Formyl-6,7-dihydro-5H-pyrrolizine 77

This is one of the hairpencil secretions of male butterflies of *Danaiae*.[352] Its simple synthesis was reported by means of a cycloaddition to propargylic aldehyde (Scheme 218).[353]

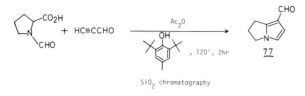

Scheme **218**

Methyl 4-methylpyrrole-2-carboxylate 78

This is a volatile trail pheromone of the leaf-cutting ant (*Atta cephalotes*).[354] A synthesis by Sonnet is shown in Scheme 219.[355]

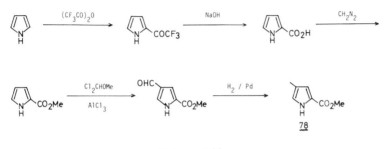

Scheme **219**

5-Methyl-3-butyloctahydroindolizine 79

This is a trail pheromone of the pharaoh ant (*Monomorium pharaonis*).[356] Synthesis of several stereoisomers of **79** was reported by Oliver and Sonnet (Scheme 220).[357]

2,3-Dihydro-2,3,5-trimethyl-6-(1-methyl-2-oxobutyl)-4H-pyran-4-one 80

This is the sex pheromone isolated from the female drugstore beetle, *Stegobium paniceum*.[358] The natural pheromone is crystalline (m.p. 52.5-53.5°) and optically active, which implies that it is a single enantiomer.[358] Its absolute stereochemistry is unknown. A synthesis of a mixture of two diastereomeric racemates was accomplished by Mori et al. as shown in Scheme 221.[359] This simple biomimetic synthesis resulted in the production of an oil that was active on the male drugstore beetle. The *trans*-isomer **80′** was removed by chromatographic purification.

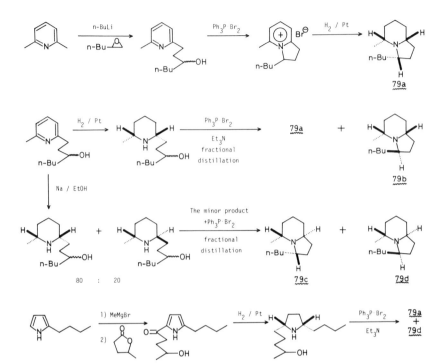

Scheme **220**

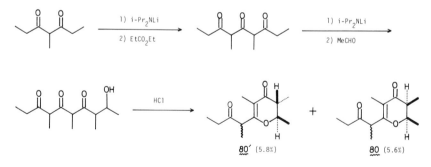

Scheme **221**

15. OPTICALLY ACTIVE PHEROMONES

Introduction

Pheromone chemistry was not discussed in three dimensions until 1973. Many pheromones are achiral aliphatic compounds. Geometrical isomerism was the chief concern of those who were working in the pheromone field.

However, there are also chiral pheromones, whose absolute stereochemistries should be established. Difficulties are often encountered in assigning the absolute stereochemistry of a natural pheromone, because it is obtainable only in minute amounts. The best solution to this problem is a synthesis starting from a compound of known absolute configuration. In 1973 Mori first demonstrated the utility of this approach by synthesizing the antipodes of the dermestid beetle pheromone artifacts **32** and **33** from (S)-$(-)$-2-methylbutan-1-ol and establishing the absolute configuration of the pheromone artifacts to be R.[360, 361] Since then a number of pheromones have been synthesized in optically active form as detailed below.

There was another reason which stimulated the synthesis of optically active pheromones. It was related to the theory of olfaction. Wright's vibration theory of olfaction predicted the unimportance of optical isomerism in pheromone perception.[362] From the generalization that no examples were known in which one of a pair of optical isomers had an odor and the other did not, Wright inferred that the primary process of olfaction must be a physical rather than a chemical interaction.[363] The slight differences reported in the odors of some optical isomers, he thought, might have resulted from different level of purity. According to Wright's theory, the vibrational frequencies of an odorous molecule in the far-infrared region (500-50 cm^{-1}) determine the quality of an odor, whereas such factors as volatility, adsorbability, and water-lipid solubility determine the strength of the odor.[364] On the other hand, Amoore emphasized the importance of molecular shape in determining the quality of odor.[365] His stereochemical theory is an example of the lock-and-key concept so well known in enzyme and drug theory. An odorous molecule must possess a stereostructure complementary to the sites of the receptors. In fact a highly significant correlation existed between molecular shape and ant alarm pheromone activity.[366] In Amoore's theory two enantiomers should be different in their odors, for they are not superimposable.

In 1971 the odor differences between extremely purified (R)-$(-)$-carvone and its (S)-$(+)$-isomer were reported simultaneously by two groups.[367, 368] The (R)-$(-)$-isomer had the odor of spearmint, while the antipode was of caraway odor. Anyway, both enantiomers were odorous, a clearer result is preferable to settle the dispute between Wright and Amoore. If we study chiral insect pheromones, then the relationship between stereostructure and olfaction might be clarified more quantitatively than by employing human noses.

In 1974 three groups reported their work along this line. Silverstein investigated the alarm pheromone of *Atta texana*.[369,370] Marumo worked on disparlure, the gypsy moth pheromone,[371,372] and Mori studied *exo*-brevicomin, the western pine beetle pheromone.[373,13,15] In all these cases, only one enantiomer of the pheromone was biologically active. This was in full accord with Amoore's stereochemical theory.

It is now evident that the syntheses of chiral pheromones are both chemically and biologically worthwhile. The results so far obtained on stereochemistry-pheromone activity relationships, which is more complex than first thought to be, will be summarized at the end of this section. Rossi recently published a review on chiral pheromones.[24]

Synthetic Strategies Towards Chiral Pheromones

If the purpose of a synthesis is the determination of the absolute configuration of a natural and optically pure pheromone with known optical rotation, the optical purity of the synthetic pheromone is not so important. What is important is the sign of optical rotation of the synthetic pheromone with known absolute configuration. However, if we want to know the relationship between absolute stereochemistry and pheromone activity, we must synthesize pheromones of high optical purity. Use of optically impure pheromone samples in bioassays may lead to the false conclusion that the both enantiomers are biologically active.

In spite of the strong research activity in asymmetric synthesis, there are only very few methods that afford highly optically pure products with known absolute configuration. It is therefore practical to adopt two classical methods, optical resolution and derivation from chiral natural products, for the synthesis of optically pure pheromones. In the synthetic sequences reactions should not be adopted that may cause racemization. The optical purity of the final product or an intermediate very near to it should be carefully determined whenever possible.

The art of optical resolution was well-reviewed recently.[374-376] The freedom of choice in the course of planning a synthesis may become wider if we adopt optical resolution instead of derivation from natural products, since we can imagine any kind of suitable intermediates for resolution. However, the resolution is not always completely successful and the absolute configuration of the resolved material must be determined by some means.

In planning a chiral synthesis starting from natural products, a wide range of knowledge concerning their stereochemistry and reactions is required. Two good reference materials are available in this area.[377,378] The merit of this approach is the satisfactory optical purities of readily available natural products such as sugars, amino acids, and terpenes. So this is a safe path leading to highly optically pure pheromones. However, the structural limitations among abundant natural products restrict the planning of syntheses and often necessitates a lengthy route for the completion of the synthesis. Another demerit is that the natural products are seldom available in two enantiomeric forms, so it is often difficult to prepare both enantiomers of the pheromone starting from a single enantiomeric form of the natural product.

Determination of Optical Purity

Remarkable progress was recorded recently concerning the techniques for determining optical purities.[374, 376, 379] The most popular present-day method is the NMR or GLC (or HPLC) analysis of (R)-(+)- or (S)-(−)-α-methoxy-α-trifluoromethylphenylacetic acid ester (MTPA ester) of an alcohol with unknown optical purity.[380] The diastereomeric esters are often separable by GLC or HPLC. In the NMR the diastereomers show a pair of signals due to methoxy protons. The chemical shift difference may be increased by the addition of NMR shift reagents such as $Eu(fod)_3$. An alternative method is to examine the NMR spectrum of a chiral compound in the presence of chiral shift reagents such as $Eu(facam)_3$ = tris [3-trifluoromethylhydroxymethylene)-d-camphorato] europium.[381] These two methods enabled Silverstein et al. to determine the enantiomeric composition of several insect pheromones with 5-500 μg of sample. Pheromone alcohols were analyzed by the MTPA method.[382] Remarkably they were not always optically pure. The enantiomeric ratios were as follows: Sulcatol **9** 65 : 35/ (+) : (−); *trans*-verbenol **81** 60 : 40/(+) : (−); 4-methylheptan-3-ol **87** 100% (−); seudenol **83** 50 : 50/(+) : (−); ipsdienol **60** from *Ips pini* (Idaho) 100% (−). Several bicyclic ketal pheromones were analyzed by the chiral shift reagent method.[383] The enantiomeric ratios were as follows: *exo*-brevicomin **6** 100% (+); frontalin **8** from *D. frontalis* 85 : 15/(−) : (+); α-multistriatin **58** 100% (−). These findings indicated that insects do not always produce only one enantiomer. This is indeed surprising, since higher animals use only one enantiomer of steroids and higher plants produce only one enantiomer of gibberellins and abscisic acid. Individual synthesis of chiral pheromones will now be discussed in the following order: (1) alcohols and acetates; (2) aldehydes, ketones, and lactones; and (3) epoxides and ketals.

(S, Z)-(+)-14-Methyl-8-hexadecen-1-ol (S)-32 and Methyl (S, Z)-(+)-14-methyl-8-hexadecenoate (S)-33

These were isolated as the pheromones of dermestid beetle and levorotatory (see Sections 5-14).[154] Their absolute configurations, however, were unknown. Mori synthesized (S)-pheromones starting from commercially available (S)-(+)-2-methylbutanol (Scheme 222).[360, 361] The synthesis was based on an acetylenic route. The synthesized (S)-pheromones were dextrorotatory, therefore the natural pheromones possess the R-configuration. Later Rossi and Carpita repeated Mori's work with some improved procedures. Their work is shown in

Scheme 223.[384] The limitation of this approach is evident when one considers the unavailability of (R)-(+)-2-methylbutan-1-ol as a natural product. The synthesis of the natural (R)-enantiomer was carried out by quite a different route as will be described later (Scheme 254).

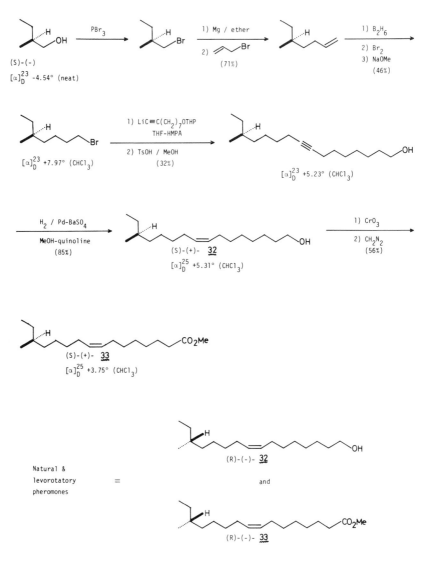

Scheme 222

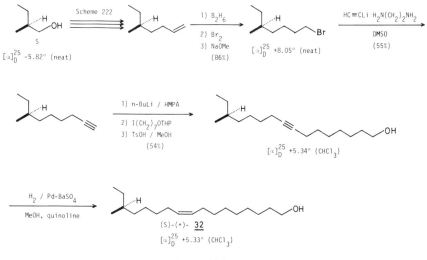

Scheme **223**

(S)-(-)-Ipsenol (S)-59 and Its Antipode (R)-59

The natural ipsenol, isolated by Silverstein et al. from the frass of *Ips paraconfusus*, was levorotatory: $[\alpha]_D^{25} - 17.5° \pm 0.7°$ (EtOH).[270, 385] Mori's first synthesis was rather lengthy employing a chiral epoxide **A** and a chiral α-methylene-γ-lactone as intermediates (Scheme 224).[386, 387] Chiral epoxides were particularly useful intermediates in chiral syntheses. In this case the epoxide was prepared from leucine. The second synthesis by Mori et al. is the chiral version of the Kondo synthesis of (±)-**59** (see Scheme 155[276]) and also employed the chiral epoxide **A** (Scheme 225).[388] Since (*S*)-leucine yielded (-)-ipsenol, the natural pheromone was of the *S*-configuration. The biological activity of both enantiomers was studied by Vité et al.[389] The five-spined engraver beetle (*Ips grandicollis*) aggregated only in response to the (*S*)-(-)-isomer. The antipode proved nearly inactive. The biological activity of the racemate was inferior to the (*S*)-(-)-isomer when released in comparable quantity, but there was no indication of definite response inhibition by the (*R*)-(+)-isomer. Ipsenol inhibits attack by *Ips pini* on ponderosa pine logs baited with male *Ips pini*.[390] Only (*S*)-(-)-ipsenol was responsible for this action as a pheromone inhibitor.[391]

(R)-(-)-Ipsdienol (R)-60 and Its Antipode (S)-60

The natural ipsdienol isolated from *Ips paraconfusus* in California was dextrorotatory: $[\alpha]_D^{20} + 10° \pm 0.9°$ (MeOH).[270, 385] Mori synthesized (*R*)-ipsdienol **60** via a chiral epoxide starting from (*R*)-(+)-glyceraldehyde acetonide as shown in

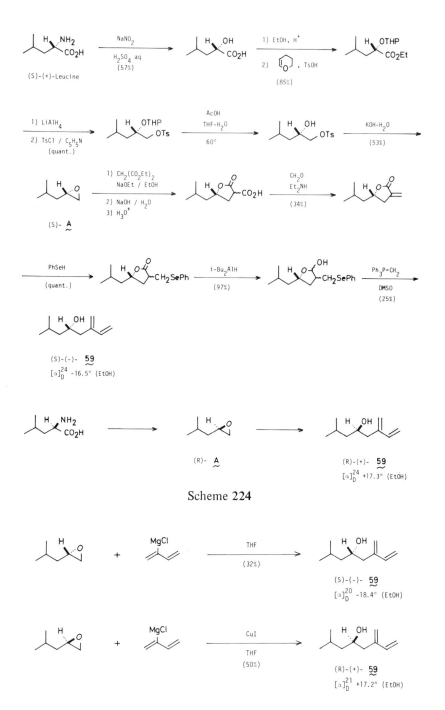

(S)-(+)-Leucine

NaNO₂
H₂SO₄ aq
(57%)

1) EtOH, H⁺
2) 〔O〕, TsOH
(85%)

1) LiAlH₄
2) TsCl / C₅H₅N
(quant.)

AcOH
THF-H₂O
60°

KOH-H₂O
(53%)

(S)- **A**

1) CH₂(CO₂Et)₂
NaOEt / EtOH
2) NaOH / H₂O
3) H₃O⁺

CH₂O
Et₂NH
(34%)

PhSeH
(quant.)

i-Bu₂AlH
(97%)

Ph₃P=CH₂
DMSO
(25%)

(S)-(-)- **59**
[α]²⁴_D -16.5° (EtOH)

(R)- **A**

(R)-(+)- **59**
[α]²⁴_D +17.3° (EtOH)

Scheme **224**

MgCl
THF
(32%)

(S)-(-)- **59**
[α]²⁰_D -18.4° (EtOH)

MgCl
CuI
THF
(50%)

(R)-(+)- **59**
[α]²¹_D +17.2° (EtOH)

Scheme **225**

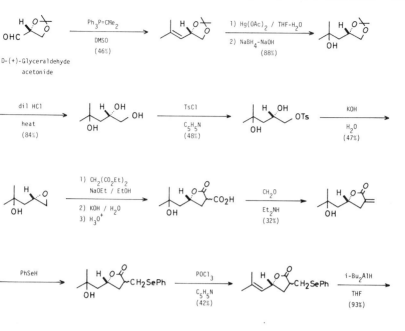

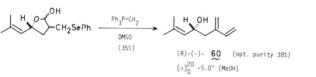

Scheme 226

Scheme 226.[392, 388] The synthetic route was similar to that employed for the synthesis of optically active ipsenol (Scheme 224) except that the extra double bond was introduced, hydrated, and reintroduced by dehydration. The product (*R*)-**60** was levorotatory. The natural and dextrorotatory pheromone was therefore of the *S*-configuration. Unfortunately in this case the synthetic pheromone was not optically pure (ca. 38% optical purity as shown later by Ohloff[393]). Since (*S*)-(−)-glyceraldehyde acetonide was not so readily available as the (*R*)-(+)-isomer, Mori et al. devised a different route to (*S*)-(+)-ipsdienol **60** starting from (*R*)-(+)-malic acid as shown in Scheme 227.[388] This synthesis also used a chiral epoxide as the key intermediate and yielded 90% optically pure (*S*)-(+)-**60**. Ohloff and Giersch reported a clever synthesis of optically active ipsdienol from verbenone in only three steps as shown in Scheme 228.[393] The key step of their synthesis is the flash-pyrolysis of 2(10)-pinen-4-ol to give ipsdienol. The

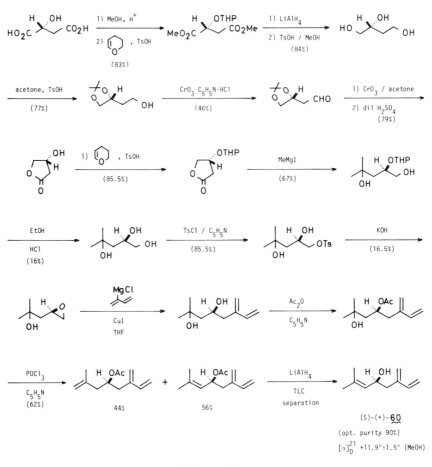

Scheme 227

oxygen-bearing chiral center of the pinenols is solely responsible for the configuration of ipsdienols, while the two bridgehead chiral centers are destroyed
during pyrolysis. Their synthetic ipsdienol showed larger $[\alpha]_D$ value than the
natural pheromone and they concluded that the optical purity of the natural ipsdienol was about 75%.

The availability of these fairly optically pure ipsdienol enantiomers prompted
Vité to study chirality-activity relationship.[394] The results were very interesting.
Both *Ips calligraphus* and *I. avulsus* responded to (R)-(−)-ipsdienol. However,
I. paraconfusus was attracted by (S)-(+)-ipsdienol. Thus there are certainly
species-specific differences in response to the ipsdienol enantiomers.

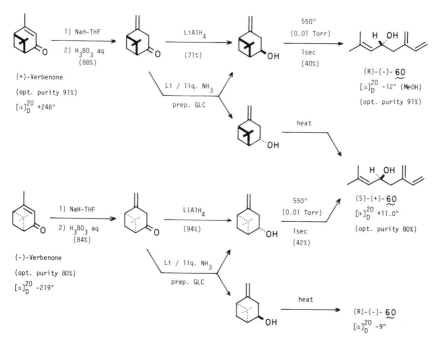

Scheme 228

(1R, 4S, 5R)-(+)-2-Pinen-4-ol[(+)-trans-verbenol, (+)-81] and Its Antipode [(−)-81]

trans-Verbenol 81 was shown to be present in the hindguts of the bark beetles, *Dendroctonus brevicomis* and *D. frontalis*[395] and later isolated from the hindguts of female *Dendroctonus ponderosae* as an aggregation pheromone.[396] Selective conversion of (+)-α-pinene to (+)-trans-verbenol was observed in *Ips paraconfusus*, while (−)-α-pinene was selectively converted to (S)-cis-verbenol 82.[397] In this insect, therefore, (+)-trans-verbenol should be optically pure. However, about 200 μg of trans-verbenol isolated from female southern pine beetles (*Dendroctonus frontalis*) was shown to be a 60 : 40 mixture of (+)- and (−)-trans-verbenol 81 by the chiral shift reagent method.[382] Mori synthesized the pure enantiomers of 81 from α-pinene (Scheme 229).[398] Fairly optically pure trans-verbenol was converted to the 3β-acetoxyetienate A, which was repeatedly recrystallized to effect purification. Removal of the steroid portion by reductive cleavage yielded optically pure trans-verbenol 81.

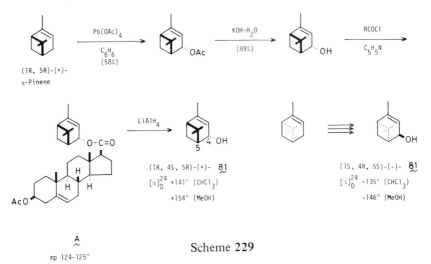

A
mp 124-125°

Scheme **229**

(1S, 4S, 5S)-2-Pinen-4-ol [(S)-cis-verbenol, (S)-82] and Its Antipode *[(R)-82]*

cis-Verbenol **82** was first isolated from male frass of a bark beetle (*Ips paracon-fusus*) as a component of the aggregation pheromone.[270, 385] Later it was found among other bark beetles such as *Ips latidens*[399] and *Ips calligraphus*.[400] (1R, 4S, 5R)-2-Pinen-4-ol (*trans*-verbenol **81**) was converted to (1R, 4R, 5R)-2-pinen-4-ol (*cis*-verbenol **82**) as shown in Scheme 230.[401] The crystalline 3β-acetoxyetienate of *cis*-verbenol was purified by repeated recrystallization. The optically pure (1R, 4R, 5R)-2-pinen-4-ol was dextrorotatory in chloroform but levorotatory in methanol or in acetone. Since the natural pheromone is known

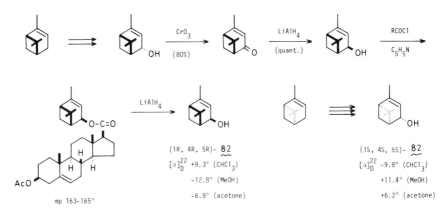

Scheme **230**

to be dextrorotatory in methanol or in acetone, it is (1*S*, 4*S*, 5*S*)-2-pinen-4-ol or (*S*)-*cis*-verbenol. The biological activity of the both enantiomers of *cis*-verbenol was tested on *Ips typographus*[402] and on *Ips calligraphus*[403]. Only (*S*)-*cis*-verbenol was active. Deploying the isomers at equal rate, (*R*)-*cis*-verbenol does not interfere with response of *Ips calligraphus* to (*S*)-*cis*-verbenol. However, released at a tenfold higher concentration, (*R*)-*cis*-verbenol strongly inhibits response to the (*S*)-isomer.[403]

(S)-(+)-Sulcatol [6-methyl-5-hepten-2-ol (S)-9] and Its Antipode [(R)-9]

This is an aggregation pheromone produced by males of *Gnathotrichus sulcatus*, an economically important ambrosia beetle in the Pacific coast of North America.[16] By detailed NMR analysis using MTPA ester method, the natural pheromone was shown to be a 65/35 mixture of the (*S*)-(+)- and (*R*)-(-)-enantiomers. The racemate was obtained by the sodium borohydride reduction of 6-methyl-5-hepten-2-one.[16] Mori synthesized both enantiomers in an optically pure state starting from glutamic acid (Scheme 231a).[17] γ-Tosyloxymethyl-γ-butyrolactone was highly crystalline and could be easily purified to an optically pure state. In laboratory and field bioassays, *G. sulcatus* responded to sulcatol only when both enantiomers were present. Response was greater to racemic sulcatol than to a mixture (65 : 35) of (*S*)-(+)- and (*R*)-(-)-enantiomers, the

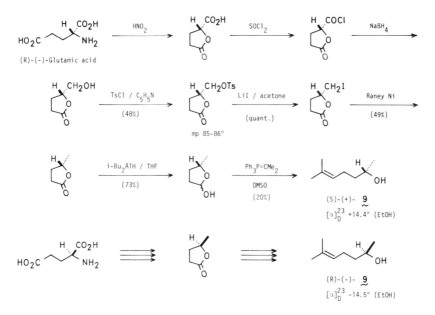

Scheme **231a**

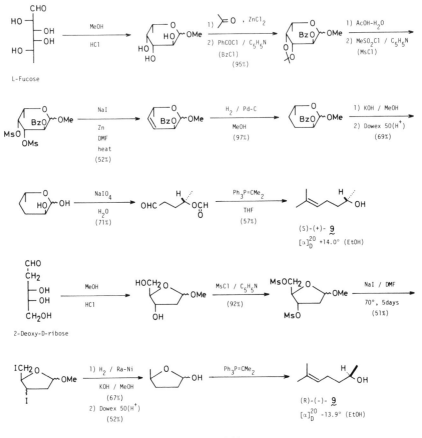

Scheme **231b**

naturally occurring isomeric ratio. This fact implies the presence of enantiomer-specific active sites on receptor proteins in the same or different cells.[18]

Another synthesis by Schuler and Slessor employed carbohydrate precursors as shown in Scheme 231b.[484] Thus (S)-(+)-sulcatol was synthesized from L-fucose by deoxygenation, carbon chain shortening, and a Wittig reaction. 2-Deoxy-D-ribose was converted to (R)-(−)-sulcatol by deoxygenation and a Wittig reaction. This synthesis, however, seems to be more complicated than that by Mori. Utilization of carbohydrates in a chiral synthesis often requires an efficient method of deoxygenation to destroy unnecessary chiral centers.[485]

(1R, 2S)-(+)-Grandisol (1R, 2S)-61 and Its Antipode (1S, 2R)-61

Natural grandisol is known to be dextrorotatory, $[\alpha]_D + 50° \pm 10°$.[284] Hobbs and Magnus accomplished the first synthesis of (1R, 2S)-grandisol starting from

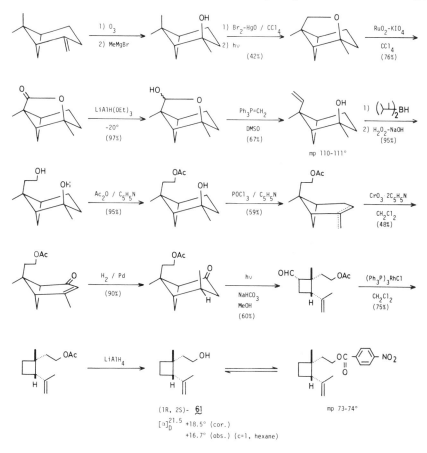

Scheme **232**

90% optically pure (1*S*, 5*S*)-(-)-β-pinene as shown in Scheme 232.[404, 405] The grandisol obtained was dextrorotatory: $[\alpha]_D^{21.5}$ + 18.5° (corrected for the optical purity of β-pinene). The natural grandisol is therefore 1*R*, 2*S*. The large $[\alpha]_D$ value of the natural pheromone might be overestimated.

Mori then published the synthesis of the both enantiomers of grandisol by resolving his intermediate (Scheme 233).[406] A bicyclic keto-acid **A** obtained by photocycloaddition was resolved with alkaloids. Then the resolved acid (-)-**A** was converted to a bicyclic ketone (-)-**B**. This yielded an acid **C** contaminated with 10-20% of the *trans*-isomer. This was purified by iodolactonization-reduction. The resulting (1*R*,2*S*)-(+)-grandisol was about 80% optically pure. Similarly (-)-grandisol was synthesized from (+)-**A**. This, too, was shown to be 80% optically pure.

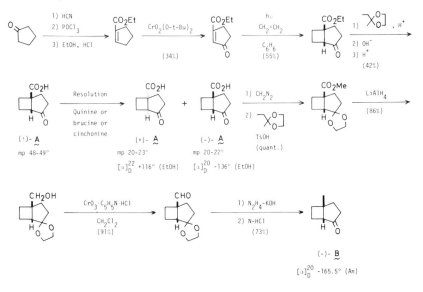

Scheme 233a

In order to secure highly optically pure (−)-grandisol, the antipode of the natural pheromone, another synthesis was carried out. The synthesis of (−)-grandisol was essential in clarifying the chirality-activity relationship of the boll weevil pheromones. Thus (±)-grandisol was oxidized to (±)-**A**. It was resolved with quinine and the resulting (−)-**A** was reduced to give (1S, 2R)-(−)-grandisol with 91-98% optical purity (Scheme 234).[407] This (−)-**61** and the previously synthesized (+)-isomer of about 80% optical purity[406] were bioassayed in the United States, and surprisingly both were fully active on female *Anthonomus grandis*.[407]

(R)-(+)-Seudenol [3-methyl-2-cyclohexen-1-ol, (R)-83] *and Its Antipode (S)-(−)-83*

Seudenol is an aggregation pheromone isolated from female hindguts of the Douglas fir beetle, *Dendroctonus pseudotsugae*.[408] The natural pheromone is racemic.[382] This posed a very interesting problem as to the pheromone activity of the pure enantiomers of seudenol: is only one enantiomer responsible for the beetle aggregation or not? To answer this question, both enantiomers of seudenol **83** were synthesized as shown in Scheme 235.[409] It should be mentioned that the resolution of (±)-seudenol **83** was unsuccessful.[409] Instead of seudenol itself, the more stable 3-iodo-2-cyclohexen-1-ol **A** was resolved, and its absolute configuration was correlated with that of 2-cyclohexen-1-ol. The iodoalcohol

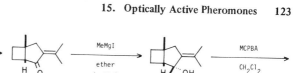

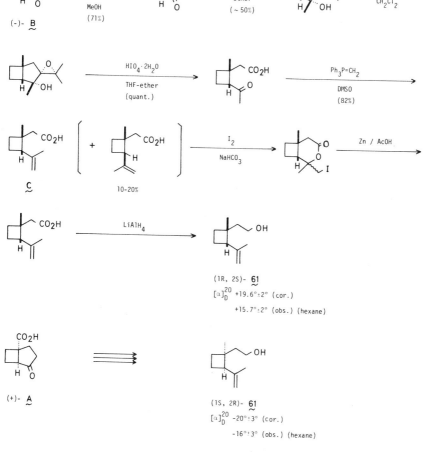

Scheme 233b

was converted to seudenol 83 with lithium dimethylcuprate. Field tests with synthetic materials revealed that (±)-83 was more active than either (−)- or particularly (+)-83 alone.[410]

erythro-3,7-Dimethylpentadec-2-yl Acetate 84 and Propionate 85

In 1976 Jewett et al. identified 3,7-dimethylpentadecan-2-ol as the alcohol in three species from two genera of pine sawflies.[411] In *Neodiprion lecontei* and

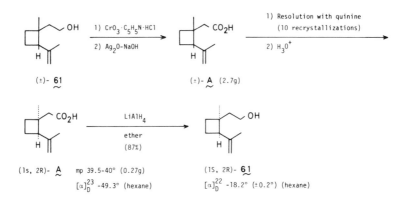

Scheme **234**

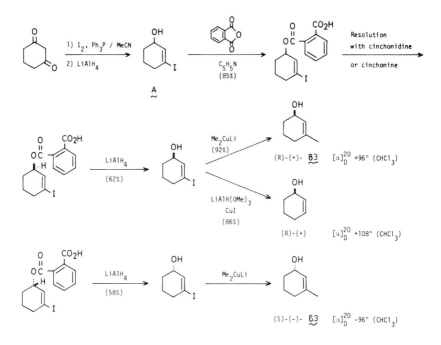

Scheme **235**

N. sertifer, the acetate **84** of this alcohol was the major component of their sex attractant, while in *Diprion similis*, it was the propionate **85**.[411] The alcohol possesses three asymmetric carbon atoms and therefore can exist in eight stereoisomeric forms. Here we also treat the syntheses of stereoisomeric mixtures so that we can understand the importance of stereoselective synthesis. The first synthesis by Jewett et al. is shown in Scheme 236[411,412] The addition of methylmagnesium iodide to an aldehyde **A** was controlled according to Cram's rule and the resulting alcohol **B** contained more of the *erythro*-isomer, which showed an identical NMR pattern in its methyl signal region with that of the natural pheromone. The relative stereochemistry at C-2 : C-3 was therefore assigned to be *erythro* and the number of the possible stereoisomers was reduced to four. Some analogs of **84** and **85** were also prepared by Jewett et al.[412]

The second synthesis by Kocienski and Ansell proceeded in satisfactory overall yield starting from 2,6-dimethylcyclohexanone (Scheme 237).[413] The crucial step was a Beckmann fragmentation of the oxime **A** to the isomeric unsaturated nitrile **B**, which proceeded in 90% yield when **A** reacted with 2 eq of *p*-TsCl in refluxing pyridine.

A synthesis by Gore et al. utilized a two-step synthetic sequence for aldehyde **D** by (1) the reaction of a vinylallenic Grignard reagent **A** with an α,β-unsaturated aldehyde **B** and (2) the oxy-Cope rearrangement in refluxing diglyme of the resulting 4-ethynylhexa-1,5-dien-3-ol **C** as shown in Scheme 238.[414] This led to a diastereomeric mixture of **84** in 47% overall yield from alcohol **C**.

Magnusson's synthesis, like Kocienski's, employed a cyclohexanone as the starting material. But he thoughtfully used *cis*-2,3-dimethylcyclohexanone **A** so that the *erythro*-configuration at C-2 : C-3 of **84** might be formed by fragmentation of the cyclohexanone. Controlling the relative configuration at C-2 : C-3 in

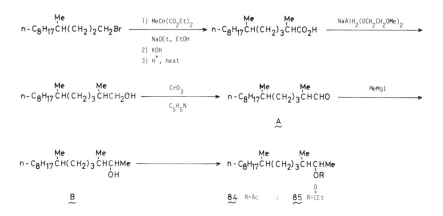

Scheme 236

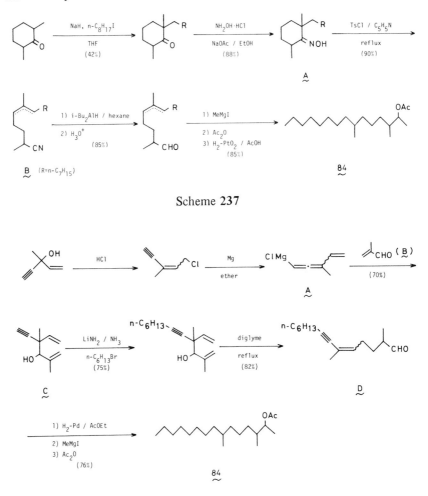

Scheme 237

Scheme 238

this manner, he obtained **84** as a mixture of two diastereomeric racemates. The stereochemistry at C-7 could not be controlled in this synthesis as shown in Scheme 239.[415, 416] He also prepared the *threo*-isomer of **84** by employing *trans*-2,3-dimethylcyclohexanone as the starting material.[416] The propionate **84** was also prepared.[416]

Tai et al. synthesized (2*R*, 3*R*)-(+)-**84** and (2*S*, 3*S*)-(−)-**84** by the sequence shown in Scheme 240.[417] They applied their new method of asymmetric hydrogenation with a modified nickel catalyst.[418] Thus, in the key-step, asymmetric (enantioface-differentiating) hydrogenation of methyl 2-methyl-3-oxobutyrate

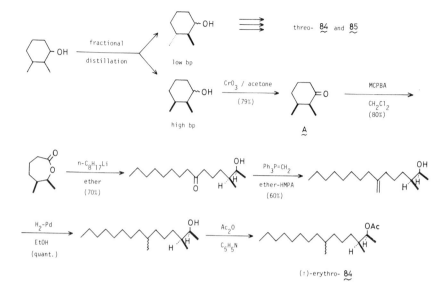

Scheme **239**

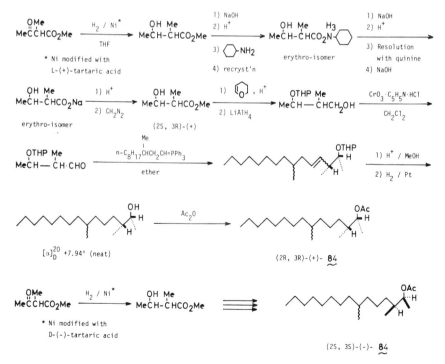

Scheme **240**

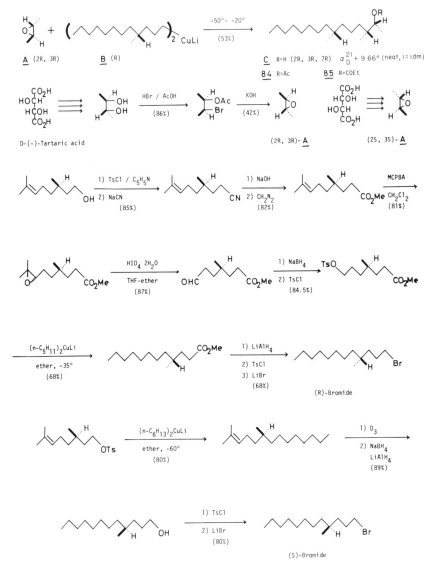

Scheme **241a**

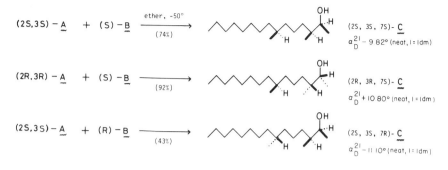

Scheme **241b**

over a nickel catalyst modified with L-(+)-tartaric acid gave methyl 3-hydroxy-2-methylbutyrate in high diastereomeric (*erythro/threo*=78/22) and enantiomeric (2S, 3R)/(2R, 3S)=84/16) excess. This asymmetric hydrogenation provided a great advantage for further optical resolution. Unfortunately, they could not control the configuration at C-7.[417, 486]

A stereocontrolled synthesis of all of the four possible stereoisomers of *erythro*-84 and 85 was accomplished by Mori and Tamada as shown in Scheme 241a,b.[419, 420] The key feature of this synthesis was a stereoselective oxirane cleavage reaction (A + B → C). The attack of a chiral organocopper reagent (B) to a chiral epoxide (A) provided the chiral alcohol C. The chiral epoxide A was prepared from tartaric acid and the chiral organocopper B was derived from (R)-(+)-citronellol. All possible stereoisomers of 84 and 85 was obtained in optically pure state by this synthesis and their pheromone activity was evaluated by field tests. *Neodiprion pinetum* in the United States was attracted only by (2S, 3S, 7S)-84.[421] On *Neodiprion sertifer* in Wales, the (2R, 3R, 7S)- and the (2S, 3S, 7S)-84 and 85 were equally effective in trapping.[422] It would, therefore, appear that the important part of these isomers is the 7S position in this case.

(R, Z)-3-Methyl-6-isopropenyl-3,9-decadien-1-yl acetate (R, Z)-68 and Its Antipode (S, Z)-68

This is one of the two pheromone components of female California red scale (*Aonidiella aurantii*). Both enantiomers of 68 were prepared from (S)- or (R)-carvone as shown in Scheme 242.[317] In each case the (Z)- and (E)-isomers were readily separable by preparative GLC (OV-1 column). Bioassays with each of the four isomers showed that only the (R, Z)-isomer attracted male red scale. Thus the natural pheromone is (R, Z)-68. The optical purity of the synthetic pheromone was checked by the HPLC analysis of an amide B derived from (R)-(+)-1-(1-naphthyl)ethylamine and an acid obtained by the oxidation of the aldehyde A and shown to be 98.4-99.0%.

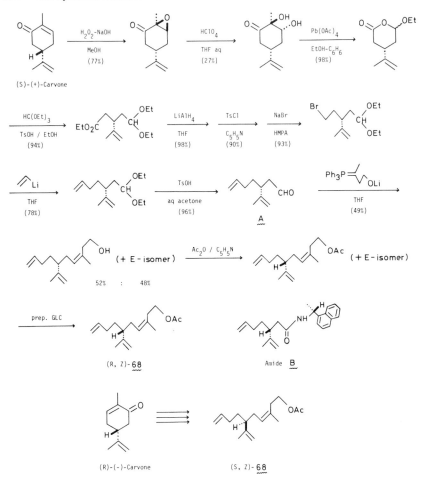

Scheme 242

(R)-(-)-10-Methyldodecyl acetate (R)-86 and Its Antipode (S)-86

Tamaki et al. isolated **86** as a minor component (2% of the total amount) of the pheromone complex of the smaller tea tortrix moth (*Adoxophyes* sp.).[423] Both enantiomers of **86** were synthesized from (R)-(+)-citronellol as shown in Scheme 243.[424] The key step was the Grignard coupling reaction. Synthesis of the intermediates from citronellol will be detailed in the sections dealing with the dermestid beetle pheromone and the German cockroach pheromone. Preliminary field tests showed that both enantiomers are biologically active.

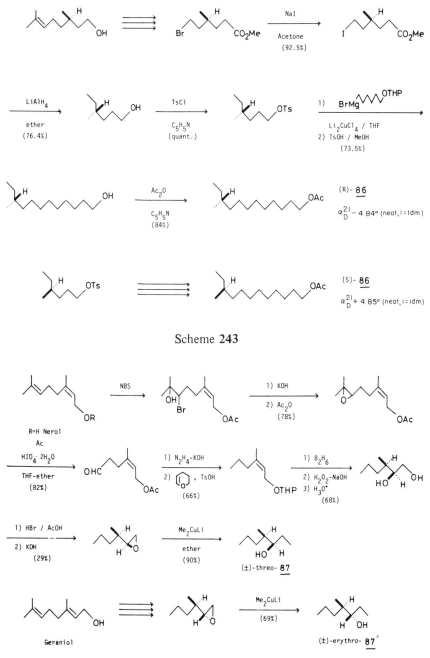

Scheme **243**

Scheme **244**

(3R, 4R)-threo-(+)-4-Methylheptan-3-ol (3R, 4R)-87

(−)-4-Methylheptan-3-ol [(−)-87] and (−)-α-multistriatin 58 are beetle-produced pheromones responsible for the aggregation of the smaller European elm bark beetles (Scolytus multistriatus).[265] Mori clarified the relative and absolute stereochemistry of (−)-87 by synthetic means. First, racemic threo-87 and erythro-87' were prepared from nerol and geraniol, respectively, as shown in Scheme 244.[425] The natural pheromone was identical with the threo-isomer on the basis of IR, NMR, and GLC. Second, (3R, 4R)-threo-4-methylheptan-3-ol 87 was synthesized from (R)-(+)-citronellic acid as shown in Scheme 245.[425] The synthesized (3R, 4R)-87 was dextrorotatory. The natural and levorotatory pheromone therefore possesses (3S, 4S)-absolute stereochemistry. The optical purity of the synthetic product was checked by its conversion to the optically pure antipode (R)-(−)-88 of the alarm pheromone of Atta texana (S)-(+)-88.

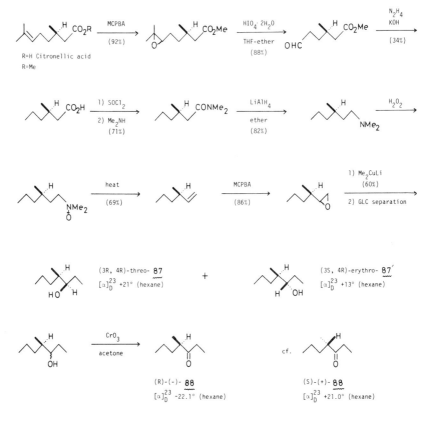

Scheme 245

(S)-(+)-4-Methyl-3-heptanone (S)-88 and Its Antipode (R)-88

This was identified as the principal alarm pheromone of the leaf-cutting ant (*Atta texana*), without specification of chirality.[426] The synthesis of the (S)-ketone employed the resolved (S)-(+)-methylallylacetic acid as the key intermediate (Scheme 246).[369,370] The (R)-ketone was prepared from (R)-(−)-2-methylpentanoic acid. The (S)-(+)-ketone 88 was about 400 times more active than the (R)-(−)-enantiomer on workers of *Atta texana*.[369] The (−)-isomer showed no inhibition of the activity of the (+)-enantiomer. The natural pheromone isolated from *Atta texana* or *Atta cephalotes* was dextrorotatory and hence assigned (S)-configuration. A synthesis of (R)-(−)-88 from (R)-(+)-citronellic acid is shown in Scheme 245.

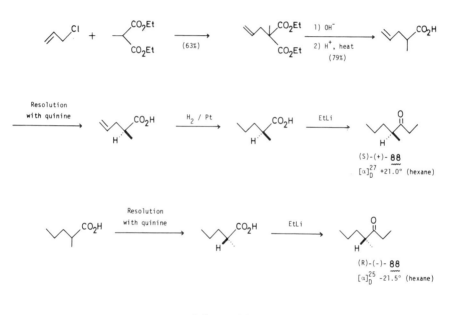

Scheme **246**

(R)-(−)-6-Methyl-3-octanone (R)-89 and Its Antipode (S)-89

This was identified as the alarm pheromone of ants in the genus *Crematogaster*.[427] Rossi et al. synthesized both enantiomers of 89 as shown in Scheme 247.[428] (R)-(−)-89 was prepared in ~33% overall yield starting from commercially available (S)-(−)-citronellol, while the antipodal (S)-(+)-89 was synthesized from (S)-(+)-2-methyl-1-butanol. Their biological activities are not yet published.

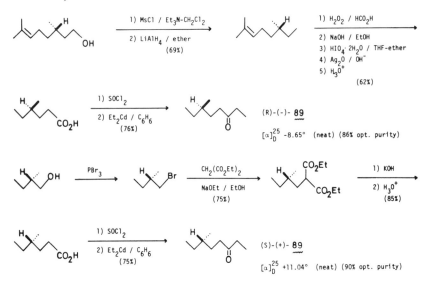

Scheme 247

(S)-(+)-Manicone (S)-50 and Its Antipode (R)-50

Manicone **50** is one of the alarm pheromones of ants. The structure was proposed without specification of chirality.[225] (S)-(+)-Manicone (S)-**50** and its antipode were synthesized by Banno and Mukaiyama employing the titanium
tetrachloride-promoted reaction of 3-trimethylsilyloxy-2-pentene with (S)-
(+)- or (R)-(-)-2-methylbutanal.[429] The S-isomer was 97% optically pure, while
the R-isomer was of 60% optical purity (Scheme 248).[429]

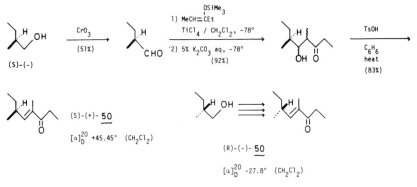

Scheme 248

(3S, 11S)-(+)-3,11-Dimethylnonacosan-2-one (3S, 11S)-55 and Its Stereoisomers

This is the sex pheromone isolated from the cuticular wax of sexually mature females of the German cockroach (*Blattella germanica*).[239, 240] Mori et al. accomplished the synthesis of all of the four possible stereoisomers of **55** in optically pure forms and established the hitherto unknown absolute configuration of the natural pheromone to be 3S, 11S.[430] The key step was the coupling of a chiral tosylate (**A**) with a chiral Grignard reagent (**B**) to give an olefin (**C**), which was converted to **55** in a standard manner (Scheme 249).[430] (R)-(+)-Citronellol was the only chiral source used in this synthesis. The preparation of (R)- and (S)-**A** is shown in Scheme 250. The intermediate alcohol **D** was obtained as crystals. The preparation of (R)- and (S)-**B** and the coupling of **A** with **B** are shown in Scheme 251. The (3S, 11S)-ketone was identified as the natural pheromone on the basis of IR, NMR, [α]$_D$ and mixture m.p. determination.

(3S, 11S)-(+)-29-Hydroxy-3,11-dimethylnonacosan-2-one (3S, 11S)-56 and Its (3R, 11S)-Isomer

This is also the sex pheromone isolated from female German cockroach (*Blattella germanica*).[239, 240] Mori and Masuda completed the synthesis of (3S, 11S)-**56** and its (3R, 11S)-isomer. The synthetic strategy was same as that used for the synthesis of **55** except that a benzyloxy group was attached at the terminal position of the long alkyl chain of the starting material (Scheme 252).[431] Comparison of the physical properties of (3S, 11S)-**56** with those of the natural pheromone strongly suggested their identity, which was proved by direct comparison.

(R, Z)-(-)-14-Methyl-8-hexadecenal (R, Z)-90 and Its Antipode (S, Z)-90

Cross et al. isolated the genuine sex pheromone of *Trogoderma inclusum* and *Trogoderma variabile* by aeration of the female beetles and identified it as (Z)-

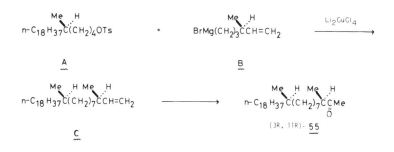

Scheme 249

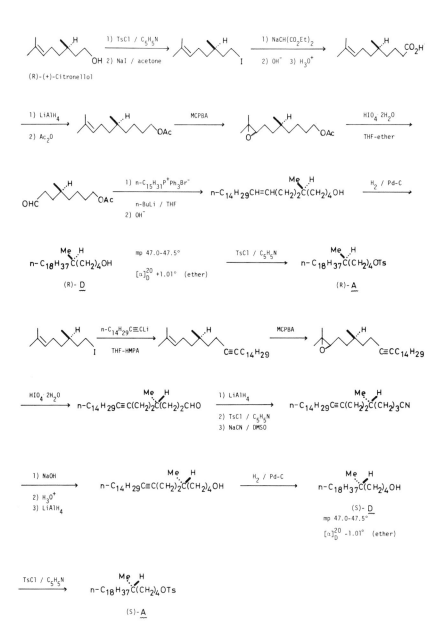

Scheme **250**

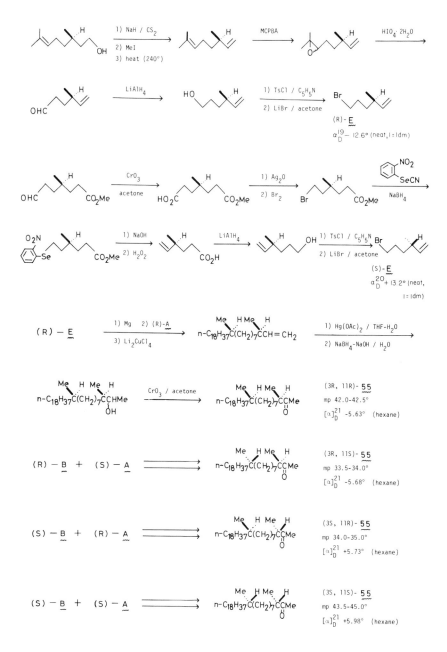

Scheme **251**

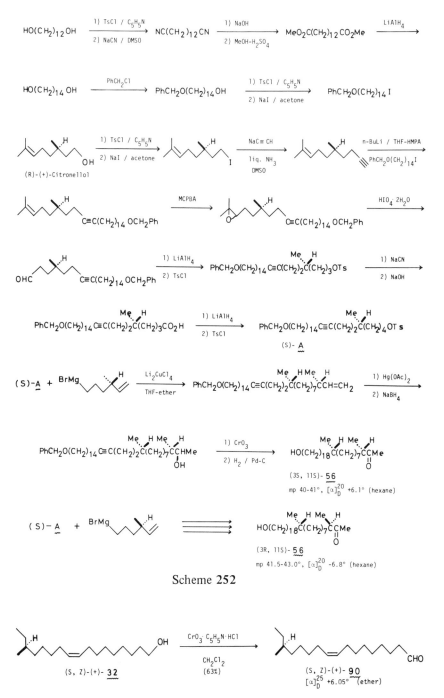

Scheme **252**

Scheme **253**

14-methyl-8-hexadecenal **90** without assigning the absolute configuration.[155] Rossi and Carpita synthesized (*S, Z*)-**90** by oxidizing (*S, Z*)-(+)-**32** (Scheme 253, cf. Scheme 223).[384] Rossi and Niccoli claimed that (*S, Z*)-(+)-**90**, when mixed with a small amount (8%) of its *E*-isomer, was highly attractive to Khapra beetle (*Trogoderma granarium*).[432] Mori et al. synthesized both enantiomers of **90**, employing (*R*)-(+)-citronellol as the common starting material, in a highly optically pure state (Scheme 254).[433] The key intermediate was an optically active

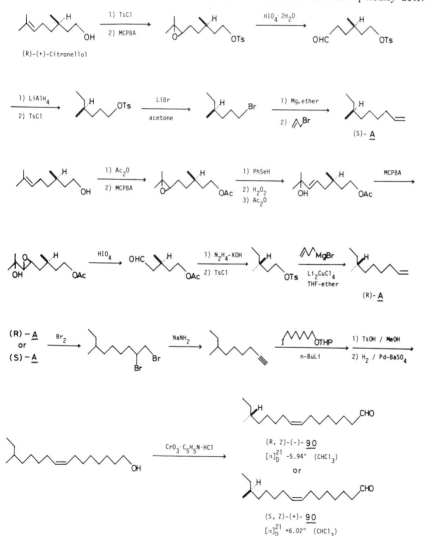

Scheme **254**

olefin **A**, which was converted to a chiral acetylene and coupled with an achiral intermediate. The (R)-isomer was about 250 times more active than its antipode when tested on dermestid beetle (*Trogoderma inclusum*).[433]

(R, E)-(−)-14-Methyl-8-hexadecenal (R,E)-91 and Its Antipode (S, E)-91

This is the pheromone of *Trogoderma glabrum* and *Trogoderma granarium* (Khapra beetle).[155] Rossi and Carpita synthesized (S, E)-(+)-**91** as shown in Scheme 255.[384] Recently both enantiomers were synthesized from (R)-(+)-citronellol as an extension of the synthesis of **90** (Scheme 256, cf. Scheme 254).[434] The R-enantiomer was 10 to 100 times more active than the S-isomer on *Trogoderma glabrum, T. inclusum*, and *T. variabile*.[435]

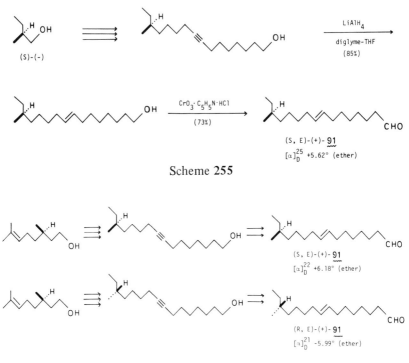

Scheme **255**

Scheme **256**

(R)-(+)-γ-Caprolactone (R)-92 and Its Antipode (S)-92

γ-Caprolactone (4-hydroxyhexanoic acid lactone) is a pheromone component of a dermestid beetle, *Trogoderma glabrum*.[436] Silverstein et al. synthesized both enantiomers of this lactone starting from glutamic acid enantiomers as shown in Scheme 257. (cf. Scheme 231).[437,438] The optical purity of (S)-(-)-92 was shown to be very high by using Pirkle's chiral solvating agent.[439] The optically pure solvating agent (R)-(-)-2,2,2-trifluoro-1-(9-anthryl)ethanol (TFAE) can induce NMR spectral nonequivalence between enantiomeric lactones. Substituents on either face of the lactone respond differently to the shielding effect of the anthryl substituent of TFAE. *T. granarium* responds to the (R)-(+)-enantiomer of γ-caprolactone but to neither the (S)-(-)-enantiomer nor the racemic form.[440]

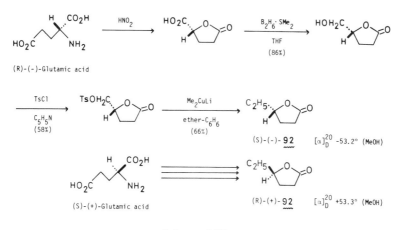

Scheme 257

(R)-(+)-δ-n-*Hexadecalactone* (R)-93 *and Its Antipode* (S)-93

δ-n-Hexadecalactone (5-hydroxyhexadecanoic acid lactone) has been found in the mandibular glands of the oriental hornet (*Vespa orientalis*).[441] This plays the role of a queen substance. The (R)-(+)- and (S)-(-)-enantiomers of 93 were synthesized by Coke and Richon as shown in Scheme 258.[442] A chiral epoxide was used as the key intermediate in the present synthesis.

(R, Z)-4-*Hydroxy-5-tetradecenoic Acid Lactone* (R, Z)-94 *and Its Antipode* (S, Z)-94

This is the pheromone isolated from the female Japanese beetle (*Popillia japonica*).[443] Both enantiomers of 94 were synthesized by Tumlinson et al. from glutamic acid enantiomers (Scheme 259).[443] The final products were obtained in highly optically pure states. They employed the Wittig reaction to construct the olefinic linkage. This caused 10-15% contamination by the unwanted (E)-isomer

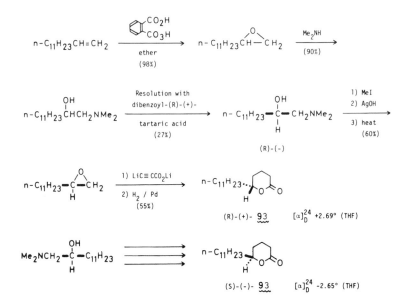

Scheme **258**

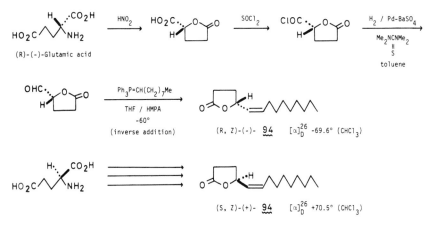

Scheme **259**

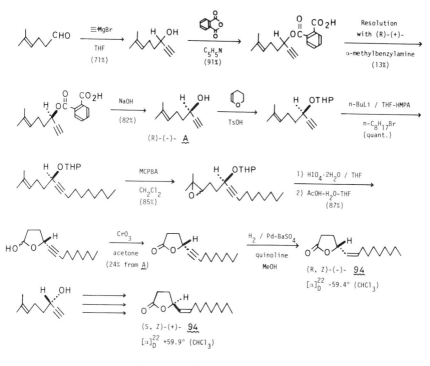

Scheme **260**

of **94** in the final product, which had to be removed by preparative HPLC and GLC. Bioassay showed that the natural pheromone is (R, Z)-**94**. Male response was strongly inhibited by small amounts of (S, Z)-**94**. The racemate of **94** was inactive.

Another synthesis by Mori et al. used resolved intermediates, and the double bond was generated by partial hydrogenation of an acetylenic lactone, which was more stereoselective than the Wittig olefination.[444] However, the optical purity of the final product was lower than the first synthesis by Tumlinson et al. The route is shown in Scheme 260.[444]

(2R, 5S)-2-Methyl-5-hydroxyhexanoic Acid Lactone (2R, 5S)-95 and Its Stereoisomers (2S, 5S)-95, (2S, 5R)-95, (2R, 5R)-95

cis-2-Methyl-5-hydroxyhexanoic acid lactone **95** was isolated as the major volatile component of the sex pheromone of the carpenter bee (*Xilocopa hirutissima*).[445] Its racemate was prepared by Baeyer-Villiger oxidation of 2.5-dimethylcyclopentanone.[445]

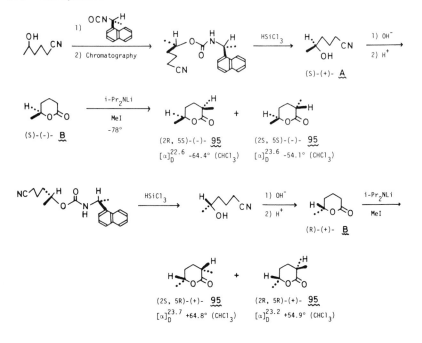

Scheme 261

Pirkle and Adams synthesized all four stereoisomers of 2-methyl-5-hydroxy-hexanoic acid lactone as shown in Scheme 261[446]. 5-Cyanopentan-2-ol was re-solved by chromatographic separation of the diastereomeric carbamate deriva-tives. Hydrolysis and lactonization of each enantiomer afforded optically pure δ-methyl-δ-valerolactone, which was methylated to give the *cis* and *trans* isomers of **95**, which were separated by preparative GLC.

Methyl (R, E)-(-)-2,4,5-tetradecatrienoate (R,E)-48 and Its Antipode (S, E)-48

This is the pheromone produced by the male dried bean beetle (*Acanthoscelides obtectus*).[217] The first synthesis of the chiral pheromone **48** was accomplished by Pirkle and Boeder as shown in Scheme 262.[447] The key intermediates, (*R, R*)- and (*S, R*)-1-ethynyl-3-carbomethoxypropyl *N*-[1-(1-naphthyl)ethyl] car-bamates, (*R, R*)-A and (*S, R*)-A, were separable using liquid chromatography. Subsequently the carbamate moiety of **A** served as the leaving group in the next step. The low-temperature reaction of (*S, R*)-A with lithium di-*n*-octylcuprate gave the (*R*)-(-)-form of methyl 4,5-tetradecadienoate [(*R*)-(-)-B] and this allene was then converted to (*R*)-(-)-**48**, [α]$_D$ -98° (hexane). The natural phero-

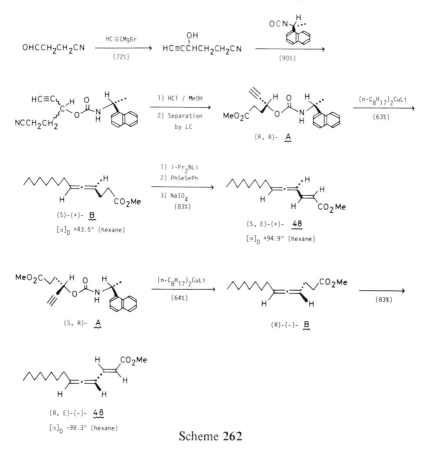

Scheme 262

mone was also levorotatory, $[\alpha]_D$ $-128°$ (hexane). Pirkle's synthetic **48** therefore had 77% the rotatory power of the naturally occurring material.

The second synthesis by Mori et al. is shown in Scheme 263.[448] The orthoester Claisen rearrangement reaction (**A** → **B**) was the key step, which proceeded in a highly stereoselective manner to give an (*S*)-(+)-allene **B**. The chirality of the starting (−)-**A** determined the chirality of **B**. The absolute configuration of (−)-**A** was proved to be *S* by correlating (+)-**A** with (*S*)-(+)-undecan-3-ol. The antipode of the natural pheromone was obtained by this synthesis. Its $[\alpha]_D$ value was + 160° (hexane). Therefore it had 125% the rotatory power of the natural pheromone. This implies that the natural pheromone was impure either chemically or enantiomerically. The synthesis of chiral allenes was reviewed.[449]

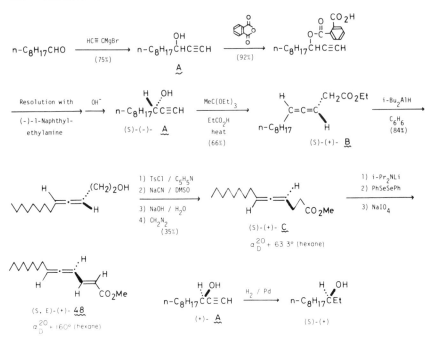

Scheme **263**

(7R, 8S)-(+)-Disparlure (7R, 8S)-21 and Its Antipode (7S, 8R)-21

This is the pheromone of the gypsy moth (*Porthetria dispar*).[121] The amount of the isolated natural pheromone was so small that its optical rotation could not be determined. The first synthesis by Marumo et al. started from L-(+)-glutamic acid as shown in Scheme 264.[371] The synthesis, however, was not stereoselective and required the tedious separation of diastereomers at the hydroxylactone **A** stage. The optical purity of the hydroxylactone was checked by the NMR method (MTPA ester plus a shift reagent). The hydroxylactone **A** was shown to contain 5.8% of its enantiomer. Both enantiomers of the *trans*-analog were also prepared. EAG and behavioral responses of the gypsy moth to these stereo-isomers revealed that *cis*-(+)-disparlure was the most effective. Racemic dispar-lure came second, while *cis*-(−)-disparlure inhibited the activity of *cis*-(+)-isomer. Enantiomers of *trans*-disparlure were not significantly different from the control.[372]

The second synthesis by Mori et al. was stereoselective and started from L-(+)-tartaric acid as shown in Scheme 265.[450,451] An intermediate in the brevicomin synthesis (Scheme 268) was also used in this case. Some intermediates, **A**, **A′**, **B** and **B′**, were crystalline, and this was particularly favorable for the purpose of

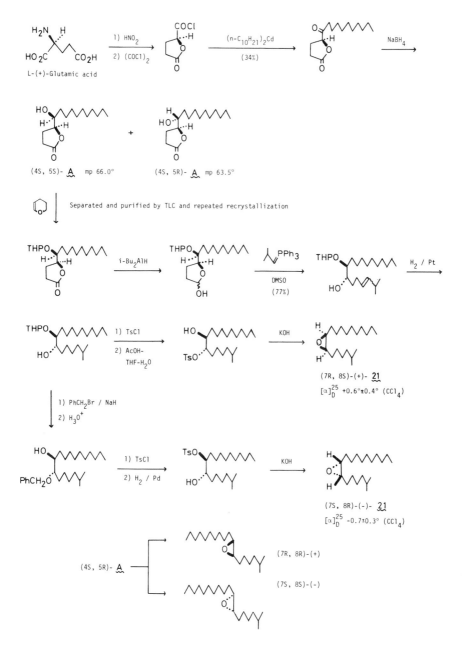

Scheme **264**

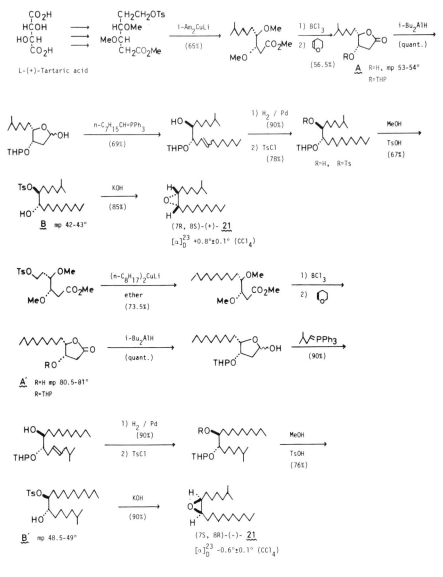

Scheme 265

obtaining highly optically pure disparlure enantiomers because they could be purified by repeated recrystallization. The optical purities of **B** and **B'** were checked by the NMR analysis of their MTPA esters in the presence of Eu(fod)$_3$ and estimated to be >98%. L-(+)-Tartaric acid was converted to both enantiomers of disparlure **21**. Under field conditions males of the nun moth (*Porthetria*

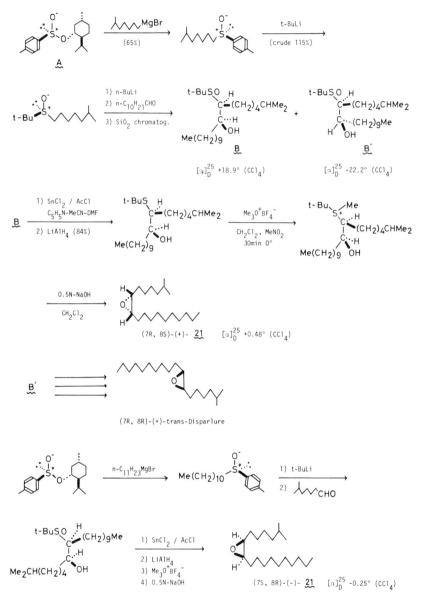

Scheme 266

149

monacha) and males of the gypsy moth (*P. dispar*) responded to (+)-disparlure. However, the addition of (−)-disparlure significantly suppressed response by *P. dispar*, while (−)-disparlure did not have such effect on the response by *P. monacha*.[452,453] Electroantennogram (EAG) studies using the differential receptor saturation technique suggest the existence of one receptor type having greatest affinity for (+)-disparlure and another type having greater affinity for (−)-disparlure than for (+)-isomer.[454]

The third synthesis by Farnum et al. was shorter than the previous routes and used for large-scale synthesis of optically active disparlure (Scheme 266).[455] Their synthesis afforded (+)-disparlure in about 15% overall yield in six steps from (−)-menthyl-*p*-toluenesulfinate **A**; 35 g of (+)-**21** was prepared by this method.[456] The key step was the chromatographic separation of a mixture of diastereomeric hydroxysulfoxides **B** and **B′**. The sulfoxide **B** yielded (+)-disparlure, while the isomer **B′** gave (+)-*trans*-disparlure. When the synthesis was conducted in a slightly modified manner (−)-disparlure was also obtained.

The fourth synthesis by Pirkle and Rinaldi was also quite effective. Enantiomerically pure (+)-*cis*-disparlure **21** was synthesized in 12% yield by a five-step sequence of reactions as shown in Scheme 267.[457] They separated the (±)-*erythro*-**A** from (±)-*threo*-**A′** by liquid chromatography. The optical resolution of (±)-*threo*-**A′** was also achieved by liquid chromatographic separation of the diastereomeric (*R*)-1-(1-naphthyl)ethyl isocyanate derived carbamates **B** and **B′**. Silanolysis of the low R_f diastereomer **B′** provided (−)-**A′**. This was stereospecifically converted to (+)-disparlure **21**. ^{13}C-NMR of (−)-**A′**, the immediate precursor to **21**, in the presence of chiral lanthanide shift reagent, allowed the determination of its enantiomeric purity.

(1R, 5S, 7R)-(+)-exo-Brevicomin (1R, 5S, 7R)-6 and Its Antipode (1S, 5R, 7S)-6

This is the aggregation pheromone of the western pine beetle (*Dendroctonus brevicomis*).[11] This pheromone possesses three asymmetric carbon atoms and therefore is a highly dissymmetric bicyclic compound. However, its 0.05% hexane solution was reported to show no optical rotation between 350 and 250 nm.[11] This observation hampered further study on the absolute stereochemistry of the pheromone until 1974 when both enantiomers were synthesized by Mori.[13] The synthesis started from tartaric acid enantiomers. The route to (1*R*, 5*S*, 7*R*)-(+)-*exo*-brevicomin **6** from D-(−)-tartaric acid is shown in Scheme 268.[13] The (+)-isomer was active on *Dendroctonus brevicomis*, while the antipode was inactive.[15] The notable feature of the synthesis was differentiation of the two carboxyl groups and demethylation by chromic acid oxidation. The demethylation was unsuccessful with boron trichloride.

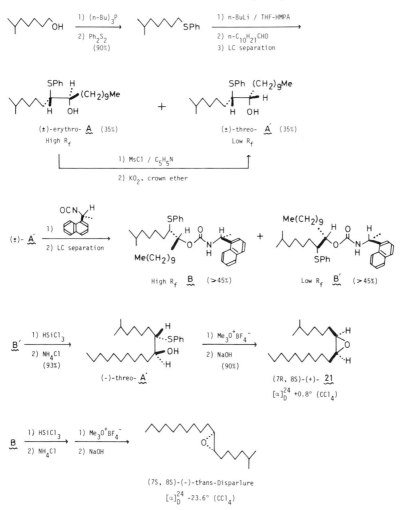

Scheme **267**

The second synthesis by Meyer also used L-(+)-tartaric acid as the starting material and (1*S*, 5*R*, 7*S*)-(−)-*exo*-brevicomin (−)-**6** of 81% optical purity was synthesized (Scheme 269).[458] The key step was the alkylation of a dithiane **B** with a bromide **A** to give a crystalline dithiane **C**. Repeated recrystallization of **C** from methanol did not increase the optical purity.

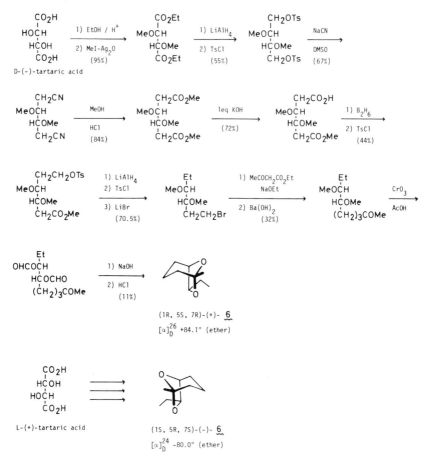

Scheme 268

(1S, 5R)-(–)-Frontalin (1S, 5R)-8 and Its Antipode (1R, 5S)-8

This is the pheromone isolated from the hindguts of the male western pine beetle and possesses two asymmetric carbon atoms.[250] Mori's first synthesis started from levulinic acid as shown in Scheme 270.[14] The lactonic acid (±)-**A** was cleanly resolved to optically pure enantiomers and hence the pheromone was obtained in an optically pure state. The absolute configuration of the resolved acids (+)-**A** and (–)-**A** was deduced by comparing their CD spectra with that of (S)-(–)-2-hydroxypentane-1,5-dioic acid of known stereochemistry.[14] Only (1S, 5R)-(–)-frontalin 8 was biologically active.[15]

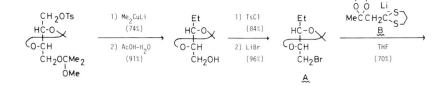

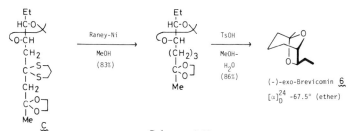

Scheme **269**

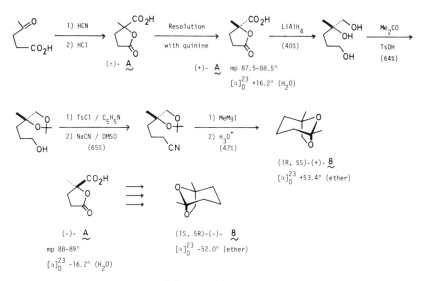

Scheme **270**

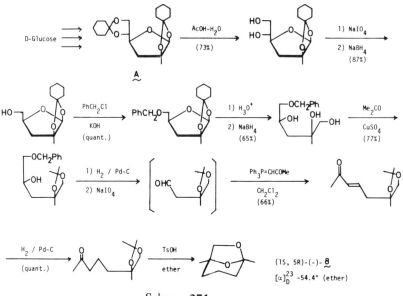

Scheme 271

Two syntheses of optically active frontalin appeared subsequently starting from D-glucose. Ohrui and Emoto converted D-glucose into (S)-(−)-frontalin 8 as shown in Scheme 271.[459] 1,2,5,6-Di-O-cyclohexylidene-3-deoxy-2-C-methyl-β-D-arabinohexofuranose A, the starting material, was synthesized from D-glucose by the methods reported by Kawana et al.[460, 461] Thus the stereochemistry of (−)-frontalin was firmly correlated to that of D-glucose. Hicks and Fraser-Reid converted methyl-α-D-glucopyranoside into frontalin as shown in Scheme 272.[462] The ketone A, derivable from methyl-α-D-glucopyranoside in four steps, was converted into one enantiomer, the other enantiomer, or a mixture of both enantiomers of frontalin in 13% overall yield, and 1 g of optically pure frontalin was obtained from 21 g of methyl-α-D-glucopyranoside.

A Unique synthesis of (R)-(+)-frontalin 8 was reported by Magnus and Roy using a new organosilicon reagent (Scheme 273).[463] The starting material was (R)-(−)-linalool. The overall yield from linalool was 23-29%.

(1S, 2R, 4S, 5R)-(−)-α-Multistriatin (1S, 2R, 4S, 5R)-58α, Its Antipode (1R, 2S, 4R, 5S)-58α, and Stereoisomers

This is a pheromone component isolated from the smaller European elm bark beetle (*Scolytus multistriatus*).[265] The natural pheromone was levorotatory ($[\alpha]_D^{25}$ −47°).[464] Silverstein et al. synthesized a mixture of (1S, 2R, 4S, 5R)-(−)-58α, (1R, 2R, 4S, 5S)-(+)-58β, (1S, 2R, 4R, 5R)-(−)-58γ and (1R, 2R, 4R, 5S)-

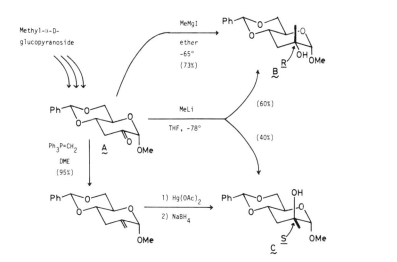

Methyl-α-D-glucopyranoside

MeMgI
ether
-65°
(73%)

B **R**

A

MeLi
THF, -78°

(60%)

(40%)

Ph₃P=CH₂
DME
(95%)

1) Hg(OAc)₂
2) NaBH₄

S
C

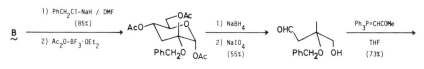

B

1) PhCH₂Cl-NaH / DMF
(85%)

2) Ac₂O-BF₃·OEt₂

1) NaBH₄
2) NaIO₄
(55%)

Ph₃P=CHCOMe
THF
(73%)

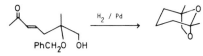

H₂ / Pd

(1R, 5S)-(+)- **8**
[α]_D +51.3° (CHCl₃)

C ⟹

(1S, 5R)-(-)- **8**
[α]_D -50.7° (CHCl₃)

Scheme 272

155

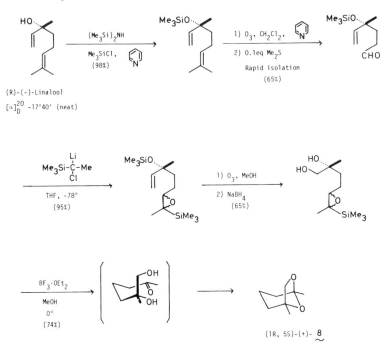

Scheme **273**

(+)-58δ starting from (S)-(+)-2-methyl-3-butenoic acid **A**.[464] This synthesis, as well as that by Mori, established the absolute configuration of natural α-multi-striatin to be 1S, 2R, 4S, 5R. Starting from (R)-(–)-**A**, (+)-58α and its isomers were also synthesized. The synthetic route is shown in Scheme 274. The mixture was separable into four components by preparative GLC. The enantiomeric composition of synthetic (–)- and (+)-58α was determined by ^{13}C-NMR with the chiral shift reagent, tris[3-(heptafluoropropylhydroxymethylene)-d-camphor-ato]europium(III).

The second synthesis by Mori started from (R)-(+)-glyceraldehyde acetonide as shown in Scheme 275.[465] In this synthesis all four stereoisomers of **58** was obtained with 1S-stereochemistry. Considerable racemization took place during this multistep synthesis.

(R)-(+)-Citronellol was used as the chiral starting material in the third synthesis of **58** by Cernigliaro and Kocienski (Scheme 276).[466]

The synthesis of (–)-α-multistriatin by Sum and Weiler was highly stereoselective, giving only pure **58**α starting from D-glucose (Scheme 277a).[467] The crucial

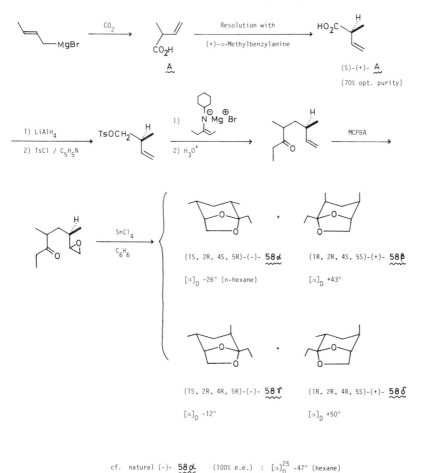

Scheme 274

step was the generation of the 1,3-diaxial methyl system of **B** from **A** by stereoselective hydrogenation using Wilkinson's catalyst. Alkylation of the dithiane **C**
was also crucial having been successful only with *tert*-butyllithium and ethyl
iodide in *n*-hexane-HMPA. The optical purity of the product **58α** was satisfactory (~98%).

Fried's synthesis of (±)-α-multistriatin was extended to a synthesis of the both
enantiomers of it by resolving (±)-**A** as shown in Scheme 277b.[483] The resolved

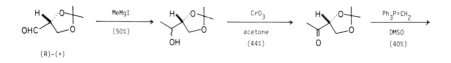

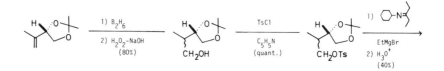

(R)-(+)

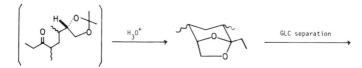

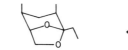

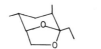

H_3O^+ GLC separation

(1S, 2R, 4S, 5R)-(-)- **58 α**

$[\alpha]_D^{23}$ -17° (ether)

(1S, 2S, 4R, 5R)-(-)- **58 β**   +

(1S, 2R, 4R, 5R)-(-)- **58 γ**   +   (1S, 2S, 4S, 5R)-(-)- **58 δ**

$[\alpha]_D^{23}$ -31° (ether)

Scheme **275**

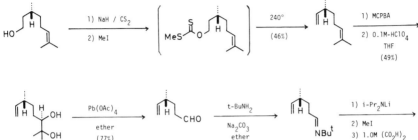

 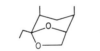

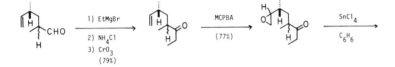

(1S, 2R, 4S, 5R)-(-)- **58 α**   (1R, 2R, 4S, 5S)-(+)- **58 β**

$[\alpha]_D$ -18.7° (hexane)

(1S, 2R, 4R, 5R)-(-)- **58 γ**   (1R, 2R, 4R, 5S)-(+)- **58 δ**

Scheme **276**

159

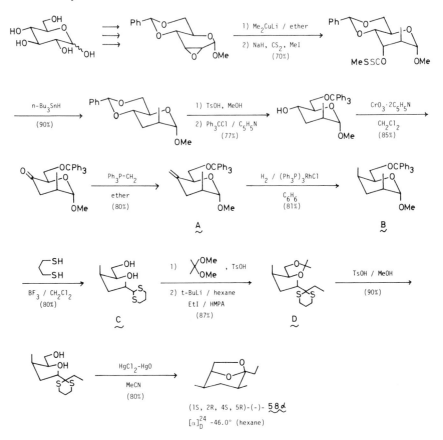

Scheme **277a**

(−)-**A** gave a mixture of (−)-α-multistriatin **58α** and (−)-γ-multistriatin **58γ** (**58α** : **58γ** = 85 : 15). The bioactive enantiomer was shown to be (−)-**58α**, while (+)-**58α** was no more active than controls in both laboratory and field tests.

A highly stereoselective synthesis of **58δ** was recently achieved by Mori and Iwasawa as shown in Scheme 278.[468] The key step was the stereoselective ring opening of an epoxy ester **A** with lithium dimethylcuprate to give **B**. Both enantiomers of **A** were prepared from tartaric acid enantiomers. (−)-δ-Multistriatin **58δ**, instead of **58α**, was attractive for the European population of *Scolytus multistriatus*.[266]

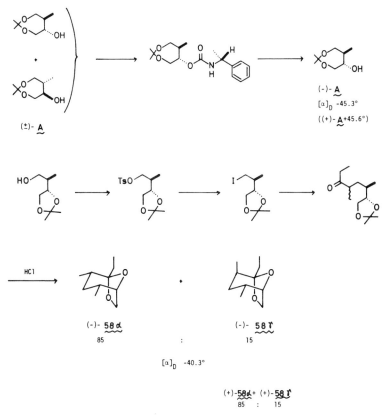

Scheme **277b**

(2R, 5RS)-Chalcogran (2R, 5RS)-57 and Its Antipode (2S, 5RS)-57

This is the pheromone of *Pityogenes chalcographus*.[249] Silverstein et al. synthe-sized (2*R*, 5*RS*)-chalcogran **57** from (*R*)-(+)-caprolactone **92** as shown in Scheme 279.[469] They were able to separate the two stereoisomers by preparative GLC and assigned stereochemistries by ^{13}C-NMR with the aid of a chiral shift reagent. The major isomer had a longer GLC retention time.

Mori et al. prepared (2*R*, 5*RS*)-**57** and (2*S*, 5*RS*)-**57** by an entirely different route. They utilized α-amino-*n*-butyric acid as the chiral source. The synthesis proceeded in a simple manner by applying the recent technique of dianion

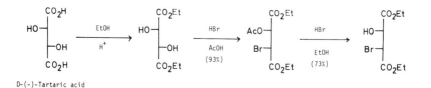

D-(-)-Tartaric acid
EtOH
H+
HBr
AcOH
(93%)
HBr
EtOH
(73%)
NaOEt
EtOH
(77%)
A
Me2CuLi
(78%)
B
TsOH
(quant.)
1) LiAlH4
2) TsOH / MeOH
(quant.)
TsOH
(76%)
1) TsCl / C5H5N
2) NaI / acetone
(74%)
i-Pr2NLi
(69%)
10% HCl
MeCN
(43%)
(1S, 2S, 4S, 5R)-(-)-58δ
[α]D20 -83.5° (pentane)
(1R, 2R, 4R, 5S)-(+)-58δ
[α]D20 +82.4° (pentane)

Scheme 278

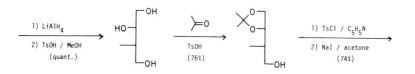

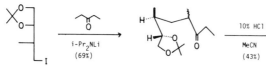

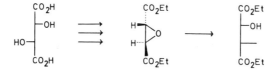

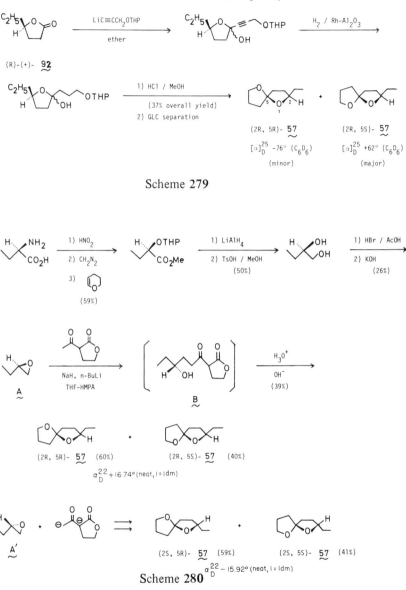

Scheme 279

Scheme 280

chemistry as shown in Scheme 280.[470,471] Thus alkylation of a dianion derived from α-acetyl-γ-butyrolactone with chiral epoxides **A** or **A**′ followed by subsequent hydrolytic decarboxylation and ketalization yielded a mixture of (2R, 5R)-**57** and (2R, 5S)-**57** or (2S, 5R)-**57** and (2S, 5S)-**57**. No separation of the mixture was attempted, since the natural pheromone itself was a mixture.

Periplanone-B 96

This is one of the two sex pheromone component of the American cockroach (*Periplaneta americana*). About 200 μg of periplanone-B was isolated by Persoons et al. and its structure was proposed to be **96** without stereochemical assignment.[472-474] A synthesis of (±)-periplanone-B was achieved by Still as shown in Scheme 281.[475] The key step was the conversion of **A** to **B** by means

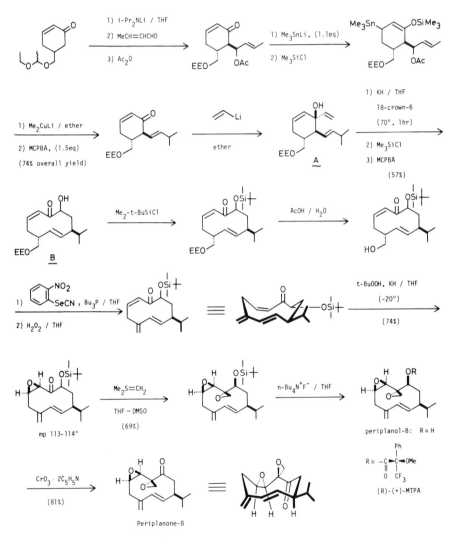

Scheme **281**

of an oxy-Cope rearrangement.[476] (±)-Periplanol-B was resolved by LC as its (R)-(+)-MTPA ester.[477] The separated esters were hydrolyzed with sodium hydroxide to give resolved periplanol-B and its enantiomer, which were oxidized with chromic anhydride-pyridine to periplanone-B **96** and its enantiomer. The threshhold activity on male cockroaches of synthetic periplanone-B was 10^{-6} to 10^{-7} μg. The antipode was devoid of activity. The absolute stereochemistry of periplanone-B was determined as **96** by Nakanishi's exciton chirality method.[477]

Stereochemistry-Pheromone Activity Relationships

The relationship between pheromone activity and absolute stereochemistry of chiral pheromones is rather complicated. In some cases only one enantiomer is

STEREOCHEMISTRY – ACTIVITY RELATIONSHIPS –1

A   Only one enantionmer is biologically active.

A – 1   The antipode does not inhibit the action of the pheromone.

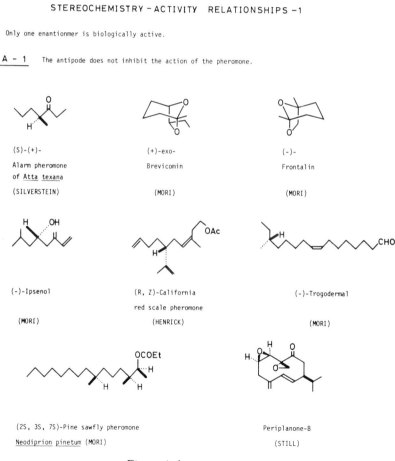

(S)-(+)-

Alarm pheromone

of Atta texana

(SILVERSTEIN)

(+)-exo-

Brevicomin

(MORI)

(-)-

Frontalin

(MORI)

(-)-Ipsenol

(MORI)

(R, Z)-California

red scale pheromone

(HENRICK)

(-)-Trogodermal

(MORI)

(2S, 3S, 7S)-Pine sawfly pheromone

Neodiprion pinetum (MORI)

Periplanone-B

(STILL)

Figure A-1

biologically active, but in others both enantiomers are fully active. The situation changes with insect by insect. The results obtained with various insects are summarized in Fig. A-D. The formula indicates the bioactive enantiomer. The name of the senior author who synthesized the pheromone is written under each formula.

In the case of those pheromones listed in Fig. A-1, only one enantiomer is biologically active. No inhibitory action was observed with the inactive antipode. Majority of chiral pheromones seems to belong to this class. Similarly, in the case of pheromones listed in Fig. A-2, only one enantiomer is biologically active. However, the inactive antipode inhibits the action of the correct enantiomer. Especially in the case of the Japanese beetle pheromone, its racemate was inactive due to the inhibitory action of the wrong enantiomer.

STEREOCHEMISTRY – ACTIVITY RELATIONSHIPS – 2

A-2   The antipode inhibits the action of the pheromone.

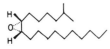

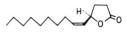

(+)-Disparlure   (S)-cis-Verbenol   (R, Z)-Japanese beetle pheromone
(MARUMO, MORI)   (MORI)   (TUMLINSON)

Figure A-2

STEREOCHEMISTRY – ACTIVITY   RELATIONSHIPS – 3

B   All stereoisomers are biologically active.

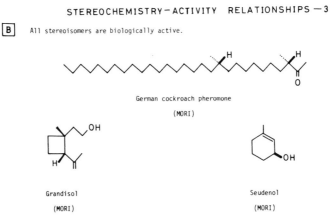

German cockroach pheromone

(MORI)

Grandisol   Seudenol

(MORI)   (MORI)

Figure B

In the case of pheromones listed in Fig. B, insects do not discriminate between stereoisomers. It is indeed surprising that the unnatural grandisol is biologically active.

Ipsdienol (Fig. C) is unique because two different species of *Ips* seem to possess enantiomeric pheromone receptor systems and the chirality of the pheromone is quite important in establishing and maintaining a particular *Ips* species.

STEREOCHEMISTRY −ACTIVITY RELATIONSHIPS − 4

C Even in the same genus, different species use different enantiomers.

(S)-(+)-Ipsdienol
Ips paraconfusus

(R)-(−)-Ipsdienol
Ips calligraphus

(OHLOFF)

Figure **C**

Sulcatol (Fig. D) is the only pheromone both of whose enantiomers are required for pheromone activity. This means that two enantioselective receptors are present in the antennae of *Gnathotrichus sulcatus*. It should be added that Chapman et al. recently demonstrated the presence of chiral dual receptors even for an achiral sex pheromone such as (*Z*)-11-tetradecenyl acetate.[480, 481] The prochiral character of (*Z*)-11-tetradecenyl acetate is used to good advantage in the receptor system. The red-banded leaf roller has evolved at least two chemo-receptors (one in common with the European corn borer) that accommodate

STEREOCHEMISTRY − ACTIVITY RELATIONSHIPS − 5

D The both enantiomers are required for the pheromone activity.

(+)-Sulcatol   (−)-Sulcatol

(MORI)

Figure **D**

different conformations of the achiral, but prochiral, pheromone. It is therefore now clear that all types of isomerism—structural, geometrical, and optical—are utilized by insects in their chemical communication system.

16. CONCLUSION

In this chapter I have reviewed the synthesis of 96 insect pheromones. The syntheses of 30 chiral pheromones were discussed with special emphasis. Advances in synthetic methodology have made it possible to synthesize both geometriacally and enantiomerically pure pheromones and has assisted progress in understanding pheromone perception. Pheromone synthesis is indeed one of the unique areas of research, which firmly ties chemistry and the life sciences. Nature's diversity as revealed by pheromone chemists will continually provide new targets for ingenious syntheses and stimulate new ideas among synthetic chemists.

The literature survey for this chapter was made up to the end of April, 1979.

ACKNOWLEDGMENTS

I thank my co-workers, whose patience and skill made it possible to achieve our own work on pheromone synthesis. They are Takashi Ebata, Hiroshi Hashimoto, Hiroko Iwasawa, Shin-ichi Kobayashi, Satoru Masuda, Noriko Mizumachi, Tsutomu Nakayama, Tomoo Nukada, Masayuki Sakakibara, Mitsuru Sasaki, Kazuo Sato, Toshio Suguro, Tetsuo Takigawa, Yoshiyuki Tachibana, Shigeharu Tamada, and Minoru Uchida. I also thank my present staff, Prof. T. Kitahara and others, for drawing the formulas. Mrs. T. Kitahara kindly typed the legends of the formulas. I am especially thankful to my wife Keiko Mori for typing the manuscript not only of this review but also almost all of my papers on synthetic chemistry.

REFERENCES

1. A. Butenandt, R. Beckmann, D. Stamm, and E. Hecker, *Z. Naturforsch.*, **14B**, 283 (1959).
2. P. Karlson and M. Lüscher, *Nature*, **183**, 55 (1959).
3. A. Butenandt and E. Hecker, *Angew. Chem.*, **73**, 349 (1961).
4. A. Butenandt, E. Hecker, M. Hopp, and W. Koch, *Liebigs Ann. Chem.*, **658**, 39 (1962).
5. E. Truscheit and K. Eiter, *Liebigs Ann. Chem.*, **658**, 65 (1962).

6. W. L. Roelofs and H. Arn, *Nature*, **219**, 513 (1968).

7. W. L. Roelofs and A. Comeau, *J. Insect Physiol.*, **17**, 435 (1971).

8. J. A. Klun, O. L. Chapman, K. C. Mattes, M. Beroza, and P. E. Sonnet, *Science*, **181**, 661 (1973).

9. M. Beroza, G. M. Muschik, and C. R. Gentry, *Nature*, **244**, 149 (1973).

10. H. E. Hummel, L. K. Gaston, H. H. Shorey, R. S. Kaae, K. J. Byrne, and R. M. Silverstein, *Science*, **181**, 873 (1973).

11. R. M. Silverstein, R. G. Brownlee, T. E. Bellas, D. L. Wood, and L. E. Browne, *Science*, **159**, 889 (1968).

12. J. P. Vité and J. A. A. Renwick, *Naturwissenschaften*, **58**, 418 (1971).

13. K. Mori, *Tetrahedron*, **30**, 4223 (1974).

14. K. Mori, *Tetrahedron*, **31**, 1381 (1975).

15. D. L. Wood, L. E. Browne, B. Ewing, K. Lindahl, W. D. Bedard, P. E. Tilden, K. Mori, G. B. Pitman, and P. R. Hughes, *Science*, **192**, 896 (1976).

16. K. J. Byrne, A. A. Swigar, R. M. Silverstein, J. H. Borden, and E. Stokkink, *J. Insect Physiol.*, **20**, 1895 (1974).

17. K. Mori, *Tetrahedron*, **31**, 3011 (1975).

18. J. H. Borden, L. Chong, J. A. McLean, K. N. Slessor, and K. Mori, *Science*, **192**, 894 (1976).

19. R. Baker and D. A. Evans, *Ann. Reports (B)*, **72**, 347 (1975).

20. R. Baker and D. A. Evans, *Ann. Reports (B)*, **74**, 367 (1977).

21. J. A. Katzenellenbogen, *Science*, **194**, 139 (1976).

22. C. A. Henrick, *Tetrahedron*, **33**, 1845 (1977).

23. R. Rossi, *Synthesis*, 817 (1977).

24. R. Rossi, *Synthesis*, 413 (1978).

25. J. G. MacConnell and R. M. Silverstein, *Angew. Chem. Int. Ed.*, **12**, 644 (1973).

26. P. Karlson and D. Schneider, *Naturwissenschaften*, **60**, 113 (1973).

27. D. A. Evans and C. L. Green, *Chem. Soc. Rev.*, **3**, 74 (1974).

28. K. Eiter, *Pure Appl. Chem.*, **41**, 201 (1975).

29. M. S. Blum, Pesticide Chemistry in the 20th Century (J. R. Plimmer, Ed.), ACS Symposium Series No 37, American Chemical Society, Washington, 1976, 209-236.

30. M. Jacobson, *Insect Sex Pheromones*, Academic Press, New York, 1972.

31. M. C. Birch, Ed., *Pheromones*, North-Holland, Amsterdam, 1974.

32. M. Beroza, Ed., *Pest Management with Insect Sex Attractants*, ACS Symposium Series No. 23, American Chemical Society, Washington (1976).

33. H. Z. Levinson, *Naturwissenschaften*, **62**, 272 (1975).

34. H. Z. Levinson, *Z. Angew. Entomol.*, **84**, 1 (1977).

35. H. Z. Levinson, *Z. Angew. Entomol.*, **84**, 337 (1977).

36. J. P. Vité and J. A. A. Renwick, *Z. Angew. Entomol.*, **82**, 112 (1976).

37. J. P. Vité and W. Francke, *Naturwissenschaften*, **63**, 550 (1976).

38. M. C. Birch, *Calif. Agr.*, **31** (11), 4 (1977).

39. J. P. Vité, *Biol. Zeit*, 8 (4), 112 (1978).

40. M. C. Birch, *Am. Sci.*, **66**, 409 (1978).

41. H. Z. Levinson, *Naturwissenschaften*, **59**, 477 (1972).

42. W. S. Bowers, *Lipids*, **13**, 736 (1978).

43. D. Schneider, *Sci. Am.*, July, p. 28 (1974).

44. D. J. Faulkner, *Synthesis*, 175 (1971).

45. J. Reucroft and P. J. Sammes, *Quart. Rev. Chem. Soc.*, **25**, 135 (1971).

46. K. Mori, *Recent Developments in the Chemistry of Natural Carbon Compounds,* **9**, 11-209, Akademiai Kiado, Budapest, 1979.

47. M. Schwarz and R. M. Waters, *Synthesis*, 567 (1972).

48. J. D. Warthen, Jr. and M. Jacobson, *Synthesis*, 616 (1973).

49. J. Attenburrow, A. F. B. Cameron, J. H. Chapman, R. M. Evans, B. A. Hems, A. B. A. Jansen, and T. Walker, *J. Chem. Soc.*, 1094 (1952).

50. R. Rossi and A. Carpita, *Synthesis*, 561 (1977).

51. A. Claesson and C. Bogentoft, *Synthesis*, 539 (1973).

52. K. Kondo, A. Negishi, and D. Tunemoto, *Angew. Chem. Int. Ed.*, **13**, 407 (1974).

53. T. Hayashi and H. Midorikawa, *Synthesis,* 100 (1975).

54. M. Schlosser and K. F. Christmann, *Angew. Chem. Int. Ed.*, **5**, 126 (1966).

55. M. Schlosser and K. F. Christmann, *Liebigs Ann. Chem.*, **708**, 1 (1967).

56. M. Schlosser, *Top. Stereochem.*, **5**, 1 (1970).

57. S. Baba, D. E. van Horn, and E. Negishi, *Tetrahedron Lett.*, 1927 (1976).

58. G. Zweifel and C. C. Whitney, *J. Am. Chem. Soc.*, **89**, 2753 (1967).

59. S. Warwel, G. Schmitt, and B. Ahlfaenger, *Synthesis*, 632 (1975).

60. E. J. Corey and R. H. Wollenberg, *J. Org. Chem.*, **40**, 2265 (1975).

61. H. Yatagai, Y. Yamamoto, K. Maruyama, A. Sonoda, and S. Murahashi, *J. Chem. Soc. Chem. Commun.* 852 (1977).

62. E. Negishi and T. Yoshida, *J. Chem. Soc. Chem. Commun.*, 606 (1973).

63. N. Okukado, D. E. VanHorn, W. L. Klima, and E. Negishi, *Tetrahedron Lett.*, 1027 (1978).

64. P. J. Kocienski, B. Lythgoe, and S. Ruston, *J. Chem. Soc. Perkin I*, 829 (1978).

65. H. Lindlar, *Helv. Chim. Acta*, **35**, 446 (1952).

66. H. Lindlar and R. Dubuis, *Org. Synth. Col. Vol.*, **5**, 880 (1973).

67. M. D. Chisholm, W. F. Steck, and E. W. Underhill, *J. Chem. Ecol.*, **4**, 657 (1978).

68. D. J. Cram and N. L. Allinger, *J. Am. Chem. Soc.*, **78**, 2518 (1956).

69. C. A. Brown and V. K. Ahuja, *J. Chem. Soc. Chem. Commun.*, 553 (1973).

70. ref. 22 p. 1848.

71. G. Goto, T. Shima, H. Masuya, Y. Masuoka, and K. Hiraga, *Chem. Lett.*, 103 (1975).

72. D. R. Hall, P. S. Beevor, R. Lester, R. G. Poppi, and B. F. Nesbitt, *Chem. Ind.*, 216 (1975).

73. H. J. Bestmann, O. Vostrowsky, and H. Platz, *Chem. Zeit.*, **98**, 161 (1974).

74. H. J. Bestmann, W. Stransky, O. Vostrowsky, and P. Range, *Chem. Ber.*, **108**, 3582 (1975).

75. H. J. Bestmann, W. Stransky, and O. Vostrowsky, *Chem. Ber.*, **109**, 1694 (1976).

76. H. C. Brown and G. Zweifel, *J. Am. Chem. Soc.*, **83**, 3834 (1961).

77. H. Yatagai, Y. Yamamoto, and K. Maruyama, *J.C.S. Chem. Commun.* 702 (1978).

78. G. Zweifel and H. Arzoumanian, *J. Am. Chem. Soc.*, **89**, 5086 (1967).

79. L. Brandsma, *Preparative Acetylenic Chemistry*, Elsevier, Amsterdam, 1971.

80. D. N. Brattesani and C. H. Heathcock, *Synth. Commun.*, **3**, 245 (1973).

81. S. Bhanu and F. Scheinmann, *J. Chem. Soc. Chem. Commun.*, 817 (1975).

82. G. Fouquet and M. Schlosser, *Angew. Chem. Int. Ed.,* **13**, 82 (1974).

83. H. Neumann and D. Seebach, *Tetrahedron Lett.*, 4839 (1976).

84. Y. Tanigawa, H. Kanamaru, A. Sonoda, and S. Murahashi, *J. Am. Chem. Soc.*, **99**, 2361 (1977).

85. A. Alexakis, J. Normant, and J. Villiéras, *Tetrahedron Lett.*, 3461 (1976).

86. H. P. Dang and G. Linstrumelle, *Tetrahedron Lett.*, 191 (1978).

87. C. A. Brown and A. Yamashita, *J. Am. Chem. Soc.*, **97**, 891 (1975).

88. C. A. Brown, *J. Chem. Soc. Chem. Commun.* 222 (1975).

89. C. A. Brown and E. Negishi, *J. Chem. Soc. Chem. Commun.*, 318 (1977).

90. W. E. Willy, D. R. McKean, and B. A. Garcia, *Bull. Chem. Soc. Japan*, **49**, 1989 (1976).

91. P. E. Sonnet, *Synth. Commun.*, **6**, 21 (1976).

92. C. C. Leznoff, *Accounts Chem. Res.* **11**, 327 (1978).

93. C. C. Leznoff and T. M. Fyles, *J. Chem. Soc. Chem. Commun.*, 251 (1976).

94. T. M. Fyles, C. C. Leznoff, and J. Weatherston, *J. Chem. Ecol.,* **4**, 109 (1978).

95. T. M. Fyles, C. C. Leznoff, and J. Weatherston, *Canad. J. Chem.*, **56**, 1031 (1978).

96. *Annual Reports in Organic Synthesis*, Academic Press, New York.

97. *Annual Reports in the Progress of Chemistry, Section B*, The Chemical Society, London.

98. E. C. Uebel, P. E. Sonnet, B. A. Bierl, and R. W. Miller, *J. Chem. Ecol.*, **1**, 377 (1975).

99. P. E. Sonnet, E. C. Uebel, R. L. Harris, and R. W. Miller, *J. Chem. Ecol.*, **3**, 245 (1977).

100. P. E. Sonnet, *J. Am. Oil Chem. Soc.*, **53**, 57 (1975).

101. J. G. Pomonis, C. F. Fatland, D. R. Nelson, and R. G. Zaylskie, *J. Chem. Ecol.,* **4**, 27 (1978).

102. D. A. Carlson, P. A. Langley, and P. Huyton, *Science*, **201**, 750 (1978).

103. A. Butenandt and N. D. Tam, *Z. Physiol. Chem.*, **308**, 277 (1957).

104. G. Pattenden and B. W. Staddon, *Ann. Entomol. Soc. Amer.*, **63**, 900 (1970).

105. R. S. Berger and T. D. Canerday, *J. Econ. Entomol.*, **61**, 452 (1968).

106. H. E. Henderson and F. L. Warren, *J. South African Chem. Inst.*, **23**, 9 (1970).

107. J. Weatherston, W. L. Roelofs, A. Comeau, and C. J. Sanders, *Canad. Entomol.*, **103**, 1741 (1971).

108. C. Hirano, H. Muramoto, and H. Horiike, *Naturwissenschaften*, **63**, 439 (1976).

109. D. S. Sgoutas and F. A. Kummerow, *Lipids*, **4**, 283 (1969).

110. (a) E. Vedejs and P. L. Fuchs, *J. Am. Chem. Soc.,* **95**, 822 (1973).

    (b) P. E. Sonnet and J. E. Oliver, *J. Org. Chem.*, **41**, 3279, 3284 (1976).

111. P. B. Dervon and M. A. Shippey, *J. Am. Chem. Soc.*, **98**, 1266 (1976).

112. D. A. Carlson, M. S. Mayer, D. L. Silhacek, J. D. James, M. Beroza, and B. A. Bierl, *Science*, **174**, 76 (1971).

113. K. Eiter, *Naturwissenschaften*, 59, 468 (1972).

114. R. L. Cargill and M. G. Rosenblum, *J. Org. Chem.*, 37, 3971 (1972).

115. R. L. Carney, R. J. Scheible, and J. W. Baum, Zoecon Corporation, unpublished results, cited in Ref. 22, p. 1882.

116. T. -L. Ho and C. M. Wong, *Canad. J. Chem.*, 52, 1923 (1974).

117. G. W. Gribble and J. K. Sanstead, *J. Chem. Soc. Chem. Commun.*, 735 (1973).

118. K. Abe, T. Yamasaki, N. Nakamura, and T. Sakan, *Bull. Chem. Soc. Japan*, 50, 2792 (1977).

119. E. C. Uebel, P. E. Sonnet, R. W. Miller, and M. Beroza, *J. Chem. Ecol.*, 1, 195 (1975).

120. R. L. Carney and J. W. Baum, Zoecon Corporation, unpublished results, cited in Ref. 22, p. 1883.

121. B. A. Bierl, M. Beroza, and C. W. Collier, *Science*, 170, 88 (1970).

122. R. T. Cardé, W. L. Roelofs, and C. C. Doane, *Nature*, 241, 474 (1974).

123. H. J. Bestmann and O. Vostrowsky, *Tetrahedron Lett.*, 207 (1974).

124. H. J. Bestmann, O. Vostrowsky, and W. Stransky, *Chem. Ber.*, 109, 3375 (1976).

125. T. H. Chan and E. Chang, *J. Org. Chem.*, 39, 3264 (1974).

126. W. Mychajlowskij and T. H. Chan, *Tetrahedron Lett.*, 4439 (1976).

127. K. Eiter, *Angew. Chem. Int. Ed.*, 11, 60 (1972).

128. A. A. Shamshurin, M. A. Rekhter, and L. A. Vlad, *Khim, Prir. Soedin*, 9, 545 (1973).

129. H. Klünenberg and H. J. Schäfer, *Angew. Chem. Int. Ed.*, 17, 47 (1978).

130. G. A. Tolstikow, V. N. Odinokov, P. J. Galeeva, and R. S. Bekeeva, *Tetrahedron Lett.*, 1857 (1978).

131. G. A. Tolstikov, V. N. Odinokov, R. I. Galeeva, R. S. Bakeeva, and S. R. Rafikov, *Dokl. Akad. Nauk SSSR.*, 239, 1377 (1978).

132. H. J. Bestmann, O. Vostrowsky, K. -H. Koschatsky, H. Platz, T. Brosche, I. Kantardjiew, M. Rheinwald, and W. Knauf, *Angew. Chem. Int. Ed.*, 17, 768 (1978).

133. R. S. Berger, *Ann. Entomol. Soc. Am.*, 59, 767 (1966).

134. W. Seidel, J. Knolle, and H. J. Schäfer, *Chem. Ber.*, 110, 3544 (1977).

135. W. L. Roelofs, A. Comeau, and R. Selle, *Nature*, 224, 723 (1969).

136. G. Holan and D. F. O'Keefe, *Tetrahedron Lett.*, 673 (1973).

137. Insecticide Chemistry Group, Acad. Sin., Peking, *Hua Hsueh Hsueh Pao*, 35, 221 (1977).

138. K. Mori, M. Uchida, and M. Matsui, *Tetrahedron*, 33, 385 (1977).

139. W. L. Roelofs, J. P. Tette, E. F. Taschenberg, and A. Comeau, *J. Insect Physiol.*, 17, 2235 (1971).

140. H. Arn., S. Rauscher, H. R. Buser, and W. L. Roelofs, *Z. Naturforsch. (C)*, 31, 499 (1976).

141. C. A. Henrick and B. A. Garcia, Zoecon Corporation, unpublished results, cited in Ref. 22, p. 1846.

142. R. Rossi, *Chim. Ind. (Milan)*, 60, 652 (1978).

143. T. Ando, S. Yoshida, S. Tatsuki, and N. Takahashi, *Agric. Biol. Chem.*, 41, 1485 (1977).

144. H. J. Bestmann, O. Vostrowsky, H. Platz, Th. Brosche, K. H. Koschatzky, and W. Knauf, *Tetrahedron Lett.*, 497 (1979).

145. A. A. Sekul and A. N. Sparks, *J. Econ. Entomol.*, **60**, 1270 (1967).

146. Y. Tamaki, H. Noguchi, and T. Yushima, *Appl. Entomol. Zool.*, **6**, 139 (1971).

147. D. Warthen, *J. Med. Chem.*, **11**, 371 (1968).

148. M. Jacobson and C. Harding, *J. Econ. Entomol.*, **61**, 394 (1968).

149. H. J. Bestmann, P. Range, and R. Kunstmann, *Chem. Ber.*, **104**, 65 (1971).

150. B. F. Nesbitt, P. S. Beevor, D. R. Hall, R. Lester, and V. A. Dyck, *Insect Biochem.*, **6**, 105 (1976).

151. H. J. Bestmann, O. Vostrowsky, K. H. Koschatzky, H. Platz, A. Szymanska, and W. Knauf, *Tetrahedron Lett.*, 605 (1978).

152. B. F. Nesbitt, P. S. Beevor, D. R. Hall, R. Lester, and V. A. Dyck, *J. Insect Physiol.*, **21**, 1883 (1975).

153. H. Fukui, F. Matsumura, M. C. Ma, and W. E. Burkholder, *Tetrahedron Lett.*, 3563 (1974).

154. J. O. Rodin, R. M. Silverstein, W. E. Burkholder, and J. E. Gorman, *Science*, **165**, 904 (1969).

155. J. H. Cross, R. C. Byler, R. F. Cassidy, Jr, R. M. Silverstein, R. E. Greenblatt, W. E. Burkholder, A. R. Levinson, and H. Z. Levinson, *J. Chem. Ecol.*, **2**, 457 (1976).

156. J. J. DeGraw and J. O. Rodin, *J. Org. Chem.*, **36**, 2902 (1971).

157. (a) M. L. Roumestand, P. Place, and J. Gore, *Tetrahedron Lett.*, 677 (1976). (b) *Idem.*, *Tetrahedron*, **33**, 1283 (1977).

158. W. L. Roelofs, J. Kochansky, R. Cardé, H. Arn, and S. Ruscher, *Mitt. Schweiz. Entomol. Ges.*, **46**, 71 (1973).

159. K. M. Nicholas and R. Pettit, *Tetrahedron Lett.*, 3475 (1971).

160. C. Descoins and D. Samain, *Tetrahedron Lett.*, 745 (1976).

161. C. Descoins, D. Samain, B. Lalanne-Cassou, and M. Gallois, *Bull. Soc. Chim. France (2)*, 941 (1978).

162. J. N. Labovitz, C. A. Henrick, and V. L. Corbin, *Tetrahedron Lett.*, 4209 (1975).

163. E. Negishi and A. Abramovitch, *Tetrahedron Lett.*, 411 (1977).

164. G. Cassani, P. Massardo, and P. Piccardi, *Tetrahedron Lett.*, 633 (1979).

165. W. L. Roelofs, A. Comeau, A. Hill, and G. Milicevic, *Science*, **174**, 297 (1971).

166. C. Descoins and C. A. Henrick, *Tetrahedron Lett.*, 2999 (1972).

167. K. Mori, *Tetrahedron*, **30**, 3807 (1974).

168. C. A. Henrick and J. B. Siddall, U.S. Pat. 3, 818, 049; *Chem. Abstr.*, **81**, 63136s (1974).

169. C. A. Henrick, R. J. Anderson, and L. D. Rosenblum, Zoecon Corporation, unpublished results (1974), cited in Ref. 22, p. 1858.

170. D. Samain, C. Descoins, and A. Commerçon, *Synthesis*, 388 (1978).

171. H. J. Bestmann, J. Süss, and O. Vostrowsky, *Tetrahedron Lett.*, 3329 (1978).

172. B. F. Nesbitt, P. S. Beevor, R. A. Cole, R. Lester, and R. G. Poppi, *J. Insect Physiol.*, **21**, 1091 (1975).

173. B. F. Nesbitt, P. S. Beevor, R. A. Cole, R. Lester, and R. G. Poppi, *Tetrahedron Lett.*, 4669 (1973).

174. S. Tanaka, A. Yasuda, H. Yamamoto, and H. Nozaki, *J. Am. Chem. Soc.*, **97**, 3252 (1975).

175. J. H. Babler and M. J. Martin, *J. Org. Chem.*, **42**, 1799 (1977).

176. T. Mandai, H. Yasuda, M. Kaito, J. Tsuji, R. Yamaoka, and H. Fukami, *Tetrahedron*, **35**, 309 (1979).

177. R. E. Doolittle, W. L. Roelofs, J. D. Solomon, R. T. Cardé, and M. Beroza, *J. Chem. Ecol.*, **2**, 399 (1976).

178. Y. Tamaki, H. Noguchi, and T. Yushima, *Appl. Entomol. Zool.*, **8**, 200 (1973).

179. Y. Tamaki and T. Yushima, *J. Insect Physiol.*, **20**, 1005 (1974).

180. B. F. Nesbitt, P. S. Beevor, R. A. Cole, R. Lester, and R. G. Poppi, *Nature, New Biol.*, **244**, 208 (1973).

181. R. L. Carney and J. W. Baum, Zoecon Corporation, unpublished results (1975), cited (by C. A. Henrick) in Ref. 22, p. 1864.

182. R. M. Silverstein, J. O. Rodin, W. E. Burkholder, and J. E. Gorman, *Science*, **157**, 85 (1967).

183. A. Butenandt, R. Beckmann, and E. Hecker, *Z. Physiol. Chem.*, **324**, 71 (1961).

184. A. Butenandt, R. Beckmann, and D. Stamm, *Z. Physiol. Chem.*, **324**, 81 (1961).

185. H. J. Bestmann, O. Vostrowsky, H. Paulus, W. Billmann, and W. Stransky, *Tetrahedron Lett.*, 121 (1977).

186. E. Negishi, G. Lew, and T. Yoshida, *J. Chem. Soc. Chem. Commun.*, 874 (1973).

187. J. F. Normant, A. Commerçon, and J. Villieras, *Tetrahedron Lett.*, 1465 (1975).

188. (a) G. Kasang, K. E. Kaissling, O. Vostrowsky, and H. J. Bestmann, *Angew. Chem. Int. Ed.*, **17**, 60 (1978).
      (b) K. E. Kaissling, G. Kasang, H. J. Bestmann, W. Stransky, and O. Vostrowsky, *Naturwissenschaften*, **65**, 382 (1978).

189. W. L. Roelofs, J. P. Kochansky, R. T. Cardé, C. A. Henrick, J. N. Labovitz, and V. L. Corbin, *Life Sci.*, **17**, 699 (1975).

190. S. Voerman and G. H. L. Rothchild, *J. Chem. Ecol.*, **4**, 531 (1978).

191. A. Alexakis, G. Cahiez, and J. F. Normant, *Tetrahedron Lett.*, 2027 (1978).

192. M. Jacobson, R. E. Redfern, W. A. Jones, and M. H. Aldridge, *Science*, **170**, 542 (1970).

193. Y. Kuwahara, H. Hara, S. Ishii, and H. Fukami, *Science*, **171**, 801 (1971).

194. U. E. Brady, J. H. Tumlinson, R. G. Brownlee, and R. M. Silverstein, *Science*, **171**, 802 (1971).

195. H. C. F. Su, P. G. Manhany, and U. E. Brady, *J. Econ. Entomol.*, **66**, 845 (1973).

196. H. J. Bestmann, O. Vostrowsky, and A. Plenchette, *Tetrahedron Lett.*, 779 (1974).

197. H. J. Bestmann, J. Süss, and O. Vostrowsky, *Tetrahedron Lett.*, 245 (1979).

198. J. Kochansky, J. Tette, E. F. Taschenberg, R. T. Cardé, K. -E. Kaissling, and W. L. Roelofs, *J. Insect Physiol.*, **21**, 1977 (1975).

199. J. P. Kochansky, R. T. Cardé, E. F. Taschenberg, and W. L. Roelofs, *J. Chem. Ecol.*, **3**, 419 (1977).

200. (a) W. A. Jones, M. Jacobson, and D. F. Martin, *Science*, **152**, 1516 (1966). (b) W. A. Jones and M. Jacobson, *Science*, **159**, 99 (1968).

201. (a) P. A. Bierl, M. Beroza, R. T. Staten, P. E. Sonnet, and V. E. Alder, *J. Econ. Entomol.*, **67**, 211 (1974). (b) P. E. Sonnet, *J. Org. Chem.*, **39**, 3793 (1974).

202. H. C. F. Su and P. G. Mahany, *J. Econ. Entomol.*, **67**, 319 (1974).

203. H. Diesselnkötter, K. Eiter, W. Karl, and D. Weindisch, *Tetrahedron*, **32**, 1591 (1976).

204. (a) K. Mori, M. Tominaga, and M. Matsui, *Agric. Biol. Chem.*, **38**, 1551 (1974). (b) K. Mori, M. Tominaga, and M. Matsui, *Tetrahedron*, **31**, 1846 (1975).

205. H. J. Bestmann, K. H. Koschatsky, W. Stransky, and O. Vostrowsky, *Tetrahedron Lett.*, 353 (1976).

206. R. J. Anderson and C. A. Henrick, *J. Am. Chem. Soc.*, **97**, 4327 (1975).

207. C. A. Henrick, Ref. 22, pp. 1871-1875.

208. A. Hammoud and C. Descoins, *Bull. Soc. Chim. France (2)*, 299 (1978).

209. J. H. Tumlinson, C. E. Yonce, R. E. Doolittle, R. R. Heath, C. R. Gentry, and E. R. Mitchell, *Science*, **185**, 614 (1974).

210. M. Uchida, K. Mori, and M. Matsui, *Agric. Biol. Chem.*, **42**, 1067 (1978).

211. K. Yaginuma, M. Kumakura, Y. Tamaki, T. Yushima, and J. H. Tumlinson, *Appl. Entomol. Zool.*, **11**, 266 (1976).

212. T. Ebata and Mori, *Agric. Biol. Chem.*, **43**, 1507 (1979).

213. M. Uchida, K. Nakagawa, and K. Mori, *Agric. Biol. Chem.*, **43**, 1919 (1979).

214. F. Matsumura, H. C. Coppel, and A. Tai, *Nature*, **219**, 963 (1968).

215. A. Tai, F. Matsumura, and H. C. Coppel, *J. Org. Chem.*, **34**, 2180 (1969).

216. R. Yamaoka, H. Fukami, and S. Ishii, *Agric. Biol. Chem.*, **40**, 1971 (1976).

217. D. F. Horler, *J. Chem. Soc. (C)*, 859 (1970).

218. P. D. Landor, S. R. Landor, and S. Mukasa, *J. Chem. Soc. Chem. Commun.*, 1638 (1971).

219. C. Descoins, C. A. Henrick, and J. B. Siddall, *Tetrahedron Lett.*, 3777 (1972).

220. R. Baudouy and J. Gore, *Synthesis*, 573 (1974).

221. D. Michelot and G. Linstrumelle, *Tetrahedron Lett.*, 275 (1976).

222. P. J. Kocienski, G. Cernigliaro, and G. Feldstein, *J. Org. Chem.*, **42**, 353 (1977).

223. K. Mori, H. Hashimoto, Y. Takenaka, and T. Takigawa, *Synthesis*, 720 (1975).

224. H. Röller, K. Biemann, J. S. Bjerke, D. W. Norgard, and W. H. McShan, *Acta Entomol. Bohemoslav.*, **65**, 208 (1968).

225. H. M. Fales, M. S. Blum, R. M. Grewe, and J. M. Brand, *J. Insect Physiol.*, **18**, 1077 (1972).

226. J. A. Katzenellenbogen and T. Utawanit, *J. Am. Chem. Soc.*, **96**, 6153 (1974).

227. P. J. Kocienski, J. M. Ansell, and R. W. Ostrow, *J. Org. Chem.*, **41**, 3625 (1976).

228. T. Nakai, T. Mimura, and T. Kurokawa, *Tetrahedron Lett.*, 2895 (1978).

229. Y. Tamaki, K. Honma, and K. Kawasaki, *Appl. Entomol. Zool.*, **12**, 60 (1977).

230. S. Tamada, K. Mori, and M. Matsui, *Agric. Biol. Chem.*, **42**, 191 (1978).

231. R. G. Smith, G. E. Daterman, and G. D. Daves, Jr., *Science*, **188**, 63 (1975).

232. L. M. Smith, R. G. Smith, T. M. Loehr, G. D. Daves, Jr., G. E. Daterman, and R. H. Wohleb, *J. Org. Chem.*, **43**, 2361 (1978).

233. R. G. Smith, G. D. Daves, Jr., and G. E. Daterman, *J. Org. Chem.*, **40** 1593 (1975).

234. P. J. Kocienski and G. J. Cernigliaro, *J. Org. Chem.*, **41**, 2927 (1976).

235. J. N. Labovitz, V. L. Graves, and C. A. Henrick, Zoecon Corporation, Unpublished work, cited by C. A. Henrick, Ref. 22, p. 1853 (1977).

236. M. Fetizon and C. Lazare, *J. Chem. Soc. Perkin I*, 842 (1978).

237. B. Åckemark and A. Ljungqvist, *J. Org. Chem.*, **43**, 4387 (1978).

238. K. Kondo and S. Murahashi, *Tetrahedron Lett.*, 1237 (1979).

239. R. Nishida, H. Fukami, and S. Ishii, *Appl. Entomol. Zool.*, **10**, 10 (1975).

240. (*a*) R. Nishida, T. Sato, Y. Kuwahara, H. Fukami, and S. Ishii, *J. Chem. Ecol.*, **2**, 449 (1976). (*b*) R. Nishida, Y. Kuwahara, H. Fukami, and S. Ishii, *J. Chem. Ecol.*, **5**, 289 (1979).

241. T. Sato, R. Nishida, Y. Kuwahara, H. Fukami, and S. Ishii, *Agric. Biol. Chem.*, **40**, 391 (1976).

242. M. Schwarz, J. E. Oliver, and P. E. Sonnet, *J. Org. Chem.*, **40**, 2410 (1975).

243. A. W. Burgstahler, L. O. Weigel, W. J. Bell, and M. K. Rust, *J. Org. Chem.*, **40**, 3456 (1975).

244. L. D. Rosenblum, R. J. Anderson, and C. A. Henrick, *Tetrahedron Lett.*, 419 (1976).

245. P. Place, M. L. Roumestant, and J. Gore, *Tetrahedron*, **34**, 1931 (1978).

246. R. Nishida, T. Sato, Y. Kuwahara, H. Fukami, and S. Ishii, *Agric. Biol. Chem.*, **40**, 1407 (1976).

247. A. W. Burgstahler, L. O. Weigel, M. E. Sanders, C. G. Shaefer, W. J. Bell, and S. B. Vuturo, *J. Org. Chem.*, **42**, 566 (1977).

248. B. P. Mundy, K. B. Lipkowitz, and G. W. Dirks, *Heterocycles*, **6**, 51 (1977).

249. W. Francke, V. Heemann, B. Gerken, J. A. A. Renwick, and J. P. Vité, *Naturwissenschaften*, **64**, 590 (1977).

250. G. W. Kinzer, A. F. Fentiman, Jr., T. F. Page, R. L. Foltz, and J. P. Vité, *Nature*, **221**, 447 (1969).

251. B. P. Mundy, R. D. Otzenberger, and A. R. DeBernardis, *J. Org. Chem.*, **36**, 2390 (1971).

252. T. D. J. D'Silva and D. W. Peck, *J. Org. Chem.*, **37**, 1828 (1972).

253. K. Mori, S. Kobayashi, and M. Matsui, *Agric. Biol. Chem.*, **39**, 1889 (1975).

254. T. Sato, S. Yamaguchi, and H. Kaneko, *Tetrahedron Lett.*, 1863 (1979).

255. T. Sato, S. Yoshiie, T. Imamura, K. Hasegawa, M. Miyahara, S. Yamamura, and O. Ito, *Bull. Chem. Soc. Japan*, **50**, 2714 (1977).

256. T. E. Bellas, R. G. Brownlee, and R. M. Silverstein, *Tetrahedron*, **25**, 5149 (1969).

257. H. H. Wasserman and E. H. Barber, *J. Am. Chem. Soc.*, **91**, 3674 (1969).

258. J. Knolle and H. J. Schäfer, *Angew. Chem. Int. Ed.,* **14**, 758 (1975).

259. P. J. Kocienski and R. W. Ostrow, *J. Org. Chem.*, **41**, 398 (1976).

260. A. Chaquin, J.-P. Morizur, and J. Kossanyi, *J. Am. Chem. Soc.*, **99**, 903 (1977).

261. J. L. Coke, H. J. Williams, and S. Natarajan, *J. Org. Chem.*, **42**, 2380 (1977).

262. M. Look, *J. Chem. Ecol.*, **2**, 83 (1976).

263. N. T. Byrom, R. Crigg, and B. Kongkathip, *J. Chem. Soc. Chem. Commun.*, 216 (1976).

264. K. Mori, *Agric. Biol. Chem.*, **40**, 2499 (1976).

265. G. T. Pearce, W. E. Gore, R. M. Silverstein, J. W. Peacock, R. A. Cuthbert, G. N. Lanier, and J. B. Simeone, *J. Chem. Ecol.*, **1**, 115 (1975).

266. B. Gerken, S. Grüne, J. P. Vité, and K. Mori, *Naturwissenschaften*, **65**, 110 (1978).

267. W. E. Gore, G. T. Pearce, and R. M. Silverstein, *J. Org. Chem.*, **40**, 1705 (1975).

268. W. E. Gore and I. M. Armitage, *J. Org. Chem.*, **41**, 1926 (1976).

269. W. J. Elliott and J. Fried, *J. Org. Chem.*, **41**, 2468, 2475 (1976).

270. R. M. Silverstein, J. O. Rodin, and D. L. Wood, *Science*, **154**, 509 (1966).

271. C. A. Reece, J. O. Rodin, R. G. Brownlee, W. G. Duncan, and R. M. Silverstein, *Tetrahedron*, **24**, 4249 (1968).

272. O. P. Vig, R. C. Anand, G. L. Kad, and J. M. Sehgal, *J. Indian Chem. Soc.*, **47**, 999 (1970).

273. J. A. Katzenellenbogen and R. S. Lenox, *J. Org. Chem.*, **38**, 326 (1973).

274. S. R. Wilson and L. R. Phillips, *Tetrahedron Lett.*, 3047 (1975).

275. S. Karlsen, P. Frøyen and L. Skattebøl, *Acta Chem. Scand., (B)*, **30**, 664 (1976).

276. K. Kondo, S. Dobashi, and M. Matsumoto, *Chem. Lett.*, 1077 (1976).

277. J. Haslouin and F. Rouessac, *Bull. Soc. Chim. France (2)*, 1242 (1977).

278. M. Bertrand and J. Viala, *Tetrahedron Lett.*, 2575 (1978).

279. A. Hosomi, M. Saito, and H. Sakurai, *Tetrahedron Lett.*, 429 (1979).

280. R. G. Riley, R. M. Silverstein, J. A. Katzenellenbogen, and R. S. Lenox, *J. Org. Chem.*, **39**, 1957 (1974).

281. K. Mori, *Agric. Biol. Chem.*, **38**, 2045 (1974).

282. C. F. Garbers and F. Scott, *Tetrahedron Lett.*, 1625 (1976).

283. J. H. Tumlinson, D. D. Hardee, R. C. Gueldner, A. C. Thompson, P. A. Hedin, and J. P. Minyard, *Science*, **166**, 1010 (1969).

284. J. H. Tumlinson, R. C. Gueldner, D. D. Hardee, A. C. Thompson, P. A. Hedin, and J. P. Minyard, *J. Org. Chem.*, **36**, 2616 (1971).

285. R. Zurflüh, L. L. Dunham, V. L. Spain, and J. B. Siddall, *J. Am. Chem. Soc.*, **92**, 425 (1970).

286. R. C. Gueldner, A. C. Thompson, and P. A. Hedin, *J. Org. Chem.*, **37**, 1854 (1972).

287. H. Kosugi, S. Sekiguchi, R. Sekita, and H. Uda, *Bull. Chem. Soc. Japan*, **49**, 520 (1976).

288. R. L. Cargill and B. W. Wright, *J. Org. Chem.*, **40**, 120 (1975).

289. W. E. Billups, J. H. Cross, and C. V. Smith, *J. Am. Chem. Soc.*, **95**, 3438 (1973).

290. W. A. Ayer and L. M. Browne, *Canad. J. Chem.*, **52**, 1352 (1974).

291. G. Stork and J. F. Cohen, *J. Am. Chem. Soc.*, **96**, 5270 (1974).

292. J. H. Babler, *Tetrahedron Lett.*, 2045 (1975).

293. B. M. Trost and D. E. Keeley, *J. Org. Chem.*, **40**, 2013 (1975).

294. B. M. Trost, D. E. Keeley, H. C. Arndt, and M. J. Bogdanowicz, *J. Am. Chem. Soc.*, **99**, 3088 (1977).

295. F. Bohlmann, C. Zdero, and U. Faass, *Chem. Ber.*, **106**, 2904 (1973).

296. E. Wenkert, D. A. Berges, and N. F. Golob, *J. Am. Chem. Soc.*, **100**, 1263 (1978).

297. V. Rautenstrauch, *J. Chem. Soc. Chem. Commun.*, 519 (1978).

298. J. G. MacConnell, J. H. Borden, R. M. Silverstein, and E. Stokkink, *J. Chem. Ecol.*, **3**, 549 (1977).

299. K. Mori and M. Sasaki, *Tetrahedron Lett.*, 1329 (1979).

300. K. Mori and M. Sasaki, *Tetrahedron*, **36**, 2197 (1980).

301. G. Gowda and T. B. H. McMurry, *J. Chem. Soc. Perkin I*, 274 (1979).

302. R. H. Bedoukian and J. Wolinsky, *J. Org. Chem.*, **40**, 2154 (1975).

303. J. H. Babler and T. R. Mortell, *Tetrahedron Lett.*, 669 (1972).

304. (a) S. W. Pelletier and N. V. Mody, *J. Org. Chem.*, **41**, 1069 (1976). (b) J. P. deSouza and A. M. R. Gonçalves, *J. Org. Chem.*, **43**, 2068 (1978).

305. P. C. Traas, H. Boelens, and H. J. Takken, *Rec. Trav. Chim. Pay-Bas*, **95**, 308 (1976).

306. T. Nakai, T. Mimura, and A. Ari-izumi, *Tetrahedron Lett.*, 2425 (1977).

307. R. H. Wollenberg and R. Peries, *Tetrahedron Lett.*, 297 (1979).

308. L. M. McDonough, D. A. George, B. A. Butt, J. M. Ruth, and K. R. Hill, *Science*, **177**, 177 (1972).

309. M. P. Cooke, Jr., *Tetrahedron Lett.*, 1281, 1983 (1973).

310. S. B. Bowlus and J. A. Katzenellenbogen, *Tetrahedron Lett.*, 1277 (1973).

311. S. B. Bowlus and J. A. Katzenellenbogen, *J. Org. Chem.*, **38**, 2733 (1973).

312. J. P. Morizur, G. Muzard, J. -J. Basselier, and J. Kossanyi, *Bull. Soc. Chim. France*, 257 (1975).

313. C. Ouannés and Y. Langlois, *Tetrahedron Lett.*, 3461 (1975).

314. M. Obayashi, K. Utimoto, and H. Nozaki, *Tetrahedron Lett.*, 1807 (1977).

315. K. Utimoto, M. Obayashi, and H. Nozaki, *J. Org. Chem.*, **41**, 2940 (1976).

316. W. L. Roelofs, M. J. Gieselmann, A. M. Cardé, H. Tashiro, D. S. Moreno, C. A. Henrick, and R. J. Anderson, *Nature*, **267**, 698 (1977).

317. W. Roelofs, M. Gieselmann, A. Cardé, H. Tashiro, D. S. Moreno, C. A. Henrick, and R. J. Anderson, *J. Chem. Ecol.*, **4**, 211 (1978).

318. B. B. Snider and D. Rodini, *Tetrahedron Lett.*, 1399 (1978).

319. W. C. Still and A. Mitra, *J. Am. Chem. Soc.*, **100**, 1927 (1978).

320. M. J. Gieselmann, D. S. Moreno, J. Fargerlund, H. Tashiro, and W. L. Roelofs, *J. Chem. Ecol.*, **5**, 27 (1979).

321. R. J. Anderson and C. A. Henrick, cited in Ref. 320, p. 30.

322. T. Suguro and K. Mori, unpublished results (1979); presented at the Annual Meeting of the Agric. Chem. Soc. Japan, April 3, 1979.

323. J. Meinwald, A. M. Chalmers, T. E. Pliske, and T. Eisner, *Tetrahedron Lett.*, 4893 (1968).

324. F. W. Sum and L. Weiler, *J. Chem. Soc. Chem. Commun.*, 985 (1978).

325. J. Meinwald, Y. C. Meinwald, and R. H. Mazzocchi, *Science*, **164**, 1174 (1969).

326. J. Meinwald, W. R. Thompson, T. Eisner, and D. F. Owen, *Tetrahedron Lett.*, 3485 (1971).

327. J. -P. Morizur, G. Bidan, and J. Kossanyi, *Tetrahedron Lett.*, 3019 (1972).

328. G. Bidan, J. Kossanyi, V. Meyer, and J. -P. Morizur, *Tetrahedron*, **33**, 2193 (1977).

329. J. Meinwald, A. M. Chalmers, T. E. Pliske, and T. Eisner, *J. Chem. Soc. Chem. Commun.*, 86 (1969).

330. D. H. Miles, P. Loew, W. S. Johnson, A. F. Kluge, and J. Meinwald, *Tetrahedron Lett.*, 3019 (1972).

331. B. M. Trost, L. Weber, P. Strege, T. J. Fullerton, and T. J. Dietsche, *J. Am. Chem. Soc.*, **100**, 3426 (1978).

332. B. M. Trost, M. J. Bogdanowicz, W. J. Frazee, and T. N. Salzmann, *J. Am. Chem. Soc.*, **100**, 5512 (1978).

333. A. J. Birch, W. V. Brown, J. E. T. Corrie, and B. P. Moore, *J. Chem. Soc. Perkin I*, 2653 (1972).

334. V. D. Patil, U. R. Nayak, and Sukh Dev, *Tetrahedron*, **29**, 341 (1973).

335. M. Kodama, Y. Matsuki, and S. Itô, *Tetrahedron Lett.*, 3065 (1975).

336. Y. Kitahara, T. Kato, T. Kobayashi, and B. P. Moore, *Chem. Lett.,* 219 (1976).

337. H. Takayanagi, T. Uyehara, and T. Kato, *J. Chem. Soc. Chem. Commun.,* 359 (1978).

338. C. G. Butler, R. K. Callow, and N. C. Johnson, *Nature,* **184,** 1871 (1959).

339. C. G. Butler, R. K. Callow, and N. C. Johnson, *Proc. Roy. Soc. (B),* **155,** 417 (1961).

340. M. Barbier, E. Lederer, and T. Nomura, *Compt. Rend.,* **251,** 1133 (1960).

341. H. J. Bestmann, R. Kunstmann, and H. Schulz, *Liebigs Ann. Chem.,* **699,** 33 (1966).

342. R. E. Doolittle, M. S. Blum, and R. Boch, *Ann. Entomol. Soc. Am.,* **63,** 1180 (1970).

343. B. M. Trost and T. N. Salzmann, *J. Org. Chem.,* **40,** 148 (1975).

344. B. M. Trost, T. N. Salzmann, and K. Hiroi, *J. Am. Chem. Soc.,* **98,** 4887 (1976).

345. B. M. Trost and K. Hiroi, *J. Am. Chem. Soc.,* **98,** 4313 (1976).

346. J. Tsuji, K. Masaoka, and T. Takahashi, *Tetrahedron Lett.,* 2267 (1977).

347. Y. Tamaru, Y. Yamada, and Z. Yoshida, *Tetrahedron Lett.,* 919 (1978).

348. Y. Tamaru, Y. Yamada, and Z. Yoshida, *Tetrahedron,* **35,** 329 (1979).

349. C. S. Subramaniam, P. J. Thomas, V. R. Mamdapur, and M. S. Chadha, *Indian J. Chem.,* **16B,** 318 (1978).

350. T. A. Hase and K. McCoy, *Synth. Commun.,* **9,** 63 (1979).

351. J. Meinwald and Y. C. Meinwald, *J. Am. Chem. Soc.,* **88,** 1305 (1966).

352. J. A. Edgar, C. C. J. Culvenor, and L. W. Smith, *Experientia,* **27,** 761 (1971).

353. M. T. Pizzorno and S. M. Albonico, *Chem. Ind.,* 349 (1978).

354. R. G. Riley, R. M. Silverstein, B. Carroll, and R. Carroll, *J. Insect Physiol.,* **20,** 651 (1974).

355. P. E. Sonnet, *J. Med. Chem.,* **15,** 97 (1972).

356. F. J. Ritter, I. E. M. Rotgans, E. Talman, P. E. J. Verwiel, and F. Stein, *Experientia,* **29,** 530 (1973).

357. J. E. Oliver and P. E. Sonnet, *J. Org. Chem.,* **39,** 2662 (1974).

358. Y. Kuwahara, H. Fukami, R. Howard, S. Ishii, F. Matsumura, and W. E. Burkholder, *Tetrahedron,* **34,** 1769 (1978).

359. M. Sakakibara and K. Mori, *Tetrahedron Lett.,* 2401 (1979).

360. K. Mori, *Tetrahedron Lett.,* 3869 (1973).

361. K. Mori, *Tetrahedron,* **30,** 3817 (1974).

362. Review: W. R. Roderick, *J. Chem. Educ.,* **43,** 510 (1966).

363. R. H. Wright, *Nature,* **198,** 782 (1963).

364. R. H. Wright, *The Science of Smell,* George Allen & Unwin, London, 1964.

365. J. E. Amoore, *Molecular Basis of Odor,* Charles C. Thomas, Springfield, Ill., 1970.

366. J. E. Amoore, G. Palmieri, E. Wanke, and M. S. Blum, *Science,* **165,** 1266 (1969).

367. G. F. Russell and J. I. Hills, *Science,* **172,** 1043 (1971).

368. L. Friedman and J. G. Miller, *Science,* **172,** 1044 (1971).

369. R. G. Riley, R. M. Silverstein, and J. C. Moser, *Science,* **183,** 760 (1974).

370. R. G. Riley and R. M. Silverstein, *Tetrahedron,* **30,** 1171 (1974).

371. S. Iwaki, S. Marumo, T. Saito, M. Yamada, and K. Katagiri, *J. Am. Chem. Soc.,* **96,** 7842 (1974).

372. M. Yamada, T. Saito, K. Katagiri, K. Iwaki, and S. Marumo, *J. Insect Physiol.,* **22,** 755 (1976).

373. K. Mori, 9th IUPAC International Symposium on Chemistry of Natural Products, Ottawa, 1974, *Abstracts* 32G.

374. S. H. Wilen, *Top. Stereochem.*, **6**, 107 (1971).

375. S. H. Wilen, *Tables of Resolving Agents and Optical Resolutions*, University of Notre-Dame Press, Notre Dame, Indiana, 1972.

376. S. H. Wilen, A. Collet, and J. Jacque, *Tetrahedron*, **33**, 2725 (1977).

377. W. Klyne and J. Buckingham, *Atlas of Stereochemistry*, Vols. 1 and 2, Chapman and Hall, London, 1974, 1978.

378. J. H. Brewster, "Assignments of Stereochemical Configuration by Chemical Methods," in *Techniques of Chemistry*, Vol. IV, Part III, Wiley-Interscience, New York, 1972.

379. A. Gaudemer, "Determination of Optical Purity and Absolute Configuration by NMR," in *Stereochemistry*, H. B. Kagan, Ed., Vol I, Georg Thieme, Stuttgart, 1977, pp. 117-136.

380. J. A. Dale and H. S. Mosher, *J. Am. Chem. Soc.*, **95**, 512 (1973).

381. H. L. Goering, J. N. Eikenberry, G. S. Koerner, and C. J. Lattimer, *J. Am. Chem. Soc.*, **96**, 1493 (1974).

382. E. L. Plummer, T. E. Stewart, K. Byrne, G. T. Pearce, and R. M. Silverstein, *J. Chem. Ecol.*, **2**, 307 (1976).

383. T. E. Stewart, E. L. Plummer, L. L. McCandless, J. R. West, and R. M. Silverstein, *J. Chem. Ecol.*, **3**, 27 (1977).

384. R. Rossi and A. Carpita, *Tetrahedron*, **33**, 2447 (1977).

385. R. M. Silverstein, J. O. Rodin, and D. L. Wood, *J. Econ. Entomol.*, **60**, 944 (1967).

386. K. Mori, *Tetrahedron Lett.*, 2187 (1975).

387. K. Mori, *Tetrahedron*, **32**, 1101 (1976).

388. K. Mori, T. Takigawa, and T. Matsuo, *Tetrahedron*, **35**, 933 (1979).

389. J. P. Vité, R. Hedden, and K. Mori, *Naturwissenschaften*, **63**, 43 (1976).

390. M. C. Birch and D. M. Light, *J. Chem. Ecol.*, **3**, 257 (1977).

391. M. C. Birch, D. M. Light, and K. Mori, *Nature*, **270**, 738 (1977).

392. K. Mori, *Tetrahedron Lett.*, 1609 (1976).

393. G. Ohloff and W. Giersch, *Helv. Chim. Acta*, **60**, 1496 (1977).

394. J. P. Vité, G. Ohloff, and R. F. Billings, *Nature*, **272**, 817 (1978).

395. J. A. A. Renwick, *Contrib. Boyce Thompson Inst.*, **23**, 355 (1967).

396. G. B. Pitman, J. P. Vité, G. W. Kinzer, and A. F. Fentiman, Jr., *Nature*, **218**, 168 (1968).

397. J. A. A. Renwick, P. R. Hughes, and I. S. Krull, *Science*, **191**, 199 (1976).

398. K. Mori, *Agric. Biol. Chem.*, **40**, 415 (1976).

399. D. L. Wood, R. W. Stark, R. M. Silverstein, and J. O. Rodin, *Nature*, **215**, 206 (1967).

400. J. A. A. Renwick and J. P. Vité, *J. Insect Physiol.*, **18**, 1215 (1972).

401. K. Mori, N. Mizumachi, and M. Matsui, *Agric. Biol. Chem.*, **40**, 1611 (1976).

402. S. Krawielitzki, D. Klimetzek, A. Bakke, J. P. Vité, and K. Mori, *Z. Angew. Entomol.*, **83**, 300 (1977).

403. J. P. Vité, D. Klimetzek, G. Loskant, R. Hedden, and K. Mori, *Naturwissenschaften*, **63**, 582 (1976).

404. P. D. Hobbs and P. D. Magnus, *J. Chem. Soc. Chem. Commun.*, 856 (1974).

405. P. D. Hobbs and P. D. Magnus, *J. Am. Chem. Soc.*, 98, 4595 (1976).

406. K. Mori, *Tetrahedron*, 34, 915 (1978).

407. K. Mori, S. Tamada, and P. A. Hedin, *Naturwissenschaften*, 65, 653 (1978).

408. J. P. Vité, G. B. Pitman, A. F. Fentiman, Jr., and J. W. Kinzer, *Naturwissenschaften*, 59, 469 (1972).

409. K. Mori, S. Tamada, M. Uchida, N. Mizumachi, Y. Tachibana, and M. Matsui, *Tetrahedron*, 34, 1901 (1978).

410. R. C. McKnight, cited in Prof. J. P. Vité's personal communication to K. M. dated June 15, 1978.

411. D. M. Jewett, F. Matsumura, and H. C. Coppel, *Science*, 192, 51 (1976).

412. D. Jewett, F. Matsumura, and H. C. Coppel, *J. Chem. Ecol.*, 4, 277 (1978).

413. P. J. Kocienski and J. M. Ansell, *J. Org. Chem.*, 42, 1102 (1977).

414. P. Place, M. -L. Roumestant, and J. Gore, *J. Org. Chem.*, 43, 1001 (1978).

415. G. Magnusson, *Tetrahedron Lett.*, 2713 (1977).

416. G. Magnusson, *Tetrahedron*, 34, 1385 (1978).

417. A. Tai, M. Imaida, T. Oda, and H. Watanabe, *Chem. Lett.*, 61 (1978).

418. T. Harada, S. Onaka, A. Tai, and Y. Izumi, *Chem. Lett.*, 1131 (1977).

419. K. Mori, S. Tamada, and M. Matsui, *Tetrahedron Lett.*, 901 (1978).

420. K. Mori and S. Tamada, *Tetrahedron*, 35, 1279 (1979).

421. F. Matsumura, private communication to K. M. dated Oct. 26, 1978.

422. R. Baker, private communication to K. M. dated Nov. 15, 1978.

423. Y. Tamaki, H. Noguchi, H. Sugie, R. Sato, and A. Kariya, *Appl. Entomol. Zool.*, 14, 101 (1979).

424. T. Suguro and K. Mori, *Agric. Biol. Chem.*, 43, 869 (1979).

425. K. Mori, *Tetrahedron*, 33, 289 (1977).

426. J. C. Moser, R. G. Brownlee, and R. M. Silverstein, *J. Insect Physiol.*, 14, 529 (1968).

427. R. M. Crewe, M. S. Blum, and C. A. Collingwood, *Comp. Biochem. Physiol., [B]*, 43, 703 (1972).

428. R. Rossi and P. A. Salvadori, *Synthesis*, 209 (1979).

429. K. Banno and T. Mukaiyama, *Chem. Lett.*, 279 (1976).

430. K. Mori, T. Suguro, and S. Masuda, *Tetrahedron Lett.*, 3447 (1978).

431. K. Mori, S. Masuda, and T. Suguro, *Tetrahedron,* in press.

432. R. Rossi and A. Niccoli, *Naturwissenschaften*, 65, 259 (1978).

433. K. Mori, T. Suguro, and M. Uchida, *Tetrahedron*, 34, 3119 (1978).

434. T. Suguro and K. Mori, *Agric. Biol. Chem.*, 43, 409 (1979).

435. T. Shapas and W. E. Burkholder, unpublished results.

436. R. G. Yarger, R. M. Silverstein, and W. E. Burkholder, *J. Chem. Ecol.*, 1, 323 (1975).

437. U. Ravid and R. M. Silverstein, *Tetrahedron Lett.*, 423 (1977).

438. U. Ravid, R. M. Silverstein, and L. R. Smith, *Tetrahedron*, 34, 1449 (1978).

439. W. H. Pirkle, D. L. Sikkenga, and M. S. Pavlin, *J. Org. Chem.*, 42, 384 (1977).

440. H. Z. Levinson, Ref. 24a in Ref. 438.

441. R. Ikan, R. Gottlier, E. D. Bergmann, and J. Ishai, *J. Insect Physiol.*, 15, 1709 (1969).

442. J. L. Coke and A. B. Richon, *J. Org. Chem.*, **41**, 3516 (1976).

443. J. H. Tumlinson, M. G. Klein, R. E. Doolittle, T. L. Ladd, and A. T. Proveaux, *Science*, **197**, 789 (1977).

444. K. Sato, T. Nakayama, and K. Mori, *Agric. Biol. Chem.*, **43**, 57 (1979).

445. J. W. Wheeler, S. L. Evans, M. S. Blum, H. H. V. Velthius, and J. M. F. deCamargo, *Tetrahedron Lett.*, 4029 (1976).

446. W. H. Pirkle and P. E. Adams, *J. Org. Chem.*, **43**, 378 (1978).

447. W. H. Pirkle and C. W. Boeder, *J. Org. Chem.*, **43**, 2091 (1978).

448. K. Mori, T. Nukada, and T. Ebata, *Tetrahedron,* in press

449. R. Rossi and P. Diversi, *Synthesis*, 25 (1973).

450. K. Mori, T. Takigawa, and M. Matsui, *Tetrahedron Lett.*, 3953 (1976).

451. K. Mori, T. Takigawa, and M. Matsui, *Tetrahedron*, **35**, 833 (1979).

452. J. P. Vité, D. Klimetzek, G. Loskant, R. Hedden, and K. Mori, *Naturwissenschaften*, **63**, 582 (1976).

453. D. Klimetzek, G. Loskant, J. P. Vité, and K. Mori, *Naturwissenschaften*, **63**, 581 (1976).

454. J. R. Miller, K. Mori, and W. L. Roelofs, *J. Insect Physiol.*, **23**, 1447 (1977).

455. D. J. Farnum, T. Veysoglu, A. M. Cardé, B. Duhl-Emswiler, T. A. Pancoast, T. J. Reitz, and R. T. Cardé, *Tetrahedron Lett.*, 4009 (1977).

456. D. J. Farnum, cited in Ref. 457 (its Ref. 5).

457. W. H. Pirkle and P. L. Rinaldi, *J. Org. Chem.*, **44**, 1025 (1979).

458. H. H. Meyer, *Liebigs Ann. Chem.*, 732 (1977).

459. H. Ohrui and S. Emoto, *Agric. Biol. Chem.*, **40**, 2267 (1976).

460. M. Kawana and S. Emoto, *Tetrahedron Lett.*, 3395 (1975).

461. M. Kawana, H. Ohrui, and S. Emoto, *Bull. Chem. Soc. Japan*, **41**, 2199 (1968).

462. D. R. Hicks and B. Fraser-Reid, *J. Chem. Soc. Chem. Commun.*, 869 (1976).

463. P. D. Magnus and G. Roy, *J. Chem. Soc. Chem. Commun.*, 297 (1978).

464. G. T. Pearce, W. E. Gore, and R. M. Silverstein, *J. Org. Chem.*, **41**, 2797 (1976).

465. K. Mori, *Tetrahedron*, **32**, 1979 (1976).

466. G. J. Cernigliaro and P. J. Kocienski, *J. Org. Chem.*, **42**, 3622 (1977).

467. Phaik-Eng Sum and L. Weiler, *Canad. J. Chem.*, **56**, 2700 (1978).

468. K. Mori and H. Iwasawa, *Tetrahedron,* **36**, 87 (1980).

469. L. R. Smith, H. J. Williams, and R. M. Silverstein, *Tetrahedron Lett.*, 3231 (1978).

470. K. Mori, M. Sasaki, S. Tamada, T. Suguro, and S. Masuda, *Heterocycles*, **10**, 111 (1978).

471. K. Mori, M. Sasaki, S. Tamada, T. Suguro, and S. Masuda, *Tetrahedron,* **35**, 1601 (1979).

472. C. J. Persoons, P. E. J. Verwiel, F. J. Ritter, E. Talman, P. J. F. Nooijen, and W. J. Nooijen, *Tetrahedron Lett.*, 2055 (1976).

473. E. Talman, P. E. J. Verwiel, F. J. Ritter, and C. J. Persoons, *Israel J. Chem.*, **17**, 227 (1978).

474. C. J. Persoons, P. E. J. Verwiel, E. Talman, and F. J. Ritter, *J. Chem. Ecol.*, **5**, 221 (1979).

475. W. C. Still, *J. Am. Chem. Soc.*, **101**, 2493 (1979).

476. W. C. Still, *J. Am. Chem. Soc.*, **99**, 4186 (1977).

477. M. A. Adamas, K. Nakanishi, W. C. Still, E. V. Arnold, J. Clardy, and C. J. Persoons, *J. Am. Chem. Soc.*, **101**, 2495 (1979).

478. R. D. Clark, *Synth. Commun.*, **9**, 325 (1979).

479. A. Marfat, P. R. McGuirk, and P. Helquist, *J. Org. Chem.*, **44**, 1345 (1979).

480. O. L. Chapman, J. A. Klun, K. C. Mattes, R. S. Sheridan, and S. Maini, *Science*, **201**, 926 (1978).

481. O. L. Chapman, K. C. Mattes, R. S. Sheridan, and J. A. Klun, *J. Am. Chem. Soc.*, **100**, 4878 (1978).

482. G. Leadbetter and J. R. Plimmer, *J. Chem. Ecol.*, **5**, 101 (1979).

483. W. J. Elliott, G. Hromnak, J. Fried, and G. N. Lanier, *J. Chem. Ecol.*, **5**, 279 (1979).

484. H. R. Schuler and K. N. Slessor, *Canad. J. Chem.*, **55**, 3280 (1977).

485. S. Hanessian, *Accounts Chem. Res.*, **12**, 159 (1979).

486. F. Matsumura, A. Tai, H. C. Coppel, and M. Imaida, *J. Chem. Ecol.*, **5**, 237 (1979).

The Total Synthesis of Cannabinoids

RAJ K. RAZDAN

SISA Incorporated,
Cambridge, Massachusetts

1. INTRODUCTION

The illicit use of marijuana, which started in a substantial way in the early 1960s, has continued to increase. By 1975 over 36 million Americans had tried the drug, and among the 20-24 age group over 10% were using it on a daily basis. This has caused grave concern to society, and marijuana has become the subject of intense sociopolitical controversy. However, the recent use of marijuana and its active constituent Δ^1-THC for glaucoma and as an antinauseant in patients undergoing cancer chemotherapy has revived public interest in its therapeutic potential. In other cultures, particularly Chinese, Indian, and Middle Eastern, the therapeutic use of cannabis preparations has been well documented for centuries, and folklore has recorded its use in insomnia, neuralgia, migraine pain, rheumatism, asthma, bronchitis, loss of appetite, and gynecological and obstetrical problems such as dysmenorrhea. Even today its use is widely practiced in the Ayurvedic and Tibbi systems of Indian medicine. The recent discovery of its use in the treatment of glaucoma and as an antinauseant was serendipitous and came out of the drug culture. It was well known among drug users that while "high" on marijuana one could chop onions without the eyes watering. Encouraged by anecdotes like this, ophthalmologists Hepler and Frank[1] in 1971 found that smoking marijuana did indeed decrease lacrimation and also decreased the intraocular pressure (IOP) of the eye. This led to clinical trials with smoked marijuana or Δ^1-THC in glaucomatous patients.

Similarly, "pot" smokers who had cancer and were undergoing cancer chemotherapy noticed a relief from the nausea that traditionally accompanies this treatment. This was clinically confirmed by Sallan and co-workers[2] in 1975. These therapeutic aspects are being vigorously pursued at present.

It is unlikely that the natural material marijuana itself will become a marketable product because it is a complex mixture of over 35 known cannabinoids, various terpenes, nitrogen bases, phenolic compounds, sugars, etc. It is possible that its main active constituent Δ^1-THC could become a viable drug, but it is most likely that the marketable products will come from the synthetic analogs, where the undesirable side effects and physical characteristics of Δ^1-THC have been modified or eliminated.

The term "cannabinoids" is used for the typical C_{21} groups of compounds present in *Cannabis sativa* L. (family Moraceae) and includes their analogs and transformation products.

The following two different numbering systems, dibenzopyran and monoterpenoid, are generally used for cannabinoids. The former is used by *Chemical Abstracts*. In the present article the monoterpenoid numbering system will be used, as it can be most easily adapted for ring-opened cannabidiol derivatives and isotetrahydrocannabinols (iso-THCs).

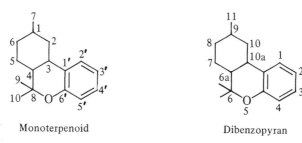

Monoterpenoid   Dibenzopyran

$(-)$-Δ^1-3,4-*trans*-Tetrahydrocannabinol (THC)[3] is the main physiologically active constituent of hashish or charas, which is the resinous exudate from the female flowers of *Cannabis sativa* L. Numerous preparations of *Cannabis* are known, and they are given different names depending on the country of origin and their mode of preparation. Thus the most potent preparation is the unadulterated resin, which is known as hashish in the Middle East and charas in India. Marijuana (pot, grass), mostly used in the United States, refers to the dried flowering tops of the plant, which are smoked in a pipe or a cigarette. Bhang, which is used in India, is a concoction made with milk and water from the flowering tops of the plant and is ingested by mouth.

$(-)$-Δ^1-3,4-*trans*-THC (**1**) is a resin that is optically active and is generally referred to as Δ^1-THC. It is also known as Δ^9-THC based on dibenzopyran numbering system. The other physiologically active isomer is Δ^6-THC (**2**; alternate name Δ^8-THC) and is found only in a few varieties of the plant. The isomers with a 3,4-*cis* ring junction are *cis*-Δ^1-THC (**3**) and *cis*-Δ^6-THC (**4**) both of which have been synthesized, but only **3** has been found in the plant so far. On theoretical grounds the *trans* compounds (**1** and **2**) are expected to be more thermodynamically stable compared to the *cis* compounds **3** and **4**. In the *trans* series Δ^6-THC (**2**) is more stable than Δ^1-THC (**1**), since **1** is easily isomerized to (**2**) on treatment with acids. The main interest pharmacologically therefore centers around the thermodynamically less stable *trans*-Δ^1-THC (**1**) and its various derivatives and metabolites. This has posed many synthetic problems because during chemical reactions the more stable derivatives of *trans*-Δ^6-THC (**2**) are mostly formed.

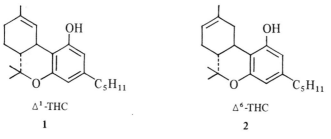

Δ^1-THC Δ^6-THC

1   **2**

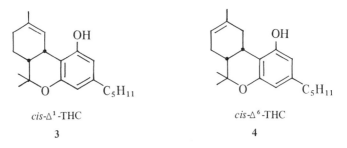

cis-Δ^1-THC

3

cis-Δ^6-THC

4

With this background let us now examine the various syntheses for $(-)$-Δ^1- and Δ^6-THCs (**1** and **2**). No attempt is made to cover the subject exhaustively in this chapter. The examples are chosen to illustrate the strategy used in the synthesis of various cannabinoids.

2. STRATEGY IN THE SYNTHESIS OF $(-)$-Δ^1- AND Δ^6-THCS AND THEIR METABOLITES

A. Synthesis of Δ^1- and Δ^6-THCs

The syntheses under discussion may be divided into two main categories: stereospecific syntheses and other approaches.

Stereospecific Syntheses

Since the structure of Δ^1-THC is not complex, the basic strategy can be envisioned as joining of the aromatic and the alicyclic parts of the molecule (as shown) by condensation of olivetol with an optically active monoterpene. Mainly, it is the selection of the monoterpene and the reaction conditions that dictate or control the position of the double bond in the Δ^1- or Δ^6- position in the final product.

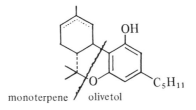

monoterpene   olivetol

From Verbenols

In 1967 Mechoulam et al.[4] described a synthesis of $(-)$-Δ^6-THC (**2**) from a pinane derivative, verbenol (**5**) and olivetol (**6**) in the presence of acid catalysts

(Chart 1.1). They visualized that the attack by the resorcinol will be favored from the side opposite the bulky dimethylmethylene bridge in verbenol and will thus provide stereochemical control of the reaction to give mainly *trans* products. Thus in the presence of BF$_3$·Et$_2$O, (–)-*cis*- or (–)-*trans*-verbenol (5) condensed with olivetol (6) to give (–)-Δ⁶-THC (2) in 44% yield. The purification of this material proved to be tedious. A better procedure was found to be a two-step sequence. When 5 was allowed to react with 6 in the presence of *p*-toluenesulfonic acid (*p*-TSA) in CH$_2$Cl$_2$, a mixture containing 7 (60%), 8, and 9 was formed from which 7 was isolated by chromatography. Treatment of 7 with BF$_3$·Et$_2$O in CH$_2$Cl$_2$ cleanly formed 2 in 80% yield.

The final conversion of Δ⁶-THC (2) to Δ¹-THC (1) was achieved by the addition of gaseous hydrochloric acid to the double bond of 2 to form 10 followed

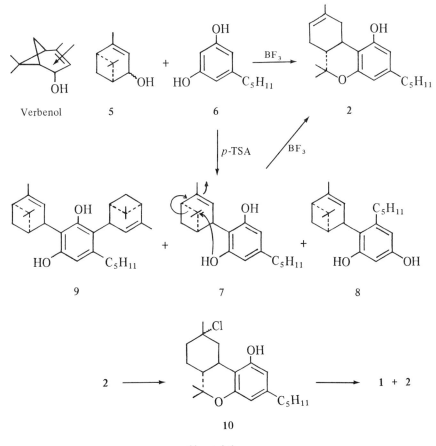

Chart **1.1**

by dehydrochlorination with sodium hydride in THF. A mixture of **1** and **2** was thus obtained, which was separated by careful chromatography.

Following the same sequence of reactions the unnatural (+)-Δ^6-THC and (+)-Δ^1-THC were similarly prepared from (+)-verbenol.[5] Both the isomers were found to be relatively inactive compared to (−)-Δ^1- and (−)-Δ^6-THCs in monkeys[6] and dogs.[7]

Utilizing this procedure various side-chain homologs of (−)-Δ^1- and (−)-Δ^6-THC's were synthesized including deuterium-labeled [3-^{2}H]-Δ^6-THC, [3-^{2}H]-Δ^1-THC,[4] and other cannabinoids[8] with tritium labeled at unspecified positions.

From Chrysanthenol

It is apparent that the mechanism of the verbenol route (Chart 1.1) is likely to involve a common allylic cation, since both *cis*- and *trans*-verbenols give the same products. However, on mechanistic grounds Razdan et al.[9] reasoned that, by virtue of the position of the double bond, verbenol can lead only to Δ^6-THC, since the double bond has to migrate into that position during the ring opening of the cyclobutane ring. On the other hand, on the basis of similar arguments they thought chrysanthenol should lead directly to Δ^1-THC. This was indeed found to be the case albeit the yield was moderate[9] (Chart 1.2). Thus (−)-verbenone, on irradiation in cyclohexane gave (−)-chrysanthenone,[10,11] which on LiAlH$_4$ reduction formed the (+)-*cis*-chrysanthenol.[10,12] Treatment with equimolar quantity of olivetol (**6**) in the presence of 0.1% BF$_3$·Et$_2$O in methylene chloride gave a resin containing ∿25% [gas-liquid chromatography (GLC)] Δ^1-THC, which was separated by chromatography and was found to be identical to the natural material in all respects. The direct synthesis from chrysanthenol

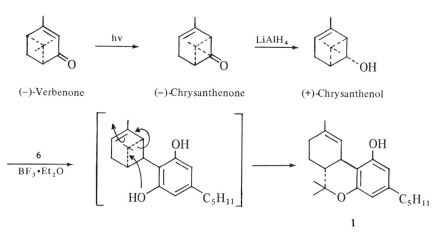

(−)-Verbenone   (−)-Chrysanthenone   (+)-Chrysanthenol

1

Chart 1.2

has some biogenetic implications, especially since chrysanthenone and chrysanthenyl acetate[12] have been found to occur naturally, and many publications report[13] the complete absence of cannabidiol in *Cannabis sativa*. It is therefore suggested that an alternative biogenetic scheme proceeding via the "pinane route"[14] might be operative in certain subspecies of *Cannabis*.

From p-*Mentha-2,8-dien-1-ol*

About the same time when Mechoulam et al.[4] reported their synthesis of (−)-Δ⁶-THC from verbenol, Petrzilka et al.[15] in 1967 demonstrated a facile entry into cannabinoids utilizing (+)-*cis*- or *trans-p*-mentha-2,8-dien-1-ol (11). By condensing with olivetol (6) in the presence of weak and strong acids they obtained (−)-cannabidiol (12) and (−)-Δ⁶-THC (2), respectively (Chart 1.3). The yield of 12 was 25%, but with a strong acid, *p*-TSA, no cannabidiol (12) was isolated, and Δ⁶-THC (2) was obtained in 53% yield. Presumably 2 was being formed via the intermediates cannabidiol → Δ¹-THC → Δ⁶-THC, since both cannabidiol and Δ¹-THC are known to yield Δ⁶-THC in nearly quantitative yields on treatment with *p*-TSA. The enhanced yield of 2 was rationalized on the postulate that abnormal cannabidiol (*abn*-CBD, 14), which always accompanies cannabidiol (12), undergoes a "retrocondensation" to give "ion *c*," which in turn forms more cannabidiol and hence Δ⁶-THC.

The Δ⁶-THC obtained by this procedure is accompanied by many by-products, which is typical of all THC syntheses and was purified by very careful column chromatography. Conversion to Δ¹-THC was carried out by the usual procedure of addition and elimination of HCl (Chart 1.1). But these authors improved the yield of Δ¹-THC in the dehydrochlorination step, previously reported by Fahrenholtz et al.[16] in the synthesis of *dℓ*-THCs (NaH/THF) and later applied by Mechoulam et al.[4] in their synthesis of (−)-Δ¹-THC by the verbenol route. By using potassium-*t*-amylate in benzene, Petrzilka and co-workers[15] reported a 100% yield of 1 from the chloro compound 10.

Because of the commercial availability of the starting terpene, (+)-*cis*/*trans-p*-mentha-2,8-dien-1-ol (11), this route was further developed by Razdan and co-workers[17] for the preparation of kilogram quantities of 1 and 2 at Arthur D. Little, Inc., Cambridge, Massachusetts for the National Institutes of Health. They reported that the dehydrochlorination procedure was very sensitive to reaction conditions and showed (GLC) that under the best conditions a mixture of 95% 1 and 5% (−)-Δ¹⁽⁷⁾-THC (13) was always obtained. This new THC (13) was completely chracterized[18] by isolation from this mixture by chromatography on silver nitrate-silica gel.

Based on the observation that cannabidiol (12) gives mainly Δ¹-THC (1) on treatment with $BF_3 \cdot Et_2O/CH_2Cl_2$, whereas under the same conditions 11 and 6 give Δ⁶-THC (2), Razdan and co-workers concluded that the formation of 2 and not 1 must be due to the acid being generated from a mole of H_2O and

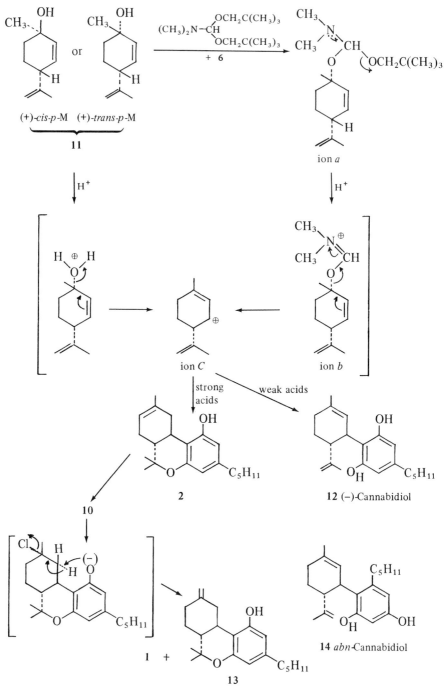

(+)-*cis-p*-M (+)-*trans-p*-M

11

ion *a*

ion *C*

ion *b*

strong acids

weak acids

2

12 (−)-Cannabidiol

10

1 +

13

14 *abn*-Cannabidiol

Chart **1.3**

BF$_3$·Et$_2$O in the reaction mixture. With this reasoning Razdan et al.[19] developed a modification of Petrzilka's cannabinoid synthesis utilizing BF$_3$·Et$_2$O/CH$_2$Cl$_2$ in the presence of MgSO$_4$ as the dehydrating agent. By this process Δ^1-THC (1) of very high optical purity was formed in a simple one-step synthesis in 50% yield (GLC) and was isolated in 31% yield after a simple and quick column chromatography.[19] The purity of Δ^9-THC was >96% by GLC and in contrast to Petrzilka's process no Δ^8-THC (2) is formed under the new conditions. Furthermore, by a slight change in the reaction conditions of the new process, (–)-cannabidiol (12) was obtained on a preparative scale.

On studying this reaction in greater detail, Razdan et al.[19] found (Chart 1.4) that normal cannabidiol (n-CBD, 12) and abnormal cannabidiol (abn-CBD, 14) were formed first and in a ratio of 1:2. The reaction stopped at this stage if <0.5% BF$_3$·Et$_2$O or wet p-TSA was used. This was followed by conversion of cannabidiols 12 and 14 into normal and abnormal THC's (1 and 16) and iso-THCs (15 and 17). Significantly, the ratio of normal to abnormal products was now greater than 3:1, indicating that apparent transformation of abn-CBD (14) to normal products was taking place. To elucidate the mechanism they studied the conversion of individual compounds with BF$_3$·Et$_2$O under the reaction conditions and arrived at the following interpretation of their results: (1) reaction rate 2 = twice rate 1, since 12 and 14 are formed in a ratio of 1:2; (2) reaction rate –2 ⩾ rate 4 because abn-CBD (14) is converted into more normal than abnormal products; (3) rate –1 ≪ rate –2 because 14 gives 50% normal products, whereas 12 gives no abnormal products and only a small amount of olivetol; and (4) rate 1 > rate 3 and rate 2 > rate 4, since the reaction can give cannabidiols 12 and 14 exclusively.

The different behavior of cannabidiols 12 and 14 can be explained on steric grounds. An examination of Dreiding models shows that in 14, unlike 12, ring closure to abnormal THC 16 results in a large steric interaction between the benzylic methylene of the C$_5$H$_{11}$ group and the C-2 vinylic proton. This interaction retards the formation of compound 16 and also increases the propensity of 14 to undergo cleavage, probably by forcing a larger contribution of trans-diaxial conformation of the cyclohexene ring substituents. This conformation favors ring closure to compound 17 or cleavage to ion c and ion d.

The importance of the steric effect of the 5-alkylresorcinol side chain has been further illustrated by the observation that when the n-C$_5$H$_{11}$ chain in 6 was substituted by a more sterically hindered CH(CH$_3$)CH(CH$_3$)C$_5$H$_{11}$ group, the reaction gave a 3:1 normal to abnormal cannabidiol ratio and subsequent ring closure gave only traces of abnormal products.

The above discussion provides an understanding of the formation of Δ^1-THC (1) as the main product from 11 and 6.

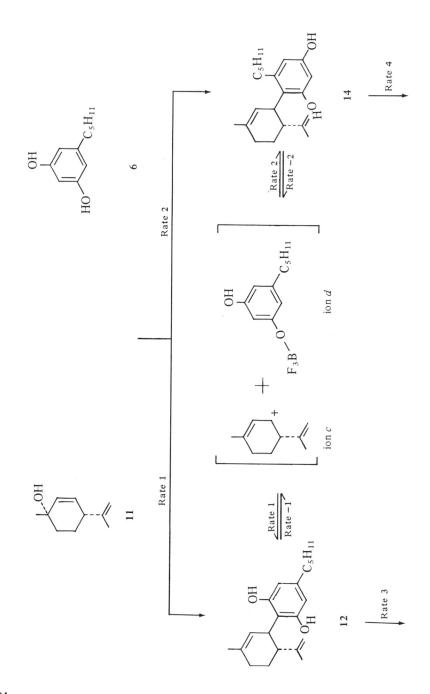

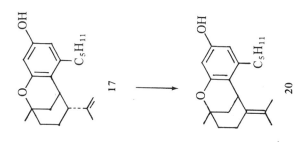

17 → **20**

+

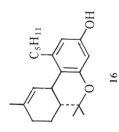

16 →

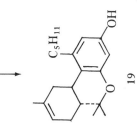

19

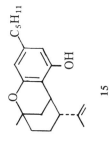

15 →

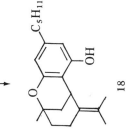

18

+

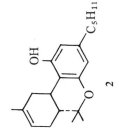

1 → **2**

Chart 1.4

195

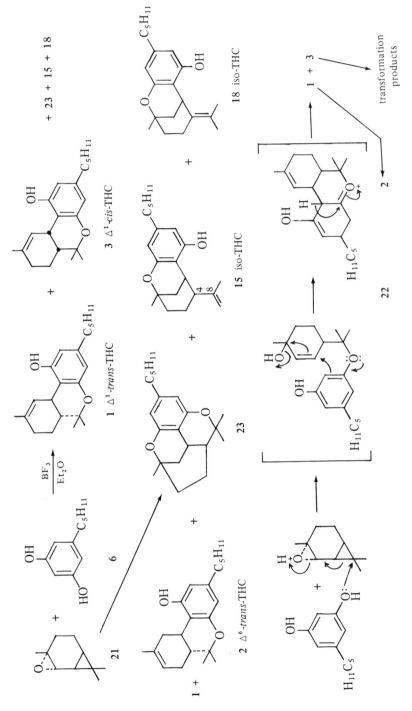

Chart 1.5

196

From Carene Epoxides

In 1970 Razdan and Handrick[20] reported an entry into cannabinoids from carane derivatives. Treatment of (+)-*trans*-2-carene epoxide (**21**) and **6** (Chart 1.5) in the presence of $BF_3 \cdot Et_2O$ gave mainly a mixture of (−)-Δ¹-*trans*-THC and (−)-Δ¹-*cis*-THC (**3**) from which the former was isolated by preparative gas chromatography. As expected, other transformation products like iso-THCs were also formed during this reaction, but no cannabidiol (**12**) was detected. Similar results were obtained by using *p*-TSA when the molar ratio of **21** was increased. However an equimolar quantity of **21** and olivetol (**6**) in the presence of *p*-TSA gave mainly the expected transformation products of Δ¹-*trans*- and Δ¹-*cis*-THCs, that is, Δ⁶-*trans*- (**2**) and iso-THCs. These results were intrepreted as suggesting that the mechanism is different from the *p*-menthadienol route and that *trans*- and *cis*-Δ¹-THCs are first formed (Chart 1.5) and are then converted into their transformation products 2 and 23 → 15 → 18, respectively.

Recently, Montero has reported in a thesis[21] that (+)-3-carene epoxide (**24**) and olivetol (**6**) give Δ⁶-THC (**2**) and the corresponding diadduct, on refluxing in benzene with *p*-TSA for 12 hours. He did not isolate any other product from this reaction and postulated the mechanism as proceeding via ion *c*, which seems reasonable under the conditions he carried out the reaction (Chart 1.6).

From p-Menth-2-ene-1,8-diol

Handrick et al.[22] (Chart 1.7) have recently used another readily available mono-terpene *p*-menth-2-ene-1,8-diol[23] (**25**) in the synthesis of (−)-Δ¹-THC (**1**). This

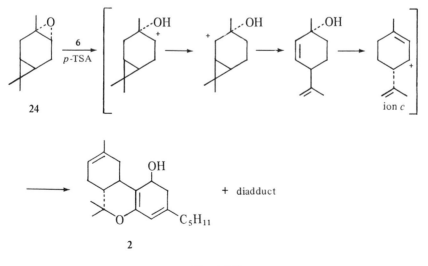

Chart 1.6

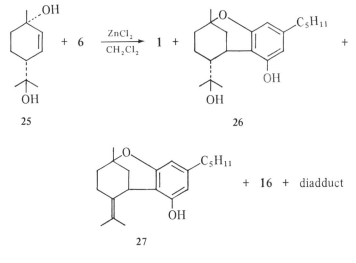

Chart 1.7

synthon was selected to facilitate the ring formation at C-8 by the presence of a hydroxyl group rather than a double bond as in p-mentha-2,8-diene-1-ol (11). A variety of catalysts were studied, and the best yield of 1 with the least amount of by-products was found to be with anhydrous $ZnCl_2/CH_2Cl_2$. The material was purified by preparative high-pressure liquid chromatography (HPLC). Although the quality of 1 was excellent, the isolated yield provided no advantage over the p-mentha-2,8-dien-1-ol route. It is interesting to note that with Zn halides the reaction stopped at the cannabidiol (12) stage with 11 but proceeded to the THC stage with 25 with apparently no isomerization of Δ^1- to Δ^6-THC, even during an extended reaction time.

Of all the procedures described above, as stated earlier, Petrzilka's process as developed by Razdan and co-workers[17] is presently used in the large-scale preparation of Δ^1-THC. Modification of Petrzilka's process ($BF_3 \cdot Et_2O/MgSO_4$) by Razdan et al.[19] has also been developed[24] to produce 50 g lots of 1 of very high purity. The purification is simplified by using Prep HPLC thus avoiding large-scale column chromatography.

Chart 1.8 summarizes the various monoterpenes that have been used for the synthesis of 1 and 2.

Other Approaches

Various other approaches have been used for the synthesis of Δ^1- and Δ^6-THCs and some of these are described below to illustrate this objective.

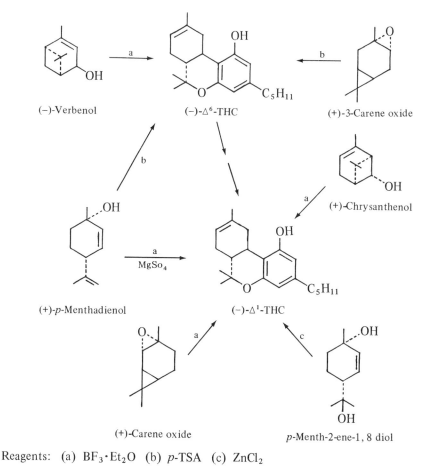

(−)-Verbenol   (−)-Δ⁶-THC   (+)-3-Carene oxide

(+)-p-Menthadienol   (−)-Δ¹-THC

(+)-Carene oxide   p-Menth-2-ene-1,8 diol

Reagents: (a) $BF_3 \cdot Et_2O$ (b) p-TSA   (c) $ZnCl_2$

Chart 1.8

Diels-Alder Reaction

An entirely different approach, which utilized a Diels-Alder reaction on an appropriately substituted cinnamic acid derivative (Chart 1.9), was developed by Jen et al.[25] This approach was originally reported by Adams and co-workers[26] but was later abandoned presumably because they failed to demethylate **29**. Jen et al.[25] deduced the *anti* configuration of **28** from its mode of preparation (Knoevenagel reaction of malonic acid with the corresponding aldehyde) and confirmed it by NMR. Based on the rule governing the retention of configuration of the dienophile constituents in the Diels-Alder reaction, compound **29**

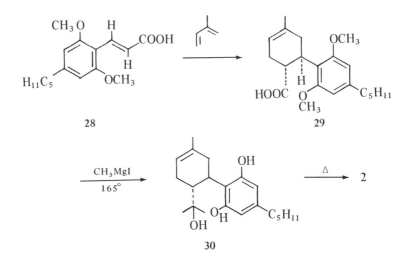

28

29

30

Chart **1.9**

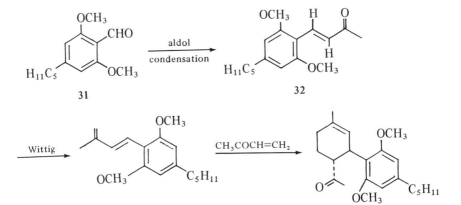

31

32

33

34

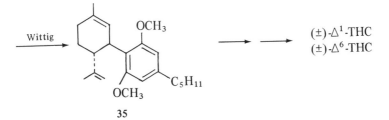

35

(±)-Δ¹-THC
(±)-Δ⁶-THC

Chart **1.10**

showed the *trans* configuration analogous to that found in THCs. After resolution of racemic **29** and treatment with excess CH_3MgI at 165°, the dimethoxy groups were successfully removed and the carboxyl was converted to a 2-hydroxypropyl group to give the triol **30**. Distillation furnished (−)-Δ⁶-THC (**2**). Similarly the (+)-isomer was also prepared.

Another variation[27,28] on the Diels-Alder approach is shown in Chart 1.10.

From Citral

A synthesis of (±)-THCs, which is of historical interest and patterned along the suggested biogenetic pathway, was published by Mechoulam and Gaoni[29] (Chart 1.11). It was the first synthesis of (±)-Δ¹-THC and was achieved from citral (**36**) and the lithium derivative of olivetol dimethyl ether (**37**). A mixture was obtained that was tosylated to yield (±)-**35**. The success of the synthesis was due to the fact that **35** could be demethylated under near-neutral conditions to give **12**, since it is known that **12** is sensitive to both acidic and strongly basic conditions. As described earlier (Chart 1.9) the same procedure was used by Jen et al.[25] in their synthesis. Acid treatment of **12** gave **1** and **2**.

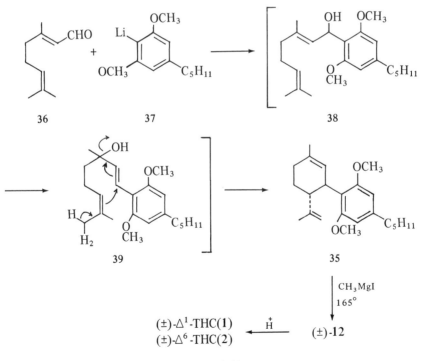

Chart **1.11**

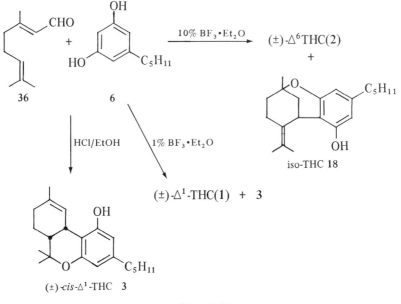

Chart 1.12

A very facile synthesis and somewhat related to the above synthesis of (±)-THCs was reported by Taylor et al.[30,31] In their approach olivetol (6) and citral were condensed with 10% $BF_3 \cdot Et_2O$ (Chart 1.12). These authors also found that with 0.0005 N HCl/C_2H_5OH, (±)-cis-Δ^1-THC (3) was formed in about 12% yield together with small amounts of (±)-trans-Δ^1-THC (1). Mechoulam et al.[29] later improved the yield of 1 to 20% by using 1% $BF_3 \cdot Et_2O$ in CH_2Cl_2. The mechanism of formation of these compounds has not yet been clearly established.

Pechmann Condensation

A very versatile route based on Von Pechmann condensation was developed by Fahrenholtz et al.[16] By this sequence (Chart 1.13) they prepared racemic Δ^1-, Δ^6-, and $\Delta^{1(7)}$-THCs. In addition the synthesis of (±)-cis-Δ^1-THC and (±)-cis-$\Delta^{1(7)}$-THC (48) was achieved from 46. The Li/NH_3 reduction of 43 gave both the cis- and trans-ketones 46 and 44, respectively, which were separated. The major product was the trans-isomer as expected. The cis-ketone 46 was also prepared by catalytic reduction (Raney nickel) under high pressure and temperature of the ketal of 42 followed by treatment with CH_3MgI and acid hydrolysis.

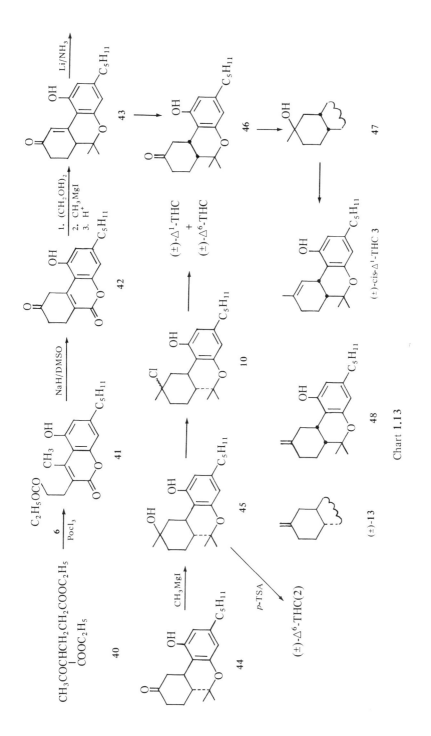

Chart 1.13

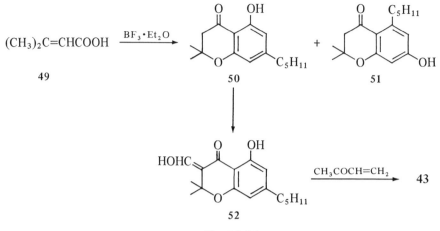

$(CH_3)_2C=CHCOOH$ $\xrightarrow{BF_3 \cdot Et_2O}$

49

50   +   **51**

52 $\xrightarrow{CH_3COCH=CH_2}$ **43**

Chart 1.14

Alternatively the key intermediate **43** was also prepared from **49** (Chart 1.14). The alicyclic ring was built by using the standard Robinson annelation procedure.

B. Synthesis of *cis*-THCs

Of the two *cis*-isomers, that is, *cis*-Δ^1-THC (**3**) and *cis*-Δ^6-THC (**4**) only **3** has been found in the plant and that only recently.[32] The Δ^1-isomer **3** is more thermodynamically stable than the Δ^6-isomer **4**. The former is well characterized, and its formation was observed during some synthetic sequences such as from citral-olivetol reaction (Chart 1.12) as reported by Taylor et al.[30] the carene oxide synthesis (Chart 1.5) by Razdan and Handrick[20] and a total synthesis by Fahrenholtz et al.[16] (Chart 1.13). Recently another synthesis, utilizing olivetol bis(tetrahydropyranyl ether) homocuprate and dehydrolinalool acetate, has been reported by Luteijn and Spronck.[32a]

In contrast, the *cis*-Δ^6-isomer **4** has had a confusing history. The earlier claims[30] regarding its synthesis has not been substantiated,[31] since at one time $\Delta^{4,8}$-iso-THC **18** was erroneously regarded as *cis*-Δ^6-THC (**4**).

Synthesis of (±)-cis-Δ^6-THC

In 1975 the synthesis of authentic (±)-**4** by three routes was achieved by Uliss et al.[33]

In the first synthesis they utilized the stereospecific intramolecular epoxide cleavage by phenolate anion.[34] Thus (Chart 1.15) epoxidation with *m*-chloroperbenzoic acid at 0° formed **54** from the acetate of (±)-*cis*-Δ^1-THC **53**. Under

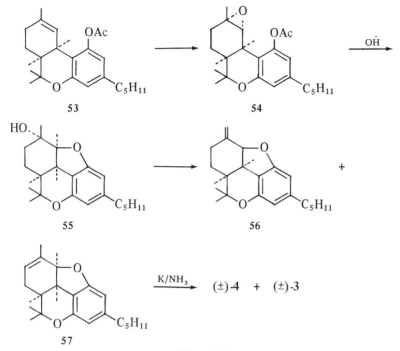

Chart 1.15

basic hydrolytic conditions **54** gave the benzofuran **55** in 90% yield. Dehydration with HMPA (240°, 0.25 h) gave a 3:2 mixture of **57** and **56**. Treatment of this mixture with K/NH₃ gave a mixture of the two *cis*-THCs, which were separated as their acetates by HPLC and subsequently hydrolyzed to give (±)-**4**.

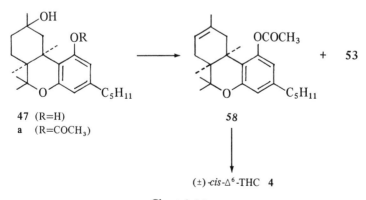

Chart 1.16

In a second synthesis (Chart 1.16), acetylation of the known cis-tertiary alcohol 47[16] gave compound 47a. Treatment of 47a with thionyl chloride/ pyridine furnished a mixture of 58 and 53 (2:3 by GLC), which was separated by HPLC. 58 was then hydrolyzed to yield (±)-4.

It was found that Δ^1-cis-THC acetate (53) under acid catalysis (p-TSA in boiling benzene) provided substantial quantities of the thermodynamically less stable 58 at equilibrium (ratio of 77:23, respectively). This is contrary to results reported in the literature[31b] and constitutes on hydrolysis the third synthesis of (±)-4.

Synthesis of (+)-cis-Δ^1- and Δ^6-THC's

Razdan and Handrick[20] had earlier reported a one-step stereospecific synthesis of (−)-Δ^1-THC (1) from a carene derivative. In this reaction (Chart 1.5) (+)-*trans*-2-carene oxide (21) and olivetol (6) in the presence of $BF_3 \cdot Et_2O$ gave a complex mixture containing optically active *trans*- and cis-Δ^1-THCs (1 and 3). Since separation of these two isomers proved difficult by chromatography on a pre-parative scale, a chemical reaction sequence was carried out (Chart 1.17) which allowed the separation of the cis-isomers from the *trans*-products. It takes advantage of the observation that *trans*-compounds do not undergo the stereo-specific intramolecular epoxide cleavage by phenolate anion because of ring strain. On this basis Uliss et al.[35] acetylated the crude mixture and allowed it to react with m-chloroperbenzoic acid to presumably give a mixture containing optically active epoxides 54 and 59. Hydrolysis with base was accompanied by an intramolecular opening of the epoxide ring (as in the racemic case, Chart 1.15), which occurred exclusively with the cis-isomer 54, resulting in the dihydrofuran derivative 55. After base treatment 55 was easily isolated as a neutral fraction and was then dehydrated, reductively cleaved and purified to give pure (+)-cis-Δ^1-THC and (+)-cis-Δ^6-THC. The natural cis-Δ^1-THC (3) has the opposite configuration to Δ^1-THC at C_4.[32]

C. Metabolites of Tetrahydrocannabinols

Extensive literature[3b,3c,36] has appeared in recent years describing the various metabolites of THC's isolated from *in vivo* or *in vitro* studies. These have been carried out on a wide variety of animal species and in some cases different animal organ homogenates have been utilized. Important sites of metabolism of Δ^1- and Δ^6-THCs are shown in Chart 1.18 with a listing of some specific meta-bolites. The hydroxylation at the 7-position appears to be the major initial point of attack in nearly every species tested including man. This metabolite, 7-hydroxy-Δ^1-THC, is pharmacologically equiactive with Δ^1-THC, and still others are active to different degrees. This has complicated the understanding of

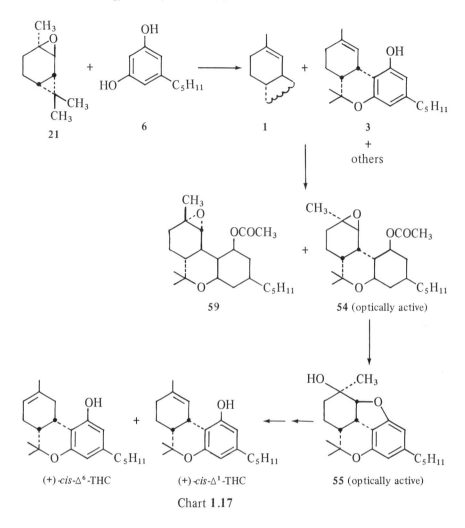

Chart 1.17

marijuana activity in man. The metabolites of Δ^1-THC so far identified in man[37] are shown in Chart 1.19.

D. Metabolites Functionalized in the Alicyclic Ring

The most important memeber of this class is the 7-hydroxy-Δ^1-THC because it is pharmacologically equiactive with Δ^1-THC. The problems associated with the synthesis of these metabolites in the Δ^1-series are somewhat similar to those

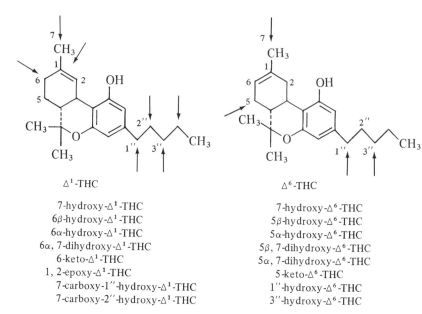

Δ¹-THC

7-hydroxy-Δ¹-THC
6β-hydroxy-Δ¹-THC
6α-hydroxy-Δ¹-THC
6α, 7-dihydroxy-Δ¹-THC
6-keto-Δ¹-THC
1, 2-epoxy-Δ¹-THC
7-carboxy-1″-hydroxy-Δ¹-THC
7-carboxy-2″-hydroxy-Δ¹-THC

Δ⁶-THC

7-hydroxy-Δ⁶-THC
5β-hydroxy-Δ⁶-THC
5α-hydroxy-Δ⁶-THC
5β, 7-dihydroxy-Δ⁶-THC
5α, 7-dihydroxy-Δ⁶-THC
5-keto-Δ⁶-THC
1″-hydroxy-Δ⁶-THC
3″-hydroxy-Δ⁶-THC

Chart **1.18**

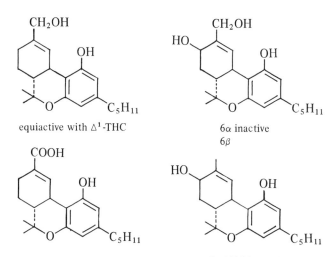

equiactive with Δ¹-THC

6α inactive
6β

6α 1/10th activity
6β 1/4 th activity

Chart **1.19**

encountered in the synthesis of Δ^1-THC itself as has already been discussed in the Introduction. Thus a number of synthetic procedures that work satisfactorily in the Δ^6 series prove to be inadequate in the Δ^1 series.

7-Substituted Metabolites

A detailed discussion of the Δ^1-metabolites, which result from the natural material Δ^1-THC, is appropriate. Therefore various approaches to the synthesis of 7-hydroxy-Δ^1-THC, the most important member of this class, are described, and in addition some examples in the Δ^6-series are included to illustrate other synthetic strategies.

Δ^1-Derivatives

Pitt et al.[38] have developed a regioselective route to some of these metabolites by a base induced epoxide-allylic alcohol rearrangement followed by SN' displacement. This procedure provides a new method of derivatizing the allylic 7-methyl group of Δ^1-THC and the synthetic sequence to 7-hydroxy-Δ^1-THC **64** from Δ^1-THC (**1**) is shown in Chart 1.20. Δ^1-THC acetate was converted to the known α-epoxide **59** which was isomerized to a mixture of the allylic alcohols **60** and **61** in excellent yield by treatment with the lithium salt of an amine in

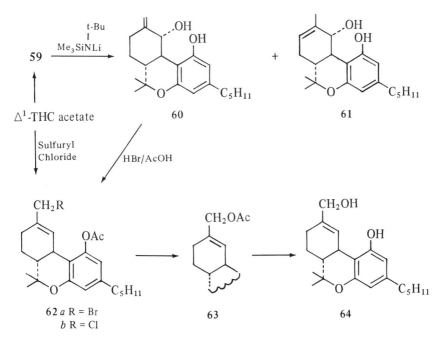

Chart 1.20

ether. The ratio of **60/61** was controlled by the size of the substituent on the amine. Thus the use of Me_3SiNH-t-Bu gave the best yield of **60**, which was converted to **62a** by 5% HBr/AcOH taking advantage of the thermodynamically controlled SN′ isomerization of allylic bromides. Acetolysis with tetramethyl ammonium acetate in acetone followed by reduction with $LiAlH_4$ converted **62a** to the desired metabolite **64** in ∿20% yield. In an alternative procedure[38] the phenol in **60** was selectively protected as ethoxymethyl ether and then treated with $SOCl_2$/Py to give the corresponding 7-chloro derivative. Acetolysis as before, followed by acid-catalyzed removal of the acetyl and ethoxymethyl groups formed the metabolite **64**. The yield was not as good as in the other procedure.

Direct allylic halogenation or oxidation of Δ^1-THC acetate has also been carried out, but the yields have been unsatisfactory mainly because of the lack of selectivity of attack at the primary and secondary allylic sites of Δ^1-THC. Pitt et al.[39] reported that treatment of Δ^1-THC acetate with sulfuryl chloride (Chart 1.20) gave a mixture containing **62b**, which, with silver acetate in acetic acid followed by hydrolysis with base, formed **64** albeit in poor yield (5%). The other metabolite 6β-hydroxy-Δ^1-THC was also isolated (14%) from this synthesis. Reagents like N-chloro- and N-bromosuccinimide, predominantly halogenate in the C-6 position, and these have been utilized in the synthesis of various 6-substituted metabolites.

Ben-Zvi et al.[40] reported the SeO_2 oxidation of Δ^1-THC acetate (Chart 1.21) followed by reduction with $LiAlH_4$ to give **64** in poor yield (1%), the main product being 7-hydroxycannabinol (**65**). This procedure has proved more successful in Δ^6-series (Chart 1.25).

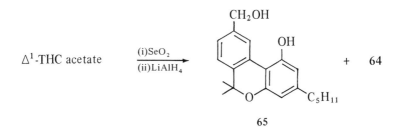

Δ^1-THC acetate $\xrightarrow{\text{(i)SeO}_2}{\text{(ii)LiAlH}_4}$ 65   +   **64**

Chart **1.21**

In a different approach, Razdan et al.[41] (Chart 1.22) oxidized the exocyclic double bond of (−)-$\Delta^{1(7)}$-THC acetate (**66**) with m-chloroperbenzoic acid to the epoxide **67**, which was hydrolyzed (basic conditions were used to avoid forming the Δ^6-dehydration products) with 0.3N KOH/DMSO to form the triol **68**. After acetylation, the diacetate alcohol **69** was treated with $SOCl_2$/Py to give a

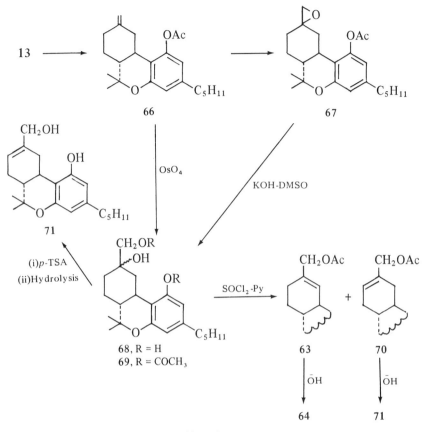

Chart 1.22

mixture of the two metabolites as their diacetates **63** and **70**. These were separated by HPLC and then hydrolyzed with base to yield 7-hydroxy-Δ^1-THC (**64**). The overall yield of **64** from $\Delta^{1(7)}$-THC (**13**) was 13%. Alternatively **69** was obtained from **66** by hydroxylation of the exocyclic double bond with OsO$_4$ in ether followed by acetylation. These syntheses were developed because **13** became available in large quantities from the kilogram synthesis of Δ^1-THC (see Chart 1.3).[17,18] The intermediate **69** also provided a facile route to the 7-hydroxy-Δ^6-THC **71**.[42] This was achieved by treatment with p-TSA followed by hydrolysis in an overall yield of 75% from **13**.

Uliss et al.[43] have recently reported a versatile route to 7-substituted Δ^1-THCs from the novel synthons **75a** and **75b**. This scheme (Chart 1.23) is based on the principal of reversal of reactivity of carbonyl compounds when masked as dithioacetals (i.e., umpolung). Interestingly, this route has resulted in the synthesis of

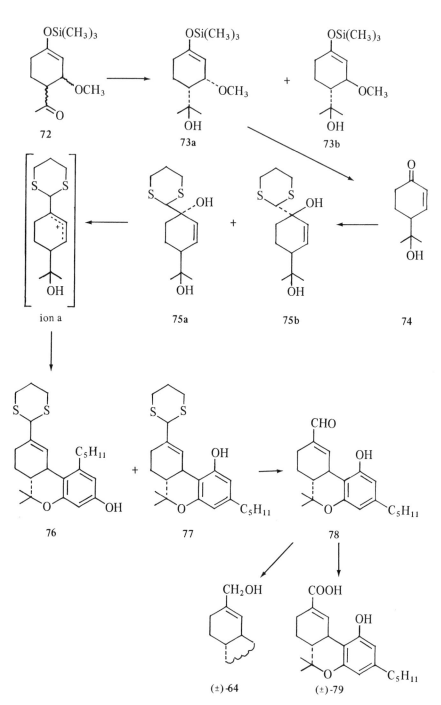

Chart 1.23

Δ^1-derivatives specifically, since by introduction of the dithiane moiety into the THC structure, isomerization of the normally labile Δ^1 unsaturation to the Δ^6 isomer is effectively inhibited. The synthons 75a and 75b were readily obtained by carrying out a Grignard reaction on the Diels-Alder adduct 72, thus extending the usefulness of Danishefsky's novel Diels-Alder diene.[44] This was followed by hydrolysis of the mixture (73a + b) to 74 with CCl_3COOH and addition of the Li anion of 1,3-dithiane to 75. Treatment with olivetol (6) in the presence of p-TSA gave a mixture of Δ^1-compounds 76 and 77, which was separated. The dithiane masking group in 77 was readily removed by $HgO/BF_3 \cdot Et_2O$ to give the metabolite (±)-78.[45] This was converted by $LiAlH_4$ reduction to the metabolite (±)-64 or oxidized with MnO_2/CH_3OH containing acetone cyanohydrin, to the metabolite (±)-79.[38]

(–)-Perillyl aldehyde was converted into perillyl alcohol acetate (80) and then to 82 via 81. On condensation with olivetol Lander et al.[46] obtained the (+)-7-hydroxy-Δ^1-THC (Chart 1.24) in low yield.

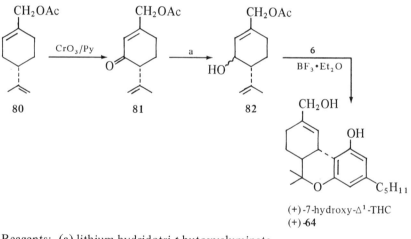

Reagents: (a) lithium hydridotri-t-butoxyaluminate

Chart 1.24

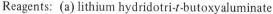

Δ^6-Derivatives

In the case of 7-substituted Δ^6- derivatives the syntheses in general are more straightforward. Thus as described earlier (Chart 1.22) 7-hydroxy-Δ^6-THC (71) is prepared[41] in excellent yield from $\Delta^{1(7)}$-THC (13). Oxidation of Δ^6-THC acetate (Chart 1.25) with SeO_2 under controlled conditions followed by acetylation gives directly the 7-acetoxy-Δ^6-THC acetate (70), which on reduction provides the metabolite 71.[40] By slightly changing the oxidation conditions and

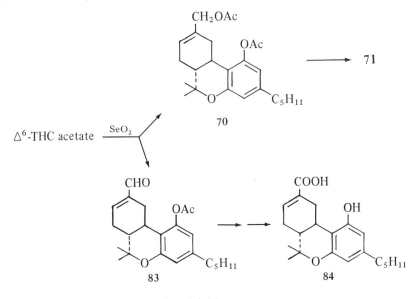

Chart 1.25

refluxing Δ^6-THC acetate with SeO_2 for a prolonged period in ethanol, the aldehyde 83 is isolated.[47,48] On further oxidation with $MnO_2/NaCN$ in CH_3OH followed by hydrolysis 83 was converted to the metabolite 84.[47] These reactions of Δ^6-THC acetate with SeO_2 are in contrast to similar oxidation of Δ^1-THC acetate (cf. Chart 1.21).

In other syntheses of 7-hydroxy-Δ^6-THC (71), the epoxide 85 of Δ^6-THC acetate with $HClO_4$ gave 86 (Chart 1.26), which after acetylation and dehydration ($SOCl_2/Py$) formed 87. After hydrolysis and allylic rearrangement 87 was converted to the metabolite 71.[3b]

The base-induced epoxide-allylic alcohol rearrangement was applied by Petrzilka and Demuth[49] on 88 to give a mixture of allylic alcohols 89 (Chart 1.27). These were acetylated after removal of the tetrahydropyranyl (THP) protecting group to give 87 and then subjected to allylic rearrangement and reduction to form 71. The allylic alcohols 89 were also obtained from Δ^6-THC acetate by a photochemical rearrangement followed by reduction.

A direct conversion of $\Delta^{1(7)}$-THC acetate 66 to 7-bromo-Δ^6-THC acetate (90) was achieved by Weinhardt et al.[50] by using N-bromoacetamide in the presence of 70% $HClO_4$. Treatment with silver acetate/AcOH gave 70, which on alkaline hydrolysis furnished 71 (Chart 1.28).

The metabolites 84 and 71 were also synthesized by Pitt et al.[39] from 91 (Chart 1.29), which has been previously prepared by Wildes et al.[51] The morpho-

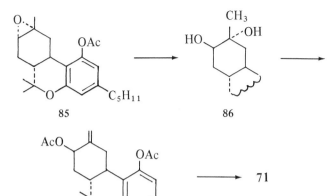

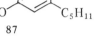

Chart **1**.26

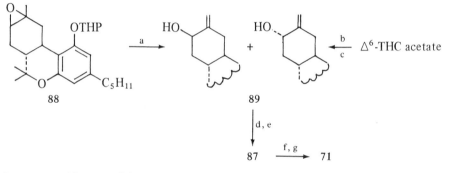

Reagents: (a) BuLi (b) irradiation, O_2-sensitizer (c) $NaBH_4$
(d) H^+ (e) Ac_2O-Py (f) Δ, 290°C (g) $LiAlH_4$

Chart **1**.27

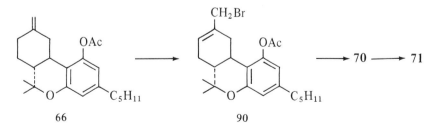

Chart **1**.28

linoenamine of **91** treated with CCl₃COOH gave the α-chloramide **92**. After removal of the benzyl protecting group and conversion to the phenoxide anion to eliminate HCl, only the Δ^6-amide **93** was obtained. Saponification of **93** gave the metabolite **84**, which was reduced with LiAlH₄ to **71**.

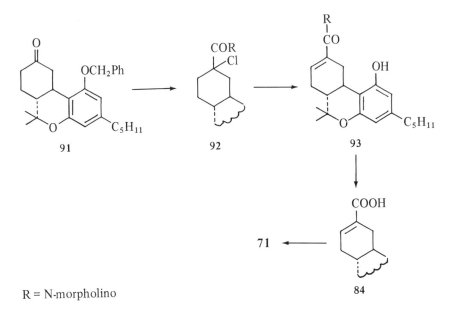

91   92   93

71 ←   84

R = N-morpholino

Chart 1.29

6-Substituted Metabolites

The 6α- and 6β-hydroxy-Δ^1-THC's are two metabolites that have been identified in man. The latter, **95**, was directly prepared (Chart 1.30) by bromination of Δ^1-THC acetate with N-bromosuccinimide[38] or sulfuryl chloride[39] followed by acetolysis and hydrolysis. The 6α-metabolite **97** was prepared[38] from **95** by selective acetylation of the phenolic group and MnO₂ dioxide oxidation to **96** followed by LiAlH₄ reduction to **97**. It is interesting to note that **96** was also obtained[52] by SeO₂ oxidation of Δ^1-THC acetate (cf. Chart 1.21).

5-Substituted Metabolites

The metabolites[53] known in this class were synthesized[52] from Δ^6-THC acetate as shown in Chart 1.31. Oxidation with t-butylchromate gave **98**. It was reduced with LiAlH₄ to give a mixture of the 5α- and 5β-hydroxy-Δ^6-THCs, **99** and **100**, respectively. These were isolated by chromatography.

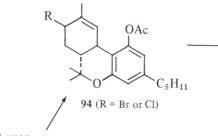

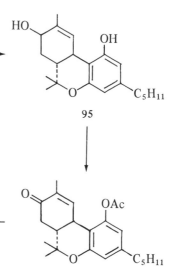

Δ¹-THC acetate

94 (R = Br or Cl)

95

97

96

Chart **1.30**

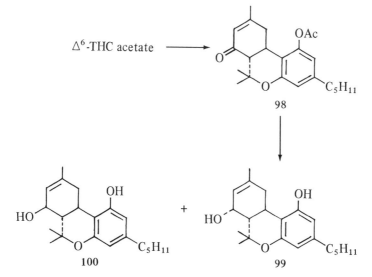

Δ⁶-THC acetate

98

100   +   99

Chart **1.31**

Other Metabolites

The two dihydroxy metabolites **103** and **104** have been found in humans[37a] and were synthesized by Pitt et al.[38] Δ^6-THC epoxide **85** (Chart 1.32) on treatment with BuLi rearranged to **101** as a 1:1 mixture of epimers at C-6. Diacetylation of **101** and treatment with OsO_4 gave **102** (R = H). Acetylation of the primary 7-hydroxyl group of **102**, dehydration with $SOCl_2$/Py followed by saponification gave **103** and a lesser amount of **104**. The latter was obtained in much better yield by using the same hydroxylation-dehydration sequence with 7-hydroxy-Δ^6-THC (**71**).[38]

Gurney et al.[53] reported the isolation of 1,2-epoxyhexahydrocannabinol as a metabolite from an *in vitro* preparation of squirrel monkey tissue. It was synthesized from Δ^1-THC acetate by epoxidation with *m*-chloroperbenzoic acid.[52,53] This metabolite has also been identified in *in vitro* preparations of dog tissue.[54]

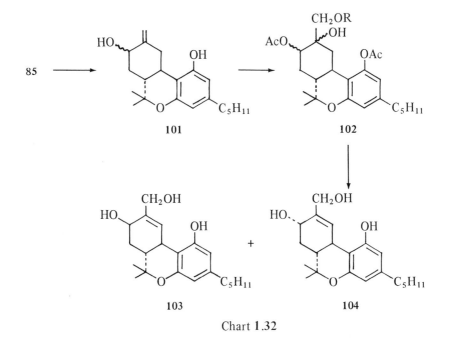

Chart 1.32

E. Metabolites Functionalized in the Aromatic Ring

The *n*-pentyl side chain of the aromatic ring is the main site of metabolic attack.[55] Thus Δ^1-THC, cannabidiol, and cannabinol give microsomal hydroxylations at carbons 1″-5″. In addition carboxylic acid metabolites, formed by oxidative cleavage of the side chain, have also been identified.[56]

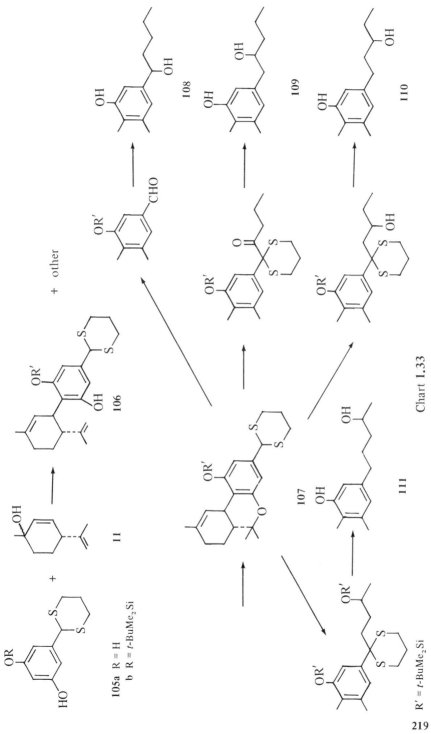

Chart 1.33

R' = t-BuMe₂Si

219

Until very recently none of the Δ^1-THC metabolites hydroxylated at the *n*-pentyl side chain had been synthesized. Pitt et al.[57] have now reported an elegant general synthesis of side-chain derivatives of Δ^1-THC as shown in Chart 1.33. The poorly soluble resorcinol derivative **105a** was made more soluble in organic solvents by converting it to **105b**. This was best achieved by quantitative conversion to the diether and then selective monodesilylation with fluoride ion. Condensation of **105b** with *p*-mentha-2,8-dien-1-ol (**11**) in $CH_2Cl_2/BF_3 \cdot Et_2O/$ $MgSO_4$, using the procedure of Razdan et al.[19] (Chart 1.4), the cannabidiol analog **106** was obtained together with other compounds. Treatment of this mixture with $BF_3 \cdot Et_2O/CH_2Cl_2$ at $-20°$ for 48 h gave the Δ^1-analog **107**, which was purified by chromatography. This key intermediate **107** was then converted to the various side-chain hydroxylated Δ^1-THCs.

It is interesting to note that Razdan et al.[58] had previously reported the reaction of the five-membered heterocyclic analog of **105a** [i.e., 2-(3′,5′-dihy-droxyphenyl)-1,3-dithiolane] with **11** in benzene/*p*-TSA and found that the Δ^6-analog was formed (see Chart 4.4). It appears that the reaction products are very sensitive to structure and/or reaction conditions in cannabinoids.

The specific synthesis of the metabolite **110** was also achieved (Chart 1.34) recently by Handrick et al.[22] by the preparation and condensation of 3′-ace-toxyolivetol (**112**) with the monoterpene *p*-menth-2-ene-1,8-diol (**25**). They essentially demonstrated the usefulness of this synthon (**25**, Chart 1.7) for the synthesis of Δ^1-THC and its metabolites by preparing **110**. 3″-Hydroxy-Δ^1-THC (**110**) was also reported to be about 3 times more active than Δ^1-THC in pre-liminary pharmacological tests in mice.[22] Christie et al.[58a] reported the synthesis of (±)-**110** by condensing 1-(3,5-dihydroxyphenyl)pentan-3-one ethylene thioacetal with citral in the presence of $BF_3 \cdot Et_2O$ followed by removal of the thioacetal and reduction with $NaBH_4$.

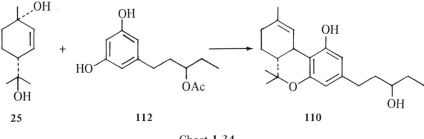

25   112   110

Chart **1.34**

The synthesis in the Δ^6-series is generally carried out (Chart 1.35) by con-densing the appropriate resorcinol or a suitable thioketal derivative thereof, with the monoterpene *p*-menthadienol[59] **11** or verbenol[55a] **5**. The direct coupling of

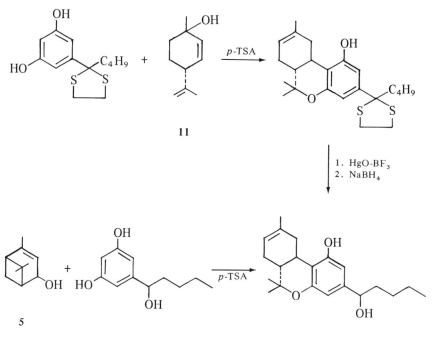

Chart 1.35

$1'$- or $3'$-hydroxyolivetol or $1'$- or $3'$-oxoolivetol with **11** failed but worked as their thioketal derivatives.[59]

Recently Lotz et al.[60] have reported that resorcylalkyl esters directly condense with p-menthadienol **11** to give the corresponding Δ^6-THC derivatives. These on LiAlH$_4$ reduction provide Δ^6-THC analogs with a hydroxyl group, substituted on the terminal carbon of the aromatic side chain. The $5''$-hydroxy-Δ^6-THC was synthesized using this procedure.

In the Δ^6-THC series the biological activity has been shown to vary with the position of the hydroxyl group in the pentyl side chain. The order of activity[55] is $3''$- $>$ $4''$- $>$ $2''$- $>$ $1''$-hydroxy-Δ^6-THC. Full details of this work have now been published.[55a]

3. SYNTHESIS OF OTHER THCS AND RELATED CANNABINOIDS

In this section the synthetic approaches to some of the "unnatural" THC's and the important cannabinoids that occur in the plant *Cannabis sativa* L. will be discussed.

A. "Unnatural" THCs

$(-)$-$\Delta^{1(7)}$-THC (**13**, Chart 1.3) was isolated during Petrzilka's dehydrochlorination procedure by Razdan et al.[18] It was also produced by photoisomerization of Δ^6-THC[42] and by dehydrochlorination of 1-chlorohexahydrocannabinol methyl ether with a bulky base followed by demethylation of the ether with potassium thiophenoxide.[51]

Cardillo et al.[61] reported a synthesis (Chart 2.1) of cannabidiol (**12**) from *p*-mentha-1,8-dien-3-ol (**113**) and olivetol (**6**) in aqueous acid. These mild condensation conditions were applied in the preparation of "unnatural" Δ^4-THC derivatives. Thus *p*-menth-4-en-3-ol (**114**) and olivetol gave **115**, which formed the novel cannabinoid **116**. This procedure led to the synthesis of Δ^4-THC (**118**) by the condensation of olivetol with *p*-menth-4-ene-3,8-diol (**117**).[62]

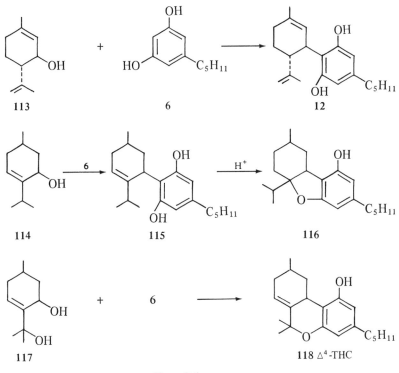

Chart 2.1

Δ^5-THC (**120**), another "unnatural" THC, was synthesized[63] (Chart 2.2) by hydroboration of Δ^6-THC to give **119a**, which was tosylated to **119b** and then treated with K-*t*-butoxide in benzene.

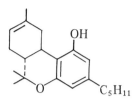

Δ^6-THC

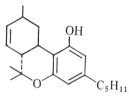

119a R = H
  b R = Ts

120 Δ^5-THC

Chart **2.2**

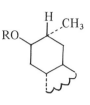

121

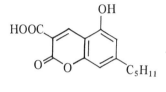

122

$\rightleftharpoons$

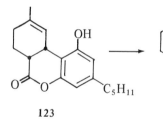

123

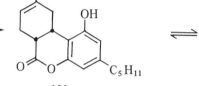

124a, R = H
  b, R = Ac

125

1. SOCl$_2$-Py
2. OH$^-$

42

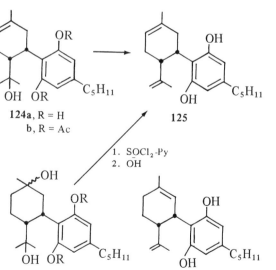

127

128a R = H
  b R = Ac

126

Chart **2.3**

B. Cannabidiols

(−)-Cannabidiol (12, Chart 1.3) is one of the major constituents of the plant. The natural material has a 3,4-*trans* ring junction with a double bond at the Δ^1 position. As discussed earlier, it was synthesized from *p*-menthadienol 11 by the Petrzilka procedure[15] (Chart 1.3) or by the modification of Razdan et al.[19] of Petrzilka's procedure (Chart 1.4). It is also prepared as shown in Chart 2.1 and by condensing 11 and olivetol in the presence of wet *p*-TSA.[19] For the synthesis of (±)-cannabidiol see Charts 1.10 and 1.11.

The Δ^6-isomer of cannabidiol is not known. The two Δ^1- and Δ^6-cannabidiols with a 3,4-*cis* junction are "unnatural" and were synthesized (Chart 2.3) recently by Handrick et al.[64] Thus lactone 122 was prepared from isoprene and 121 by a Diels-Alder reaction accompanied by decarboxylation.[64,65] Reaction of 122 with CH_3MgI gave the triol 124a; its diacetate 124b was dehydrated with $SOCl_2$/Py and then saponified to give (±)-Δ^6-3,4-*cis*-cannabidiol 125. The Δ^1-isomer 126 was obtained by equilibrating the lactone 122 to a 1:1 mixture of 122 and 123. The identical procedure then gave a mixture of Δ^6- and Δ^1-*cis*-cannabidiol diacetates. These were easily separated by HPLC and the Δ^1-isomer was saponified to give 126. In an alternative synthesis by Handrick et al.[64] the keto lactone 42 (Chart 1.14) was ketalized, reduced with Raney nickel at high pressure,[16] and hydrolyzed to give 127. Reaction with CH_3MgI furnished the tetrol 128a. Its acetate 128b was dehydrated and then saponified to form 125.

C. Cannabinol

This was one of the first cannabinoids to be synthesized in early 1940, as it established the basic skeleton of the THC structure. Δ^1-THC is readily oxidized to cannabinol on exposure to air.[18]

The first synthesis (Chart 2.4) by Adams et al.[66] condensed 129 with dihydro-olivetol 130 to form the pyrone 131 which on dehydrogenation gave 132. Grignard reaction followed by acid treatment furnished cannabinol 136. It is interesting to note that a similar reaction of 129 with olivetol (6) formed the "abnormal" isomer of 132 resulting from an attack on the 4 position of olivetol.[67]

In a different approach, based on Pechmann condensation, both the British[68] and American groups[69] independently arrived at the synthesis of cannabinol using the same sequence. Thus the keto ester 133 and olivetol formed the pyrone 134, which on treatment with CH_3MgI, followed by acid treatment, gave Δ^3-THC (135). Sulfur dehydrogenation converted 135 to cannabinol. Both groups discovered that the physiological activity of Δ^3-THCs was similar to the natural material. The American group led by Roger Adams carried out extensive structure activity work in this series, which formed the basis for drug develop-

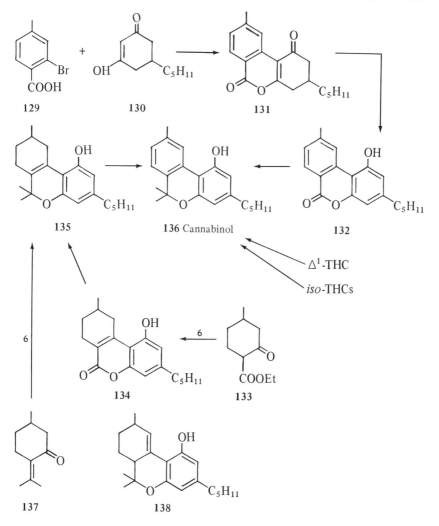

Chart 2.4

ment in cannabinoids during the late 1960s. This particular aspect will be discussed in a later section. The Pechmann condensation was reinvestigated by Claussen and Korte[70] and they isolated from this reaction the other isomer Δ^2-THC (138).

Another synthesis of cannabinol was achieved from pulegone (137) by both Adams et al.[71] and Ghosh et al.[72] (Chart 2.4). Cannabinol is also formed easily by chloranil dehydrogenation of Δ^1-THC.[73] In contrast, Δ^6-THC and cis-Δ^1-THC do not form cannabinol. This has been explained on the basis of stereo-electronic factors.[73] The various iso-THCs are easily converted to 136.[31b]

D. Cannabinoid Acids[3]

They form an important group of compounds present in the plant.[3] These acids are inactive by themselves but are readily decarboxylated on heating (smoking) or on GLC to the parent active THCs. The two methods which have been developed for their synthesis are shown in Chart 2.5. Methyl magnesium carbonate (MMC) carboxylated Δ^1-THC. Similarly cannabidiolic acid and cannabigerolic acid were synthesized. Petrzilka's procedure[15] was used to prepare 141, but the acid itself could not be isolated, as during hydrolysis it decarboxylated to give 12.

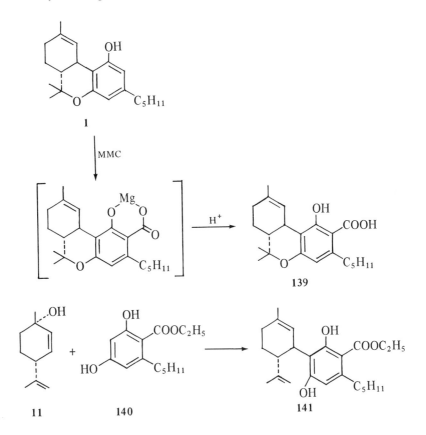

Chart 2.5

E. Cannabigerol

It is a minor constituent of the plant and was synthesized by condensation of geraniol with olivetol in the presence of p-TSA in CH_2Cl_2.[75] It was also prepared by a direct aklylation of ethoxycarbonyl olivetol with geranyl bromide followed by hydrolysis.[76]

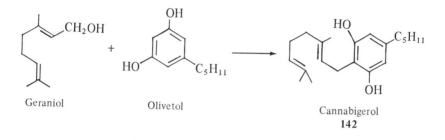

Geraniol   Olivetol   Cannabigerol
                                                            142

F. Cannabichromene

It was previously considered to be a minor constituent of the plant, but recent advances in analytical techniques have established that it is no longer a minor cannabinoid. In fact in some variants it is more abundant that cannabidiol.[77] Two main approaches to the synthesis of cannabichromene have been reported (Chart 2.6). One is based on the dehydrogenation of cannabigerol (142) with chloranil[73] or dichlorodicyanobenzoquinone.[76] With both these reagents the reaction probably proceeds via hydride abstraction but the O-quinone methide intermediate 143 has also been proposed.[76] The other synthesis of 144 was simultaneously and independently reported by Crombie and Ponsford[78] and Kane and Razdan.[79] These two groups achieved a one-step total synthesis of cannabichromene 144 and cannabicyclol (148) by heating citral (36) and olivetol (6) in the presence of pyridine. From this reaction the tetracyclic ether (147, also called "cannabicitran") and iso-THCs were also isolated. They showed that the reaction between substituted resorcinols or phloroglucinol and citral in pyridine to form ethers of type 147 is a general one which leads to substituted iso-THC derivatives.[78b, 79b] The most likely mechanism[78b, 80] for the pyridine catalyzed reaction appears to be the initial condensation to 145 followed by cyclization to cannabichromene (144). This can either form cannabicyclol (148) or proceed via an intramolecular [2 + 4] cycloaddition within a quinone-methide tautomer 146 of the phenol, to 147. The mechanism of the cyclization of 144 to 147 was previously considered to be ionic but Crombie and co-workers[81] have now shown it to be electrocyclic.

The yield of cannabichromene from citral reaction has been improved by using t-butylamine in place of pyridine and refluxing the mixture in toluene.[82]

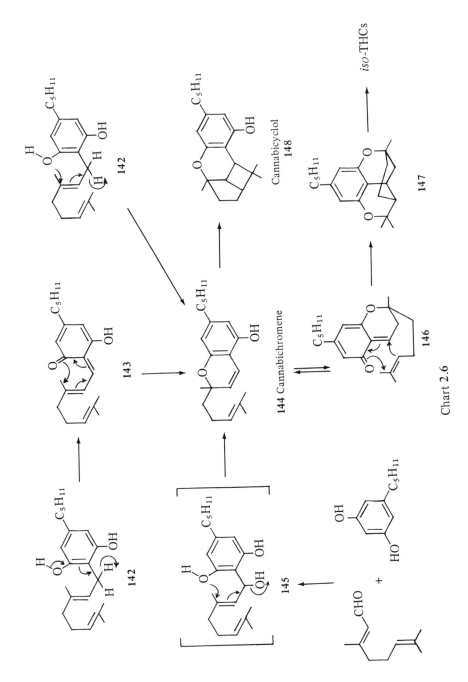

Chart 2.6

G. Cannabicyclol

It is a very minor compound of the plant and was initially assigned structure
149. Crombie and Ponsford[78a] revised the structure to **148** mainly on the basis
of NMR spectrum where the C-3 proton appears as a doublet being coupled to
one proton only. This structure has since been firmly established on the basis of
X-ray results published by Begley et al.[83] However, reading their paper gives the
incorrect impression that we (Kane and Razdan)[79a] preferred the original structure **149** for cannabicyclol in spite of conclusive evidence. These authors placed
our work completely out of context and ignored the fact that most of the
evidence they gave appeared after the publication of our paper.[79a] This clarifica-
tion* is necessary, as it can be misleading, particularly to those unfamiliar with
the field.

The first synthesis of cannabicyclol was achieved by the pyridine catalyzed
condensation of citral and olivetol as described before (Chart 2.6). Treatment of
144 with either $BF_3 \cdot Et_2O$[84] or photolysis in the presence of a sensitizer[85]
(*t*-butanol-acetone) also formed cannabicyclol.

---

*In 1968 when Crombie and Ponsford's paper[78a] appeared, in which they had suggested a
revision of cannabicyclol structure from **149** to **148**, our paper[79a] had already been submitted and in a footnote we suggested "that in the absence of further experimental data the
structure and stereochemistry as suggested by Korte and Mechoulam should not be dis-
carded *at the present time*." In our subsequent paper[79b] in early 1969, we depicted
structures **149** and **148** for cannabicyclol as either of the structures were acceptable to us
in the absence of more definitive evidence. We based our position on the fact that we were
unable to convert cannabichromene to cannabicyclol with pyridine, in contrast to Crombie
and Ponsford's results.[78a] The conversion of cannabichromene to cannabicyclol by heat was
not mentioned in the reference quoted by these authors and conversion under acid conditions appeared much later. The photochemical conversion in the presence of a sensitizer
(*tert*-butanol-acetone) supported structure **148** but did not prove it unequivocally.

Furthermore, the NMR in this series of compounds is very complex particularly in the
region where the benzylic proton appears and heavy reliance on this data can sometimes be
misleading, e.g., the benzylic proton in

shows up as a sharp singlet (220 Mc/s, $CDCl_3$ 2500 Hz sweep width) at 2.8 ppm. However,
at a 1000 Hz sweep width it shows signs of an ill-resolved multiplet.[24] Another example is
the case of iso-THC[31] which was previously misinterpreted as Δ^6-3,4-*cis*-THC. Yet another
example is the wrong assignment of the benzylic protons in Δ^6-THC [Archer et al. *J. Am.
Chem. Soc.* **92**, 5200 (1970)]. It was with this background that at the time we had indicated both the structures for cannabicyclol in the absence of definite proof such as now
provided by X-ray.

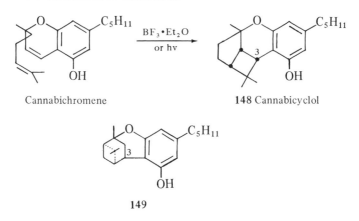

Cannabichromene   148 Cannabicyclol

149

H. Novel Cannabinoids

Cannabielsoic acids (151a, b), which have been isolated from hashish,[86] represent a novel cannabinoid structure. The decarboxylation product of these naturally occurring consititutents is cannabielsoin (152), which is also the major product obtained by pyrolysis[87] of cannabidiol in air at 70°C. The synthesis of cannabielsoic acid A (151a) was achieved by Shani and Mechoulam[86] from cannabidiolic acid (150) by a novel photooxidative cyclization process (Chart 2.7). A mixture of 151a and its isomer at C-1 was obtained.

Uliss et al.[34] reported (Chart 2.7) a stereochemically unambiguous synthesis of cannabielsoin (152) from the epoxide 156. Thus cannabidiol diacetate 153 gave a mixture of epoxides 154, 155, and 156 when allowed to react with m-chloroperbenzoic acid. These were separated and the assignment of an α-configuration to the endocyclic epoxides in 155 and 156 was established. Treatment with base at room temperature converted 156 to 152. This transformation involves an intramolecular trans diaxial cleavage of the α-epoxide at its less hindered site, which fixes the stereochemistry of the fused furan ring at C-2 and C-3 as cis and the configuration of the C-1 hydroxyl group as α (axial). This established the stereochemistry of cannabielsoin at C-1 so as to conform to structure 152. Independently Shani and Mechoulam[86] arrived at similar conclusions while working on the cannabielsoic acid series. Similar treatment with base gave the novel cannabinoids 157 and 158 from 154 and 155, respectively. These studies[34] have also suggested that the entropy of ring formation is the major factor in determining the product of an intramolecular epoxide cleavage. (−)-8β-Hydroxymethyl-Δ^1-THC (157)[88] is equiactive with Δ^1-THC in biological potency and represents the first example of functionalization in the geminal methyl part of the molecule of Δ^1-THC.

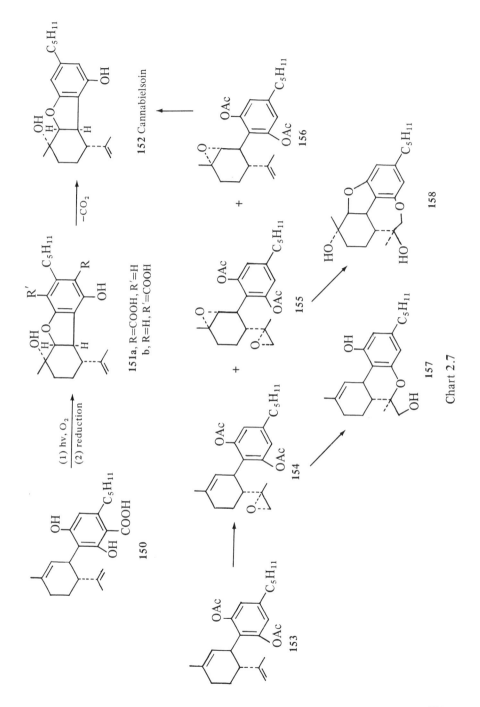

151a, R=COOH, R′=H
b, R=H, R′=COOH

Chart 2.7

231

During their studies on cannabielsoin (152) Uliss et al.[89] also synthesized (Chart 2.8) a novel cannabinoid 159 containing a 1,8-cineole moiety. It was formed in quantitative yield from 152 by an intramolecular cyclization on treatment with a catalytic amount of p-TSA. This conversion serves as confirmatory evidence for the stereochemical assignment of the C-1 hydroxyl group as α (axial) in 152.

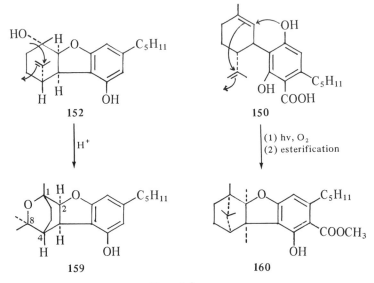

Chart 2.8

Another cannabinoid containing a camphane moiety 160 was synthesized by Kirtany and Paknikar[90] by irradiation of 150 in the presence of oxygen followed by esterification (Chart 2.8).

Ciommo and Merlini[91] reported the synthesis (Chart 2.9) of cannabinoid-like benzoxocinols 163 and 164 by condensing p-menth-3-en-8-ol (161) or p-mentha-3,8-diene (162) with olivetol (6) in the presence of HCOOH. It was shown that the reaction between 161 and 6 proceeds via dehydration of 161 to 162.

Compound 163 was also synthesized (Chart 2.9) by Houry et al.[92] from carvone and olivetol using $POCl_3$ followed by reduction. They also prepared 164 from limonene or pinene. Compound 163 was reported to be biologically active.[92]

Citronellal and phloroacetophenone were condensed in the presence of pyridine to give hexahydrocannabinoid analogs.[93] This reaction is a variation of the citral-olivetol-pyridine reaction discussed earlier (see Section 3F; Chart 2.6).

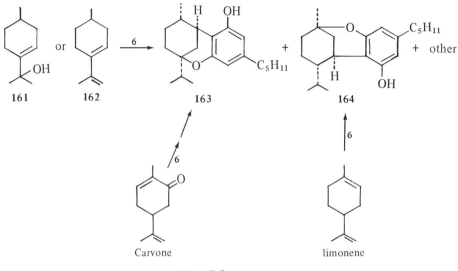

Chart 2.9

In recent years interest has focused on 1-ketocannabinoids because nabilone,[94] a member of this class, has been undergoing clinical trials as an antinausea and antiglaucoma agent. Wilson and May[95] oxidized the exocyclic double bond of (−)-$\Delta^{1(7)}$-THC (13) with OsO_4 and then cleaved it with $NaIO_3$ to give (−)-167 (R = C_5H_{11}). Archer et al.[94] prepared nabilone [167, R = $C(CH_3)_2C_6H_{13}$] from the corresponding $\Delta^{1(7)}$-THC by ozonolysis. Since the overall yield was low they developed a different route to 1-ketocannabinoids (Chart 2.10). The known (+)-apoverbenone (166)[96] was prepared from β-pinene via ozonolysis to nopinone (165) followed by bromination and dehydrobromination. Reaction of 166 with the appropriate resorcinol in the presence of anhydrous $AlCl_3$ gave (−)-nabilone. Alternatively, 165 was converted to the enol acetate 168,[97] which on oxidation with lead tetraacetate in refluxing benzene for 2 h gave 170. If the reflux time was extended to 18 h, 169 was isolated. Either of them on condensation with the resorcinol derivative and p-TSA formed 171. Stannic chloride treatment converted 171 to 167. On the other hand, 171 on treatment with p-TSA gave the optically active cis-ketone 172. The transformation of cis-172 to trans-167 was accomplished by treatment with $AlCl_3$ at 0°C.

For the preparation of racemic 1-keto cannabinoids the known optically inactive diene 173[98] was condensed with olivetol or the other appropriate resorcinol in the presence of acid catalysts (Chart 2.11). The products of the reaction varied with the nature of the catalyst, solvent, temperature, and reaction time. Compounds 174 and 175 appear to be intermediates in the formation of the cis-ketone 172 following the sequence 174 → 175 → 172. Since 167[16]

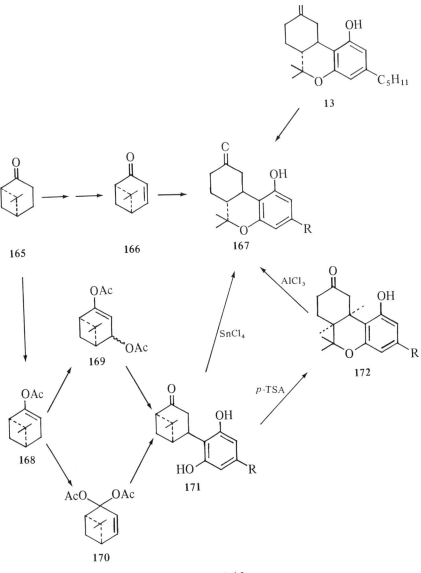

Chart 2.10

has been previously converted into (±)-Δ¹-THC (Chart 1.13) this represents a new synthesis of THCs.

A variation of this scheme included the use of "masked ketones" such as **176** and **177** instead of the diene **173**. Both formed **167** (R= C_5H_{11}) in 80% yield.

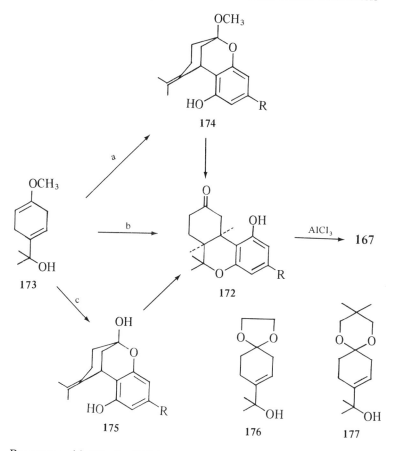

Reagents: (a) $BF_3 \cdot Et_2O/C_6H_6$, RT, 6h; (b) $SnCl_4/CH_2Cl_2$, 0°C, 7h; (c) $BF_3 \cdot Et_2O/CH_2Cl_2$, 0°C, 7h.

Chart 2.11

The C-glucosidation of Δ^6-THC in the 3'-position has been reported[99] pre-viously by treating Δ^6-THC with β-glucose pentaacetate in benzene containing $BF_3 \cdot Et_2O$. The formation of a C-glucoside in the presence of a phenolic group was unexpected but is not surprising in view of the reactivity of THCs toward electrophilic attack on the aromatic ring (e.g., formation of THC acids using methylmagnesium carbonate; Chart 2.5). Using the same procedure, the C-glucuronide of Δ^6-THC in the 3'-position was prepared.[100] On the basis of the large coupling constant observed (J = 10 Hz) for the C-1' H in the sugar moiety it was considered to be a β-glucuronide.

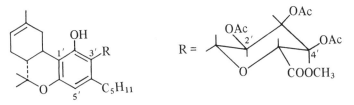

Δ^6-THC-C-3'-glucuronide methyl ester triacetate

4. NEW CANNABINOID TRANSFORMATIONS

Many interesting isomerizations and transformations have been observed in cannabinoids, due to the presence of double bonds, free phenolic groups and *cis* or *trans* ring junctions. The isomerizations and interconversions in the presence of acid catalysts of Δ^1- to Δ^6-THCs in the *trans* series and *cis*-THCs → tetra-cyclic ether (cannabictran, **147**) → iso-THCs in the *cis* series, are well established and have been discussed in detail in previous articles.[3] In this section some new transformations, which have since appeared, are described.

A. Photochemical

Bowd et al.[101] carried out the ultraviolet irradiation of cannabinol (**136**) in ethanol and showed (Chart 3.1) that cannabinodiol (**178**), a minor constituent of cannabis, is first formed by a photoinduced ring opening and hydrogen

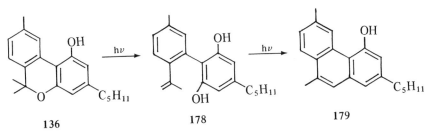

Chart 3.1

transfer. This in turn undergoes a photoinduced dehydration and ring closure to give the highly fluorescent hydroxyphenanthrene **179**. The conversion of **136** to **178** is analogous to a previously reported[102] photochemical transformation of 2,2-disubstituted chromenes.

B. cis → trans Conversion

Uliss et al.[103] have elucidated the mechanism of the conversion of 3,4-*cis*- to 3,4-*trans*-cannabinoids. In 1969 Razdan and Zitko[104] reported the first example of the conversion of a *cis*- to a *trans*-THC. They found that (±)-Δ^1-3,4-*cis*-THC (**3**) was converted to (±)-Δ^6-3,4-*trans*-THC (**2**) on treatment with BBr₃. In a more recent example Archer et al.[94] have reported the conversion of a 3,4-*cis*-1-ketocannabinoid to its *trans* counterpart using AlCl₃ (Chart 2.10).

At the time Razdan and Zitko[104] had proposed that this transformation involved cleavage of the ether bond followed by probable inversion at C-4 rather than at C-3. With the recent availability[35] of (+)-Δ^1-3,4-*cis*-THC (**3**) of known absolute configuration (3*S*, 4*R*) by synthesis from 1*S*, 2*S*, 3*R*, 6*R*-carene-2-oxide (**21**, Chart 1.17) and olivetol, the conversion of *cis* to *trans* was reinvestigated. Inasmuch as the products formed by epimerization of **3a** at C-3 (Δ^6-3*R*, 4*R*-THC, **2b**) and at C-4 (Δ^6-3*S*, 4*S*-THC, **2a**) are enantiomers, preference for epimerization at either site will be reflected in the sign and magnitude of the optical rotation of the product. Hence a sample of **3a** of known optical purity was subjected to BBr₃ treatment and from the rotation of the product (Δ^6-*trans*-THC, 33% yield) it was found that it corresponds to a mixture of 24% enantiomer **2b** and 76% enantiomer **2a**. It was thus shown that epimerization at C-4 is the favored process and is accompanied by a lesser amount of C-3 epimerization or racemization. Supportive evidence was also provided, which indicates that the first step is cleavage of the pyran ring (Chart 3.2). The resulting equilibrium is driven to the relatively stable isomer **2a** via epimerization at C-4. Friedal-Crafts cleavage of the C-3-C-1′ bond in **126a** or **180a** followed by recombination leads to C-3 epimerization in the former and racemization of the latter.

Similar studies on *cis*-hexahydrocannabinols (HHCs) were also undertaken[103] to determine if removal of the carbocyclic unsaturation affects the stereochemical outcome of the conversion. It was shown that the conversion of 3,4-*cis*- to 3,4-*trans*-HHCs proceeds with exclusive C-4 inversion.

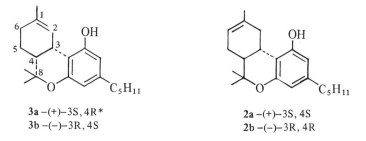

3a −(+)−3S, 4R *
3b −(−)−3R, 4S

2a −(+)−3S, 4S
2b −(−)−3R, 4R

*The **a** refers to the compound shown and **b** to its enantiomer.

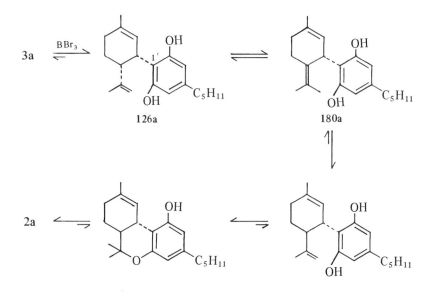

Chart 3.2

C. Pyrolysis

Since cannabis is generally ingested by smoking, the products formed on pyrolysis of cannabinoids are of general interest. Salemink and co-workers[105] have studied the pyrolysis of cannabidiol in detail and have isolated several

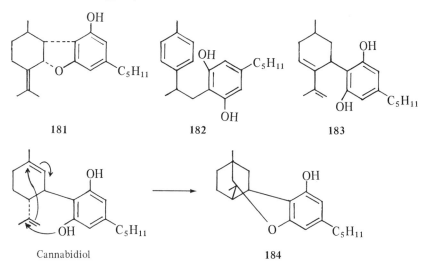

Chart 3.3

transformation products. They have observed that the nature of the gas phase considerably influences the nature of the pyrolytic products and have identified Δ^1-THC, cannabinol, several aromatic compounds, for example, olivetol, 2-methylolivetol and 2-ethylolivetol; *abn*-cannabidiol (14); cannabielsoin (152);[105a] a dibenzofuran 181[105d] in which there has been a rearrangement of the carbon skeleton (Chart 3.3); 182;[105e] a new cannabidiol isomer 183[105b] and the bicyclic cannabinoid 184.[105b] The structures of 183 and 184 were based on their mass spectral fragmentations only. Supportive NMR evidence was provided for 181 whereas 182 was confirmed by synthesis.[105f] In addition the isolation of 4-hydroxy-6-pentylbenzofuran and 2,2-dimethyl-5-hydroxy-7-pentylchromene were reported and confirmed by synthesis.[105f]

5. SYNTHESIS OF THC ANALOGS[106]

The cannabis plant, being a complex mixture, is unlikely to be used as a marketable drug; therefore, the future of therapeutic agents from cannabinoids undoubtedly lies in the synthetics.

This has led to intense interest in structural modification of Δ^1-THC and synthetic Δ^3-THCs. In addition a large variety of heterocyclic analogs have been prepared. All these modifications have resulted in a series of novel THC derivatives and analogs, which show a wide variety of enhanced activities such as antiglaucoma, antinausea, analgesic, tranquilizer, antihypertensive, etc. Like morphine, lysergic acid diethylamide, and cocaine, which have structurally related analgesics, oxytoxics, and local anaesthetics, respectively, the socially abused cannabinoids may now be on the verge of generating a family of safer and more useful therapeutic agents.

A. Carbocyclic Analogs

Compounds Related to Δ^1- and Δ^6-THCs

Some of these compounds are shown in Chart 4.1. Apart from the various metabolites of Δ^1- and Δ^6-THCs that have been synthesized, efforts have been directed to modify the THC structure to make water soluble derivatives. This is deemed necessary, since Δ^1- and other THCs are resinous materials completely insoluble in water, and to carry out pharmacological studies these materials have to be administered in various solvents such as polyethylene glycol, Tween, triton, and alcohol, which are themselves not without pharmacological activity. An aminoalkyl ester derivative of THCs is an obvious choice, but until recently conventional methods of esterification were not successful in its preparation. Razdan and co-workers[107] have shown that water-soluble esters of Δ^1- and other THCs can be easily prepared with carbodiimide as the condensing agent (Chart

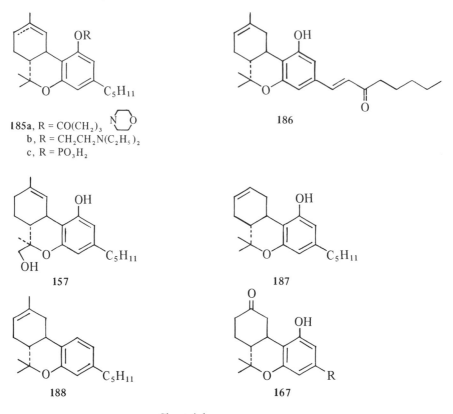

185a, R = CO(CH$_2$)$_3$ —N⟨O⟩
   b, R = CH$_2$CH$_2$N(C$_2$H$_5$)$_2$
   c, R = PO$_3$H$_2$

186

157

187

188

167

Chart 4.1

4.2). Thus the γ-morpholinobutyric ester (185a·HBr) of Δ^1-THC is a solid, freely soluble in water and is equiactive with Δ^1-THC in various pharmacological tests. In contrast, the ether derivative 185b[107a] is quite different and does not show the pharmacological profile of Δ^1-THC.

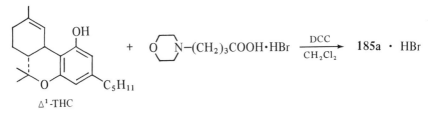

Δ^1-THC   +   O⟨N⟩—(CH$_2$)$_3$COOH·HBr $\xrightarrow[\text{CH}_2\text{Cl}_2]{\text{DCC}}$ 185a · HBr

Chart 4.2

A biologically active, water-soluble phosphate ester of Δ^6-THC (**185c** as the sodium salt) has also been reported[108] recently. It was synthesized (Chart 4.3) by treating Δ^6-THC with phosphoryl chloride in pyridine followed by mild hydrolysis with water. Treatment with alcoholic sodium hydroxide gave the sodium salt of **185c**.

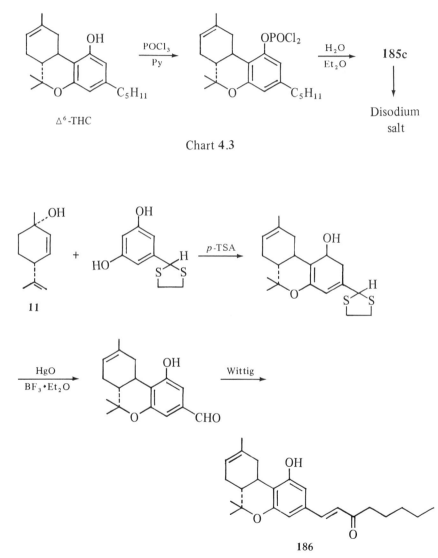

Chart 4.3

Chart 4.4

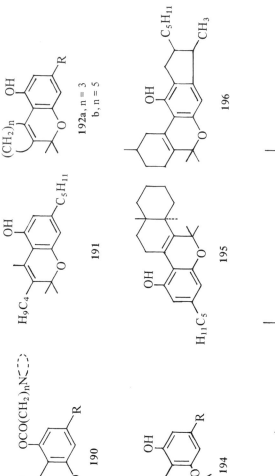

192a, n = 3
b, n = 5

191

190

189

196

195

194

193

200

199

198

197

Chart 4.5

The potent analog **186** was synthesized by Razdan et al.[58] as shown in Chart 4.4. The resorcinol thioketal was prepared from the corresponding aldehyde.

As described earlier (Section 3H; Chart 2.7) the potent Δ^1-THC analog **157** was synthesized[34] from cannabidiol diacetate.

The Δ^6-THC analog **187** was synthesized[95] from the 1-keto compound **167** (R = C_5H_{11}) by reduction and dehydration with *p*-TSA. This compound was shown to be equiactive with Δ^6-THC.[95] Compound **188** was prepared[109] from Δ^6-THC by treatment with $C\ell PO(OC_2H_5)_2$ followed by cleavage with Li/NH_3.

The synthesis of 1-keto analog **167** has already been described in detail (see Section 3H; Charts 2.10 and 2.11).

Compounds Related to Δ^3-THC

Some examples of this type are shown in Chart 4.5. These compounds differ from the natural THCs **1** and **2** in the relative position of the double bond and show typical marijuana-like activity in rodents and dogs. They were discovered[106] independently by two groups, Adams and co-workers in the United States, and Todd and co-workers in Great Britain, in the course of their work on the structure elucidation of the active constituents of hashish and marijuana. Todd and Adams in particular carried out extensive structure-activity relationship (SAR) studies in Δ^3-THCs. Some of these are depicted by the general formula **189**, **191**, and **192** and have been reviewed in detail by Mechoulam[3b,c] and Pars et al.[106a] They were synthesized according to the general scheme shown in Chart 4.6. A Pechmann condensation between the appropriate keto ester and

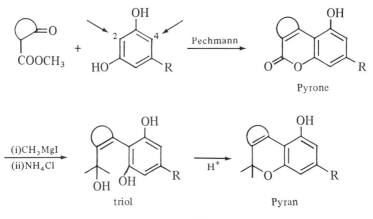

Chart 4.6

the resorcinol formed the pyrone. Generally the major product was formed by attack on the 2-position of the resorcinol. The isomers (attack on the 4-position) are also formed and their amount depends on the reaction conditions and the nature of the substituent R in the resorcinol. Treatment of the pyrone with excess CH_3MgI gives the triol, which on treatment with catalytic amounts of acids (e.g., p-TSA) ring closes to the pyran.

Various water-soluble aklyl amino esters of type **190** and of the cyclopenteno derivative **192a** were prepared[107b] from the pyrans by using the appropriate acid and carbodiimide, as in the case of Δ^1-THC. Compounds of type **193** were prepared by oxidation of the pyrane as its acetate, with ceric ammonium nitrate.[110]

The novel cannabinoids with a homopyran ring of type **194** were reported by Matsumoto et al.[111] Their synthesis is shown in Chart 4.7. The pyrone **201** was reduced with Red-Al to the triol **202a**. Under mild alkaline conditions it was converted to the dibenzyl derivative **202b**. After bromination with PBr_3 and further treatment with NaCN/DMSO it formed the nitrile, which was debenzylated under mild hydrogenation conditions to **203**. Conversion of nitrile **203** to the ester **204** proceeded smoothly with HCl/EtOH. Grignard reaction with CH_3MgBr followed by treatment with ethanolic HCl gave the oxepin **194**. Some other related oxepins were also prepared by these authors.

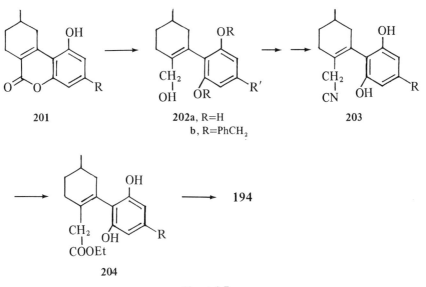

201   202a, R=H   203
                       b, R=PhCH$_2$

204

Chart 4.7

Razdan et al.[112] synthesized the steroidal analog **195**. They also prepared **196**.[113] Both these analogs were prepared using the general scheme shown in Chart 4.6. Similarly the analogs **197** and **198** were synthesized by Malik et al.[114] In the case of **197**, the corresponding pyrone was first reduced with Li/NH_3 to avoid aromatization of the alicyclic ring. Grignard reaction and ring closure with acid gave the desired pyran **197**.

The ring-opened analog **199** was prepared by Razdan et al.[115] by Li/NH_3 treatment of the corresponding pyran. The ring opening was unexpected as no reduced pyrans were formed. It shows a very low extinction coefficient in the UV spectrum, λ_{max}^{EtOH} 275 nm (ϵ 1600), in spite of the conjugation present, indicating that the alicyclic ring is out of plane from the aromatic ring. It has potent marijuana-like activity unlike other ring-opened cannabinoids, for example, cannabidiol.

Loev et al.[116] prepared the analog **200** by $LiAlH_4$ reduction of the corresponding pyrone to the allylic alcohol **202a** ($R' = C_9H_{19}$) followed by ring closure with p-TSA. The biologically active benzoxocinols **163**, **164** have already been discussed (section 3H; Chart 2.9).

B. Heterocyclic Analogs

Interest in preparing nitrogen analogs of THCs was stimulated by the observation that THC is one of the very few potent drugs that act on the central nervous system (CNS) yet has no nitrogen in its structure. As long ago as 1946 Anker and Cook[117] synthesized the nitrogen analog **205** and its dihydro derivative and found them to be without analgesic activity. Other types of CNS activity were not mentioned in their report. They prepared **205** by following the general scheme as used for the synthesis of Δ^3-THCs (Chart 4.6).

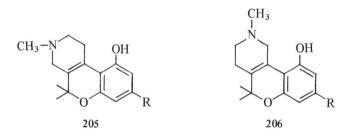

205   206

In 1966 Pars et al.[118] reported the synthesis of the nitrogen analog **206**, which showed marijuana-like pharmacological profile. Since that time various nitrogen analogs,[106, 107b, 119, 120] type **205**, **207** to **212** (Chart 4.8), and sulfur analogs[106, 120-122] (Chart 4.10), all showing varying amounts of CNS activity in laboratory animals, have been synthesized.

The sequence of reactions for the synthesis of **207** ($R = C_9H_{19}$), shown in Chart 4.9, illustrates the general procedure used in the preparation of N-substituted nitrogen analogs of type **207**. The synthetic scheme is an adaptation of the general scheme used for the synthesis of Δ^3-THCs (Chart 4.6). The pyrone

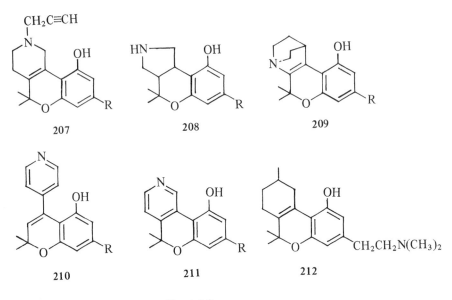

Chart 4.8

215 was obtained from the keto ester **213** and the resorcinol **214**. Best results in this Pechmann condensation were obtained by using a mixture of POCl$_3$/H$_2$SO$_4$. The Grignard addition was carried out in anisole and during workup the pyran **216** was formed. Debenzylation of **216** with Pd/C and H$_2$ formed the nor-base **217**. Alkylation with propargyl bromide-Na$_2$CO$_3$ in EtOH furnished the desired compound **207**. A similar sequence was used for the synthesis of analogs **208**, except that during the debenzylation step the double bond was also reduced.

Compounds **209**, **210**, and **212** were prepared following the general scheme in Chart 4.6. Compound **211** was prepared by dehydrogenation of **207** (R = C$_9$H$_{19}$) with 10% Pd/C in xylene.[123]

Surprisingly, all these nitrogen analogs as their acid addition salts were not very water soluble. Hence, like the carbocyclic analogs, they were made water soluble by making their alkylamino ester derivatives. A large number of water-soluble derivatives were prepared and studied by Razdan et al.[107b]

The sulfur analogs[121,122] **218** to **222** (Chart 4.10) were prepared according to the general scheme in Chart 4.6. In the sulfur series the Pechmann reaction was best carried out in the presence of HCl/EtOH. The sulfur analog **223** was prepared from **218** by dehydrogenation.

Recently Cushman and Castagnoli[124] reported the synthesis (Chart 4.11) of biologically active nitrogen analogs **230** having a *trans* ring fusion similar to that

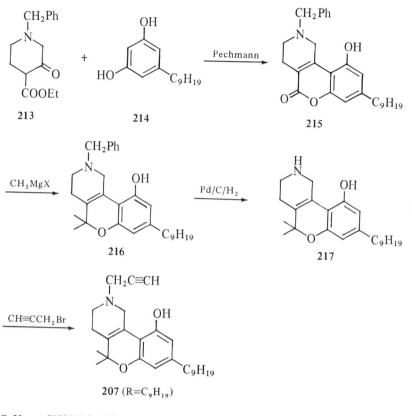

$C_9H_{19} = CH(CH_3)CH(CH_3)C_5H_{11}$

Chart 4.9

found in the natural THCs **1** and **2**. Their novel approach to the synthesis of these analogs untilizes the condensation of the Schiff base **225** with glutaric anhydride to give predominantly the *trans*-piperidone **226a**. The ester **226b** on Grignard treatment followed by demethylation with BBr_3 and subsequent dehydrohalogenation, afforded the olefin **227**. After ring closure with $BF_3 \cdot Et_2O$, compound **228** was treated with CH_3MgBr and then dehydrated to form the enamine **229**. Catalytic reduction of **229** gave a diastereomeric mixture of amines **230**. Preliminary pharmacological results showed the mixture to be active.

Several other nitrogen analogs have been prepared, and in most cases their biological activity has not been reported.

218

219

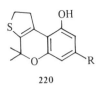

220

221

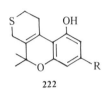

222

223

R = CH(CH₃)CH(CH₃)C₅H₁₁

Chart **4.10**

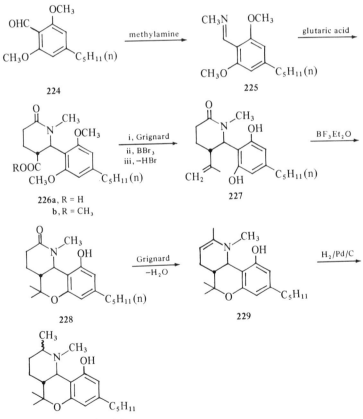

224 → methylamine → **225** → glutaric acid →

226a, R = H
  b, R = CH₃

i, Grignard
ii, BBr₃
iii, −HBr

→ **227** → BF₃·Et₂O →

228 → Grignard, −H₂O → **229** → H₂/Pd/C →

230 diastereomeric mixture

Chart **4.11**

248

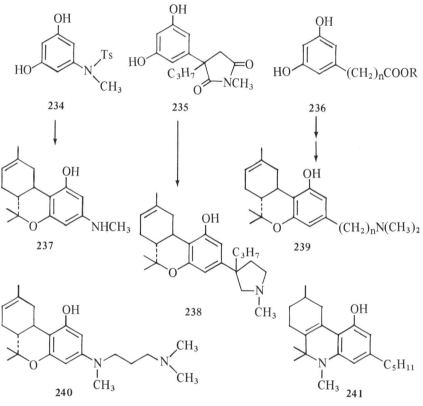

231   232   233a, X = CH
b, X = N

234   235   236

237   238   239

240   241

242

Chart **4.12**

Condensation of **231** (Chart 4.12) with benzamidine furnished **232**, whereas condensation with the appropriate diamine followed by thermal ring closure in vacuum gave **233**.[125] The resorcinols **234** and **235** were condensed with *p*-menthadienol **11** and then reduced to yield **237**[126] and **238**,[127] respectively. Similarly **236** (R = CH_3 or C_2H_5) condensed with **11** to give the corresponding ester, which on treatment with dimethylamine followed by $LiAlH_4$ reduction formed the analogs **239** (*n* = 2 to 5).[128] The analog **240**[126] was obtained from **237** on treatment with dimethylaminopropyl chloride in the presence of butyl-lithium. No biological data were reported for any of these compounds.

Another nitrogen analog **241** (Chart 4.12), where the pyran oxygen is replaced by NCH_3, was synthesized and reported to be inactive.[129] However, a related analog **242** has been reported[130] to be very active biologically, but no details of its synthesis have been described.

The novel analog **243** was synthesized[131] according to the scheme in Chart 4.13. The Diels-Alder reaction is used to give the *trans* ring junction as in the synthesis of Δ^6-THC (see Chart 1.9).

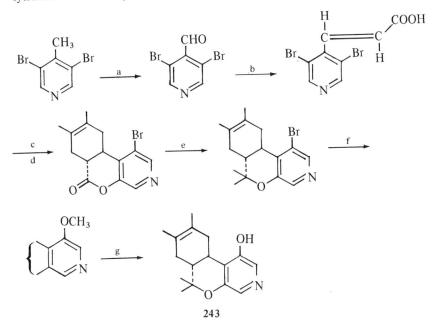

243

Reagents: (a) SeO_2-DMSO   (b) $CH_2(COOH)_2$ (c) CH_3OH/H^+
(d) , Dioxan at 120°C   (e) CH_3MgI, Δ (f) $NaOCH_3/CH_3OH$
/DMF/CuI   g, $NaSC_2H_5/DMF, H^+$

Chart 4.13

The analogs where the phenolic hydroxy group is replaced by SH (**244**) or NH$_2$ (**245**) were synthesized by Matsumoto et al.[132] (Chart 4.14). They showed that the amino analogs retained pharmacological activity, but the sulfur analogs were relatively inert.

Other sulfur analogs include **247**, **248**, and **249**. They were prepared[133] from pulegone (**137**) by a base-catalyzed reaction with the thiophenol (Chart 4.15) followed by the ring closure to **246**. Demethylation of the methoxyl gave **247**, which on dehydrogenation gave the cannabinol analog **248**. Sulfones **249** were also prepared by *m*-chloroperbenzoic acid oxidation of **246**. No biological data were reported.

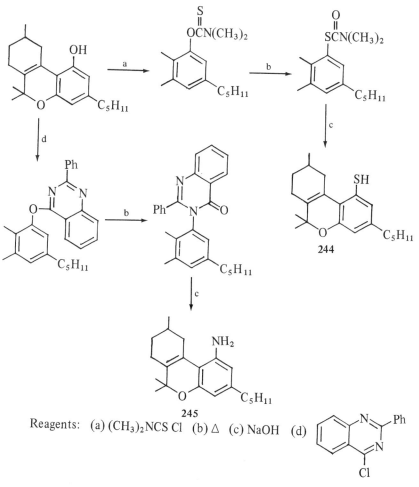

Reagents: (a) (CH$_3$)$_2$NCS Cl (b) Δ (c) NaOH (d)

Chart **4.14**

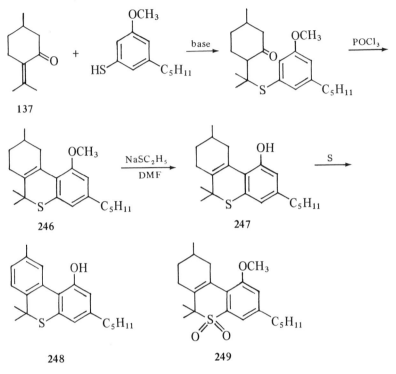

Chart 4.15

6. OVERALL STRUCTURE-ACTIVITY RELATIONSHIPS IN CANNABINOIDS

In various animals Δ^1-THC and other synthetic THCs show predominantly central nervous system (CNS) depression and ataxia, which lasts from several hours to days, depending on the dose administered. The characteristic effect of THCs, which distinguishes them from all other pyschoactive drugs, is a postural arrest phenomenon with relaxed staring and associated hypersensitivity to external stimuli.[134] For example, when Δ^1-THC is given at a dose of 0.2-0.5 mg/kg (i.v.) to dogs, they stand in a trancelike state, sway from side to side, pitch forward and backward, and overreact to a swinging object. When aroused, ataxia is evident for 3 to 4 hours after the injection. It is characteristic for the dogs to urinate and defecate soon after receiving the drug and sleep a great deal for the next 24 hours.

The synthesis and demonstration of CNS activity for a wide variety of cannabinoids has resulted in an expansion of the structure-activity (SAR) conclu-

sions originally put forth by Roger Adams and co-workers.[135] By using dog ataxia as the basis for THC activity, Adams found that the potency is increased when R is a highly branched alkyl with the 1,2-dimethylheptyl (C_9H_{19}) showing optimum activity. He also showed that when R_1 is methyl the activity is greater than when these (R_1) substituents are higher alkyls. Reduction of the double bond in the C-ring retained activity, and the C-ring could be contracted, expanded or even opened without entirely eliminating activity.
eliminating activity.

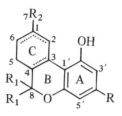

Recent studies of various metabolites and other synthetics have expanded these SAR observations. Thus based on CNS pharmacological profiles in laboratory animals, the SAR picture can be summarized as follows.

1. Essentially a benzopyran structure with an aromatic hydroxyl group at 2'-position and an alkyl or alkoxyl group on the 4'-position are a requirement for activity.

2. The position and the environment around the aromatic hydroxyl group are very important for the activity, viz.;

   a. The OH at position C-2' is in itself necessary for CNS activity.

   b. Esterification of the phenol retains activity and in some carbocyclic and heterocyclic benzopyrans, can lead to greater selectivity of action. Etherification of the phenol eliminates activity. Replacement of the OH by NH_2 retains, whereas by SH, eliminates activity.

   c. Methyl substituents at C-2 in the C-ring significantly alter the activity of both carbocyclic and (C-ring) heterocyclic benzopyrans particularly in the case of planar five-membered C-rings.

3. Substitution in the aromatic ring by electronegative groups like carboxyl, carbomethoxyl, acetyl eliminates activity, whereas alkyl groups in C-3' position retain, and in C-5' position reduce, activity.

4. A minimum length of the aromatic side chain is necessary to elicit activity. The branching of the alkyl side chain increases potency. Thus 1,2-dimethylheptyl or 1,1-dimethylheptyl gives the most potent compounds. Similarly p-fluorphenyl alkyl and side chains as shown give good activity.

$-O-CH(CH_3)(CH_2)_3Ph,$ and

5. On ring B the *gem*-dimethyl group at C-8 (R_1) is optimum for activity. Replacement of one of the R_1 substituents on the B ring with a hydroxymethyl group retains activity. Replacement of pyran O by N and ring expansion of ring B by one carbon can retain activity.

6. In the alicyclic ring C, compounds with the double bond in the Δ^1-, Δ^6-, or Δ^3-position are active. A 3,4-*trans* junction increases and a *cis* junction decreases activity. The natural THCs are active in the 3R, 4R series only. A methyl at C-1 increases activity, but metabolism to the 7-hydroxymethyl is not a prerequisite for THC activity.

7. The C-ring can be substituted by a variety of nitrogen and sulfur-containing rings without loss of CNS activity. With the nitrogen and sulfur analogs the most active CNS agents are obtained when the heteroatom is in a phenethyl orientation, e.g., inserted in place of C-1 or C-5.

8. Planarity of the C ring is not a necessary criterion for activity. See, for example, the quinuclidine analog (**209**, chart 4.8) and the benzoxocine compounds (**163, 164**; Chart 2.9).

9. In both carbocyclic and heterocyclic analogs, opening the pyran ring generally decreases activity. An exception is compound **199** (Chart 4.5), which is approximately equiactive with Adams DMHP (Chart 6.1).

7. THERAPEUTIC INDICATIONS AND POTENTIAL OF NEW DRUGS FROM CANNABINOIDS[136]

As discussed in the Introduction, the main therapeutic indications for Δ^1-THC have emerged from folklore anecdotes. The use of Δ^1-THC, as an antinauseant to patients undergoing cancer chemotherapy, and its utility as an antiglaucoma agent, is now well established clinically. In many states in the United States and in some foreign countries, a strong movement in favor of legalizing the use of Δ^1-THC for these purposes has developed. This is mainly because Δ^1-THC is more effective in controlling nausea than presently available drugs. In the case of glaucoma where patients tend to become refractory to the drug in use, the addition of a new class of drug which presumably acts by a novel mechanism is of great interest. The clinical activity of Δ^1-THC in these two fields has led investigators to use synthetics for these indications. Presently three such drugs are being actively pursued in the clinic. These are shown in Chart 6.1; a carbocyclic 1-ketocannabinoid, Nabilone,[137] and two nitrogen analogs, Nabitan,[107b, 138] a water-soluble derivative, and A-41988.[139, 140]

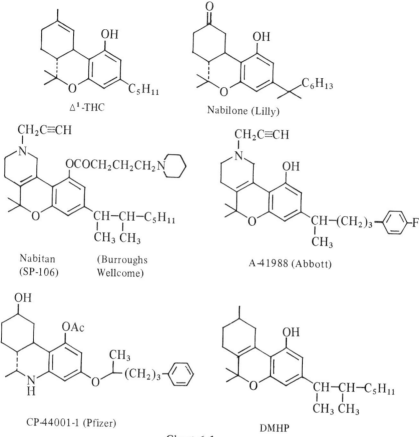

Chart **6.1**

Animal studies to date, both in heterocyclic and carbocyclic analogs, clearly point to therapeutic potential as analgesics. Preliminary studies in man have already given some indication of analgesic potential for both Δ^1-THC[141] and Nabitan (SP-106).[142-144] Since establishment of a new type of analgesic in the clinic is not simple due to subjective and placebo effects, the studies with Nabitan are continuing and as such the place of cannabinoids as analgesics has not yet been firmly established. Recently another nitrogen analog CP-44001-1 (Nantradol) has been reported[130] to be a potent analgesic in animals and is presently undergoing clinical trials.

The recent finding[145, 146] that DMHP, a carbocyclic analog originally synthesized by Roger Adams in the 1940s, shows hypotensive effects at doses where no CNS effects occur, is noteworthy and, it is hoped, will open up a fruitful area for cardiovascular drugs. In addition a preliminary study has indicated the use of Nabilone as a tranquilizing agent.[137b]

Based on a few clinical studies with Δ^1-THC there are indications of its use as an antiasthmatic[147] and as an appetite stimulant.[148]

A recent survey[149] suggests a high incidence of marijuana use among young epileptics. Their observations of its efficacy in alleviating their symptoms, like those of the glaucoma and cancer patients, may point to another area where clinical utility should be focused. There is already some indication clinically[150, 151] to supprot this view and animal studies clearly point to thereapeutic potential in this area.

Other areas of interest, supported by animal pharmacology, are antiinflammatory, antipyretic, antitussive, antispasmodic, antidiarrheal, sedative-hypnotic, antidepressant, anticancer, and treatment of alcoholism and narcotic addiction.

It is obvious that Δ^1-THC and analogs display a wide range of pharmacological action. The future development of drugs from this area will undoubtedly depend on the success achieved by structural changes to provide selectivity of pharmacological action. However, it should be emphasized that the concept of drug development from THCs and cannbinoids is based on very sound foundations, since Δ^1-THC has a remarkably low toxicity in animals and humans. In addition, it has practically no respiratory-depressant activity, no or very low physical dependence liability and finally has a unique pharmacological profile compared to other psychoactive drugs.

On reflection, it is surprising that drug development from cannabinoids did not take place sooner.

REFERENCES

1. R. S. Hepler and I. M. Frank, *J. Am. Med. Assoc.*, **217**, 1392 (1971).

2. S. E. Sallan, N. E. Zinberg, and E. Frei, *N. Engl. J. Med.*, **293**, 795 (1975).

3. Review: (a) R. K. Razdan, in W. Carruthers and J. K. Sutherland, Eds., *Progress in Organic Chemistry*, Vol. 8, Butterworths, London, 1973; (b) R. Mechoulam, Ed., *Marihuana, Chemistry, Pharmacology, Metabolism and Clinical Effects*, Academic Press, New York, 1973; (c) R. Mechoulam, N. K. McCallum, and S. Burstein, *Chem. Rev.*, **76**, 75 (1976).

4. R. Mechoulam, P. Braun, and Y. Gaoni, *J. Am. Chem. Soc.*, **89**, 4552 (1967); **94**, 6159 (1972).

5. See also Ref. 25 for a synthesis by a different route.

6. H. Edery, Y. Grunfeld, Z. Ben-Zvi, and R. Mechoulam, *Ann. N.Y. Acad. Sci.*, **191**, 40 (1971).

7. Unpublished results from our laboratory.

8. J. Idänpään-Heikkilä, G. E. Fritchie, L. F. Englert, B. T. Ho, and W. M. McIssac, *N. Engl. J. Med.*, **281**, 3129 (1969).

9. R. K. Razdan, G. R. Handrick, and H. C. Dalzell, *Experientia*, **31**, 16 (1975).

10. J. J. Hurst and G. H. Whitam, *J. Chem. Soc.*, 2864 (1960).

11. W. F. Erman, *J. Am. Chem. Soc.*, **89**, 3828 (1967).

12. J. T. Pinhey and I. A. Southwell, *Austr. J. Chem.*, **24**, 1311 (1971); P. Teisseire, P. Rouillier, and A. Galfre, *Recherches, Paris*, **16**, 68 (1967).

13. C. E. Turner and K. Hadley, *J. Pharm. Sci.*, **62**, 251 (1973) and references cited therein.

14. See, for example, L. Ruzicka, *Pure Appl. Chem.*, **6**, 493 (1963).

15. T. Petrzilka, W. Haefliger, C. Sikemeier, G. Ohloff, and A. Eschenmoser, *Helv. Chim. Acta*, **50**, 719 (1967); T. Petrzilka, W. Haefliger, and C. Sikemeier, *Helv. Chim. Acta*, **52**, 1102 (1969).

16. K. E. Fahrenholtz, M. Lurie, and R. W. Kierstead, *J. Am. Chem. Soc.*, **89**, 5934 (1967).

17. Arthur D. Little, Inc., Technical Report 3, to National Institute of Mental Health, January 1972, Contract PH-43-68-1339.

18. R. K. Razdan, A. J. Puttick, B. A. Zitko, and G. R. Handrick, *Experientia*, **28**, 121 (1972).

19. R. K. Razdan, H. C. Dalzell, and G. R. Handrick, *J. Am. Chem. Soc.*, **96**, 5860 (1974).

20. R. K. Razdan and G. R. Handrick, *J. Am. Chem. Soc.*, **92**, 6061 (1970).

21. J. L. Montero, Ph.D. Thesis, University of Lanqudoc, Montpellier, France; *Chem. Abstr.*, **80**, 44430^d (1976).

22. G. R. Handrick, D. B. Uliss, H. C. Dalzell, and R. K. Razdan, *Tetrahedron Lett.*, 681 (1979).

23. R. S. Prasad and Sukh Dev, *Tetrahedron*, **32**, 1440 (1976) and references cited therein.

24. Unpublished results from our laboratory.

25. T. Y. Jen, G. A. Hughes, and H. Smith, *J. Am. Chem. Soc.*, **89**, 4551 (1967).

26. R. Adams and T. E. Bockstahler, *J. Am. Chem. Soc.*, **74**, 5436 (1952); R. Adams and Carlin, *J. Am. Chem. Soc.*, **65**, 360 (1943).

27. F. Korte, E. Dlugosch, and U. Claussen, *Ann. Chem.*, **693**, 165 (1966).

28. H. Kochi and M. Matsui, *Agr. Biol. Chem.*, **31**, 625 (1967).

29. R. Mechoulam, P. Braun, and Y. Gaoni, *J. Am. Chem. Soc.*, **94**, 6159 (1972); R. Mechoulam and Y. Gaoni, *J. Am. Chem. Soc.*, **87**, 3273 (1965).

30. E. C. Taylor, K. Lenard, and Y. Shvo, *J. Am. Chem. Soc.*, **88**, 367 (1966).

31. (*a*) Y. Gaoni and R. Mechoulam, *J. Am. Chem. Soc.*, **88**, 5673 (1966); (*b*) *Isr. J. Chem.*, **6**, 679 (1968).

32. R. M. Smith and K. D. Kempfert, *Phytochemistry*, **16**, 1088 (1977).

32a. J. M. Luteijn and H. J. W. Spronck, *J. Chem. Soc., Perkin*, I, 201 (1979).

33. D. B. Uliss, R. K. Razdan, H. C. Dalzell, and G. R. Handrick, *Tetrahedron Lett.*, 4369 (1975).

34. D. B. Uliss, R. K. Razdan, and H. C. Dalzell, *J. Am. Chem. Soc.*, **96**, 7372 (1974).

35. D. B. Uliss, R. K. Razdan, H. C. Dalzell, and G. R. Handrick, *Tetrahedron*, **33**, 2055 (1977).

36. S. Burstein, in J. A. Vinson, Ed., *Cannabinoid Analysis in Physiological Fluids*, ACS Symposium Series 98, American Chemical Society, Washington, D.C., 1979.

37. (a) M. E. Wall, D. R. Brine, C. G. Pitt, and M. Perez-Reyes, *J. Am. Chem. Soc.*, **94**, 8579 (1972); (b) M. E. Wall and D. R. Brine, Abst. Int. Symp. on the Mass Spectroscopy of Biologial Medicine, Milano, Italy, May, 1973; (c) M. E. Wall, D. R. Brine, and M. Perez-Reyes, Int. Congress of Pharm. Sci., Stockholm, Sweden, September, 1973.

38. C. G. Pitt, M. S. Fowler, S. Sathe, S. C. Sirivastava, and D. L. Williams, *J. Am. Chem. Soc.*, **97**, 3798 (1975).

39. C. G. Pitt, F. Hauser, R. L. Hawks, S. Sathe, and M. E. Wall, *J. Am. Chem. Soc.*, **94**, 8578 (1972).

40. Z. Ben-Zvi, R. Mechoulam, and S. Burstein, *Tetrahedron Lett.*, 4495 (1970).

41. R. K. Razdan, D. B. Uliss, and H. C. Dalzell, *J. Am. Chem. Soc.*, **95**, 2361 (1973).

42. J. L. G. Nilsson, I. M. Nilsson, S. Agurell, B. A. Kermark, and I. Lagerlund, *Acta Chem. Scand.*, **25**, 768 (1971).

43. D. B. Uliss, G. R. Handrick, H. C. Dalzell, and R. K. Razdan, *J. Am. Chem. Soc.*, **100**, 2929 (1978).

44. S. Danishefsky and T. Kitahara, *J. Am. Chem. Soc.*, **96**, 7807 (1974).

45. Z. Ben-Zvi and S. Burstein, *Res. Commun. Chem. Pathol. Pharmacol.*, **8**, 223 (1974).

46. N. Lander, Z. Ben-Zvi, R. Mechoulam, B. Martin, M. Nordqvist, and S. Agurell, *J. Chem. Soc., Perkin* I, 8 (1976).

47. R. Mechoulam, Z. Ben-Zvi, S. Agurell, I. M. Nilsson, J. L. G. Nilsson, H. Edery, and Y. Grunfeld, *Experientia*, **29**, 1193 (1973).

48. S. Inayama, A. Sawa, and E. Hosoya, *Chem. Pharm. Bull.*, **22**, 1519 (1974).

49. T. Petrzilka and M. Demuth, *Helv. Chim. Acta*, **57**, 121 (1974).

50. K. K. Weinhardt, R. K. Razdan, and H. C. Dalzell, *Tetrahedron Lett.*, 4827 (1971).

51. J. W. Wildes, N. H. Martin, C. G. Pitt, and M. E. Wall, *J. Org. Chem.*, **36**, 721 (1971).

52. R. Mechoulam, H. Varconi, Z. Ben-Zvi, H. Edery, and Y. Grunfeld, *J. Am. Chem. Soc.*, **94**, 7930 (1972).

53. O. Gurney, D. E. Maynard, R. G. Pitcher, and R. W. Kierstead, *J. Am. Chem. Soc.*, **94**, 7928 (1972).

54. M. Widman, M. Nordqvist, C. T. Dollery, and R. H. Briant, *J. Pharm. Pharmacol.*, **27**, 842 (1975).

55. For a recent review, see S. Agurell, M. Binder, K. Fonseka, J. E. Lindgren, K. Leander, B. Martin, J. M. Nilsson, M. Nordqvist, A. Ohlsson, and M. Widman, in G. G. Nahas, Ed., *Marihuana: Chemistry, Biochemistry and Cellular Effects*, Springer-Verlag, New York, 1976, pp. 141-157.

55a. A. Ohlsson, S. Agurell, K. Leander, J. Dahmen, H. Edery, G. Porath, S. Levy, and R. Mechoulam, *Acta Pharm. Suec.*, **16**, 21 (1979).

56. B. R. Martin, D. J. Harvey, and W. D. Paton, *J. Pharm. Pharmacol.*, **28**, 773 (1976).

57. C. G. Pitt, H. H. Seltzman, Y. Sayed, C. E. Twine, Jr., and D. L. Williams, *J. Org. Chem.*, **44**, 677 (1979).

58. R. K. Razdan, H. C. Dalzell, P. Herlihy, and J. F. Howes, *J. Med. Chem.*, **19**, 1328 (1976).

58a. R. M. Christie, R. W. Rickards, and W. P. Watson, *Aust. J. Chem.*, **31**, 1799 (1978).

59. K. E. Fahrenholtz, *J. Org. Chem.*, **37**, 2204 (1972).

60. F. Lotz, U. Kraatz, and F. Korte, *Z. Naturforsch.*, **33B**, 349 (1978).

61. B. Cardillo, L. Merlini, and S. Servi, *Gazz. Chim. Ital.*, **103**, 127 (1973); *Tetrahedron Lett.*, 945 (1972).

62. A. Arnone, L. Merlini, and S. Servi, *Tetrahedron*, **31**, 3093 (1975).

63. R. Mechoulam, Z. Ben-Zvi, H. Varconi, and Y. Samuelov, *Tetrahedron*, **29**, 1615 (1973).

64. G. R. Handrick, R. K. Razdan, D. B. Uliss, and H. C. Dalzell, *J. Org. Chem.*, **42**, 2563 (1977).

65. R. L. Hively, Ph.D. Thesis, University of Delaware, 1966.

66. R. Adams, B. R. Baker, and R. B. Wearn, *J. Am. Chem. Soc.*, **62**, 2204 (1940).

67. R. Adams, C. K. Cain, and B. R. Baker, *J. Am. Chem. Soc.*, **62**, 2201 (1940); R. Adams, D. C. Pease, J. H. Clark, and B. R. Baker, *J. Am. Chem. Soc.*, **62**, 2197 (1940).

68. R. Ghosh, A. R. Todd, and S. Wilkinson, *J. Chem. Soc.*, 1121, 1393 (1940).

69. R. Adams and B. R. Baker, *J. Am. Chem. Soc.*, **62**, 2401, 2405 (1940).

70. U. Claussen and F. Korte, *Z. Naturforsch.*, **21B**, 594 (1966).

71. R. Adams, C. M. Smith, and S. Loewe, *J. Am. Chem. Soc.*, **63**, 1973 (1941).

72. R. Ghosh, A. R. Todd, and D. C. Wright, *J. Chem. Soc.*, 137 (1941); G. Leaf, A. R. Todd, and S. Wilkinson, *J. Chem. Soc.*, 185 (1942).

73. R. Mechoulam, B. Yagnitinsky, and Y. Gaoni, *J. Am. Chem. Soc.*, **90**, 2418 (1968).

74. R. Mechoulam and Z. Ben-Zvi, *Chem. Commun.*, 343 (1969).

75. R. Mechoulam and B. Yagen, *Tetrahedron Lett.*, 5349 (1969).

76. G. Cardillo, R. Cricchio, and L. Merlini, *Tetrahedron*, **24**, 4825 (1968).

77. J. H. Holley, K. W. Hadley, and C. E. Turner, *J. Pharm. Sci.*, **64**, 892 (1975).

78. (a) L. Crombie and R. Ponsford, *Chem. Commun.*, 894 (1968); (b) *Tetrahedron Lett.*, 4557 (1968); (c) *J. Chem. Soc., C*, 796 (1971).

79. (a) V. V. Kane and R. K. Razdan, *J. Am. Chem. Soc.*, **90**, 6551 (1968); (b) *Tetrahedron Lett.*, 591 (1969).

80. V. V. Kane and T. L. Grayeck, *Tetrahedron Lett.*, 3991 (1971); V. V. Kane, *Tetrahedron Lett.*, 4101 (1971).

81. L. Crombie, S. D. Redshaw, and D. A. Whiting, *Chem. Commun.*, 630 (1979) and references cited therein.

82. M. A. Elsohly, E. G. Boeren, and C. E. Turner, *J. Heterocyclic Chem.*, **15**, 699 (1978).

83. M. J. Begley, D. G. Clark, L. Crombie, and D. A. Whiting, *Chem. Commun.*, 1547 (1970).

84. B. Yagen and R. Mechoulam, *Tetrahedron Lett.*, 5353 (1969).

85. L. Crombie, R. Ponsford, A. Shani, B. Yagnitinsky, and R. Mechoulam, *Tetrahedron Lett.*, 5771 (1968).

86. A. Shani and R. Mechoulam, *Tetrahedron*, **30**, 2437 (1974); *Chem. Commun.*, 273 (1970).

87. F. J. E. M. Kuppers, R. J. J. Ch. Lousberg, C. A. L. Bercht, C. A. Salemink, J. K. Terlouw, W. Heerma, and A. Laven, *Tetrahedron*, **29**, 2797 (1973).

88. R. K. Razdan, J. F. Howes, D. B. Uliss, H. C. Dalzell, G. R. Handrick, and W. L. Dewey, *Experientia*, **32**, 416 (1976).

89. D. B. Uliss, G. R. Handrick, H. C. Dalzell, and R. K. Razdan, *Experientia*, **33**, 577 (1977).

90. J. K. Kirtany and S. Paknikar, *Chem. Ind.*, 324 (1976).

91. M. D. Ciommo and L. Merlini, *Gazz. Chim. Ital.*, **106**, 967 (1976).

92. S. Houry, R. Mechoulam, P. J. Fowler, E. Macko, and B. Loev, *J. Med. Chem.*, **17**, 287 (1974); S. Houry, R. Mechoulam, and B. Loev, *J. Med. Chem.*, **18**, 951 (1975).

93. S. Y. Dike, M. Kamath, and J. R. Merchant, *Experientia*, **33**, 985 (1977).

94. R. A. Archer, W. B. Blanchard, W. A. Day, D. W. Johnson, E. R. Lavagnino, C. W. Ryan, and J. E. Baldwin, *J. Org. Chem.*, **42**, 2277 (1977).

95. R. S. Wilson and E. L. May, *J. Med. Chem.*, **18**, 700 (1975); **17**, 475 (1974).

96. J. Grimshaw, J. T. Grimshaw, and H. R. Juneja, *J. Chem. Soc., Perkin* I, 50 (1972).

97. J. M. Coxon, R. P. Garland, and M. P. Hartshom, *Aust. J. Chem.*, **23**, 1069 (1970).

98. A. J. Birch, *J. Proc. R. Soc. N.S.W.*, **83**, 245 (1949); H. H. Inhoffen, D. Kampe, and W. Milkowski, *Justus Liebigs Ann. Chem.*, **674**, 28 (1964).

99. K. Bailey and D. Verner, *Chem. Commun.*, 89 (1972).

100. B. Yagen, S. Levy, R. Mechoulam, and Z. Ben-Zvi, *J. Am. Chem. Soc.*, **99**, 6444 (1977).

101. A. Bowd, D. A. Swann, and J. H. Turnbull, *Chem.Commun.*, 797 (1975).

102. A. Padwa and G. A. Lee, *Chem. Commun.*, 795 (1972).

103. D. B. Uliss, G. R. Handrick, H. C. Dalzell, and R. K. Razdan, *Tetrahedron*, **34**, 1885 (1978).

104. R. K. Razdan and B. A. Zitko, *Tetrahedron Lett.*, 4947 (1969).

105. (a) F. J. E. M. Kuppers, R. J. J. Ch. Lousberg, C. A. L. Bercht, C. A. Salemink, J. K. Terlouw, W. Heerma, and A. Laven, *Tetrahedron*, **29**, 2797 (1973); (b) F. J. E. M. Kuppers, C. A. L. Bercht, C. A. Salemink, R. J. J. Ch. Lousberg, J. K. Terlouw, and W. Heerma, *Tetrahedron*, **31**, 1513 (1975); (c) *ibid., J. Chromatogr.*, **108**, 375 (1975); (d) H. J. W. Spronck and R. J. J. Ch. Lousberg, *Experientia*, **33**, 705 (1977); (e) H. J. W. Spronck and C. A. Salemink, *Recl. Trav. Chim.*, **97**, 185 (1978); (f) J. M. Luteyn, H. J. W. Spronck, and C. A. Salemink, *Recl. Trav. Chim.*, **97**, 187 (1978).

106. For a review see (a) H. G. Pars, R. K. Razdan and J. F. Howes, in N. J. Harper and A. B. Simmonds, Eds., *Advances in Drug Research*, Vol. 11, Academic Press, London, 1977; (b) Refs. 3b and 3c.

107. (a) R. K. Razdan, G. R. Handrick, H. G. Pars, A. J. Puttick, K. K. Weinhardt, J. F. Howes, L. S. Harris, and W. L. Dewey, Committee on Problems of Drug Dependence, National Academy of Sciences/National Research Council Annual Report, p. 6860 (1970); (b) R. K. Razdan, B. Zitko-Terris, H. G. Pars, N. P. Plotnikoff, P. W. Dodge, A. T. Dren, J. Kyncyl, and P. Somani, *J. Med. Chem.*, **19**, 454 (1976); (c) B. A. Zitko, J. F. Howes, B. C. Dalzell, H. C. Dalzell, W. L. Dewey, L. S. Harris, H. G. Pars, R. K. Razdan, and J. C. Sheehan, *Science*, **177**, 442 (1972).

108. H. Yoshimura, K. Watanabe, K. Orgui, M. Fujiwara, and U. Showa, *J. Med. Chem.*, **21**, 1079 (1978).

109. U. Kraatz and F. Korte, *Tetrahedron Lett.*, 1977 (1976).

110. R. K. Razdan, H. C. Dalzell, and P. Herlihy, *J. Heterocyclic Chem.*, **13**, 1101 (1976).

111. K. Matsumoto, P. Stark, and R. G. Meister, *J. Med. Chem.*, **20**, 25 (1977).

112. R. K. Razdan, H. G. Pars, F. E. Granchilli, and L. S. Harris, *J. Med. Chem.*, **11**, 377 (1968).

113. R. K. Razdan and H. C. Dalzell, *J. Med. Chem.*, **19**, 719 (1976).

114. O. P. Malik, R. S. Kapil, and N. Anand, *Ind. J. Chem*, **14B**, 449 (1976).

115. R. K. Razdan, H. G. Pars, W. R. Thompson, and F. E. Granchelli, *Tetrahedron Lett.*, 4315 (1974).

116. B. Loev, B. Dienel, M. M. Goodman, and H. Van Hoeven, *J. Med. Chem.*, **17**, 1234 (1974).

117. R. M. Anker and A. H. Cook, *J. Chem. Soc.*, 58 (1946).

118. H. G. Pars, F. E. Granchelli, J. K. Keller, and R. K. Razdan, *J. Am. Chem. Soc.*, **88**, 3664 (1966).

119. H. G. Pars, F. E. Granchelli, R. K. Razdan, F. Rosenberg, D. Teiger, and L. S. Harris, *J. Med. Chem.*, **19**, 445 (1976) and references cited therein.

120. W. L. Dewey, L. S. Harris, J. F. Howes, J. S. Kennedy, F. E. Granchelli, H. G. Pars, and R. K. Razdan, *Nature*, **226**, 1265 (1970).

121. R. K. Razdan, B. Zitko-Terris, G. R. Handrick, H. C. Dalzell, H. G. Pars, J. F. Howes, N. Plotnikoff, P. Dodge, A. T. Dren, J. Kyncyl, L. Shoer, and W. R. Thompson, *J. Med. Chem.*, **19**, 549 (1976).

122. R. K. Razdan, G. R. Handrick, H. C. Dalzell, J. F. Howes, M. Winn, N. P. Plotnikoff, P. W. Dodge, and A. T. Dren, *J. Med. Chem.*, **19**, 552 (1976).

123. C. Lee, R. J. Michaels, H. E. Zaugg, A. T. Dren, N. P. Plotnikoff, and P. R. Young, *J. Med. Chem.*, **20**, 1508 (1977).

124. M. Cushman and N. Castagnoli, Jr., *J. Org. Chem.*, **39**, 1546 (1974).

125. W. Greb, D. Bieniek, and F. Korte, *Tetrahedron Lett.*, 545 (1972).

126. T. Petrzilka and W. G. Lusuardi, *Helv. Chim. Acta.*, **56**, 510 (1973).

127. T. Petrzilka, M. Demuth, and W. G. Lusuardi, *Helv. Chim. Acta*, **56**, 519 (1973).

128. F. Lotz, U. Kraatz, and F. Korte, *Ann. Chem.*, 1132 (1977).

129. J. F. Hoops, H. Bader, and J. H. Biel, *J. Org. Chem.*, **33**, 2995 (1968).

130. G. M. Milne, A. Weissman, B. K. Koe, and M. R. Johnson, *Pharmacologist*, **20**, 243 (1978).

131. F. Lotz, U. Kraatz, and F. Korte, *Z. Naturforsch.*, **34B**, 306 (1979).

132. K. Matsumoto, P. Stark, and R. G. Meister, *J. Med. Chem.*, **20**, 17 (1977).

133. H. Kurth, U. Kraatz, and F. Korte, *Chem. Ber.*, **109**, 2164 (1976).

134. For a detailed description of the gross behavioral effects of THC's in laboratory animals see E. F. Domino, *Ann. N.Y. Acad. Sci.*, **191**, 166 (1971).

135. R. Adams, M. Harfenist, and S. Loewe, *J. Am. Chem. Soc.*, **71**, 1624 (1949).

136. Review (*a*) H. M. Bhargva, *Gen. Pharmac.*, **9**, 195 (1978); (*b*) R. A. Archer in R. V. Heinzelman, Ed., *Annual Reports in Medicinal Chemistry*, Vol. 9, Academic Press, New York, 1974, p. 253.

137. (*a*) T. S. Herman, L. H. Einhorn, S. E. Jones, C. Nagy, A. B. Chester, J. C. Dean, B. Furans, S. D. Williams, S. A. Leigh, R. T. Dorr, and T. E. Moon, *N. Engl. J. Med.*, **300**, 1295 (1979); (*b*) L. Lemberger and H. Rowe, *Clin. Pharmacol. Ther.*, **18**, 720 (1975) and references cited therein.

138. P. A. Weber and J. R. Bianchine, private communication.

139. M. Winn, D. Arendsen, P. Dodge, A. Dren, D. Dunnigan, R. Hallas, K. Hwang, J. Kyncyl, Y. Lee, N. Plotnikoff, P. Young, H. Zaugg, H. C. Dalzell, and R. K. Razdan, *J. Med. Chem.*, **19**, 461 (1976).

140. A. Guterman, P. Somani, and R. T. Bachand, *Clin. Pharmacol. Ther.*, **25**, 227 (1979).

141. S. F. Brunk, R. Noyes, D. H. Avery, and A. Canter, *J. Clin. Pharmacol.*, **15**, 664 (1974).

142. R. W. Houde, S. L. Wallenstein, A. Rogers, and R. F. Kaiko, Committee on Problems of Drug Dependence, National Academy of Sciences/National Research Council Annual Report, p. 149 (1976); 169 (1977).

143. M. Staquet, C. Gantt, and D. Machin, *Clin. Pharmacol. Ther.*, **23**, 397 (1978).

144. P. R. Jochimsen, R. L. Lawton, K. Versteeg, and R. Noyes, *Clin. Pharmacol. Ther.*, **24**, 223 (1978).

145. L. Lemberger, R. McMahon, R. Archer, K. Matsumoto, and H. Rowe, *Clin. Pharmacol. Ther.*, **15**, 380 (1974).

146. F. R. Sidell, J. E. Pless, H. Neitlich, P. Sussman, H. M. Copelan, and V. M. Sim, *Proc. Soc. Exper. Bio. Med.*, **142**, 867 (1973).

147. (a) L. Vachon, M. X. Fitzgerald, N. H. Sotliday, I. A. Gould and E. A. Gaensler, *N. Engl. J. Med.*, **288**, 985 (1973); (b) D. P. Tashkin, B. J. Shapiro, and I. M. Frank, *N. Engl. J. Med.*, **289**, 336 (1973); (c) D. P. Tashkin, B. J. Shapiro, Y. E. Lee, and C. E. Harper, *Am. Rev. Resp. Dis.*, **112**, 377 (1975); **109**, 420 (1974).

148. See, for example, I. Greenberg, J. Kuehnle, J. H. Mendelson, and J. G. Bernstein, *Psychopharmacology*, **49**, 79 (1976).

149. D. M. Feeney, *J. Am. Med. Assoc.*, **235**, 1105 (1976).

150. P. F. Consroe, G. C. Wood, and H. Bauchsbaum, *J. Am. Med. Assoc.*, **234**, 306 (1975).

151. P. L. Moreselli, M. Rizzo, and S. Garattini, *Ann. N.Y. Acad. Sci.*, **179**, 88 (1971).

The Total Synthesis of Ionophores

WENDEL WIERENGA

The Upjohn Company, Kalamazoo, Michigan

1. INTRODUCTION

The ability of a molecule to complex an ion, and to assist in the transport of this ion through a lipophilic interface, is the general description of an *ionophore*. The term ionophore, with this functional definition, was first coined by Pressman in 1964.[1,2] A related adjective, *complexone*, perhaps accentuating more the property of complexation, is employed by Ovchinnikov and co-workers at the Shemyakin Institute to describe the same molecules. Although the ion involved in the complexation-transport process is usually an alkali or alkaline earth metal, other possibilities include iron(III), proton, and protonated organic substrates, such as biogenic amines.

The class of molecules termed ionophores encompasses a diversity of structural types. A common feature of these structures is the presence of heteroatoms capable of acting as ligands for an ion. Equally important is the 3-dimensional capability of an inward orientation of the ligands in an appropriate geometry for the ion, simultaneously exposing a relatively lipophilic exterior. This chapter is organized primarily according to structural types. Included are linear and cyclic peptides, cyclic depsipeptides, macrolides, and the antibiotic polyether monoacids. A group not often included in reviews of ionophores is the siderophores, which are iron-specific ionophores. This group is classified functionally, since it incorporates several different structural types, although many are N-hydroxylated peptides (hydroxamates). Its inclusion is based on both the fulfillment of the dual role of ionophore's complexation and transport and the accomplishment of some interesting total syntheses. Within the various groups of ionophores will be some limited descriptions of "synthetic analogs," which mimic that group of natural ionophores structurally and/or functionally.

Most of the naturally occurring ionophores were discovered because of their antibiotic properties. In fact, much of the structure proof, synthesis, and determination of the spectrum of antibiotic activity preceded their definition as ionophores. Although a correlation of specific ion complexation and transport with antibiotic activity is apparent,[3] complexation/transport does not constitute a universal mechanism of action. Besides antibiotic activity, other pharmacological effects include cardioregulatory action such as positive inotropic effects and a decrease in coronary resistance. Other effects, including distribution of radionuclides, and lowering of intraocular pressure have been noted. These have recently been reviewed by Pressman[4] and Westley.[5,6]

The phenomenon of transport through a membrane has several mechanisms: diffusion, carrier-mediated, and active transport. Diffusion-controlled passage is driven simply by thermal molecular motion. In carrier-mediated transport the introduction of a second factor, a mobile component (ionophore), assists in ionic penetration and movement through the membrane. This can occur either via a free, ion-molecular complex (carrier) or a transporting, relatively stationary complex (channel or pore). Active transport involves operating against an electrical or chemical gradient and therefore requires coupling with an energy source, such as ATP hydrolysis.

Therefore, in simplistic terms ionophore or carrier-mediated transport involves (1) complexation of the ionophore with an ion, requiring a substitution of the ionophore binding sites for the hydration sphere surrounding the ion through the lowest energy (stepwise) mechanism, (2) diffusion of the complex from one membrane interface to the other, and (3) the reverse of the complexation process [step (1)]. Several factors requiring optimization are a best-fit situation of the ion for the ligands, the 3-dimensional lipophilicity of the complex, and the difference between the energies of ligation and solvation. These phenomena and their interrelationships have been reviewed.[44,45]

The physiochemical and biological criteria for functionally defining ionophores have many facets. In the former case such techniques as bulk phase extraction (aqueous to organic phase),[46] U-tube experiments (two aqueous phases separated by an organic phase),[47,48] ionophore impregnated liquid membrane ion selective electrodes,[49] and specific ion saturated liposomes[53a] are included. All of these techniques act as models, to one degree or another, for biological membranes. Perhaps the closest model of all are bilayer lipid membranes (BLM; also known as black lipid membranes). Essentially this is a bilayer soap film oriented across an orifice suspended in aqueous media, thus creating an electrochemical cell. The BLM can be utilized to determine passive ion transport, voltage-dependent transport, and ion selectivity.[50] Oft-employed biological parameters for ionophore classification include mitochondrial swelling and uncoupling of oxidative phosphorylation,[29,51] Na^+ or K^+ depletion of erythrocytes,[53b] and such Ca^{2+} transport-dependent phenomena as histamine release from rat mast cells,[54] cortical granule rupture in sea urchin eggs,[55] serotonin release from, and aggregation of,

Table 1

Property	Li^+	Na^+	K^+	NH_4^+	Rb^+	Cs^+	Mg^{2+}	Ca^{2+}	Sr^{2+}	Ba^{2+}
Ionic radius, Å	0.66	0.95	1.33	1.44	1.48	1.69	0.65	0.99	1.13	1.35
Hydrated radius, Å	3.40	2.76	2.32		2.28	2.28				
Total hydration number	25	16	10			10				
ΔH_{hyd}, kcal/mole	124	95	76	72	69	62	459	371	353	325
Coordination number	4,6	4,6	8,10			8,10	6	6,8	6,8	8,10

blood platelets,[56] and many others. For reference to ion selectivities cited in the foregoing sections Table 1 includes data on some ionic parameters of pertinent ions.

2. THE PEPTIDES

Of the many peptide antibiotics isolated from bacteria, several have been categorized as ionophores. These include the gramacidin complex, alamethicin, and the cyclic peptide, antamanide. From the viewpoint of structural similarity, one might project related peptides as ionophore candidates such as the amphomycins,[7] the bottromycins,[8] the antimycins (both cyclic[9] and linear[10]) and probably most of the homomeric homodetic cyclic peptides (gramicidin S, J, tyrocidin A-D, polymixins, viomycins, ilamycins, cyclosporins, and others).

Interestingly, their biosynthesis appears to involve multienzyme complexes independent of the usual protein biosynthesis pathway of ribosomal-, messenger-, and transfer-RNA. The appearance of the many D-amino acids stems from an ATP-dependent racemization within the enzyme complex after amino acid incorporation.[11]

A. Gramicidins

The linear gramicidins, isolated from *Bacillus brevis*[12] along with the cyclic gramicidins S and J and the tyrocidins,[13] consist of at least six components (1a-f).[14a-c] They have in common 13 of 15 amino acids, a formyl blocked N-terminal, and ethanolamide C-terminals. They differ in the replacement of isoleucine for valine at the N-terminal and substitutions of tryptophan, phenylalanine, and tyrosine at position 11.

$$
\begin{array}{c}
\overset{O}{\overset{\|}{HC}}-\overset{1}{X}-\overset{2}{Gly}-\overset{3}{Ala}-D-\overset{4}{Leu}-\overset{5}{Ala}-D-\overset{6}{Val}-\overset{7}{Val}-D-\overset{8}{Val} \\
HOCH_2CH_2\underset{H}{N}-\underset{15}{Trp}-D-\underset{14}{Leu}-\underset{13}{Trp}-D-\underset{12}{Leu}-\underset{11\ 10}{Y}-D-\underset{}{Leu}-\underset{9}{Trp}
\end{array}
$$

		X	Y
1a	Val-gramicidin A	Val	Trp
1b	Ileu-gramicidin A	Ileu	Trp
1c	Val-gramacidin B	Val	Phe
1d	Ileu-gramacidin B	Ileu	Phe
1e	Val-gramicidin C	Val	Tyr
1f	Ileu-gramicidin C	Ileu	Tyr

Evidence has been accumulated which indicates the ion transporting capacity of gramicidin A involves the formation of ion-conducting channels in membranes.

It has been proposed[16,17] that this pentadecapeptide forms a helical, head-to-head dimeric complex which forms a pore or channel in the membrane with approximately a 4 Å diameter and 30 Å length.[18]

The relative order of transport has been determined as H > NH$_4$ > Cs > Rb > K > Na > Li.[17] Bulk phase extraction and U-tube transport experiments indicate a 1:1 K/gramicidin association ratio with weak binding and poor diffusional transport, thus further supporting the premise of non-carrier-mediated transport.[31]

The structures of gramicidins A-C were elucidated by Sarges and Witkop. They also synthesized gramicidin A(1a,b).[15] Their successful strategy involved the dicyclohexylcarbodiimide coupling of the pentapeptide, Z-Val-Gly-Ala-D-Leu-Ala-*OH*, constituting positions 1-5 of 1a, with the decapeptide, constituting positions 6-15. The respective fragments were built up from the requisite amino acids, using the mixed anhydride method (EtO$_2$CCl, Et$_3$N). The details of the synthesis will not be related here, since they have already been reviewed by Ovchinnikov and co-workers[3] and also by Johnson[19] in a previous volume of this series.

The gramicidin complex exhibits mainly Gram-positive antibacterial activity (gramicidin A constitutes ~75% of the mixture). The syntheses of shorter chain analogs led to the conclusion that excision of more than two amino acids substantially reduced activity.[20] A pentadecapeptide model, HCO-(Ala-Ala-Gly)$_5$-OH, has been proposed, based on conformation analysis, to form ion-conducting channels.[21] Recently, simplified analogs such as (Trp-D-Leu)$_7$Trp and Trp-Gly-(Trp-D-Leu)$_6$Trp have been shown to form cation-permeable channels with properties similar to gramicidin A.[136]

B. Alamethicin

A second linear polypeptide reported to enhance passive ion transport is alamethicin. Isolated from *Trichoderma viride*,[22] it was initially believed to be a cyclic octadecapeptide[23,24] and was accompanied by a closely related polypeptide (alamethicin II). However, recent work with ^{1}H-NMR spectroscopy has demonstrated alamethicin's linear nature,[25] and the combination of ^{13}C-NMR with high resolution and field desorption mass spectrometry has yielded a

```
      1    2    3    4    5    6    7    8    9    10
   Ac—Aib—Pro—Aib—Ala—Aib—Ala—Gln—Aib—Val—Aib
                             (Aib)                 |
          α
   Phol—Glu—Glu—Aib—Aib—Val—Pro—Aib—Leu—Gly
    20    19   |  (γ) 17  16   15   14   13   12   11
              OH
```

Alamethicin I and (II)

second, revised linear 20-amino acid sequence. Several rather unique features include a substantial number of α-aminoisobutyric acid residues (Aib, α-methylalanine), an acetylated N-terminal, and an amino alcohol, phenylalanol [-NHC(CH$_2$PH)CH$_2$OH, Phol], at the other terminal.

Alamethicin (I, II) exhibits a voltage-dependent, discrete conductance across lipid bilayers[27, 28]. It is postulated that alamethicin behaves as an ion-conducting channel.[28] Bulk-phase partition experiments demonstrated transport of K > Rb > Cs > Na[29]. The initially proposed linear structure[25] for alamethicin has been synthesized[32] by the solid-phase method in a block condensation approach using four fragments. In analogy to the structure shown, these would constitute residues 1-4, 5-8, 9-12, and 18-20. They found no pore-forming activity. This same group and another group[26b] have also prepared the alamethicin structures depicted above and found 30-40% of the "natural" pore-forming activity, thus suggesting other minor components exist in the alamethicin complex which contribute to its ion conductance properties.[33]

Several closely related polypeptides include antiamoebin[26], emerimicin[26], and suzukacillin A.[26, 30] The latter exhibits discrete conductance fluctuations in black lipid membranes similar to alamethicin and therefore is presumably a channel former.[30]

$$\text{Ac}-\text{Aib}-\text{Pro}-\text{Val}-\text{Aib}-\text{Val}-\text{Ala}-\text{Aib}-\text{Ala}-\text{Aib}-\text{Aib}-\text{Gln}-\text{Aib}$$
$$|$$
$$\text{L}-\text{Phol}-\text{Gln}-\text{Glu}-\text{Aib}-\text{Aib}-\text{Val}-\text{Pro}-\text{Aib}-\text{Leu}-\text{Gly}-\text{Aib}-\text{Leu}$$
$$|$$
$$(\gamma-\text{OH})$$

Suzukacillin A

C. Antamanide

A third example in the category of peptide ionophores is a cyclic decapeptide, antamanide. Isolated from the mushroom *A. phalloides*,[34] the structure has an all-L-Phe$_4$Pro$_4$ValAla composition arranged in a nearly symmetrical fashion.[35]

The decapeptide was first synthesized by Wieland[35,36] and co-workers and involved the cyclization in two different approaches of two decapeptides, one ring-closure at Pro^8-Phe^9 and the other at Phe^5-Phe^6. Numerous other syntheses of antamanide and related analogs by Wieland's[37] group and others[38] have been reported. Since some of these have previously been reviewed,[3] the details will not be covered in this review.

Antamanide, as opposed to the previously discussed linear polypeptides in this section, forms very stable complexes, particularly with Na$^+$ and Ca^{2+}. It is a carrier-type ionophore with a high Na$^+$ preference in the monovalent series (Na > Li > K > Rb > Cs).[39]

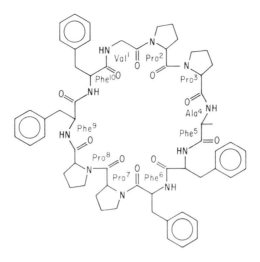

ANTAMANIDE

Synthetic cyclopeptides that complex and transport ions have been reported. For example, Blout and co-workers[40] have synthesized cyclo-(Pro-Gly)$_3$ and demonstrated selectivity for the smaller cations in both the monovalents (Li$^+$ and Na$^+$) and the divalents (Mg^{2+}, Ca^{2+}). In a slightly different approach Schwyzer's group, who had previously synthesized gramicidin S (cyclic),[41] prepared a cysteine (S—S) bridged cyclopeptide.[42] The strategy was to construct two face-to-face cyclic homodetic peptide rings to act in a sandwich ion-binding fashion. Therefore, they synthesized (Chart 2.1) S,S'-bis-cycloglycyl-L-hemicystyl-glycyl-glycyl-L-proline in 15% overall yield. Ion transport studies employing the liquid membrane electrode method of Simon[49] established a specificity order of K > Na > Li > Ca.

Other recent entries into the area of synthetic cyclopeptides include cyclo-(Gly-Pro-Gly-D-Ala-Pro),[43] which exhibits binding with divalent ions such as

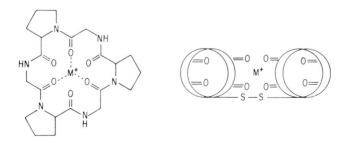

CYCLO-(PRO-GLY)^3

Mg^{2+}, Mn^{2+} and Ca^{2+}, as well as with Li^+, cyclo-[L-Tyr(OBzl)-Gly]$_3$ which exhibits K and Na complexation[74], and cyclo [Glu-Sar-Gly-(N-decyl)Gly]$_2$, which, by virtue of two ionizable $Glu\text{-}CO_2H$ groups and a lipophilic chain, exhibits excellent Ca^{2+} selectivity.[75]

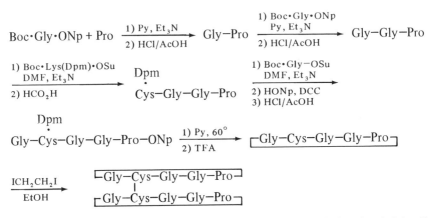

Chart 2.1.   ONp, p-nitrophenyloxy; OSu, N-succinimidoxy; Dpm, diphenylmethyl (on S); BOC, t-butoxycarbonyl (on N).

3. CYCLIC DEPSIPEPTIDES

Depsipeptides are characterized by the inclusion of at least one non-amino acid residue (heteromeric). Many of this class of natural products are cyclic in nature, and of these, three have been exploited as ionophores: valinomycin, enniatins, and monamycins.

A. Valinomycin

Valinomycin is a cyclic, regular dodecadepsipeptide of the structure (L-valine-D-hydroxyisovaleric-D-valine-L-lactic acid)$_3$, produced by several *Streptomyces* sp. Even though valinomycin exhibited rather broad antimicrobial activity including G+ bacteria, yeast, and fungi, it has received most of its attention because of its extremely high K/Na selectivity and carrier-mediated transport kinetics (approx. 10^4 K$^+$/valinomycin/sec in BLM). It also forms stable complexes with Rb$^+$ and Cs$^+$. Numerous studies have been carried out to elucidate the various conformers of valinomycin in solution and in its metal-complexed state.[3, 58, 61] The evidence indicates that only the ester carbonyls are involved directly as ligands for the metal in a bracelet conformation, stabilized by intramolecular H-bonding between the amide linkages. The X-ray crystal structure of valinomycin has also been elucidated.[59]

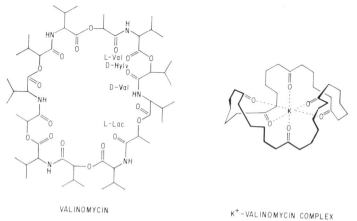

VALINOMYCIN   K⁺-VALINOMYCIN COMPLEX

The structure of valinomycin was originally proven by total synthesis by Shemyakin and co-workers.[60] The preparation of the 36-membered ring began by building up the repeating tetradepsipeptide unit Z-D-Val-L-Lac-L-Val-D-HyIv-O-t-Bu by amide bond formation between Z-D-Val-L-Lac-OH and H-L-Val-D-HyIv-O-t-Bu (SOCl₂, Et₃N). The initial acylations of t-butyl-L-lactate and D-hydroxyisovalerate with D- and L-valine (Z) were carried out with benzene sulfonyl chloride in pyridine. In the original synthesis the initial tetradepsipeptide was then coupled to the next didepsipeptide; however, in a later approach,[3] the tetrapeptides were fashioned into the linear dodecadepsipeptide by amide bond formations with PCl₅/Et₃N followed by ring closure under high dilution conditions with SOCl₂/Et₃N (25% yield). Since activation of the hydroxy acid (Lac, HyIv) C-terminal residues with thionyl chloride does not cause extensive racemization, these rarely employed conditions could be utilized. Other syntheses, including Merrifield solid-phase methods, have been reported.[62] The Shemyakin group has prepared a large number of analogs, most of which have similar or substantially inferior complexing properties, except for one analog that substitutes L-MeAla for L-Lac, exhibiting an even more stable K-complex than valinomycin. Interestingly, there appears to be a very good correlation between ion transport and antibiotic activities with these analogs.[65] Since the syntheses of valinomycin has been reviewed several times elsewhere,[3,63,64] we will not report the details here.

The syntheses of four cyclic peptides related to valinomycin have recently been completed[52] employing the Merrifield solid-phase methodology: cyclo-[Pro-Ala-D-Pro-D-Val]₃, cyclo-[Ala-Val-D-Pro-D-Val]₃, cyclo-[Pro-Val-D-Pro-D-Val]₂-Pro-D-Val, and cyclo-[Pro-Val-D-Pro-D-Val]₂. The first 12-membered oligopeptide ring was ~10³ > valinomycin for K⁺ complexation with a selectivity sequence (bulk-phase extraction) of K ~ Rh > Cs > Na > Li. This analog was substantially better than the other three and was also effective in increasing membrane conductance of sheep red blood cell bilayers in the presence of KCl.

B. Enniatins

Again, as was the case with the previously described gramicidins, a mixture of closely related structures make up the enniatins. Isolated from several *Fusarium* species, three compounds, enniatins, A, B, and C, are the most studied. A closely related cyclodepsipeptide, beauvericin, was isolated from the fungus *Beauveria bassiana*.

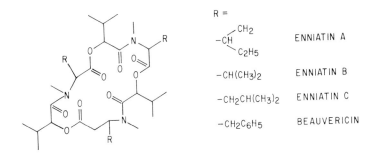

R =	
$-CH \begin{smallmatrix} CH_2 \\ C_2H_5 \end{smallmatrix}$	ENNIATIN A
$-CH(CH_3)_2$	ENNIATIN B
$-CH_2CH(CH_3)_2$	ENNIATIN C
$-CH_2C_6H_5$	BEAUVERICIN

The enniatins are cyclic hexadepsipeptides, one half the ring size of valinomycin. They do not exhibit the ion selectivity seen with valinomycin but form disc-shaped complexes $(1:1)^{67}$ with most alkali and alkaline earth metals, ammonium ion, and even some transition metals.[3, 66] Alternatively, they exhibit a better antimicrobial spectrum of activity, including *M. tuberculosis*, and are significantly less toxic than valinomycin.

As with valinomycin, their structure was first confirmed by total synthesis (enniatin B, (D-HyIv-L-MeₙVal)₃).[68, 69] The general strategy involved the coupling of the $Z(-NO_2)$-L-MeVal-D-HyIv-Ot-Bu units together via amide bond formation with PCl_5/Et_3N. The ester linkage of the depsipeptide was formed with the benzene sulfonyl chloride, pyridine mixed anhydride method, or diimidazole carbonyl as exemplified in the beauvericin synthesis (Chart 3.1). Analogous to the valinomycin synthesis, ultimate ring closure was effected with $SOCl_2/Et_3N$. Numerous analogs were prepared with similar chemistry.[3]

C. Monamycins

The monamycin family of antibiotics, isolated from *Streptomyces jamaicensis*, exhibit G+ antibacterial activity and form monovalent ion complexes.[71] Although they are structurally related to the enniatins in that they are 18-membered ring, cyclohexadepsipeptides,[72, 73] they are considerably more com-

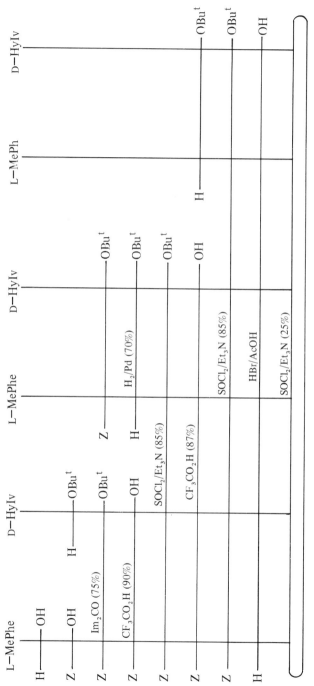

Chart **3.1.** Beavericin.

273

plicated and have, as yet, not been prepared by total synthesis. They are composed of some unusual amino acids such as $(3R)$-piperizinic acid, $(3S,5S)$-5-hydroxypiperizinic acid, $(3R,5S)$-5-chloropiperizinic acid, D-isoleusine, and trans-4-methyl-L-proline in 15 structural variations.

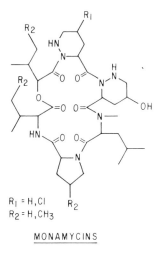

$R_1 = H, Cl$
$R_2 = H, CH_3$

MONAMYCINS

4. SIDEROPHORES

One of the unique aspects of nutrient uptake by microorganisms is the transport of iron utilizing chelating agents.[76a] Originally termed *siderochromes*, these "ferric ionophores,"[76b] or *siderophores*, have been grouped into two categories: secondary hydroxamates and catechols.[77] They are biosynthesized under iron-deficient conditions, secreted into the medium to complex, and thus "solubilize" any iron present, and then apparently transported into the cells to provide iron for the electron transport processes and other requirements. Their coordination chemistry involves high-spin ferric complexes of octahedral geometry.

A. Hydroxamates

The naturally occurring hydroxamic acids represent by far the larger of the two categories of siderophores.[76c] They range in structural complexity from the simplest, hadacidin, to perhaps the most complex, the mycobactins.

The group of monohydroxamic acids include, among others, hadacidin, actinonin, trichostatin A, and the aspergillic acids. Aspergillic acid structurally represents five closely related N-hydroxypyrizinones which vary at R from hydroxylated butyryl to propyl side chains. Aspergillic acid was isolated from

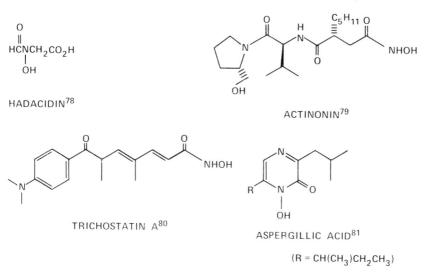

HADACIDIN[78]

ACTINONIN[79]

TRICHOSTATIN A[80]

ASPERGILLIC ACID[81]

(R = CH(CH₃)CH₂CH₃)

Aspergillus flavus by White and Hill[82] and structure established by Dutcher in 1947.[81] Apparently derived from leucine and isoleucine, the oxidation of the isoleucine-derived nitrogen occurs after cyclization to the deketopiperazinone.[83]

The parent structure was synthesized by Masaki et al. in 1965.[84] Their preparation of racemic aspergillic acid began with the alkylation of *N*-leucyl-*o*-benzylhydroxylamine with 1-chloro-3-methyl-2-oximinopentane (Chart 4.1). The oxime was unmasked to the ketone with 3 *N* hydrochloric acid with strict temperature control (41-43°). Catalytic hydrogenation released the hydroxamic acid, which underwent ammonia-catalyzed cyclization with aromatization to the desired product.

Ollis and co-workers reported a synthesis of actinonin,[79] isolated from *Streptomyces roseopallidus*, in 1974. As depicted in Chart 4.2, regioselective addition of *O*-benzylhydroxylamine on pentylmaleic anhydride was exclusive yielding only isomer 1. Cyclization, however, afforded two isomers with 3 predominating.

Nucleophilic addition of L-valyl-L-prolinol to isomaleimide-3 regioselectively afforded 4, wherein the double bond had apparently isomerized. This was subsequently deprotected and reduced to afford actinonin in unspecified yield.

There are only several examples of the dihydroxamic acid group. By far the most studied is rhodotorulic acid. Isolated from the yeast *Rhodotorula pilimanae*, and characterized by Atkins and Neilands,[85] this diketopiperazine is biosynthesized from ornithine and δ-*N*-acetyl-δ-*N*-hydroxyornithine.[86] Studies of its coordination chemistry have established a binding constant at neutral pH of 31.1 (log K_f) with the predominant Fe(III) species as Fe₂(RA)₃.[87]

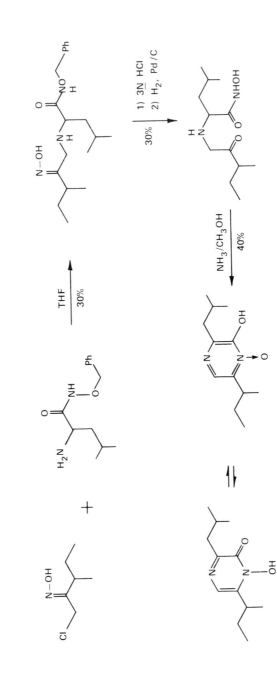

Chart **4.1**

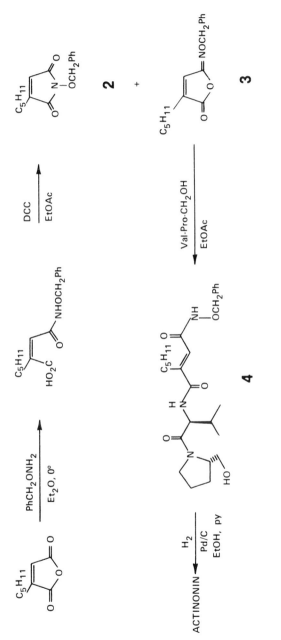

Chart 4.2

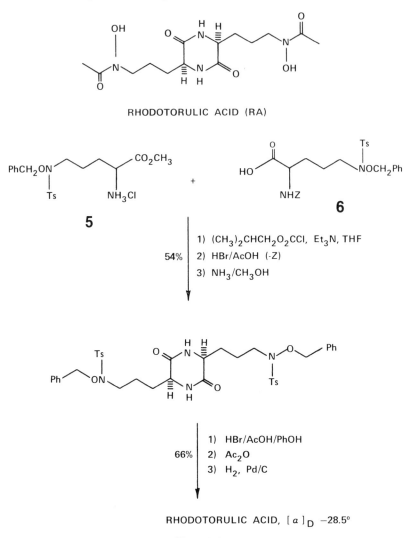

RHODOTORULIC ACID (RA)

Chart **4.3**

Rhodotorulic acid was synthesized by Y. Isowa et al.[88] in 1971. Preparation of the requisite ornithine dipeptide intermediate is accomplished by the mixed anhydride coupling of the appropriately protected L-amino acids, **5** and **6**, in 98% yield. Deprotection of the benzyloxycarbonyl protected amine followed by treatment with ammonia yielded the diketopiperazine. Removal of the tosyl groups was effected with acid in the presence of phenol at room temperature, and benzyl group cleavage, subsequent to acetylation, afforded the unmasked secondary hydroxamates in good overall yield. These workers also prepared the optical antipode of rhodotorulic acid, starting with the D-amino acid. Two other structurally interesting dihydroxamates are schizokinen and aerobactin; however, these have not as yet been prepared by total synthesis. Aerobactin, coordinatively produced by *Aerobacter aerogenes* with the catechol siderophore, enterobactin (vide infra), coordinates Fe(III) with the terminal N-acetylhydroxyl amines and the central citric group with a binding constant of 31.7.[120]

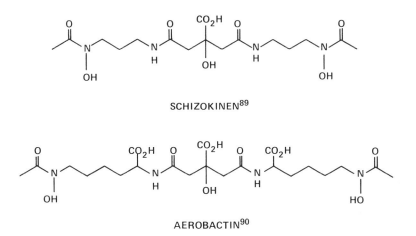

SCHIZOKINEN[89]

AEROBACTIN[90]

The trihydroxamic acids encompass the largest subgroup of the hydroxamate siderophores. These include the cyclic ferrioxamines (e.g., nocardamin), Desferal, the ferrichromes, and the mycobactins. Some of these are depicted below as their ferric complexes.

Ferrichrome, isolated from the fungus *Sphaerogena*, was the first trihydroxamate siderophore to have been characterized.[94, 98] It is biosynthetically derived form ornithine and glycine in a sequence involving (1) oxidation to ornithine hydroxylamine, (2) acetylation, (3) condensation with glycine, and (4) the incorporation of Fe(III).[76b] Ferrichrome exhibits an iron binding constant of 29.1,[95] which competes very favorably with rhodotorulic acid (31.1) and the clinically employed Desferal (30.6).

This cyclic hexapeptide was synthesized by Keller-Schierlein and Maurer[96] in 1969 and by Isowa and co-workers[97] in 1974. The procedure employed by the earlier group (Chart 4.4) began with L-δ-nitronorvaline [$NO_2-(CH_2)_3$-$CH(NH-)CO-$], a masked form of the required ornithine component, and its condensation with the dipeptide, Gly-Gly. The linear hexapeptide was assembled Gly-(L-δ-nitronorval)$_3$-Gly-Gly-OCH$_3$ by the mixed anhydride method using t-butyloxycarbonyl amine protection. Cyclization of the resulting hexapeptide

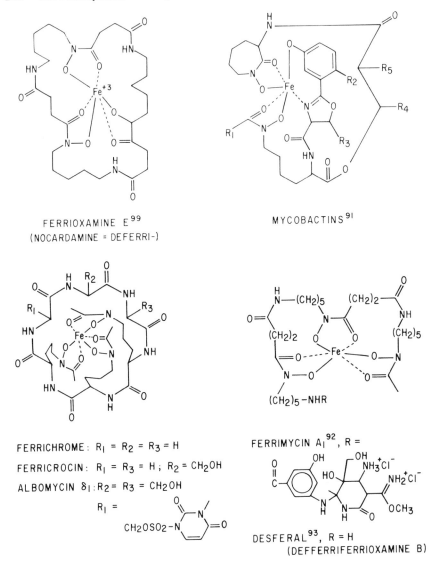

FERRIOXAMINE E[99]
(NOCARDAMINE = DEFERRI-)

MYCOBACTINS[91]

FERRICHROME: $R_1 = R_2 = R_3 = H$
FERRICROCIN: $R_1 = R_3 = H$; $R_2 = CH_2OH$
ALBOMYCIN δ_1 : $R_2 = R_3 = CH_2OH$

$R_1 = $

CH_2OSO_2-N

FERRIMYCIN A_1[92], R =

DESFERAL[93], R = H
(DEFFERRIFERRIOXAMINE B)

was carried out via the active ester, *p*-nitrophenyl, derived by carbonate mediated hydrolysis of the methyl ester precursor and reesterification. Final modifications required reduction of the nitronorvaline residues to the ornithine hy-
droxylamines and acetylation. An alternate coupling strategy to the linear
hexapeptide stage was also carried out to confirm the optical integrity of the
deferri-ligand.

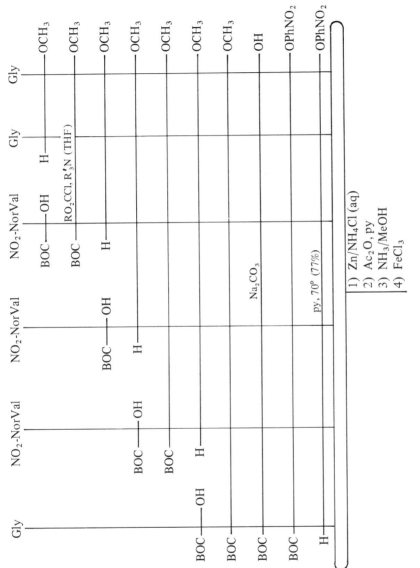

Chart **4.4** Ferrichrome.

281

The second group employed a similar synthetic strategy except they employed δ-*N*-tosyl-δ-*N*-benzyloxy-L-ornithine, previously utilized in their rhodotorulic acid synthesis, instead of the nitronorvaline. Amino acid coupling (~80% yields) and cyclization of the linear hexapeptide (41% yield) were carried out with dicyclohexylcarbodiimide/*N*-hydroxy succinimide. Acid-catalyzed detosylation (36% HBr/AcOH/PhOH) followed by acetylation (Ac$_2$O, py) and catalytic hydrogenation afforded deferriferrichrome in 47% yield.

The mycobactins, isolated from various *Mycobacteria*, are a large group of rather unique siderophores. Iron is bound unsymmetrically with two different hydroxamate ligands and an *o*-hydroxyphenyl isoxazoline; thus, strictly speaking, they are not trihydroxamate siderophores. Incorporating six centers of asymmetry and combinations of five different substituents, these complex ligands have not as yet been synthesized.

The syntheses of several ferrioxamines have been reported by Prelog[122] and Keller-Schierlein.[123] An example of this chemistry is the synthesis of ferrioxamine B [Desferal minus Fe(III)] depicted in Chart 4.5. Beginning with 1-benzyl-

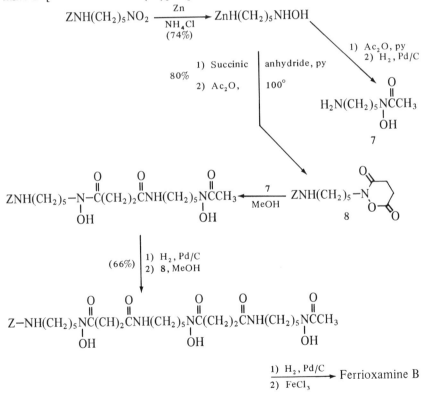

Chart 4.5

oxycarbonylamino-5-nitropentane, the ε-amino-protected hydroxylamine was generated by a zinc reduction. The novel use of succinylation followed by cyclo-dehydration yielded the internal active ester **8** with the necessary two new carbons. This could be condensed with **7** to give the 15-carbon fragment. Alternatively, condensation with 1-nitropentylamine followed by zinc reduction and acetylation yielded the same intermediate. Reuse of this 15-carbon interme-diate in condensation with **8** yielded the benzyloxycarbonyldeferri ligand. Hydrogenolysis and the introduction of iron(III) gave ferrioxamine B.

B. Catechols

The catechol or dihydroxybenzoyl is a much smaller class[76d] of natural sidero-phores than the hydroxamates, and examples to date are 2,3-dihydroxybenzoyl-glycine (*Bacillus subtilis*)[100], 2-N,6-N-di-(2,3-dihydroxybenzoyl)-L-lysine (*Azoto-bacter vinelandii*),[101] 2,3-dihydroxybenzoyl derivative of spermidine[124], agrobactin[236], and enterobactin.

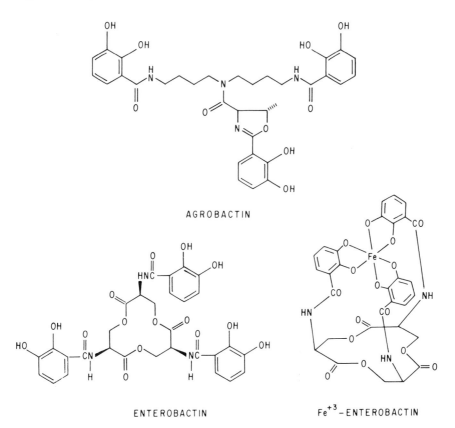

AGROBACTIN

ENTEROBACTIN

Fe^{+3}-ENTEROBACTIN

Enterobactin (or enterochelin), isolated from, and characterized as the main iron-transporting species of, *E. coli*,[102] *S. typhimurium*,[103] and *A aerogenes*[76d] is a cyclic trimer of 2,3-dihydroxy-*N*-benzoyl-L-serine. One molecule of enterobactin binds one Fe(III) to give a red complex with a binding constant (52) that is greater than the hydroxamates.[107] Its biosynthesis incorporates the elements of aromatic biosynthesis to give the 2,3-dihydroxybenzoyl grouping through the chorismic acid-isochorismic acid pathway, as well as the peptide derivation from L-serine.

Enterobactin has been recently synthesized by Corey and Bhattacharyya.[104] Beginning with *N*-benzyloxycarbonyl-L-serine, **1** (Chart 4.6), the hydroxyl was protected as the THP-ether which required, in turn, a carboxyl protection step via the phenacyl ester. Coupling of the deprotected serine residue with **2** was accomplished using the active imidazole thioester. Reductive removal of the phenacyl ester and repetition of this coupling process yielded the linear triester

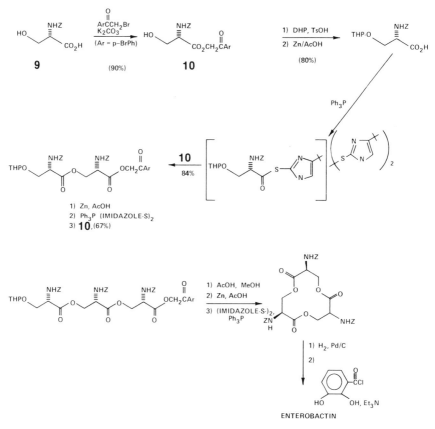

Chart 4.6

in 45% overall yield from **1**. Cyclization was carried out with the same active thioester procedure in 40% yield after deprotection of the termini (75%). Finally, removal of the N-benzyloxycarbonyl groups and replacement with the required 2,3-dihydroxybenzoyl groups[125] gave the target molecule in unspecified yield.

The severe anemia (Cooley's anemia) associated with the genetic disease β-thalassemia major can only be treated by transfusion therapy. This leads to an iron-overload state, secondary hemochromatosis, since the body retains most of the iron from transfused erythrocytes. Attempts to prevent excessive iron accumulation have involved the use of such siderophores as desferrioxamine (Desferol), rhodotorulic acid, and 2,3-dihydroxybenzoic acid.[108] However, because of several drawbacks with these agents, efforts are underway to synthesize enterobactin mimics. One example is the report of the synthesis and examination of **11** as an isosteric equivalent.[105] This was prepared from 1,3,5-benzenetrialdoxime, generated from the corresponding tri(methyl)carboxylate by diisobutylaluminum hydride reduction, pyridinium chlorochromate oxidation back to the aldehyde stage, and condensation with hydroxylamine. The benzenetrialdoxime was reduced to the tribenzylamine and subsequently benzoylated to give **11**. Binding experiments confirm an iron affinity in the range of enterobactin.

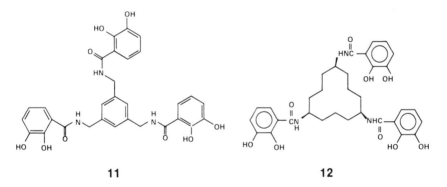

11   **12**

Corey and Hurt have synthesized a carbocyclic enterobactin analog, **12**. Not only was the question relating to iron complexation addressed with this analog, but also the question of the possibility of ester hydrolysis of the cyclic triester, enterobactin, was asked as it relates to a mechanism of iron dissociation after the complex passes into the microorganism. Obviously this mechanism is not possible with **12**.

Chart 4.7 depicts the preparation of this analog.[106] The all-*cis*-cyclodecan-1,5,9-triol, prepared from all-*trans*-cyclodecatriene,[109] was converted to the all-*cis*-triamino analog. Similar chemistry was previously documented by Collins et al.[110] Acylation with the protected 2,3-dihydroxybenzoyl chloride, itself prepared from catechol, afforded the desired 12 after acid hydrolysis of the aceto-nides. This carbocyclic analog exhibited approximately 75% of the bacterial iron-transporting capacity of enterobactin.

Raymond and co-workers have also recently reported a synthesis of 11[111], analogous to that previously described, as well as a preparation of the trisdi-hydroxybenzoyl spermidine derivative 13 (Chart 4.8).[121] Tosylation of spermi-dine afforded the tritosyl amide (84%), which was converted to the cyclic triamine via alkylation of the terminal dianion (NaH/DMF; 76%) with 1,3-ditosylpropane. Deprotection with concentrated sulfuric acid (100°, 40 h), benzoylation with 2,3-dioxomethylene benzoyl chloride, and unmasking of the catechol (BCl$_3$, CH$_2$Cl$_2$) gave the hexadentate spermidine catecholate ligand 13 (30%).

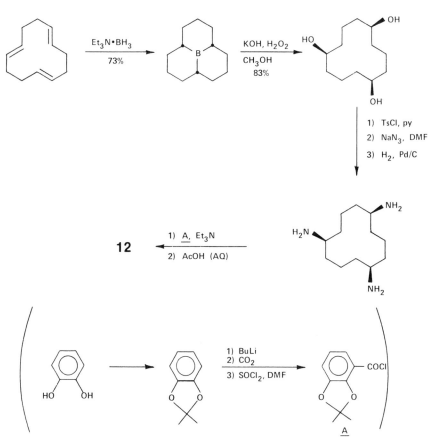

Chart 4.7

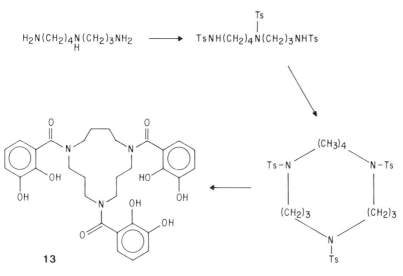

Chart 4.8

5. OXYMACROCYLES

This somewhat nondescript title for molecular classification is in fact distinguished by only three structures in the ionophore domain thus far: the macrotetrolides, the synthetic crown ethers, and a macrocyclic dilactone (boromycin). Since the crown ethers share many common properties with the naturally occurring macrotetrolides, we will consider them as synthetic analogs thereof, although they were not originally prepared for that reason.

A. Macrotetrolides

This group of macrotetrolide antibiotics, of which several are represented by the structure below, was originally isolated from various actinomycetes and structures were elucidated as 32-membered macrocyclic tetraesters (cyclodepsides).[113] Nonactin was isolated in 1955 as the original member of the actin family.[112] The compounds exhibit antimicrobial activity (G+, fungi), which increases with increasing length of R, and good insecticidal activity against mites.[114,115]

The actins afford one to one complexes with many alkali/alkaline earth metal ions. Specifically, nonactin exhibits a selectivity sequence of $NH_4^+ > K^+ \approx Rb^+ > Cs^+ > Na^+ > Ba^{2+}$.[116] The K/Na selectivity is between that observed for valinomycin and the enniatins. The first crystal structure of an ionophore-metal complex was that of the nonactin-potassium thiocyanate complex.[117] The potassium is completely dehydrated with its coordination sphere comprised of the four tetrahydrofuran oxygens and the four ester carbonyl oxygens. This arrangement creates an overall "tennis ball seam" configuration of the carbon framework as it wraps around the metal.

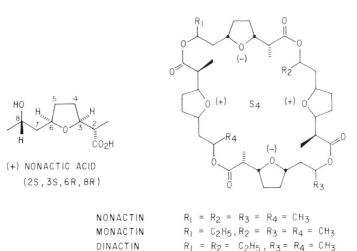

(+) NONACTIC ACID
(2S, 3S, 6R, 8R)

NONACTIN	$R_1 = R_2 = R_3 = R_4 = CH_3$
MONACTIN	$R_1 = C_2H_5, R_2 = R_3 = R_4 = CH_3$
DINACTIN	$R_1 = R_2 = C_2H_5, R_3 = R_4 = CH_3$
TRINACTIN	$R_1 = R_2 = R_3 = C_2H_5, R_4 = CH_3$
TETRANACTIN	$R_1 = R_2 = R_3 = R_4 = C_2H_5$

Nonactin is composed of four nonactic acid units with overall *meso* configuration due to the alternating sequence of (+) and (−) nonactic acid diastereomers. Four groups have independently reported syntheses of the basic nonactic acid unit, two of whom have converted the monomer into the cyclic tetramer, nonactin.

Gerlach and co-workers investigated[118] two different routes to the nonactic acid building block (Chart 5.1). In the first approach 2-acetonylfuran was functionalized electrophilically at the 5-position employing chemistry introduced by Eschenmoser et al.[119] with N-cyclohexyl-N-propenylnitrosonium ion. Hydrolysis afforded the 2,5-disubstituted furan with all the required carbons for nonactic acid. After oxidation, the required *cis*-disubstituted tetrahydrofuran was obtained by catalytic hydrogenation over rhodium (palladium affords *cis/trans* mixtures). Sodium borohydride reduction yielded mostly the 8-epinonactic acid.

An alternative approach involved the early establishment of the desired relative stereochemistry at C-6 and C-8 (Chart 5.2). The desired *threo*-1-octen-5,7-diol, generated by terminal alkylation of the dianion of acetylacetone followed by reduction and chromatographic separation, was acetylated and ozonized. Horner-Wittig reaction on the resultant aldehyde afforded an isomeric mixture

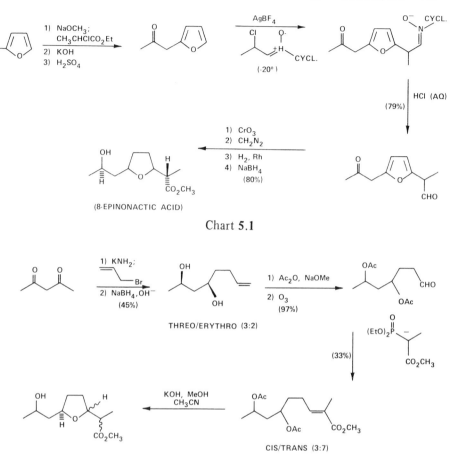

Chart **5.1**

Chart **5.2**

of α,β-unsaturated esters. Base-catalyzed cyclization produced the methyl ester of nonactic acid as the main product (60%).

The "oligomerization" of nonactic acid into the linear tetraester, 6 (Chart 5.3), was accomplished through mixed anhydride methodology. Appropriately protected monomers, 1 and 2, were prepared by transesterification of the *t*-butyl ester (1), and monobenzylation of the dianion to the 8-*O*-benzyl-2. Coupling was achieved with trimethylbenzenesulfonyl chloride in pyridine in good yield. The selective removal of benzyl on a portion of substrate 3 afforded the deprotected diester, 5, and acid hydrolysis on the remainder of 3 yielded the acid 4. Coupling and final deprotection gave the penultimate, isolated intermediate, 6. This linear tetramer was cyclized by the pyridyl thioester method[126] employing triphenyl-

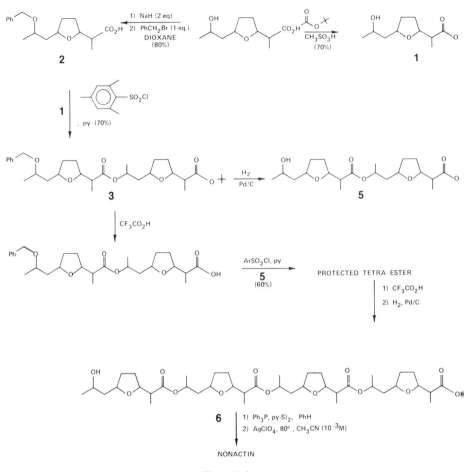

Chart 5.3

phosphine/dipyridyl disulfide[127]. The thioester intermediate was cyclized with
Ag(I) assistance. Since the synthesis was carried out with *rac*nonactic acid, four
possible cyclic configurations are possible. Two *meso* forms, the natural configuration with S_4 symmetry as well as Ci, and the enantiomeric pairs with C_1
and C_4 symmetry. Only three of the four possible were isolated and separated
chromatographically (40% yield) in 1:5:2 relative proportions. The last component represented authentic nonactin as determined by physical, biological, and
spectral properties. The authors assigned the Ci configuration to the initial
component and C_1 symmetry to the major isomer, employing primarily NMR
characteristics.

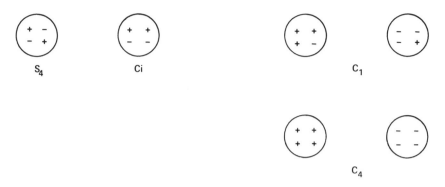

The initial synthesis of the nonactic acid diastereomers was carried out by Beck and Henseleit[128] in 1971. Alkylation of 2-methoxycarbonylmethyl furan with methyl iodide gave the α-methyl derivative in 60% yield (Chart 5.4). The C_7-C_9-alcohol fragment was introduced via a Friedel-Crafts alkylation of 2-methylbut-1-en-3-one. After reduction to the *cis*-tetrahydrofuran, this five-carbon fragment was truncated to the required three-carbon group employing Baeyer-Villiger oxidation of the methyl ketone. Hydrolysis afforded the diastereomeric nonactic acids in an overall five-step synthesis.

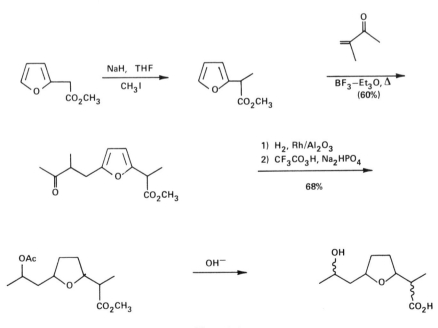

Chart 5.4

Schmidt and co-workers[129] have also prepared nonactic acid and carried it on to the cyclic tetraester (Chart 5.5). The reaction of 2-lithiofuran with propylene oxide affords the expected secondary alcohol. The introduction of the three-carbon carboxylic acid unit begins with a Villsmeyer formylation. A Wittig reaction followed by a Rh(III) catalyzed carbonylation completes the assembly. Silver(I)oxide-mediated oxidation to the required acid followed by hydrogenation produces nonactic acid. If (−)-propylene oxide is employed [from (+)-lactic acid], the 8-epimonactic acid is obtained, which can be cycled to the (+)-nonactic acid by an efficient inversion at C-8 with potassium acetate (DMSO) displacement of the 8-tosylate.

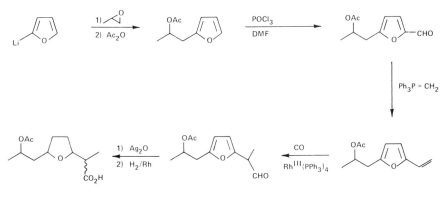

Chart 5.5

The synthesis of nonactin was completed by coupling potassium (−)-nonactinate with the 8-epitosylate of (+)-nonactic acid benzyl ester (70% yield) to produce (−)-nonactinyl-(+)-nonactic acid benzyl ester (**1**). The other "diester," 8-epi-(−)-nonactinyl-(+)-nonactic acid benzyl ester (**2**), was similarly prepared. Hydrogenolysis of **1**, conversion to the potassium salt with $KHCO_3$, and reaction with the 8-tosylate of **2** afforded the linear tetraester [(−)·(+)·(−)·(+)·benzyl ester]. Hydrogenolytic deprotection and cyclization with the conditions employed by Gerlach (20%) yielded natural nonactin.

White and coworkers[130] reported two approaches to racemic nonactic acid (Chart 5.6). Acylation of 1-(2-furyl)-2-propanol, derived via the same chemistry as the Schmidt approach, produced the desired 2,5-disubstituted furan. The incorporation of the last carbon (2-carboxyl) was accomplished via a Wittig reaction and the correct oxidation state by diborane followed by a Jones oxidation. Reduction with lithium tri(sec-butyl)borohydride yielded primarily the 8-epinonactic acid, which could inverted at C-8 with the Mitsunobu reaction.

All the approaches to nonactic acid described thus far have utilized the rhodium-catalyzed hydrogenation of furan to derive the necessary cis-2,5-disubsti-

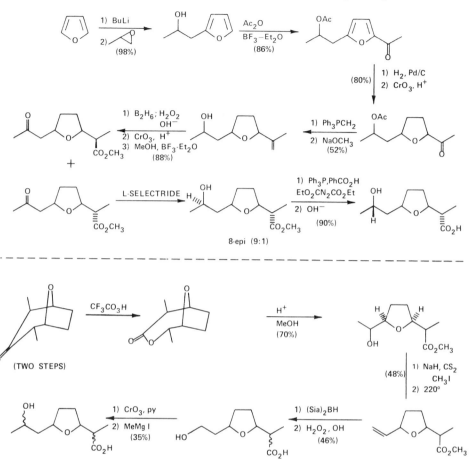

Chart 5.6

tuted tetrahydrofuran. The second approach by White et al. derives the *cis*-tetrahydrofuran via preincorporation into a bicyclic structure. Employing the chemistry developed separately by Noyori and Hoffman, 2,5-dibromo-3-penta-none was cyclo-added to furan (Zn/Cu). After hydrogenation of the remaining double bond the *cis*-2,5-disubstituted tetrahydrofuran was liberated from the (3.2.1)-bicyclooxipane by a Baeyer-Villiger oxidation followed by a methano-lysis. In contrast to the former approach, this route now required the elabora-tion of the hydroxyl side chain with an extra carbon. Pyrolysis of the methyl xanthate followed by hydroboration with diisoamyl borane yielded the primary alcohol, which had unfortunately lost configurational fidelity at C-2 during this operation. Collins oxidation followed by a Grignard on the resultant aldehyde gave the diastereomeric nonactic acids.

B. Crown Ethers

Cyclic polyethers, or crown ethers, and macrobicyclic polyethers, or cryptands, in a general sense resemble the macrotetrolides in their complexing capabilities. In many cases through the precise tailoring of the cavity size to cation diameter the stability of metal-crown complexes surpasses that of the natural ionophores. This is even more apparent with the three-dimensional ligand cryptands. Because of these aspects the crown ethers and cryptands are superb complexation-solubilization agents of a metal cation in an organic phase, but are poor transport agents[139] because of high binding constants due to the lack of a stepwise-low energy pathway for the decomplexation process,[131a] a feature also in common with siderophores.

Several examples of crown ethers and cryptands are depicted below. The pioneering work of Pederson reported in 1967 first opened up this extremely interesting area. Most of the synthetic operations employed for assembling the cyclic ethers involved modified Williamson ether syntheses. Many cyclization

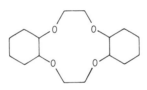

DICYCLOHEXYL-12-CROWN-4

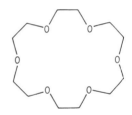

18-CROWN-6

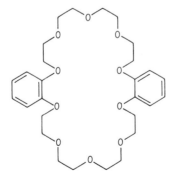

DIBENZO-30-CROWN-10

BICYCLO-2,2,2-CRYPTAND

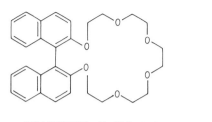

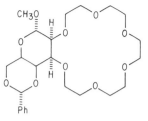

BINAPTHTHYL-18-CROWN-6

yields are remarkably high with the appropriate choice of the counter metal cation, presumably through a "template" effect. Since most of this work has been reviewed we will not incorporate it here.[3,131-136] Dicyclohexyl-12-crown-4 indicates a complexation preference of Na > K whereas the larger dicyclohexyl-18-crown-6 is more selective for potassium (K > Rb > NH$_4$ > Na ~ Li)[131,137]. The concept of chiral crown ethers, such as the binaphthyl-18-crown-6, as resolving agents for amino acids and chiral catalysts for asymmetric synthesis,[140] has been notably exploited by Cram's group,[141] Stoddart and co-workers,[142] and others.[143]

Although having investigated the properties of some crown ethers,[138] Lehn and colleagues have been primarily responsible for fashioning the interesting bicyclic ligands, the cryptands.[144] With the tertiary nitrogen as the focal point of the three-dimensional array of ligands, the synthetic chemistry involves amide bond formation to this incipient amide nitrogen followed by reduction to the amine. Also prepared have been the cylindrical and spherical molecular cryptands 7 and 8. These represent excellent topology for binuclear cryptates in the cylindrical case and complete envelopment in the spherical case yielding activation energies for dissociation > 16 kcal/mol. The bicyclo-2,2,2-cryptand exhibits a selectively ordering of K > Rb > Na > Cs > Li.

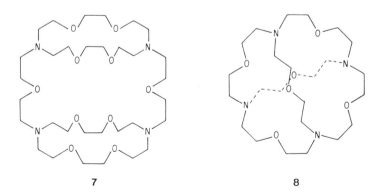

7   8

C. Boromycin

In 1971 Dunitz et al. reported the structure of the first boron-containing natural product,[190] boromycin. Isolated from *Streptomyces antibioticus*,[192] its structure was established by X-ray crystallography of the Rb⁺ salt of the compound after the removal of D-valine. Boromycin forms a cup-shaped ligand around the rubidium with interactions with the two hydroxyls as well as the three ether oxygens of the tetrahydrofuran/pyrans and the valine esterified oxygen. Boromycin is a monovalent ionophore which induces K⁺ loss from cells[193] and induces a cortical granule discharge in hamster eggs to prevent spermatozoa binding.[194] It also exhibits G+ antibacterial and antifungal activity.

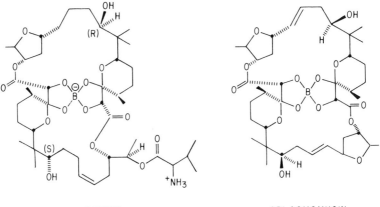

BOROMYCIN   APLASMOMYCIN

A closely related dilactone from a marine isolate belonging to *Streptomyces griseus* was recently identified. Named aplasmomycin,[191] since it exhibited *in vivo* antiplasmodium activity as well as G+ antibacterial activity, it is a symmetrical macrocyclic ligand encompassing boron. The structure was ascertained by X-ray crystallography of the Ag⁺ salt. Coordination included the two hydroxyls as well as two oxygens of the borate and one water molecule.

An approach to a total synthesis of boromycin with its 17 asymmetric centers by White and co-workers[194] involves a strategy of preparing the two halves resulting from lactone bond disconnections. Specifically (Chart 5.7), the preparation of the "southwestern" unit proceeds from the morpholine enamine of isobutyraldehyde (9). After acylation, the unmasked aldehyde was protected as the dimethyl acetal and the acyl ketone was reduced and protected. Application of the tiglate dianion, reported by Katzenellenbogen and Crumrine,[195] in an aldol fashion at the β-methyl of tiglic acid on the deprotected aldehyde gave 10. After hydrogenation of the double bond, 10 was converted to the δ-lactone, 11 (this was using "old" PtO₂; a "new" lot also removed the benzyl group). Introduction of the second hydroxyl group for ultimate boronate formation, as well the adjacent macrolide lactone carbon, was achieved on the δ-lactone carbonyl by the chemistry recently reported by Meinwald and co-workers.[196] Condensation with the protected α-lithioglycollate followed by acetonide protection afforded **12**. This unit was subsequently elaborated to the complete carbon framework of the lower half of boromycin, **18**, through alkylation of the ketone **17**, derived via debenzylation and oxidation of **12**, with the allyl bromide **16**.

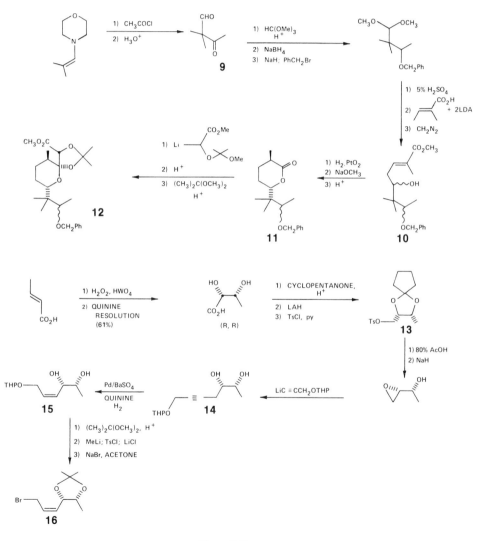

Chart 5.7

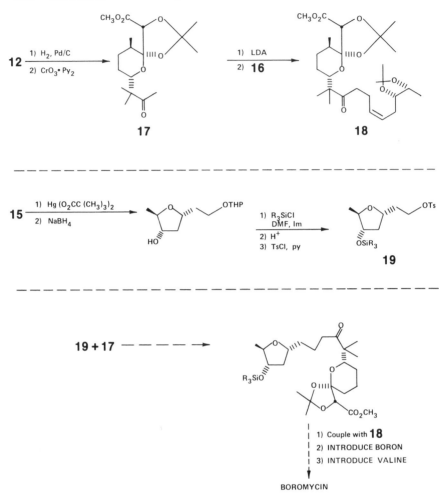

Chart 5.7 (Continued)

This allylic bromide, **16**, was prepared by oxidation of crotonic acid followed by quinine resolution. The R,R-isomer diol was protected and the acid converted to the tosylate. Acetic acid hydrolysis followed by base afforded the terminal epoxide, which was opened regioselectively with the anion of propargyl alcohol tetrahydropyranyl ether. It was noted that this regioselective epoxide opening was only successful with the adjacent hydroxyl unprotected. Limited hydrogenation afforded the cis olefin which was then converted to the allyl bromide **16** with standard chemistry. The synthesis of **15** actually served two purposes in that it also provided the precursor for the furan grouping (**19**) in the upper half of boromycin. Olefin-**15** could be converted to a mixture of *cis*- and *trans*-2,5-disubstituted tetrahydrofurans in 85% yield by treatment with Hg(II) followed by borohydride reduction of the organomercury. The reaction with the bulkier mercuric pivalate was more stereoselective than with the acetate of Hg(II).

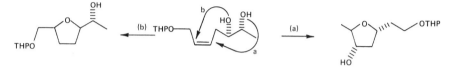

The preferred regioselectivity results from pathway a (*exo*-5-*trig*) as opposed to the alternative b pathway (*endo*-5-*trig*; disfavored). The THP-ether can be hydrolyzed and converted to the tosylate leaving group as depicted in structure **19**.

The strategy for completion of the synthesis involves the alkylation of **17** with **19** to give the upper half and its subsequent coupling with the lower half, perhaps with the assistance of boronate formation.

6. POLYETHER ANTIBIOTICS

One of the most recent entries into the field of ionophores is the polyether class. Although the first example of this class was isolated as early as 1950 (nigericin), structural elucidations have only been reported in the last ten years. This most recent class of ionophores has been growing dramatically both in numbers (now over 40) and uses.

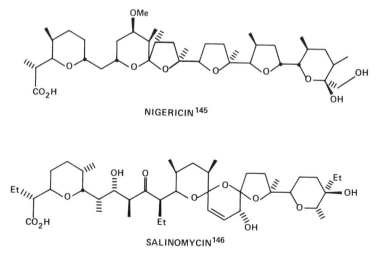

NIGERICIN [145]

SALINOMYCIN [146]

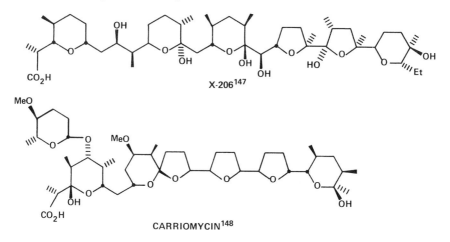

X-206[147]

CARRIOMYCIN[148]

As a class they are distinguished by a linear carbon framework containing many tetrahydrofurans and pyrans, multiple centers of asymmetry, and a structure often terminating in a carboxylic acid. Several representative structures are depicted below: a more complete, though by now dated, list is given in Westley's review on polyether antibiotics (1977).[6] Another feature distinguishing these ionophores from those discussed earlier is their excellent transport of divalent ions. In many cases these are 2:1 ligand to metal complexes. As is evident from their complex structures, the process of structure elucidation and, particularly, synthesis is difficult, and only very recently have successful synthetic approaches been completed with three representatives: lasalocid, monensin, and calcimycin.

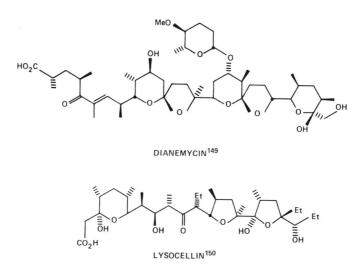

DIANEMYCIN[149]

LYSOCELLIN[150]

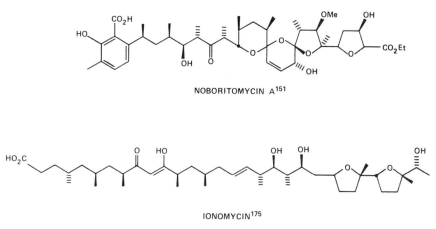

NOBORITOMYCIN A[151]

IONOMYCIN[175]

A. Lasalocid

Coisolated with X-206 and nigericin from *Streptomyces* in 1951 by Berger[152] et al., the structure of lasalocid (X-537A) was determined in 1970 by X-ray crystallography.[153,154] Relative to the other polyether antibiotics at the time, lasalocid was the smallest and unique in that it possessed an aromatic ring. Four homologs have also been identified as minor components of the lasalocid complex (4-16%).[156] The X-ray structure revealed an unsymmetrical 2:1 complex (Ba^{2+}), wherein one lasalocid forms a cyclic ligand donating the carboxyl oxygen and terminal hydroxyl (tetrahydropyran) together with a molecule of water for coordination. The other lasalocid ligand employs both of these heteratoms as well as the other four "internal" oxygens for the complex. This unusual sandwich complex (with Ba^{2+}) also appears to be preferred in solution as well, as determined by NMR studies.[155]

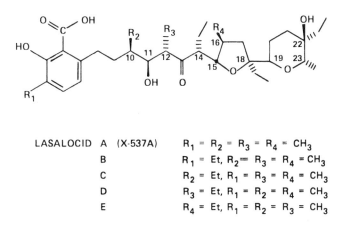

LASALOCID	A (X-537A)	$R_1 = R_2 = R_3 = R_4 = CH_3$
	B	$R_1 = Et, R_2 = R_3 = R_4 = CH_3$
	C	$R_2 = Et, R_1 = R_3 = R_4 = CH_3$
	D	$R_3 = Et, R_1 = R_2 = R_4 = CH_3$
	E	$R_4 = Et, R_1 = R_2 = R_3 = CH_3$

The complexes of lasalocid that have been examined (Ag^+, Na^+, Ba^{2+}) exhibit irregular coordination. Lasalocid is unremarkable in its ion specificity, for it complexes and transports all alkali/alkaline earth metals, lanthanides, Th(IV), and organic amines,[29,48,157] with Cs^+ being the most preferred ($Cs > Rb \approx K > Na > Li; Ba > Sr > Ca > Mg$).[158]

Biogenetically, lasalocid is derived from acetate, propionate, and butyrate in the pattern depicted below.[6] It is speculated that isolasalocid, a co-occurring isomer, is produced by a regioselective opening of an intermediate epoxide to produce the five-membered tetrahydrofuran ring in conjunction with the six-membered tetrahydropyran of lasacocid.

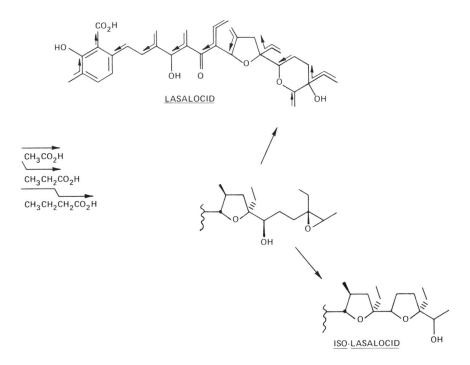

Besides its antimicrobial activity, lasalocid also exhibits coccidiostat activity (for which use FDA approval has been obtained) and improved ruminant feed utilization in cattle.[5,6] Additionally, an extensive amount of investigation has been carried out primarily by Pressman and co-workers[159] and, independently, Schwartz[160] et al. on the inotropic effects of lasalocid. Beneficial effects in stimulating myocardial contractility include a decrease in the coronary resistance and the peripheral vasculature. Initially attributed to Ca^{2+} mobilization or catecholamines, the mechanism of action now appears considerably more complex.[4]

The total synthesis of lasalocid A[161] reported in 1978 by Kishi and co-workers represents the first synthesis of a polyether antibiotic. Some synthetic work had preceded this chemistry, including mostly aromatic analogs derived from lasalocid, as well as some degradative work demonstrating a *retro*-aldol cleavage,[162] the mode of final bond formation chosen by Kishi.

The synthesis began with the assembly of an appropriately substituted 1,5-diene as the substrate to form the right half (tetrahydrofuran and pyran) of the molecule. The preparation of the first 1,5-diene depicted in Chart 6.1 employed the Claisen rearrangement approach, popularized by Johnson[163] and co-workers, to *trans*-trisubstituted olefins. The initial orthoester gave a monoolefin which was converted to the second allylic alcohol and then elaborated through a second orthoester-Claisen to 1.

In chemistry presumably similar to that employed in the Claisen rearrangement sequence to prepare it, 1 was converted to the aldehyde (Chart 6.1) and then to the phenyl ketone via a Grignard followed by a Jones oxidation. A highly stereospecific reduction with a complexed aluminum hydride recently reported by Mukaiyama et al.,[164] lithium aluminum hydride and 2-(o-toluidino-methyl)pyrrolidine, gave the alcohol shown, in a preponderence (as predicted by Cram's rule) of 10:1 over its isomer. Resolution of the 1-α-methylbenzyl urethane via HPLC gave the (−)-alcohol, 2.

Elaboration of the trisubstituted olefins to the tetrahydrofurans was accomplished with the Sharpless epoxidation procedure to give epoxide 3.[165] This

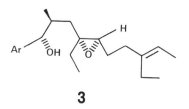

3

epoxide was opened with acetic acid to the tetrafuran 4 as an 8:1 ratio of stereoisomers. Reepoxidation afforded the undesired epoxide of the terminal double bond, so inversion was required. Acid-catalyzed opening to the diol followed by formation of the secondary tosylate and base-catalyzed closure gave the inverted epoxide. This was transformed with acetic acid to the ditetrahydrofuran **6**, after protection as the methoxymethyl ether, with a stereoisomer preference of 5:1.

The aromatic ring carried along thus far served as a latent α-ethyl, ethyl ketone required for the right half. Birch reduction afforded the cyclohexa-1,4-diene. This was ring-opened via epoxidation and periodate cleavage. Reduction produced bishomoallylic alcohol **7**. Conversion to alkene was accomplished via reduction of the ditosylate. Diborane followed by Jones oxidation served to introduce the ketone from the olefin.

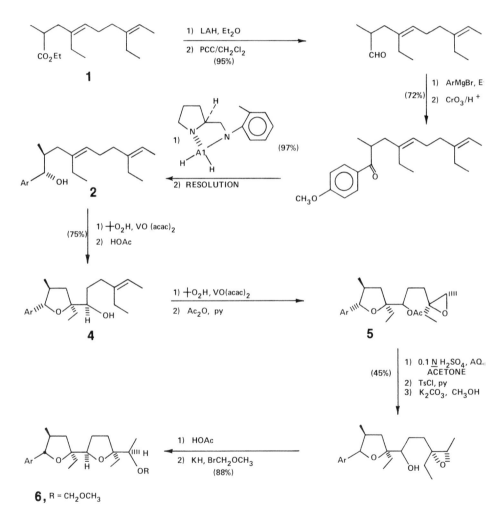

Chart 6.1. Right half of lasalocid.

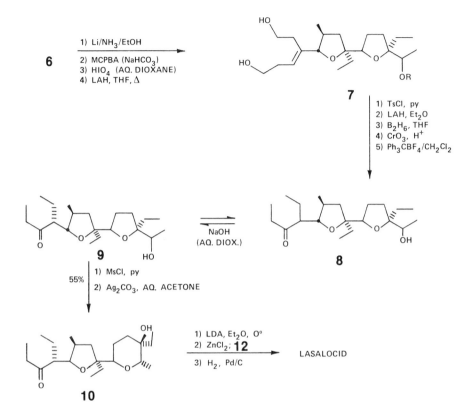

6

1) Li/NH₃/EtOH

2) MCPBA (NaHCO₃)
3) HIO₄ (AQ. DIOXANE)
4) LAH, THF, Δ

7

1) TsCl, py
2) LAH, Et₂O
3) B₂H₆, THF
4) CrO₃, H⁺
5) Ph₃CBF₄/CH₂Cl₂

9

NaOH
(AQ. DIOX.)

8

55%

1) MsCl, py
2) Ag₂CO₃, AQ. ACETONE

10

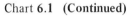

1) LDA, Et₂O, 0°
2) ZnCl₂; **12**
3) H₂, Pd/C

LASALOCID

Chart **6.1** (Continued)

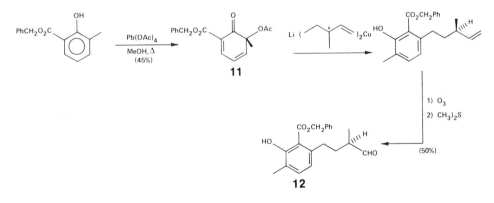

11

12

1) O₃
2) CH₃)₂S

(50%)

Pb(OAc)₄
MeOH, Δ
(45%)

Chart **6.2.** Left half of lasalocid.

305

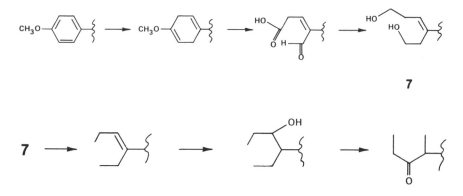

7

Deprotection of the methoxymethyl ether with the Barton procedure afforded **8** in 13% overall yield from **6**. Ketone **8** gave a 1:1 equilibrium mixture under basic conditions with the desired isomer, **9**. Ketone **9** was shown to be identical with the right half of isolasalocid, derived via a *retro*-aldol of the natural product. Silver(I) mediated solvolysis to the lasalocid series, **10**. The preparation of the salicylate right half (Chart 6.2) began with the *m*-methylsalicyclic acid, benzyl ester, which was oxidized to the cyclohexadienone, **11**.[166] Conjugate addition with lithium di(ℓ-3-methyl-4-pentenyl) cuprate and *in situ* rearomatization followed by oxidative cleavage of the double bond yielded the left-half aldehyde, **12**, required for the final aldol bond connection with the right half, **10**.

After surveying a number of conditions, Kishi's group found that the zinc-enolate chemistry of House provided four aldol products (67% yield) in a 40:10:7:3 ratio. The major product, after debenzylation, was identical with natural lasalocid. Better stereoselectivity was found with DME as solvent but the efficiency of condensation was much lower.

An alternate preparation of the isolasalocid ketone, **8**, was concurrently reported out of Kishi's group (Chart 6.3).[167] Beginning with enone **13** this alternate approach served to prepare the terminal tetrahydrofuran first. After the epoxidation, reduction, and cyclization conversion of **13** to the tetrahydrofuran, **14**, it was resolved at this stage, in a manner similar to that described in Chart 6.1. Addition of the optically active Grignard to the aldehyde introduced the required carbon skeleton for the adjacent tetrahydrofuran. Ozonolysis produced the cyclic hemiacetal to which was added the magnesium enolate of 3-hexanone. Acid-catalyzed recyclization followed by debenzylation gave **8** in a 10% overall yield, 11-step sequence from **13** (compared to 23-step-3.3% yield in the earlier approach starting from **1**).

Ireland and colleagues have embarked on a total synthesis of lasolocid A.[231] The overall synthetic plan is convergent at two points: the joining of the tetra-

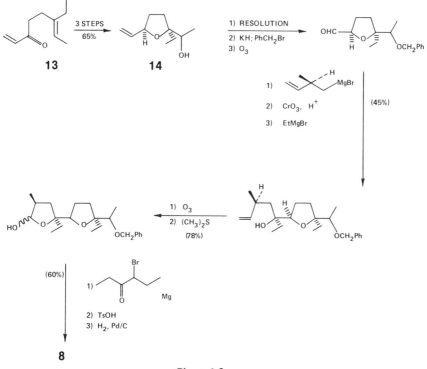

Chart 6.3

hydropyran A-ring to the tetrahydrofuran B-ring and the coupling of this assemblage to the aromatic left half. The approach relies on a stereoselective ester enolate Claisen rearrangement to accomplish the former and the precedented aldol condensation for the latter. Specifically, the tetrahydrofuran/pyran bond formation involves a Claisen rearrangement of allyl ester **15** to a stereochemically correct precursor of the A and B rings, **16**. To secure the required tetrahydrofuran/pyran building blocks in chiral form for **15**, the Ireland group utilized monosaccharide starting materials.

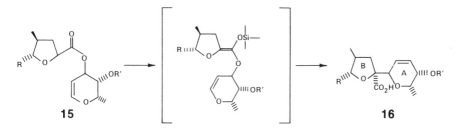

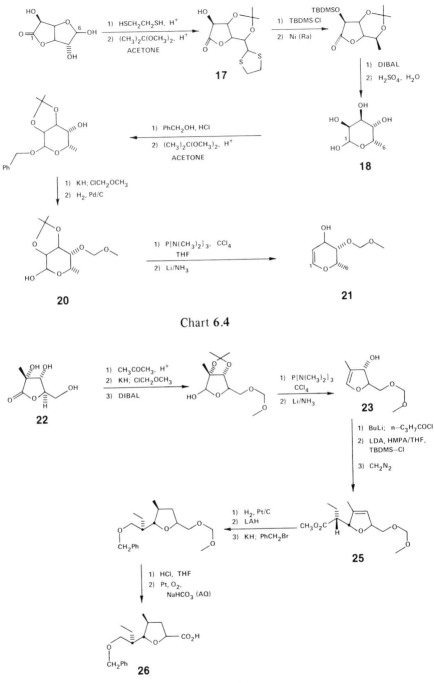

Chart 6.4

Chart 6.5

pentane. Subsequently, a four-step sequence gave the desired **10**, apparently without any epimerization at C-14 (for example, isolasalocid epimerization, Chart 6.1).

The left-half, aromatic portion of lasalocid was secured by synthesis from nonaromatic precursors (Chart 6.7). Pentenyl bromide-**31**, prepared in four steps from 2-buten-1-ol, underwent a conjugate addition as its Grignard salt to pyrone-**32**. The resulting dihydropyrone was oxidized to pyrone-**33**. An regiospecific ynamine cycloaddition with extrusion of CO_2 yielded the dibenzylaniline, **34**. Vicinal hydroxylation of the terminal olefin with osmium tetroxide/N-methyl-morpholine-N-oxide interestingly afforded the diol acetonide directly.[233] After hydrogenolytic N,N-debenzylation and deesterification of the methyl ester with n-propyl mercaptide, the resulting amino acid was esterified and diazotized. Subsequent periodate treatment of the liberated diol yielded the left-half aldehyde,

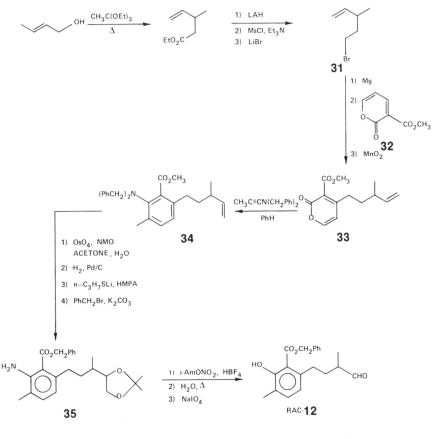

Chart 6.7

12, currently in racemic form. The chemistry remaining for the completion of the total synthesis involves preparation of optically active 31 and the aldol coupling of the two halves. The latter has actually been accomplished with the *retro*-aldol fragments from natural lasalocid employing the zinc enolate of the right half in dimethoxyethane at 0° and a 4-min reaction time with aldehyde 12.

B. Monensin

Isolated in 1967 from *Streptomyces cinnamonensis*, the structure of monensin was established by X-ray crystallographic analysis of its silver salt[168] and later of the free acid.[16] Monensin, like lasalocid, is accompanied by several homologs in

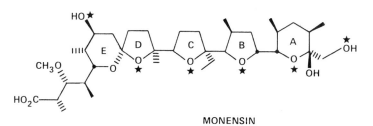

MONENSIN

the fermentation isolate; however, in contrast, it can be produced in unusually high titer by this streptomycete. The X-ray data depicts an irregular hexacordinate ligand array (starred oxygens) around silver and hydrogen bonding between both oxygens of the carboxylic acid and the two hydroxyls of the A-ring. Examination of the solution phase conformation both potentiometrically and spectroscopically[170] have revealed a structure nearly identical to the solid state.[171]

Monensin gave quite a high Na/K selectivity ratio[47,172], varying from 7 to 13 as a function of solvent, and exhibited a selectivity sequence of Na $\gg$ K > Rb > Li > Cs.[157] The biosynthetic origins of monensin parallel lasalocid in that it is derived from five acetates, seven propionates, and one butyrate. The methoxymethyl originates from methionine.[173]

Although monensin exhibits weaker antimicrobial activity than its structurally similar relative, nigericin, it also displays activity as a coccidiostat. It has been a successful drug in the control of coccidia infections in poultry since its 1971 market introduction. However, even greater sales have been realized as a feed additive for the improvement in the efficiency of feed utilization in cattle. Monensin has also been popular in the study of polyether-mediated cardioregulatory effects. It exhibits a fivefold greater inotropic potency than lasolocid.[174] Interestingly, its Ca^{2+} transporting capability relative to lasalocid is only 10^{-4} and norephinephrine transport is 1/300th, yet its sodium transport is 30 times greater.[4]

In a structural comparison with lasalocid, monensin contains 17 asymmetric centers compared to lasalocid's 10, and it is devoid of an aromatic chromophore and a formal aldol unit. However, utilizing methodology developed in their synthesis of lasalocid, Kishi's group has successfully completed a total synthesis of monensin.[178] This remarkable effort in fact involves a convergence of a right half (rings A-C) and a left half, in an aldol condensation to form the dioxospirane-D,E rings, the deketalization of which reveals the aldol unit.

The left half[176] was prepared starting from 2-furylacetonitrile (Chart 6.8). As will become evident, the furan is a latent methylcarboxylate group. The 2-furylacetonitrile was converted to 2-(2-furyl)propionaldehyde, seen earlier as an intermediate in the Schmidt nonactic acid synthesis, which was subjected to a Wittig reaction with carbethoxyethylidene triphenylphosphorane to give the *trans* (*E*) ester with approx. 5% *cis* (*Z*). The first diasteromeric centers were

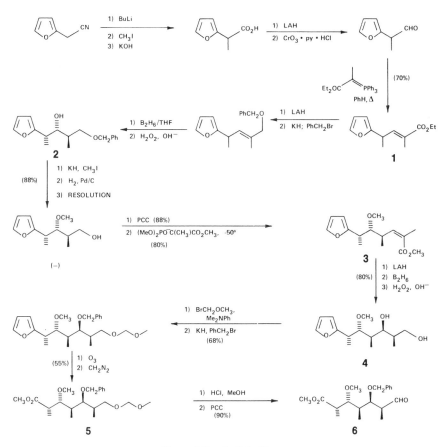

Chart 6.8. Left half.

introduced by an unusual hydroboration of the protected alcohol analog of **1**. This hydroboration proceeded with excellent stereospecificity from the sterically less hindered α-face (furyl vs methyl) of the double bond to produce **2**, in 85% overall yield from **1**, in a ratio of 8:1 relative to the other diastereomer. At this point structure assignment was based on literature precedent.[179] After protection of the 2°-alcohol and deprotection of the 1°-alcohol, resolution was achieved, as in the lasalocid synthesis, via HPLC separation of the diasteromeric urethanes.

Cognizant that the Horner-Emmons modification of the Wittig reaction produces *cis*-stereoselectivity as a function of solvent, anion, and temperature, they converted the alcohol to the aldehyde and secured the *cis*-ester, **3**, in 64% yield from the alcohol (<98% *cis*) with the dimethylphosphonate anion at −50° in THF. Repetition of the hydroboration sequence on olefin-**3** gave the diastereomeric diol, **4**, as a 12:1 mixture. After differential protection of the 1°- and 2°-alcohols, the furan was oxidatively cleaved to the carboxylate. Removal of the methoxymethyl protecting group and pyridinium chlorochromate oxidation produced aldehyde **6** ready for the aldol with the right half. The structure was confirmed by comparison with a sample obtained via alternate chemistry wherein an intermediate was correlated with a degradation product of monensin. This alternate approach to **6** involved an aldol between a furylbutyraldehyde and

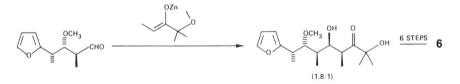

(1.8:1)

2-methyl-2-hydroxy-3-pentanone zinc enolate. However, since the diastereomeric selectivity was much poorer in this approach, the approach in Chart 6.8 was chosen.

The preparation of the 19-carbon backbone right half[177] (Chart 6.9) employs 2-allyl-1,3-propane diol, **7**, prepared from malonate and an allylic halide, as the building block. Monobenzylation was achieved by reduction of the benzyl acetal with the mixed alane, LiAlH$_4$-AlCl$_3$. Urethane resolution gave the S(−) alcohol, correlated in four steps with (−)2-methylpentanoic acid. Elaboration of **8** was through the Johnson orthoester Claisen chemistry on substrate **9**. After reduction of the ester and oxidation back to the aldehyde, a Grignard with *p*-methoxyphenylmagnesium bromide followed by oxidation and deprotection gave the homoallylic alcohol, **10**. Because of its symmetry, the S(+) alcohol could also be used in the preparation of **10** by a salvage process of debenzylation of **8** after protecting the other alcohol and then carrying through the same chemistry as on S(−)**8**.

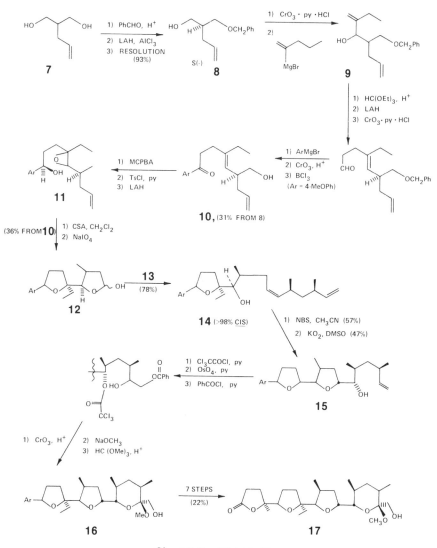

Chart 6.9. Right half.

Epoxidation with *m*-chloroperbenzoic acid produced only the mono-β-epoxide, presumably due to alcohol directed oxidation from the β-face of the proximate, more nucleophilic double bond, wherein the least steric repulsion exists between the vinyl ethyl group and the terminal allyl unit. Since the alcohol was no longer needed, it was removed by LAH reduction of its tosylate ester to give the benzylic alcohol **11**. Camphor sulfonic acid catalysis produced the tetrahydrofuran in a 7:2 diastereomeric ratio and periodate cleavage of the terminal olefin yielded the internal hemiacetal, 12.

This was reacted with ylid 13 to give the diene 14 containing the skeleton for the monensin A-ring. The preparation of the required phosphonium salt originated (below) from 3,5-dimethylphenol. Raney nickel hydrogenation followed by sodium dichromate gave cis-dimethylcyclohexanone. Baeyer-Villeger oxidation followed by basic hydrolysis and resolution via the α-methylbenzylamine

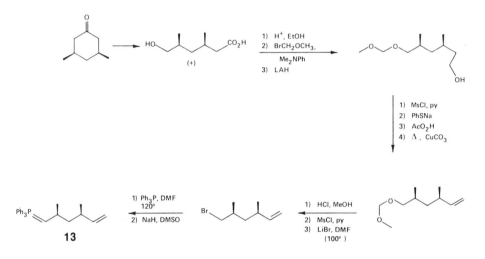

salt gave the (+)-hydroxy acid. After protecting the alcohol, the acid group was converted to a vinyl group via pyrolysis of the terminal sulfoxide. The alcohol was then deprotected and converted to the phosphonium salt from which ylid 13 was generated by NaH in DMSO.

Since the bishomoallylic alcohol, 14, could not be epoxidized with the Sharpless procedure employed in the lasalocid synthesis, the tetrahydrofuran was formed with N-bromosuccinimide and the resultant bromide converted with inversion to the alcohol 15 with superoxide in 27% overall yield. The terminal tetrahydropyran A-ring was then formed by osmium tetroxide oxidation of the terminal olefin sandwiched between two protecting procedures. The resulting 2°-alcohol was oxidized, methoxide liberated the protecting groups affording a single hemiketal, and trimethylorthoformate gave the cyclic ketal, 16, in 53% yield from 15.

In contrast to the lasalocid syntheses where the p-methoxyphenyl functioned as a latent α-ethyl, ethyl ketone, this aromatic residue in 16 was converted in a series of seven steps to a γ-methylfuranone. This was accomplished by first a Birch reduction from which the resultant 1,4-dienol ether was converted to the dimethyl ketal. Ozonolysis and subsequent treatment with magnesium bromide

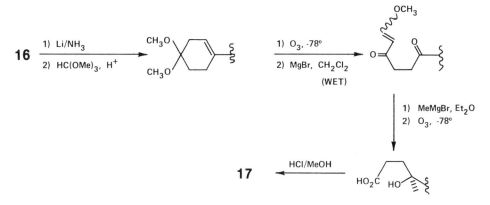

afforded the enol ether of the β-keto aldehyde. Stereospecific addition of CH_3MgBr to the ketone adjacent to the furan followed by a second ozonolysis afforded the precursor to **17** from which two aromatic ring carbons had been excised. Acid catalysis gave the furanone **17**, which was converted in quantita-

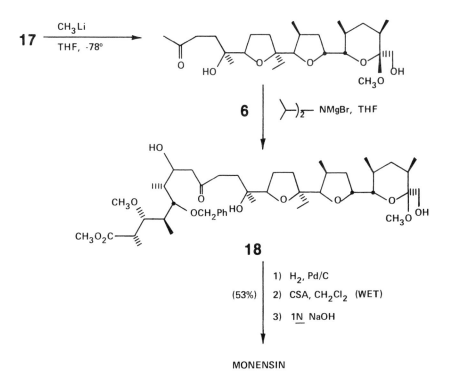

MONENSIN

Chart **6.10**

tive yield with CH_3Li to the methyl ketone required for the aldol joining of the right and left halves.

This critical coupling reaction was accomplished with diisopropylamine magnesium bromide in 21% yield at $-78°$ with the desired aldol, 18, in an 8:1 preponderance over the undesired diastereomer, whereas a 71% yield was obtained at $0°$ with a 1:1 ratio. Hydrogenolytic debenzylation liberated dihydroxy ketal. When exposed to acid, the desired dioxospirane, the isomer with the thermodynamically more stable tetrahydrofuran C-O bond (D-ring) of the spiroketal axial, was produced. These conditions also served to hydrolyze the A-ring methoxy ketal. Base hydrolysis of the methyl ester completed the synthesis of monensin.

As noteworthy as the first total synthesis of monensin has been, with its structural and stereochemical complexity, an equally notable, yet distinctive second total synthesis has just recently been successfully completed by Still and co-workers.[234] This group pursued a different strategy regarding the choice of various subfragments for a convergent total synthesis, although the final bond connection was an aldol analogous to the Kishi approach. It was envisioned that left, central, and right portions could be combined in the manner depicted to

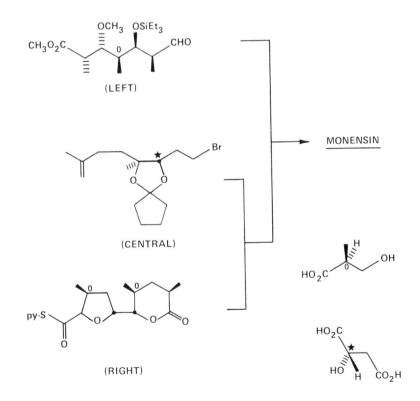

produce monensin and that these could be derived *retro*-synthetically from two optically active starting materials, (+)-β-hydroxy isobutyric acid and ℓ-malic acid.

Still undertook a multipronged degradation of monensin to secure key intermediates for the purposes of identification and structural proof of intermediates in the synthetic sequence and understanding the chemical manipulability and stability of these intermediates. These degradations of the natural substrate are summarized in Chart 6.11.

The utility of these degradation fragments becomes more apparent as the synthesis sequences unfold in the following charts. The preparation of the left portion (Chart 6.12) began with an appropriately protected and partially reduced derivative of (+)-β-hydroxisobutyric acid, itself derived from a micro-

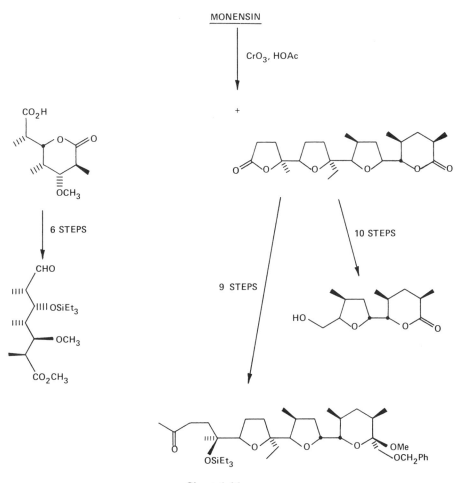

Chart 6.11

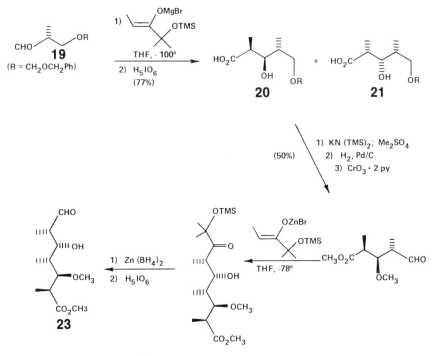

Chart **6.12**. **Left fragment.**

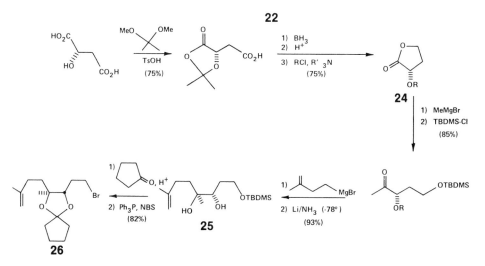

Chart **6.13**. **Central fragment.**

biological oxidation of isobutyric acid. This aldehyde, **19**, was condensed with Heathcock's enolate to yield, after oxidative cleavage of the resultant α-hydroxy-ketone, a 5:1 mixture of **20** and **21**. The postulated rationale for the predomi-nance of the anti-Cram product, **20**, is the chelation control effect of the proximate benzyloxymethoxy group with the magnesium salt directing the attack of the E-enolate syn to the α-methyl group. A later example of this same phenomenon was found in the synthesis of the central fragment (vide infra). The major isomer **20** was di-*O*-methylated, deprotected, and oxidized to an aldehyde, which in turn was subjected to a second aldol with the *E*-zinc enolate this time. The expected ("Cram") product **22** was reduced and oxidatively cleaved to provide the second three-carbon addition needed for the left fragment, **23**. Since conversion of **22** to **23** proceeded in less than acceptable yield, an alternative procedure was devised. In a maner conceptually related to the allyl chromium chemistry of Hiyama, *et. al.*,[238] reaction of the aldehyde derived from **20** with 4(2-*cis*-pentenyl)diethylaluminum afforded upon workup and flash chromato-graphy, **22a**, in high yield as the predominant (3:1) isomer. This could in turn be converted to silyl-**23** by hydrolysis, esterification, and silylation followed by ozonolysis to generate the required aldehyde.

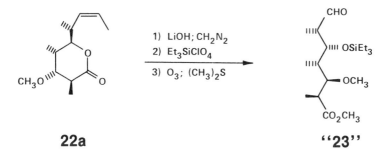

1) LiOH; CH$_2$N$_2$
2) Et$_3$SiClO$_4$
3) O$_3$; (CH$_3$)$_2$S

22a   **"23"**

The route to the central fragment (Chart 6.13) proceeded from the acetonide of ℓ-malic acid to the γ-lactone **24** (R = CH$_2$OCH$_2$Ph). After CH$_3$MgBr and silylation, a second highly stereospecific Grignard with isopentenyl magnesium

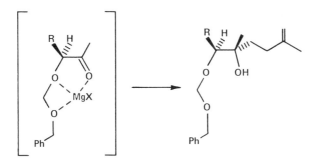

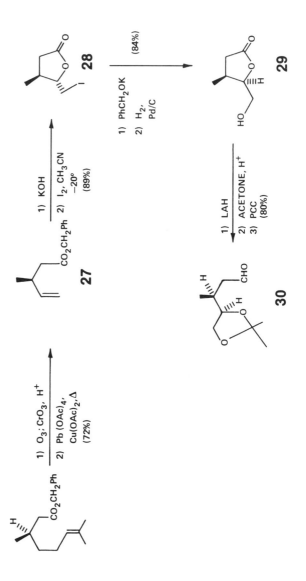

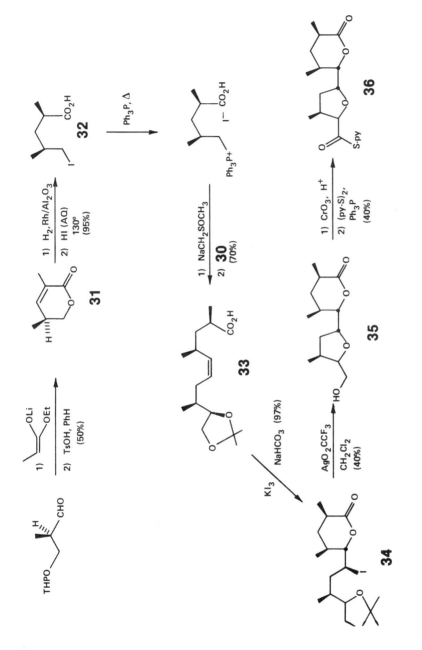

Chart 6.14. Right fragment.

bromide afforded diastereomer **25** in a 50:1 ratio over the alternate isomer. This is the second example of "chelation control," wherein in this case addition is directed *cis* to the α-hydrogen with the benzyloxymethoxy group one carbon closer than the earlier example. The diol was then protected as the labile cyclopentylidene derivative with concomitant desilylation. Conversion of the primary alcohol to a bromide gave **26** as the central fragment.

The right fragment resulted from a convergent approach beginning with benzyl citronellate and another form of β-hydroxyisobutyric acid (Chart 6.14). The former was oxidatively degraded to olefin **27** which upon iodolactonization afforded the γ-lactone **28** with a 20:1 preponderance of the isomer depicted. The newly created chiral center was inverted to the *erythro* stereochemistry of

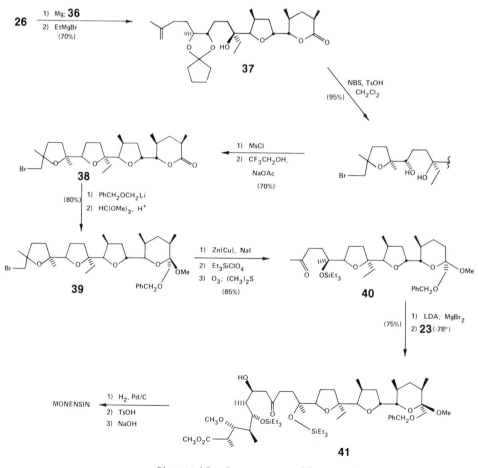

Chart **6.15**. Convergence of fragments.

29 via an intermediate epoxybenzyl ester. Standard chemistry gave synthetic intermediate **30**. A fragment to be condensed with **30** to afford the right-hand fragment was prepared from another form of the chiral isobutyric acid building block. Condensation with the lithium enolate of ethylpropanoate followed by acid-catalyzed cyclization yielded a δ-lactone. Reduction of the double bond yielded the isomer shown in a 10:1 preponderance and the resultant lactone was opened to iodide **32** with brief HI treatment. Direct alkylation of Ph_3P was achieved without solvent and the phosphonium salt condensed with aldehyde **30** in DMSO to give the *cis*-olefin **33**. A second iodolactonization followed by a silver(I)-mediated solvolysis produced the D- and E-ring portions of monensin, **35**. For coupling to the central fragment **35** was converted to the active ester **36**.

Sequential treatment of **36** with the magnesium Grignard of **26** and then ethylmagnesium bromide gave only isomer **37**, a product of chelation stereocontrol during the second Grignard. The terminal olefin was functionalized as the bromoether and the C-ring tetrahydrofuran formed by buffered solvolysis of the 2°-mesylate to give **38**. Addition of benzyloxymethyllithium followed by trimethylorthoformate gave the thermodynamic methyl ketal **39**. The required methyl ketone for the forthcoming aldol coupling was liberated by reductive elimination to the terminal isopropene followed by ozonolysis. It was found that the subsequent aldol with the left fragment **23** required protection of the hindered 3°-alcohol of **40**, so this was achieved via silylation with the perchlorate. Condensation with **23** proceeded in good yield to give a 3:1 ratio of the desired to undesired diostereomers. Deprotection, acid-catalyzed spiroketal formation, and hydrolysis, as described previously in the Kishi approach, afforded monensin.

C. Calcimycin (A-23187)

One of the more recent entries into the polyether-carboxylic ionophores is A-23,187 or calcimycin. Isolated from *Streptomyces chartreusensis*, its structure was divulged in 1974 as a benzoxazole and an α-ketopyrrole bridged by a 1,7-dioxospiro(5.5)undecane ring system.[180] This ionophore is unique not only

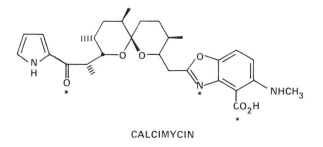

CALCIMYCIN

from a structural perspective but also because of its selectivity for divalent over monovalent cations. Particularly because of its excellent Ca^{2+} specificity and transport, its use has been cited in well over two thousand publications during its relative short history.[181]

Calcimycin forms a 2:1 complex with Ca^{2+} in a distorted octahedral array (starred atoms are liganding sites).[182] Interligand H-bonds occur in the crystalline state between the pyrrole and carboxyl, and an intraligand H-bond exists between the methylamino and carboxyl. Its structure in solution is apparently very similar.[183] This is the only known polyether-carboxylic acid-type ionophore to employ a nitrogen in a binding site, although recently a new pyrrole-containing ionophore (X-14547A) was reported,[186] which transports divalent cations, forms a dimeric ammonium salt, and, in addition to its antibacterial properties, exhibits hypotensive activity in animals.

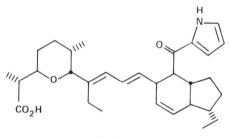

X-14547A

Its ion selectivity is reported to be Mn > Ca > Mg > Sr > Ba and Li > Na > K,[184,185] however, as in all cases previously cited, this is a function of which model is employed, ionophore concentration, pH, and other factors. For example, the transport kinetics for Ca^{2+} are different than for Mg^{2+} in isolated mitochondria with a turnover number of 45/sec for Ca^{2+} (calcimycin)$_2$.[181]

Calcimycin has been extremely instrumental in helping to understand the role of Ca^{2+} as a cellular control mediator and second messenger in hormonal processes. Essentially any Ca^{2+}-dependent phenomenon ranging from histamine release in mast cells to platelet aggregation to release of biogenic amino neurotransmitters has been examined with the aid of this ionophore. It also exhibits inotropic effects on isolated heart but appears to depress overall cardiovascular functions in the intact animal.

The first total synthesis of calcimycin was recently completed by Evans and co-workers.[187] Their approach was based on the proposition that the acyclic ketone diol form of the spiroketal would close under thermodynamic, dehydrative conditions to the required arrangement of calcimycin dioxospirane.

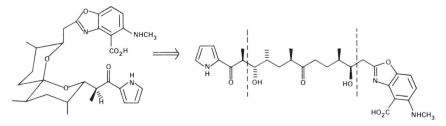

This belief, initially based on the axial vs equatorial stabilizations of the anomeric effect (2-3 kcal/mol), was born out in some initial model studies[188] (Chart 6.16). Selective ring-cleavage of the dihydroanisoles with ozone followed by alkylation with benzyl chloride afforded the protected bishomoallylic alcohols. Hydroboration (90%) gave a monoalcohol, which was then converted to the bromide. An interesting bisalkylation of the carbonyl anion equivalent, potassium methylthiomethylsulfoxide, yielded the vinyl ethers (E- and Z-iso-mers) in good yield as a presumed 1:1 diastereomeric mixture. This reaction, presumably involving bisalkylation of the central carbon followed by syn-elimination of CH_3SOH, was preferred over 1,3-dithiane methodology in this case. Although hydrolysis of the mixture could lead to four possible dioxospi-ranes, only two were produced (R = H) and the desired isomer predominated. These results were indicated by the expected C_2-symmetry of the predominant isomer as evidenced by ^{13}C-NMR and confirmed by X-ray crystallography. Analogous results were obtained with the dinor-model as depicted. Sondheimer and co-workers in an independent model study also reported the predominance of the desired dioxospiroundecane system. A double Grignard on methyl formate followed by a Jones oxidation gave the symmetrical ketone (Chart 6.16). Bromohydrin formation with acid-catalyzed ring closure yielded a 2.5:1 mixture of spiroketals.

With this background, the ultimately successful strategy employed by the Evans' group in the total synthesis involved aldol-type bond disconnections as depicted in the acyclic precursor. The benzoxazole chromophore was elaborated from the protected 2-amino-5-hydroxycarboxylate, 2 (Chart 6.17). Nitration afforded a 2/1 mixture of nitrophenols with the desired isomer predominating. After reduction to the amine, the oxazole was formed with acetyl chloride and the amine methylated in standard fashion.

The ketodiol-functionalized carbon framework for the spiroketal was con-structed from a central acetone unit and two 3-carbon alcohol units. The latter units were prepared from fermentation derived (S)-(+)-β-hydroxyisobutyric acid, a building block previously noted in Still's synthesis of monensin. One 3-carbon unit (4) was generated by silylation followed by reduction with diisobutylalumi-num hydride. The alcohol was then converted to iodide 4. Enroute to 5, benzyl-

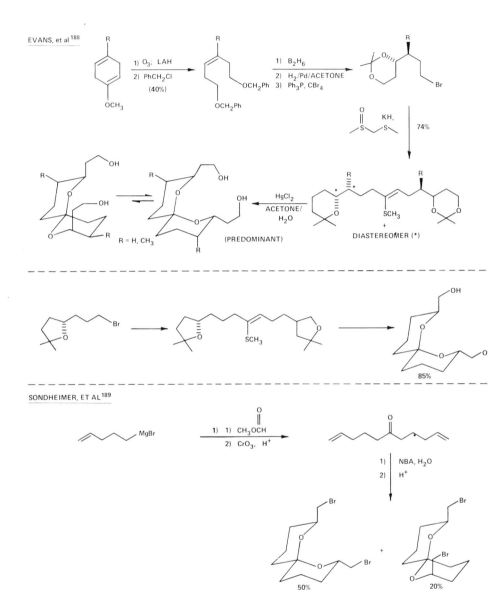

EVANS, et al [188]

R = H, CH₃   (PREDOMINANT)   DIASTEREOMER (*)

SONDHEIMER, ET AL [189]

50%   20%

Chart **6.16**

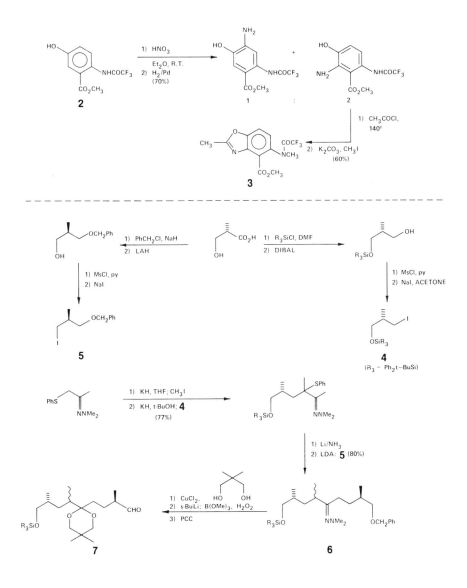

Chart **6.17**

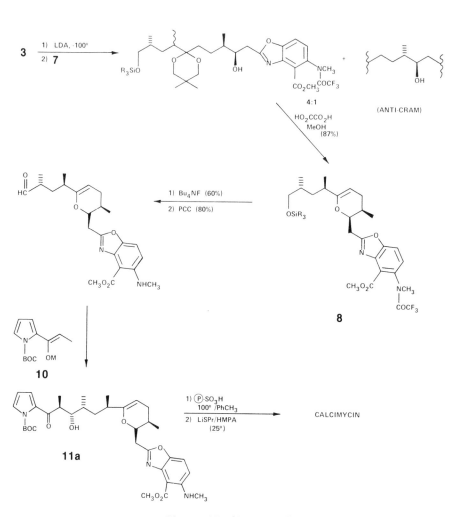

Chart 6.17   (Continued)

ation of the acid produced the dibenzyl derivative which afforded the monoalcohol upon reduction with lithium aluminum hydride. Hydroxyl for iodide interchange gave **5**.

Using the dimethylhydrazone of α-phenylthioacetone, successive monoalkylations were performed on the more substituted potassium salt first with methyl iodide and then by **4**. Reductive cleavage of the thiophenylether followed by alkylation with **5** on the other acetone terminus gave the desired central carbon framework **6** in greater than 60% overall yield.

After exchange of the dimethylhydrazone for the dimethylpropanediol ketal, the terminal benzyl ether was cleaved by forming the benzylic anion with butyllithium followed by oxidation to the labile hemiacetal. The resultant alcohol was oxidized to the aldehyde with the pyridinium chlorochromate. Due to the relative instability of the benzoxazole α-methyl anion (**3**), the aldol with **7** required low temperatures. These conditions afforded the desired diastereomer (predicted by Cram's rule) as the predominant isomer (33% from **6**). Oxalic acid treatment served to convert the ketal to the internal enol ether (**8**) as well as isomerize the central ketone α-methyl to the desired, thermodynamic β-epimer. Removal of both the diphenyl-*t*-butylsilyl group and the trifluoroacetyl group was achieved with tetrabutylammonium fluoride.

The aldehyde, **9**, from pyridium chlorochromate oxidation, was treated with the *N*-*t*-butyloxycarbonyl-protected pyrole enolate **10** to yield the aldol **11a** as the predominant diastereomer among three others. The zinc enolate afforded a predominance of the desired *E*-isomer which in turn leads to threo (thermodynamic) products, whereas the lithium enolate slightly favored the *Z*-isomer. Resin

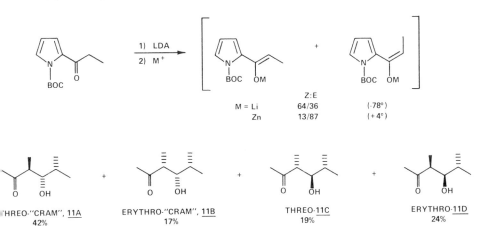

sulfonic acid treatment generated the dioxaspirane system and the methyl ester was cleaved with mercaptide to give the natural product in 23% overall yield from **9**.

An alternate approach to the total synthesis of calcimycin by Grieco and colleagues[228] originates with a suitably functionalized bicyclo[2.2.1]heptane. The strategy involves the incorporation of pyrrole and C_8-methylbenzoxazole at the end of the synthesis. The required C_9-C_{19} carbon backbone (calcimycin numbering) with the appropriate stereocenters is assembled by a series of oxidations and one- to three-carbon additions, beginning with a functionalized equivalent of norbornenone-14, as depicted in the *retro*-synthetic diagram below.

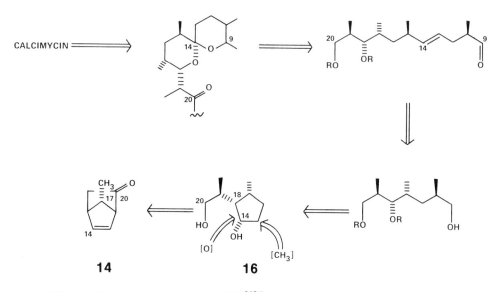

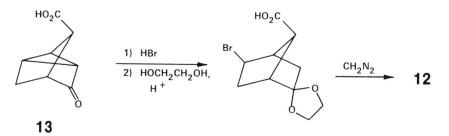

The synthesis and elaboration of 14[229a] in the total synthesis of calcimycin is detailed in Chart 6.18. Bicycloheptane-12, derived from the tricycloheptane carboxylic acid 13, was dehydrohalogenated and reduced to the epimeric-7-methanol. Carboxylic acid 13 is available in racemic or optically active form and

readily obtained from norbornadiene in two steps $((CH_2O)_n, HCO_2H; CrO_3/H^+)^{229b}$. Reduction of the corresponding tosylate was achieved in good yield with lithium triethylborohydride. After deprotection 14 was submitted to Baeyer-Villeger oxidation conditions to give a bicyclo[3.2.1]lactone, which underwent Lewis acid-catalyzed rearrangement to γ-lactone 15. After a highly stereoselective methylation, the lactone was reduced to diol 16. Selective protection of the primary alcohol allowed for a second Baeyer-Villeger oxidation on the cyclopentanone analog of 16 to afford a δ-lactone.

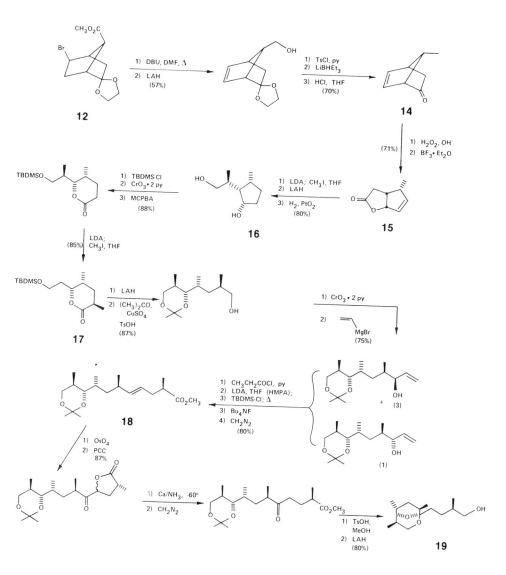

Chart **6.18**

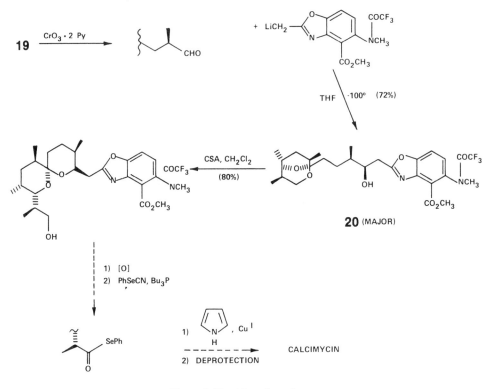

Chart 6.18   (Continued)

Here again α-methylation proceeded with a high degree of stereoselectivity to give **17**, a stereoisomer of which was previously employed by Grieco to synthesize the Djerassi-Prelog lactone and methynolide.[229a] After reduction and protecting group exchange to the 1,3-acetonide, the resulting primary alcohol was oxidized with Collins' reagent and converted with vinyl Grignard to a 3:1 mixture of allylic alcohols. These were elaborated by three carbons to ester-**18** employing the ester enolate Claisen rearrangement. This is another example of the chemistry of Ireland[230] involving solvent effects on enolate geometry (vide supra: lasalocid synthesis). Specifically, the erythro alcohol in THF yields the Z-enolate which rearranges to the β-methyl ester. The minor *threo*-alcohol isomer can also be employed by generating the E-enolate with HMPA cosolvent to afford the same β-methyl ester.

Reaction of **18** with osmium tetroxide followed by pyridinium chlorochromate served to introduce the "C-14" ketone. Reductive cleavage of the oxygen of the γ-lactone α to the ketone was effected with calcium in liquid ammonia. Internal transketalization and reduction then gave the bicyclic ketal **19**.

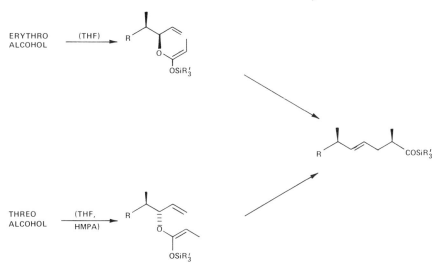

Compound **19**, after conversion to the corresponding aldehyde, was condensed with the lithium salt of the benzoxazole to afford alcohol-**20** as the major product.

The heterocycle was prepared in an analogous fashion to the previously described Evan's synthesis except that nitration was carried out in nitromethane (92%). Acid catalyzed equilibration of **19** yielded the thermodynamically more stable dioxospirane system of calcimycin. It is anticipated that conversion to the natural product can be achieved, as depicted, by oxidation and acylation of pyrrole.

D. Synthetic Analogs

A strategy for preparing noncyclic ligand systems, which might particularly mimic the polyether ionophores, must include a recognition that there is a greater than 2.3×10^4 reduction in binding capacity between the cyclic 18-crown-6 and its acyclic counterpart.[197] Thus, in approaching an appropriate balance between efficient complexation and transport, other structural constraints must be introduced to overcome the large entropy barrier in ion envelopment by an acyclic ligand framework.

The initial work in the area of noncyclic ligands was by Simon[198] and co-workers, who reported on a variety of diglycolic acid diamide ligands as ionophores, determined by incorporation in liquid membrane electrodes.[199] Several of these neutral ligands, including **1-3**, exhibited excellent selectivity for Ca^{2+} over other ions.[200] These amides were prepared[201] by the bisalkylation of the appropriate glycol with ethyl diazoacetate. The diester was hydrolyzed,

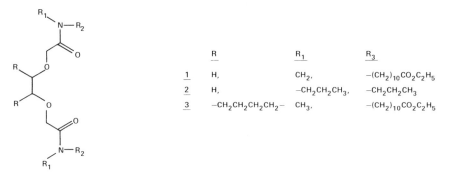

	R	R_1	R_3
1	H,	CH_2,	$-(CH_2)_{10}CO_2C_2H_5$
2	H,	$-CH_2CH_2CH_3$,	$-CH_2CH_2CH_3$
3	$-CH_2CH_2CH_2CH_2-$	CH_3,	$-(CH_2)_{10}CO_2C_2H_5$

converted to the acid chlorides and condensed with the appropriate secondary amines.

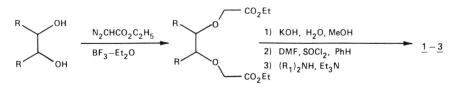

Recently, very similar ligands were reported wherein R = H and R_1 was a variety of lipophilic residues.[202] These workers also reported on a series of diglycolamic acids, **4**, which exhibited excellent Ca^{2+} transport in U-tube experiments.

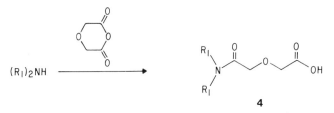

Vögtle, Maass, and co-workers have prepared and examined a large number of "open chain crown" ligands.[203] These directed, polyether ligand systems often contain rigid end groups, which can function as additional donor centers. Several

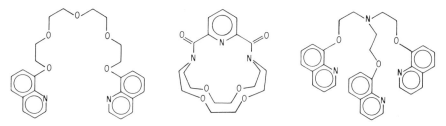

examples are shown below. A variety of alkali metal ion complexes have been prepared and the ion translocation processes of some of the ligands have been examined in isolated heart cells.[204]

Relatively modest ligands have proven capable of alkali metal complexation. For example, the permanganate oxidation of geranyl acetate affords a diol,[225] which, when treated with potassium hydroxide, yields an isolable potassium complex.[226] This interesting chemistry is currently being exploited to prepare

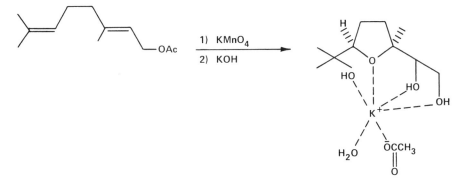

stereospecifically from 1,5-dienes, variously substituted tetrahydrofurans,[227] which are ubiquitous units in the polyether ionophore antibiotics.

Two examples of polyether carboxylic acid ligand systems are the work of Gardner and Beard[205] and Wierenga and co-workers.[206] The former workers described the synthesis of a variety of analogs of 5, however, they failed to form

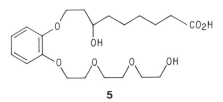

5

organic soluble salts and no properties were reported which would distinguish these compounds as ionophores. Wierenga's attempts culminated in two structures (6, 7), which compared favorably to lasalocid and calcimycin in transporting Ca^{2+} in the U-tube experiment.

The synthesis of 7 (Chart 6.19) is representative. Employing the Noyori cycloaddition chemistry, the bicyclo[3.2.1]octenone was prepared from cyclopentadiene. Stereoselective reduction to the *endo*-alcohol of this bicyclic building unit was achieved with lithium tri-*sec*-butyl borohydride. Sodium borohydride in methanol afforded a 4:1 *endo* to *exo* ratio. After protection of the alcohol, the

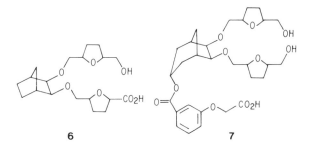

6   7

olefin was transformed to the *exo*-diol with catalytic osmium tetroxide and *N*-methylmorpholine-*N*-oxide (18% overall yield from cyclopentadiene). The *cis*-diol was bisalkylated (88%) with the difunctional furyl chloride, 8, prepared in 78% yield from hydroxymethylfurfural. Deprotection followed by acylation (90%) with the metasubstituted benzoyl chloride (prepared in three steps from *m*-hydroxybenzaldehyde) served to incorporate the third, directing ligand system. Pyruvic acid removed the acetal protecting groups and reduction to a diastereomeric mixture of *cis*-tetrahydrofurans was achieved with hydrogen over Raney nickel. Selective hydrolysis of the relatively unhindered methyl ester gave 7. Examination of the C-P-K models of 7 indicated a directional tetrahydrofuran-ligand array with 2.0-2.6 Å diameter and the third ligand juxtaposed beneath the "noncyclic crown" to provide pole-dipole interactions with the metal ion.

It should be noted that certain crown ethers have been modified to afford ligands for divalent ion complexation and transport. Two examples are 8, prepared by Cram and co-workers,[207] and 9, of Smid et al.[208] However, at least

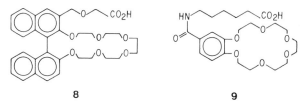

8   9

in the former case, these are not as efficient as 7 in Ca^{2+} transport.[206] Also, Brown and Foubister recently reported some anticoccidial activity for several related benzo-15-crown-5 polyethers.[237]

7. MISCELLANEOUS IONOPHORES

The inclusion of several structural classes in the last section is either for the reason that no chemical synthesis has been achieved or that the classification as an ionophore is not unambiguously established.

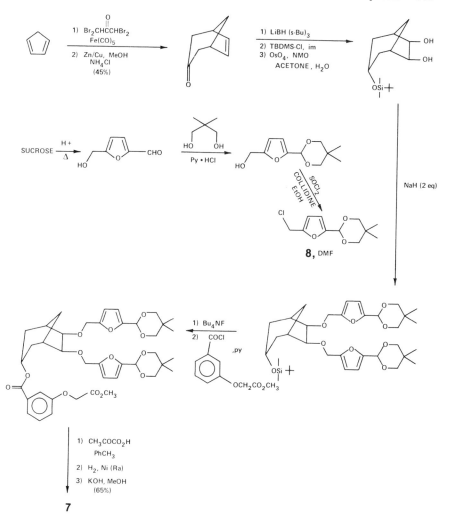

Chart 6.19

In the former category are the polyene antibiotics. These antifungal antibiotics include amphotericin B, filipin, nystatin, lienomycin, candicidin, pimaricin, etruscomycin, and others. Although the absolute and relative configurations are not known, the basic structures have been ascertained and two examples are depicted below. There has been a substantial amount of work on their effects on membrane permeability[209] and ionophoric properties.[210] All of the structures include a macrocyclic lactone of various sizes with a polyene half and a polyol half. Attached to the macrolide is usually the amino sugar, mycosamine. The

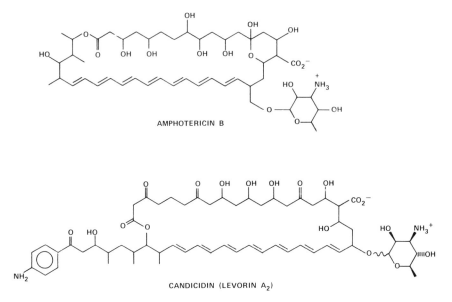

AMPHOTERICIN B

CANDICIDIN (LEVORIN A$_2$)

effects of these ionophores have suggested both a carrier and a pore-forming mechanism for ion transport.

A second structure class includes the phospholipids such as lecithin, cephalin, and cardiolipin.[211] These phosphoglycerides, constituents of cellular membranes, particularly nerve tissue, transport Ca^{2+} in the U-tube system with good efficiency.[212]

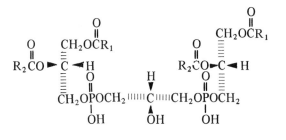

Cardiolipin (R$_1$ = steric, R$_2$ = oleic)

Related to these are the prostaglandins. There have been reports both claiming ionophoric properties for certain prostaglandins (such as PGB$_2$[213] and PGA$_2$[214]) as well as disclaiming ionophoric action.[215] Effects of prostaglandins on intestinal zinc absorption have been attributed to a PGE$_2$ liganding effect.[216] The concept of thromboxane A_2 (TxA$_2$) transport of Ca^{2+} as an integral feature of platelet aggregation has been reported based on studies which include TxA$_2$-assisted Ca^{2+} transport in aqueous to organic phases.[235]

Several isolated structures have been reported to exhibit ionophore-like properties but have received little study. These include streptolydigin,[212] several isomeric octadecadienoic acids (e.g., 13-hydroxyoctadecadienoic acid),[217] and avenaciolide.[218] Although avenaciolide has been synthesized by three different

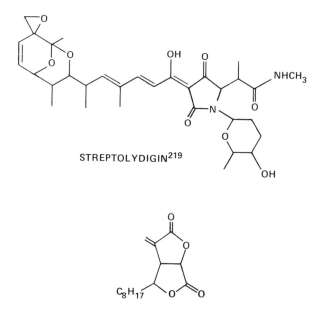

STREPTOLYDIGIN[219]

AVENACIOLIDE

groups,[220-222] its lack of structural similarity with any of the well established ionophores and absence of multiple binding sites for metal ions argues against its incorporation at this time.[181]

Lastly, work is ongoing in the search for and identification of *in vivo*, "natural" ionophores. For example, mitochondrial Ca^{2+} transport appears to be carrier-mediated[223] and $Ca^{2+} + Mg^{2+}$-ATPase from sarcoplasmic reticulum has an ionophore component. Recent results describe the isolation of a low molecular weight (ca. 3000) Ca^{2+} carrier protein from the inner mitochondrial membrane, called calciphorin.[224] Such results as these strongly suggest such future possibilities as the design and synthesis of "natural" ionophores to effect tissue and ion-selective transport in vertebrates to correct ionic imbalances or enhance the biochemical understanding of energy transduction through membranes.

ACKNOWLEDGMENTS

The author expresses his gratitude to Professors D. A. Evans, P. A. Grieco, R. E. Ireland, W. C. Still, and J. D. White for allowing the incorporation of work prior to publication. The reviewing of this work by Drs. P. A. Aristoff, J. Szmuczkovicz, and J. E. Pike, the encouragement of Drs. J. C. Babcock and J. E. Pike, technical assistance with the manuscript preparation by Ms. L. Riley and D. Piper, and the support of The Upjohn Company is gratefully acknowledged.

ADDENDUM

In 1980 Bartlett and Jernstedt disclosed an alternative synthesis of (±) nonactic acid,[239] the subunit of nonaction (Section 5A). This approach intersects with the second synthesis of Gerlach[118] at the octene diol (Chart 5.2); however, the Bartlett route affords the *erythro* diol exclusively, starting with dimethyl-1,7-octadien-4yl phosphate. This is converted regio- and stereospecifically to the terminal monoepoxide by treatment with iodine followed by methoxide (intramolecular phosphate cyclization). Lithium aluminum hydride gives the diol. This diol is converted to the aldehyde analogous to the Gerlach chemistry and then departs by performing a $TiCl_4$-mediated aldol with $CH_3CH=C(OCH_3)$-$OSi(CH_3)_3$ followed by Jones oxidation to give the β-ketoester equivalent of the Gerlach α,β-unsaturated ester. The β-ketoester is cyclized with dehydration to the epi-2,3-dehydrofuran which is reduced (Rh/Al_2O_3, H_2) to give a high yield of the 8-epinonactic acid methyl ester. Inversion of hydroxyl as described by White[130] gave (±)-nonactate in 25% overall yield.

An interesting synthetic model for nonactin was described by Samat, et al.[240] This was a dimeric, 32-membered macrocycle composed from the $-CH_2CH_2$-$OCOCH_2CH_2OCH_2CH_2CH_2OCH_2CH_2CO_2CH_2-$ subunit.

Ireland and coworker have reported a chiral synthesis of the aromatic left-half of lasalocid[241]. The basic route follows the chemistry depicted in Chart 6.7 for the earlier racemic synthesis. The required absolute stereochemistry is derived from optically active pentenyl bromide-**31**. Both enantiomers were independently derived from (−) citronellene. A discrepancy was noted between the optical rotations observed and those reported in the earlier Kishi synthesis[161] (Chart 6.2). These were attributed to incomplete resolution in the Kishi chemistry. A further complication is that the final aldehyde (**12**) exhibits instability problems. A reinvestigation of the isomeric outcome of the final aldol coupling with **12** to lasalocid is in progress.

REFERENCES

1. C. Moore and B. C. Pressman, *Biochem. Biophys. Res. Commun.*, **15**, 562 (1964).

2. B. C. Pressman, E. J. Harris, W. S. Jagger, and J. H. Johnson, *Proc. Natl. Acad. Sci. USA*, **58**, 1949 (1967).

3. See Yu. A. Ovchinnikov, V. T. Ivanov, and A. M. Shkrob, *Membrane-Active Complexones*, BBA Library, Vol. 12, 1974, Elsevier, New York; Yu. A. Ovchinnikov, *Frontiers in Bioorganic Chemistry and Molecular Biology*, Yu. A. Ovchinnikov and M. N. Kolosov, Eds., Elsevier/North Holland Biomedical Press, 1979, pp. 129-165.

4. B. C. Pressman, *Ann. Review Biochem.*, **45**, 925 (1976).

5. J. W. Westley, *Ann. Reports Med. Chem.*, **10**, 246 (1975).

6. J. W. Westley, *Adv. Appl. Microbiol.*, **22**, 177 (1977).

7. M. Bodanszky, G. F. Sigler, and A. Bodanszky, *J. Am. Chem. Soc.*, **95**, 2352 (1973).

8. S. Nakamura, T. Yajima, Y-C. Lin, and H. Umezawa, *J. Antibiot.*, **20**, 1 (1967).

9. T. Takita, Y. Muraoka, A. Yoshioka, K. Fujii, K. Maeda, and H. Umezawa, *J. Antibiot.*, **25**, 755 (1972).

10. G. Schilling, D. Berti, and D. Kluepfel, *J. Antibiot.*, **23**, 81 (1970).

11. M. Bodanszky and D. Perlman, *Science*, **163**, 352 (1969).

12. R. J. Dubos, *J. Exp. Med.*, **70**, 1, 11 (1939).

13. R. D. Hotchkiss, *Adv. Enzymol.*, **4**, 153 (1944).

14. (a) R. Sarges and B. Witkop, *J. Am. Chem. Soc.*, **86**, 1862 (1964); (b) *idem., ibid.*, **87**, 2027 (1965); (c) *idem., Biochemistry*, **4**, 2491 (1965).

15. R. Sarges and B. Witkop, *J. Am. Chem. Soc.*, **87**, 2020 (1965).

16. D. W. Urry, M. C. Goodall, J. D. Glickson, and D. F. Mayers, *Proc. Natl. Acad. Sci. USA*, **68**, 1907 (1971).

17. S. B. Hladky and D. A. Haydon, *Biochem. Biophys. Acta*, **274**, 294 (1972).

18. A. Finkelstein, *Drugs and Transport Processes*, B. A. Callingham, Ed., University Park Press, Baltimore, Md., 1973, pp. 241-250.

19. F. Johnson, *The Total Synthesis of Natural Products*, Vol. 1, J. ApSimon, Ed., pp. 331-466, Wiley-Interscience, New York, 1973.

20. S. Rambhav and L. Ramachandron, *Ind. J. Biochem. Biophys.*, **9**, 21 (1972).

21. D. W. Urry, *Proc. Nat. Acad. Sci. USA*, **69**, 1610 (1972).

22. C. E. Meyer and F. Reusser, *Experientia*, **23**, 85 (1967).

23. J. W. Payne, R. Jakes, and B. S. Hartley, *Biochem. J.*, **117**, 757 (1970).

24. Yu. A. Ovchinnikov, A. A. Kiryushkin, and I. V. Koshevnikova, *Zh. Obshch. Chim. (USSR)*, **41**, 2085 (*J. Gen. Chem.*, p. 2105) (1971).

25. D. R. Martin and R. J. P. Williams, *Biochem. J.*, **153**, 181 (1976).

26. (a) R. C. Pandy, J. C. Cook, Jr., and K. L. Rinehart, Jr., *J. Am. Chem. Soc.*, **99**, 8469 (1977); (b) see footnote 28 of 26(a).

27. P. Mueller and D. O. Rudin, *Nature*, **217**, 713 (1968).

28. L. G. M. Gordon, *Drugs and Transport Processes*, B. A. Callingham, Ed., University Park Press, Baltimore, Md., 1973, pp. 251-264 (and references cited therein).

29. B. C. Pressman, *Fed. Proc.*, **27**, 1283 (1968).

30. G. Boheim, K. Janko, D. Leibfritz, T. Ooka, W. A. König, and G. Jung, *Biochem. Biophys. Acta*, **433**, 182 (1976).

31. S. Byrn, *Biochemistry*, **13**, 5186 (1974).

32. B. F. Gisin, S. Kobayashi, and J. E. Hall, *Proc. Natl. Acad. Sci.*, **74**, 115 (1977).

33. Private communication from K. L. Rinehart; see also footnotes 18b and 28 of Ref. 26(*a*).

34. Th. Wieland and J. X. deVries, *Liebigs Ann. Chem.*, **700**, 174 (1966).

35. Th. Wieland, G. Lüben, H. Ottenheym, J. Faesel, J. X. DeVries, W. Konz, A. Prux, and J. Schmid, *Ang. Chem.*, **80**, 209 (1968).

36. Th. Wieland, J. Faesel, and W. Konz, *Liebigs Ann. Chem.*, **722**, 197 (1969).

37. C. Birr, F. Flor, P. Fleckenstein, and Th. Wieland, *Pept., Proc. Eur. Pept. Symp., 11th*, H. Nesvadba, Ed., North Holland, Amsterdam, 1973, pp. 175-84.

38. See, for example, Yu. A. Ovchinnikov, V. T. Ivanov, A. Miroshnikov, K. Khaliluline, and N. N. Uvarova, *Khim. Prir. Soedin (USSR)*, 469, (1971) and J. Halstrom, T. Christensen, and K. Brunfeldt, *Z. Physiol. Chem.*, **357**, 999 (1976).

39. Th. Wieland, H. Faulstich, W. Burgermeister, W. Otting, W. Mohle, M. M. Shemyakin, Yu. A. Ovchinnikov, V. T. Ivanov, and G. G. Malenkov, *FEBS Lett.*, **9**, 89 (1970).

40. V. Madison, M. Atreyi, C. M. Deber, and E. R. Blout, *J. Am. Chem. Soc.* **96**, 6725 (1974); see also C. M. Deber, V. Madison, and E. R. Blout, *Accts. Chem. Res.*, **9**, 106 (1976) and *J. Am. Chem. Soc.*, **99**, 4788 (1977).

41. R. Schwyzer and P. Sieber, *Helv. Chim. Acta*, **40**, 624 (1957).

42. R. Schwyzer, A. Tun-Kyi, M. Caviezel, and P. Moser, *Helv. Chim. Acta*, **53**, 15 (1970).

43. L. G. Pease and C. Watson, *J. Am. Chem. Soc.*, **100**, 1279 (1978).

44. See, for example, G. Eisenman and S. J. Krasne, *The Ion Selectivity of Carrier Molecules, Membranes, and Enzymes*, Chapter 2, and W. Epstein, *Membrane Transport*, Chapter 10, in *MTP Int. Rev. of Science, Biochemistry Series*, Vol. 2, Butterworth, London, (1975).

45. H. Diebler, M. Eigen, G. Ilgienfritz, G. Moas, and R. Winkler, *Pure Appl. Chem.*, **20**, 93 (1969).

46. C. J. Pederson and H. K. Frensdorf, *Angew Chem. Int. Ed.*, **11**, 16 (1972) and references cited therein.

47. R. Ashton and L. K. Steinrauf, *J. Mol. Biol.*, **49**, 547 (1970).

48. B. C. Pressman, *Fed. Proc. (Fed. Am. Soc. Exp. Biol.)*, **32**, 1698 (1973).

49. R. P. Scholer and W. Simon, *Chimia*, **24**, 372 (1970).

50. See monograph by H. T. Tien, *Bilayer Lipid Membranes–Theory and Practice*, 1974, Dekker, New York.

51. C. Moores and B. C. Pressman, *Biochem. Biophys. Res. Commun.*, **15**, 562 (1964).

52. B. F. Gisin, H. P. T. Beall, D. G. Davis, E. Grell, and D. C. Tosteson, *Biochem. Biophys. Acta*, **509**, 201 (1978).

53. (*a*) P. J. Henderson, J. D. McGivan, and J. B. Chappell, *Biochem. J.*, **111**, 121 (1969); (*b*) *idem., ibid.*, **111**, 521 (1969).

54. J. C. Foreman, J. L. Monger, and B. D. Gomperts, *Nature*, **245**, 249 (1973).

55. R. A. Steinhardt and D. Epel, *Proc. Nat. Acad. Sci., USA*, **71**, 915 (1974).

56. R. D. Feinman and T. C. Detwiler, *Nature*, **249**, 172 (1974).

57. Yu. A. Ovchinnikov and V. T. Ivanov, *Tetrahedron*, **31**, 2177 (1975).

58. D. J. Patel and A. E. Tonelli, *Biochemistry*, **12**, 486 (1973).

59. G. D. Smith, W. L. Daux, D. A. Langs, G. T. deTitta, J. W. Edmonds, D. C. Rohrer, and C. M. Weeks, *J. Am. Chem. Soc.*, **97**, 7242 (1975).

60. M. M. Shemyakin, N. A. Aldonova, E. I. Vinogradova, and Yu. M. Feigina, *Tetrahedron Lett.*, 1921 (1963).

61. V. T. Ivanov, I. A. Laine, N. D. Abdulaev, L. B. Senyavina, E. M. Popov, Yu. A. Ovchinnikov, and M. M. Shemyakin, *Biochem. Biophys. Res. Commun.*, **34**, 803 (1969).

62. B. F. Gisin, R. B. Merrifield, and D. C. Tosteson, *J. Am. Chem. Soc.*, **91**, 2691 (1969).

63. G. R. Pettit, *Synthetic Peptides*, Vol. 3, p. 385, Academic Press, New York, 1975.

64. M. Bodanszky and M. A. Ondetti, *Peptide Synthesis*, Wiley-Interscience, New York, 1966, p. 243.

65. M. M. Shemyakin, E. I. Vinogradorva, M. Yu Feigina, N. A. Aldanova, N. F. Loginova, I. D. Ryabova, and I. A. Pavlenko, *Experientia*, **21**, 548 (1965).

66. V. T. Ivanov, A. V. Evstratov, L. V. Sumskaya, E. I. Melnick, T. S. Chumburidze, S. L. Portnova, T. A. Balashova, and Yu. A. Ovchinnikov, *FEBS Lett.*, **36**, 65 (1973).

67. M. Dobler, J. D. Dunitz, and J. Krajewski, *J. Mol. Biol.*, **42**, 603 (1969).

68. M. M. Shemyakin, Yu. A. Ovchinnikov, A. A. Kiryushkin, and V. T. Ivanov, *Tetrahedron Lett.*, 885 (1963).

69. Pl. A. Plattner, K. Vogler, R. O. Studer, P. Quitt, and W. Keller-Schierlein, *Helv. Chim. Acta*, **46**, 927 (1963).

70. Yu. A. Ovchinikov, V. T. Ivanov, and I. I. Mikhaleva, *Tetrahedron Lett.*, 159 (1971). See also R. W. Roeske, S. Isaac, L. K. Steinrauf, and T. King, *Fed. Proc.*, 30 (1971).

71. M. J. Hall, *Biochem. Biophy. Res. Commun.*, **38**, 590 (1970).

72. K. Bevan, J. Davies, M. J. Hall, C. H. Hassall, R. B. Morton, D. A. S. Phillips, Y. Ogihara, and W. A. Thomas, *Experientia*, **26**, 122 (1970).

73. C. H. Hassall, R. B. Morton, Y. Ogihara, and D. A. S. Phillips, *J. Chem. Soc. (C)*, 526 (1971).

74. J. Al-Hassan and J. S. Davies, *Tetrahedron Lett.*, 3843 (1978).

75. C. M. Deber, P. D. Adawadker, and J. Tom-Kun, *Biochem. Biophys. Res. Commun.*, **81**, 1357 (1978).

76. (a) J. B. Neilands, "Iron and Its Role in Microbial Physiology", in *Microbial Iron Metabolism*, J. B. Neiland, Ed., Chap. 1, Academic Press, New York, 1974; (b) B. R. Byers, "Iron Transport in Gram-Positive and Acid-Fast Bacilli", Chap. 4, *ibid.*; (c) T. Emery, "Biosynthesis and Mechanism of Action of Hydroxamate-Type Siderochromes", Chap. 5, *ibid.*; (d) H. Rosenberg and I. G. Young, "Iron Transport in the Enteric Bacteria", Chap. 3, *ibid.*

77. J. B. Neilands, *Inorganic Biochemistry*, G. Eichorn, Ed., Elsevier, New York, 1973.

78. E. A. Kaczka, C. O. Gitterman, E. L. Dulaney, and K. Folkers, *Biochemistry*, **1**, 340 (1962).

79. N. H. Anderson, W. D. Ollis, J. E. Thorpe, and A. D. Ward, *J. C. S. Chem. Commun.*, 420 (1974).

80. N. Tsuji, M. Kobayashi, K. Nagashima, Y. Wakisaka, and K. Koizumi, *J. Antibiot.*, **29**, 1 (1976).

81. J. D. Dutcher, *J. Biol. Chem.*, **171**, 321, 341 (1947).

82. E. C. White and T. H. Hill, *J. Bacteriol.*, **45**, 433 (1943).

83. R. G. Micetich and J. C. MacDonald, *J. Chem. Soc.*, 1507 (1964).

84. M. Masaki, Y. Chigira and M. Ohta, *J. Org. Chem.*, **31**, 4143 (1966).

85. C. L. Atkins and J. B. Neilands, *Biochemistry*, 7, 3734 (1968).

86. H. A. Akers, M. Llinas, and J. B. Neilands, *Biochemistry*, **11**, 2283 (1972).

87. C. J. Carrano, S. R. Cooper, and K. N. Raymond, *J. Am. Chem. Soc.*, **101**, 599 (1979).

88. Y. Isowa, T. Takashima, M. Ohmori, H. Kurita, M. Sato, and K. Mori, U. S. Patent 3,772, 265 (1973).

89. K. B. Mullis, J. R. Pollack, and J. B. Neilands, *Biochemistry*, **10**, 1071 (1971).

90. F. Gibson and D. I. Magrath, *Biochem. Biophys. Acta*, **192**, 175 (1969).

91. Reviewed by G. A. Snow, *Bacteriol. Rev.*, **34**, 99 (1970).

92. H. Bickel, P. Mertens, V. Prelog, J. Seibl, and A. Walser, *Tetrahedron, Suppl. B (I)*, 171 (1966).

93. H. Bickel, G. E. Hall, W. Keller-Schierlein, V. Prelog, E. Vischer, and A. Wettstein, *Helv. Chim. Acta*, **43**, 2129 (1960).

94. J. B. Neilands, *J. Am. Chem. Soc.*, **74**, 4846 (1952).

95. G. Anderegg, F. L'Eplattenier, and G. Schwarzenbach, *Helv. Chim. Acta*, **46**, 1409 (1963).

96. W. Keller-Schierlein and B. Maurer, *Helv. Chim. Acta*, **52**, 603 (1969).

97. Y. Isowa, M. Ohmori, and H. Kurita, *Bull. Chem. Soc. Japan*, **47**, 215 (1974).

98. S. Rogers and J. B. Neilands, *Biochemistry*, **3**, 1850 (1964); see D. van der Helm, J. R. Baker, D. L. Eng-Wilmot, M. B. Hossain, and R. A. Loghry, *J. Am. Chem. Soc.*, **102**, 4224 (1980) for crystal structure.

99. W. Keller-Schierlein and V. Prelog, *Helv. Chim. Acta*, **44**, 1981 (1961).

100. T. Ito and J. B. Neilands, *J. Am. Chem. Soc.*, **80**, 4645 (1958).

101. J. L. Corbin and W. A. Bulen, *Biochemistry*, **8**, 757 (1969).

102. I. G. O'Brien and F. Gibson, *Biochem. Biophys. Acta*, **215**, 393 (1970).

103. J. R. Pollack and J. B. Neilands, *Biochem. Biophys. Res. Commun.*, **38**, 989 (1970).

104. E. J. Corey and S. Bhattacharyya, *Tetrahedron Lett.*, 3919 (1977).

105. M. C. Venuti, W. H. Rastetter, and J. B. Neilands, *J. Med. Chem.*, **22**, 123 (1974).

106. E. J. Corey and S. D. Hurt, *Tetrahedron Lett.*, 3923 (1977).

107. W. R. Harris, C. J. Carrano, and K. N. Raymond, *J. Am. Chem. Soc.*, **101**, 2213 (1979).

108. R. W. Grady, J. H. Graziano, H. A. Akers, and A. Cerarni, *J. Pharmcol. Exp. Ther.*, **196**, 478 (1976).

109. G. W. Rotermund and R. Köster, *Ann. Chem.*, **686**, 153 (1965).

110. D. J. Collins, C. Lewis, and J. M. Swan, *Aust. J. Chem.*, **28**, 673 (1975).

111. W. R. Harris, F. L. Weitl, and K. N. Raymond, *J.C.S. Chem. Comm.*, 177 (1979).

112. R. Corbaz, L. Ettlinger, E. Gäumann, W. Keller-Schierlein, F. Kradoffer, L. Neipp, V. Prelog, and H. Zähner, *Helv. Chim. Acta*, **38**, 1445 (1955).

113. W. Keller-Schielein and H. Gerlach, *Fort. Chem. Org. Naturst.*, **26**, 161 (1968).

114. K. H. Wallhäusser, G. Huber, G. Nessenmann, P. Präve, and K. Zept, *Arzneimittelforschung*, **14**, 356 (1964).

115. K. Ando, H. Oishi, S. Hirano, T. Okutani, K. Suzuki, H. Okasaki, M. Sawada, and T. Sagawa, *J. Antibiot.*, **24**, 347 (1971).

116. D. H. Haynes and B. C. Pressman, *J. Membr. Biol.*, **18**, 1 (1974).

117. M. Dobler, J. D. Dunitz, and B. T. Kilbourn, *Helv. Chim. Acta*, **52**, 2573 (1969).

118. H. Gerlach and H. Wetter, *Helv. Chim. Acta*, **57**, 2306 (1974). H. Gerlach, K. Oertle, A. Thalmann, and S. Servi, *ibid.*, **58**, 2036 (1975).

119. S. Shatzmiller, P. Gygax, D. Hall, and A. Eschenmoser, *Helv. Chim. Acta*, **56**, 2973 (1973).

120. W. R. Harris, C. J. Carrano, and K. N. Raymond, *J. Am. Chem. Soc.*, **101**, 2722 (1979). C. J. Carrano and K. N. Raymond, *ibid.*, **101**, 5401 (1979).

121. F. L. Wietl and K. N. Raymond, *J. Am. Chem. Soc.*, 2728 (1979).

122. V. Prelog and A. Walser, *Helv. Chim. Acta*, **45**, 631 (1962).

123. W. Keller-Schierlein, P. Mertens, V. Prelog, and A. Walser, *Helv. Chim. Acta*, **48**, 931 (1965).

124. A. Jacobs, G. P. White, and G. H. Tait, *Biochem. Biophys. Res. Commun.*, **74**, 1626 (1977).

125. This is reported by Weitl and Raymond[121] not to be 2,3-dihydroxybenzoyl chloride but 2,3-dioxosulfinylbenzoyl chloride (conditions: $SOCl_2$ neat at reflux for 3 hr).

126. See the accompanying chapter on Macrocyclic Synthesis by R. K. Boeckman for further details and examples.

127. T. Mukaiyama, M. Araki, and H. Takai, *J. Am. Chem. Soc.*, **95**, 4763 (1973).

128. G. Beck and E. Henseleit, *Chem. Ber.*, **104**, 21 (1971).

129. H. Zak and U. Schmidt, *Ang. Chem. Int. Ed.*, **14**, 432 (1975); J. Gombos, E. Haslinger, H. Zak, and U. Schmidt, *Tetrahedron Lett.*, 3391 (1975); *idem.*, *Monatsh. Chem.*, **106**, 219 (1975).

130. M. J. Arco, M. H. Trammell, and J. D. White, *J. Org. Chem.*, **41**, 2075 (1976).

131. R. M. Izatt and J. J. Christensen, Eds., *Synthetic Multidentate Macrocyclic Compounds*, Academic Press, New York, 1978; (*a*) G. W. Liesegang and E. M. Eyring, *ibid.*, Chap. 5, pp. 245-288.

132. C. M. Starks and C. Liotta, *Phase Transfer Catalysis*, Academic Press, New York, 1978.

133. W. P. Weber and G. W. Gokel, *Phase Transfer Catalysis in Organic Synthesis*, Springer-Verlag, New York, 1977.

134. (a) D. E. Fenton, *Chem. Soc. Rev.*, **6**, 325 (1977); (b) I. M. Kolthoff, *Anal. Chem.*, **51**, 1R-22R, (1979).

135. G. W. Gokel and H. D. Durst, *Synthesis*, 168 (1976).

136. E. Bamberg, H.-J. Apell, H. Alpes, E. Gross, I. L. Morell, J. F. Harbaugh, K. Janko, and P. Lauger, *Fed. Proc.*, **37**, 2633 (1978).

137. H. K. Frensdorff, *J. Am. Chem. Soc.*, **93**, 600 (1971).

138. B. Dietrich, J. M. Lehn, and J. P. Sauvage, *Chem. Commun.*, 15 (1973).

139. E. J. Harris, B. Zaba, and M. R. Truter, *Arch. Biochem. Biophys.*, **182**, 311 (1977).

140. Personal communication from D. J. Cram. See also J. G. DeVries and R. M. Kellogg, *J. Am. Chem. Soc.*, **101**, 2759 (1979).

141. E. P. Kyba, J. M. Timko, L. J. Kaplan, F. deJong, G. W. Kokel, and D. J. Cram, *J. Am. Chem. Soc.*, **100**, 4555 (1978) and references cited therein; see also D. J. Cram

in *Applications of Biochemical Systems in Organic Chemistry*, Part II, J. B. Jones, Ed., Wiley, New York, 1976, Chap. V, p. 852, and D. J. Cram and J. M. Cram, *Acct. Chem. Res.*, 11, 8 (1978).

142. D. A. Laidler and J. F. Stoddart, *Tetrahedron Lett.*, 453 (1979) and earlier communications cited therein.

143. See for example V. Prelog, *Pure Appl. Chem.*, 50, 893 (1978); L. Toke, L. Fenichel, P. Bako, and J. Szejtli, *Acta. Chim. Acad. Sci., Hung.*, 98, 357 (1978); J. P. Behr and J. M. Lehn, *J.C.S. Chem. Commun.*, 143 (1978).

144. J. M. Lehn, *Pure Appl. Chem.*, 49, 857 (1977); *Acct. Chem. Res.*, 11, 49 (1978).

145. L. K. Steinrauf, M. Pinkerton, and J. W. Chamberlin, *Biochem. Biophys. Res. Commun.*, 33, 29 (1968).

146. H. Kinashi, N. Otake, and H. Yonchara, *Tetrahedron Lett.*, 4955 (1973).

147. J. F. Blount and J. W. Westley, *Chem. Commun.*, 533 (1975).

148. A. Imada, Y. Nozaki, T. Hasegawa, E. Mizuta, S. Igarasi, and M. Yoneda, *J. Antibiot.*, 31, 7 (1978).

149. E. W. Czerwinski and L. K. Steinrauf, *Biochem. Biophys. Res. Commun.*, 45, 1284 (1971).

150. N. Otake, M. Koenuma, H. Kinashi, S. Sato, and Y. Saito, *J.C.S. Chem. Commun.*, 92 (1979). See also N. Otake et al., *J. Antibiot.*, 30, 186 (1972).

151. C. Keller-Justen, *J. Antiobiot.*, 31, 820 (1978).

152. J. Berger, A. I. Rachlin, W. E. Scott, L. H. Steinbach, and M. W. Goldberg, *J. Am. Chem. Soc.*, 73, 5295 (1951).

153. J. W. Westley, R. H. Evans, T. Williams, and A. Stempel, *Chem. Commun.*, 71, 1467 (1970).

154. S. M. Johnson, J. Herrin, S. J. Liu, and I. C. Paul, *J. Am. Chem. Soc.*, 92, 4428 (1970).

155. D. J. Patel and C. Shen, *Proc. Natl. Acad. Sci.*, 73, 1786 (1976).

156. J. W. Westley, W. Benz, J. Donahue, R. H. Evans, C. G. Scott, A. Stempel, and J. Berger, *J. Antibiot.*, 27, 744 (1974).

157. D. H. Haynes and B. C. Pressman, *J. Membr. Biol.*, 16, 195 (1974).

158. H. Schadt and G. Haeusler, *J. Membr. Biol.*, 18, 277 (1974).

159. N. T. de Guzman, B. C. Pressman, K. Lasseter, and P. Palmer, *Clin. Res.*, 21, 413 (1973).

160. A. Schwartz, R. M. Lewis, H. G. Hanley, R. G. Munson, F. D. Dial, and M. Y. Ray, *Circ. Res.*, 34, 102 (1974).

161. T. Nakata, G. Schmid, B. Vranesic, M. Okigawa, T. Smith-Palmer, and Y. Kishi, *J. Am. Chem. Soc.*, 100, 2933 (1978).

162. J. W. Westley, E. P. Oliveto, J. Berger, R. H. Evans, R. Glass, A. Stempel, V. Toome, and T. Williams, *J. Med. Chem.*, 16, 397 (1973).

163. W. S. Johnson, L. Wetheman, W. R. Barlett, T. J. Brocksom, T. Li, D. J. Faulkner, and M. R. Peterson, *J. Am. Chem. Soc.*, 92, 741 (1970).

164. T. Mukaiyama, M. Asami, J. Hanna, and S. Kobayashi, *Chem. Lett.*, 783 (1977). See also M. Asami and T. Mukaiyama, *Heterocycles*, 12, 499 (1979).

165. See A. O. Chong and K. B. Sharpless, *J. Org. Chem.*, 42, 1587 (1977) for mechanistic details and also T. Fukarama, B. Vranesic, D. P. Negri, and Y. Kishi, *Tetrahedron Lett.*, 2741 (1978) for further details on this chemistry of preparing tetrahydrofurans.

166. F. Wessely, E. Zbiral, and H. Sturm, *Chem. Ber.*, 93, 2840 (1960).

167. T. Nakata and Y. Kishi, *Tetrahedron Lett.*, 2745 (1978).

168. A. Agtarap, J. W. Chamberlin, M. Pinkerton, and L. K. Steinrauf, *J. Am. Chem. Soc.*, 89, 5737 (1967); M. Pinkerton and L. K. Steinrauf, *J. Mol. Biol.*, 49, 533 (1970).

169. W. K. Lutz, F. K. Winkler, and J. D. Dunitz, *Helv. Chim. Acta*, 54, 1103 (1971).

170. P. G. Gertenbach and A. I. Popov, *J. Am. Chem. Soc.*, 97, 4738 (1975).

171. M. J. O. Ateunis and G. Verhegge, *Bull. Soc. Chim. Belges*, 86, 353 (1977).

172. W. K. Lutz, H. -K. Wipf, and W. Simon, *Helv. Chim. Acta*, 53, 1741 (1970).

173. L. E. Day, J. W. Chamberlin, E. Z. Gordee, S. Cheu, M. Gorman, R. L. Hammill, T. Wess, R. E. Weeks, and R. Stroshane, *Antimicrob. Agents Chemother.*, 410 (1973).

174. B. C. Pressman and N. T. deGuzman, *Ann. N. Y. Acad. Sci.*, 264, 373 (1975); see also G. Kabell, R. K. Saini, P. Jomani, and B. C. Pressman, *J. Pharmacol. Exp. Ther.*, 211, 231 (1979).

175. B. K. Toeplitz, A. I. Cohen, P. T. Funke, W. L. Parker, and J. Z. Gougoutas, *J. Am. Chem. Soc.*, 101, 3344 (1979).

176. G. Schmid, T. Fukuyama, Y. Akasaka, and Y. Kishi, *J. Am. Chem. Soc.*, 101, 259 (1979).

177. T. Fukuyama, C. -L. J. Wang, and Y. Kishi, *J. Am. Chem. Soc.*, 101, 260 (1979).

178. T. Fukuyama, K. Akasaka, D. J. Karenewsky, C. -L. J. Wang, G. Schmid and Y. Kishi, *J. Am. Chem. Soc.*, 101, 262 (1979).

179. T. Matsumoto, Y. Hosoda, K. Mori, and K. Fukui, *Bull Chem. Soc. Japan*, 45, 3156 (1972).

180. M. O. Chaney, P. V. DeMarco, N. D. Jones, and J. L. Occolowitz, *J. Am. Chem. Soc.*, 96, 1932 (1974).

181. D. R. Pfeiffer, R. W. Taylor, and H. A. Lardy, *Annals N.Y. Acad. Sci.*, 402 (1978).

182. M. O. Chaney, N. D. Jones, and M. Debono, *J. Antibiot.*, 29, 424 (1976).

183. C. M. Deber and D. R. Pfeiffer, *Biochemistry*, 15, 132 (1976).

184. A. H. Caswell and B. C. Pressman, *Biochem. Biophys. Res. Commun.*, 49, 292 (1972).

185. D. R. Pfeiffer, P. W. Reed, and H. A. Lardy, *Biochemistry*, 19, 4007 (1974); see also *J. Biol. Chem.*, 247, 6970 (1972).

186. J. W. Westley, R. H. Evans, Jr., C. -M. Liu, T. Hermann, and J. F. Blount, *J. Am. Chem. Soc.*, 100, 6786 (1978); U.S. Patent 4,100,171 (1978).

187. D. A. Evans, C. E. Sacks, W. A. Kleischick, and T. R. Tabor, *J. Am. Chem. Soc.*, 101, 6789 (1979).

188. D. A. Evans, C. E. Sacks, R. A. Whitney, and N. G. Mandel, *Tetrahedron Lett.*, 727 (1978).

189. T. M. Cresp, C. L. Probert, and F. Sondheimer, *Tetrahedron Lett.*, 3955 (1978).

190. J. D. Dunitz, D. M. Hawley, D. Miklos, D. N. J. White, Yu. Berlin, R. Morusic and V. Prelog, *Helv. Chim. Acta*, 54, 1709 (1971).

191. H. Nakamura, Y. Iitaka, T. Kitahara, T. Okazaki, and Y. Okami, *J. Antibiot.*, 30, 714 (1977).

192. R. Hütter, W. Keller-Schierlein, F. Knüsel, V. Prelog, G. C. Rodgers, P. Suter, G. Vogel, W. Yoser, and H. Zahner, *Helv. Chim. Acta*, 50, 1533 (1976).

193. W. Pache, Boromycin, in *Antibiotics*, Vol. 3, J. W. Corcoran and F. E. Hahn, Eds., Springer-Verlag, Heidelberg, 1975, p. 585.

194. Seminar at the Upjohn Company, June 12, 1979 and private communication by J. D. White, Oregon State Univ. (collaborators are O. Dhingra and B. Sheldon); 179th ACS National Meeting, *Abstr. ORGN* 48, 1980.

195. J. A. Katzenellenbogen and A. L. Crumrine, *J. Am. Chem. Soc.*, 98, 4975 (1976).

196. A. J. Duggan, M. A. Adams, P. J. Byrnes, and J. Meinwald, *Tetrahedron Lett.*, 4323 (1978).

197. J. M. Timko, S. S. Moore, D. M. Walba, P. C. Hiberty, and D. J. Cram, *J. Am. Chem. Soc.*, 99, 4207 (1977).

198. W. Simon, W. E. Morf, and P. Ch. Meier, *Structure and Bonding*, 16, 113 (1973).

199. These ligands appear to be less effective transport agents when measuring absolute transport in the U-tube experiment as seen in our laboratories and others.[4]

200. D. Ammann, R. Bissig, M. Guggi, E. Pretsch, W. Simon, I. J. Borowitz, and L. Weiss, *Helv. Chem. Acta*, 58, 1535 (1975); see also T. Wun, R. Bittman, and I. J. Borowitz, *Biochemistry*, 16, 2074 (1977); E. Pretsch, D. Ammann, H. F. Osswald, M. Guggi, and W. Simon, *Helv. Chim. Acta*, 63, 191 (1980).

201. D. Ammann, E. Pretsch, and W. Simon, *Biochemistry*, 56, 1780 (1973).

202. M. J. Umen and A. Scarpa, *J. Med. Chem.*, 21, 505 (1978).

203. W. Rashofer, G. Oepen, and F. Vögtle, *Chem. Ber.*, 111, 419, 1108 (1978) and references cited therein to earlier communications; F. Vögtle and E. Weber, *Ang. Chem. Int. Ed.*, 18, 753-776 (1979).

204. B. Tümmler, G. Maass, E. Weber, W. Wehner, and F. Vögtle, *J. Am. Chem. Soc.*, 99, 4683 (1977); see also 101, 2588 (1979).

205. J. O. Gardner and C. C. Beard, *J. Med. Chem.*, 21, 357 (1978).

206. W. Wierenga, B. R. Evans, and J. A. Woltersom, *J. Am. Chem. Soc.*, 101, 1334 (1979).

207. J. M. Timko, R. C. Helgeson, and D. J. Cram, *J. Am. Chem. Soc.*, 100, 2828 (1978).

208. R. Ungaro, B. El Haj, and J. Smid, *J. Am. Chem. Soc.*, 98, 5198 (1976).

209. W. B. Pratt, "Antibiotics that Affect Membrane Permeability," in *Fundamentals of Chemotherapy*, Oxford University Press, 1973, Chap. 6.

210. See, for example, A. M. Spielvogel and A. W. Norman, *Arch. Biochem. Biophys.*, 167, 335 (1975) and T. Zieniawa, J. Popinigis, M. Wozniak, B. Cybulska, and E. Borowski, *FEBS Lett.*, 76, 81 (1977).

211. G. B. Ansell, J. N. Hawthorne, and R. M. C. Dawson, Eds., *Form and Function of Phospholipids*, Vol. 3, BBA Library, Elsevier, Amsterdam, 1973.

212. R. J. Seely, unpublished results, The Upjohn Co., 1975. See also Refs. 206 and 214.

213. M. E. Carsten and J. D. Miller, *Arch. Biochem. Biphys.*, 185, 282 (1978); S. T. Ohnishi and T. M. Devlin, *Biochem. Biophys. Res. Commun.*, 89, 240 (1979).

214. R. J. Kessler, C. A. Tyson, and D. E. Green, *Proc. Nat. Acad. Sci. (USA)*, 73, 3141 (1976).

215. F. Hertelendy, H. Tood, and R. J. Narconis, Jr., *Prostaglandins*, 15, 575 (1978).

216. M. K. Song and N. F. Adham, *Am. J. Physiol.*, 234, E99 (1978).

217. G. T. Carter, *Dissert. Abstr. Int.*, B37 (2), 766 (1976); G. A. Blondin and D. E. Green, *Chem. Eng. News*, p. 26-42, Nov. 10, 1975.

218. E. J. Harris and J. M. Wimhurst, *Nature New Biol.*, 245, 271 (1973); *idem., Arch. Biochem. Biophys.*, 162, 426 (1974).

219. C. DeBoer, A. Dietz, T. E. Eble, and C. M. Large, US Patent 3160560 (1964).

220. W. L. Parker, *J. Am. Chem. Soc.*, **91**, 7208 (1969). [See full account in *J. Org. Chem.*, **38**, 2489 (1973)].

221. J. L. Herman, M. H. Berger, and R. H. Schlessinger, *J. Am. Chem. Soc.*, **95**, 7923 (1973). (See full account, *ibid.*, **101**, 1344 (1979).

222. R. C. Anderson and B. Fraser-Reid, *J. Am. Chem. Soc.*, **97**, 3870 (1975).

223. E. Carafoli and G. Sottocasa, in *Dynamics of Energy-transducing Membranes*, L. Ernster, R. W. Estabrook, and E. C. Slater, Eds., 1974, Elsevier, Amsterdam, pp. 455-469. A. E. Shamoo and T. J. Murphy, *Curr. Topics Bioenergetics*, **9**, 147-177 (1979).

224. A. Y. Jeng, T. E. Ryan, and A. E. Shamoo, *Proc. Natl. Acad. Sci. USA*, **75**, 2125 (1978).

225. E. Klein and W. Rojahn, *Tetrahedron*, **21**, 2353 (1965).

226. R. Hackler, *J. Org. Chem.*, **40**, 2978 (1975).

227. D. M. Walba, M. D. Wand, and M. C. Wilkes, *J. Am. Chem. Soc.*, **101**, 4396 (1979); D. M. Walba and P. D. Edwards, *Tetrahedron Lett.*, **21**, 3531 (1980).

228. Private communication from P. A. Grieco, Univ. of Pittsburg (July, 1979); see also P. A. Grieco, E. Williams, H. Tanaka, and S. Gilman, *J. Org. Chem.*, **45**, 3537 (1980).

229. (*a*) The diastereoisomer of **14** was conveniently utilized in a synthesis of methylnolide, P. A. Grieco, Y. Ohfune, Y. Yokoyama, and W. Owens, *J. Am. Chem. Soc.*, **101**, 4750 (1979); (*b*) J. S. Bindra, A. Grodski, T. K. Schaaf, and E. J. Corey, *ibid.*, **95**, 7522 (1973).

230. R. E. Ireland and A. K. Willard, *Tetrahedron Lett.*, 3975 (1975); R. E. Ireland, R. H. Mueller, and A. K. Willard, *J. Am. Chem. Soc.*, **98**, (1976).

231. R. E. Ireland, S. Thaisrivongs, C. S. Wilcox, and C. G. McGarvey, private communication (July, 1979). (The synthesis of **10** was just communicated in *J. Am. Chem. Soc.*, **102**, 1155 (1980) and related chemistry described in *J. Org. Chem.*, **45**, 48, 197 (1980)).

232. R. E. Ireland, C. S. Wilcox, S. Thaisrivongs, and N. R. Vanier, *Can. J. Chem.*, **57**, 1743 (1979).

233. V. VanRheenan, R. C. Kelly, and D. Y. Cha, *Tetrahedron Lett.*, 1973 (1976) report that in acetone/water as the solvent combination of choice the diol is normally recovered underivitized.

234. Private communication from W. C. Still, Columbia University (August, 1979, J. McDonald and D. Collum and co-worker; just communicated in *J. Am. Chem. Soc.*, **102**, 2117, 2118, 2120 (1980)).

235. J. M. Gerrard, J. G. White, and D. A. Peterson, *Thrombox. Haemostas. (Stuttg)*, **40**, 224 (1978) and references cited therein.

236. S. A. Ong, T. Peterson, and J. B. Neilands, *J. Biol. Chem.*, **254**, 1860 (1979).

237. G. R. Brown and A. J. Foubister, *J. Med. Chem.*, **22**, 997 (1979).

238. Y. Okude, J. Hirano, T. Hiyama, and H. Nozaki, *J. Am. Chem. Soc.*, **99**, 3179 (1977).

239. P. A. Bartlett and K. K. Jernstedt, *Tetrahedron Lett.*, **21**, 1607 (1980).

240. A. Samat, J. Elguero, and J. Metzger, *J. C. S. Chem. Commun.*, 1182 (1979).

241. R. E. Ireland, G. J. McGarvey, R. C. Anderson, R. Badoud, B. Fritzsimmons, and S. Thaisrivongs, *J. Am. Chem. Soc.*, **102**, 6178 (1980).

The Synthesis of Prostaglandins

JASJIT S. BINDRA

Central Research,
Pfizer Inc., Groton, Connecticut

1. INTRODUCTION

Since the review of this topic in a previous volume[1], there has been a flood of literature on the synthesis of prostaglandins. However, the outpouring witnessed during the 1970s has begun to abate, and it appears to be an appropriate time to review the progress made during the decade.

It is difficult to imagine that only a decade ago the prostaglandins were scarce and difficultly obtainable materials, and that in the 1960s a veritable arsenal of modern ultramicroanalytical techniques had to be deployed by Bergström to unravel the structure of these substances, which were then available only in milligram quantities. The advances in total synthesis have altered all that, and today the natural prostaglandins are relatively abundantly available. Much of the

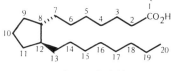

Prostanoic Acid

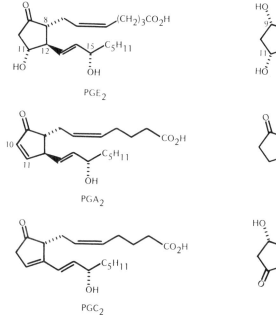

PGE$_2$

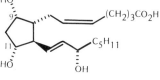

PGF$_{2a}$

PGA$_2$

PGB$_2$

PGC$_2$

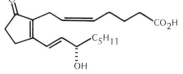

PGD$_2$

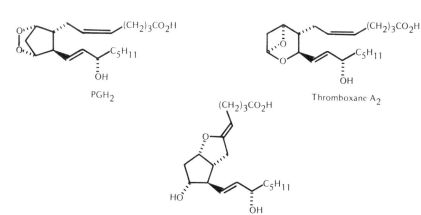

PGH$_2$

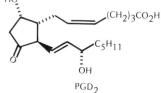

Thromboxane A$_2$

Prostacyclin

Chart 1.1. Structure and nomenclature of prostaglandins and thromboxanes.

credit for synthetic accomplishments in this field belongs to Professor E. J. Corey, who pioneered the early successes, and whose synthetic approaches continue to dominate the field.

The arachidonic acid cascade, to which the prostaglandins belong, has yielded several new members, some with striking biological activity. Notable among these are the thromboxanes and prostacyclins, which play an important role in the regulation of platelet aggregation. Though not strictly prostaglandins, the syntheses of these substances have been included in this chapter for review.

During the late 1970s there has been a clear shift away from devising syntheses of natural prostaglandins toward the synthesis of analogs possessing greater stability and selectivity of activity. Several analogs are in advanced stages of development. No attempt is made in this chapter to catalog the chemistry of these modified prostaglandins, or of the isomeric prostaglandins. Prostaglandin analogs are legion. Interested readers are referred elsewhere for a detailed review.[2] This chapter attempts to update the earlier article[1] by reviewing some of the major advances in prostaglandin chemistry from 1971 to the end of 1979.

Several books,[2-5] reviews, and proceedings of symposia[6-10] are available and provide details on many aspects of prostaglandin chemistry that may have been excluded or are only briefly mentioned here. The nomenclature of prostaglandins has been reviewed in detail (Chart 1.1).[2-11]

2. PROSTAGLANDIN SYNTHESES

A. Approach to Prostaglandin Synthesis

Two major problems encountered in the synthesis of prostaglandins are stereochemistry and sensitivity of the functional groups present in these molecules. For example, prostaglandin E_2 has four chiral centers, a labile β-ketol function in the five-membered ring, and two double bonds in the side chains. Three of the four chiral centers, those at positions 8, 11, and 12, are contiguous on the cyclopentane nucleus and are mutually *trans*. Since this is the more stable preferred arrangement, it does not offer any obstacles, and is relatively easily established. Stereochemical control of the fourth center at C-15, however, placed in a conformationally mobile side chain, and at a considerable distance from influence of the ring, poses a much more difficult problem.

The F prostaglandins possess an additional center of asymmetry at C-9. However, this additional complexity is more than offset by the much greater stability of the F prostaglandins, making them quite attractive synthesis targets. Oxidation of F-type precursors selectively protected at the 11- and 15-hydroxyl functions, followed by deprotection affords a convenient access to PGEs. Several protecting groups (e.g., tetrahydropyranyl and silyl ethers) are available, which can be removed under conditions to which E prostaglandins are stable. Since the E prostaglandins in turn can be readily converted to any other members of the same class, the E and F prostaglandins are primary synthetic targets of the more important and promising general approaches to prostaglandins.

The strategic use of blocking or masking groups and latent functionality, as well as the use of neighboring groups for influencing or establishing chiral centers elsewhere in the molecule is all important in prostaglandin synthesis. Effective employment of these strategies is most evident in the syntheses devised by E. J. Corey and R. B. Woodward, and in the 1,4-addition approaches, notably in the work of C. J. Sih, G. Stork, and the Syntex group.

B. Classification of Syntheses

Prostanoid syntheses may be divided into three major classes (Chart 2.1): (1) syntheses in which the cyclopentane ring is formed from acyclic precursors, (2) syntheses which commence from precursors containing a preformed cyclopentane nucleus, and (3) syntheses in which a bicyclic system is used to introduce an additional element of stereochemical control. The first two approaches are characteristically beset by difficulties in controlling stereochemistry and problems of selective protection and manipulation of functional groups. There is one exception, however: the 1,4-addition approach, which is completely stereospecific and allows for excellent control over the C-15 center. The third general class of syntheses relies on the relatively easier proposition of controlling stereochemistry in rigid cagelike structures. Bulky reagents are effectively employed to attack from predetermined directions, and one or more rings are subsequently

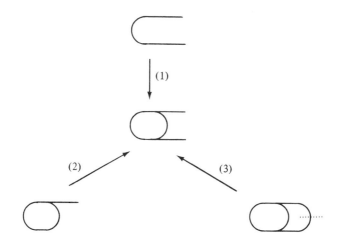

Chart 2.1.   Classification of prostanoid syntheses.

cleaved during the synthesis to unravel latent functionality, to establish the desired stereochemistry. An excellent illustration of this approach is Corey's bicycloheptane synthesis discussed separately in Section 8.

C. Commercial Feasibility

A desirable prostaglandin synthesis would be adaptable for production of all the primary prostaglandins and analogs in optically active form from a single precursor. Resolution must be accomplished at an early stage, as has been done in the Corey synthesis, or optical activity must be conferred by the use of chiral reagents. Mircrobial transformations offer a promising alternative for production of optically active compounds without resorting to chemical resolution. Alternatively, the synthesis could commence from readily available, optically active starting materials.

From the commercial point of view the synthesis should be of the converging type, involving a relatively small number of steps. Furthermore, the synthesis should be adaptable to large scale, although in the synthesis of such complex polyfunctional molecules it is very difficult to avoid using sophisticated and expensive reagents, some of which can be quite hazardous and difficult to handle even at the pilot-plant scale. Many of the syntheses that have been developed thus far require chromatographic separations at one stage or another. However, chromatographic separations, which are perfectly acceptable in the laboratory, often make the process quite undesirable on a larger scale and therefore, syntheses involving a minimum number of chromatographic separations are preferred. In spite of the various handicaps of the present approaches to prostaglandins, there is no doubt in the author's mind that a practical synthesis can be speedily tailored today for production of a specific prostaglandin analog, once it is identified.

D. Outline of Discussion

In the discussion that follows, syntheses of the primary prostaglandins, the E and the F prostaglandins, are discussed first in Sections 3 to 11. Of special interest to the reader looking for practical laboratory approaches to prostaglandins are the 1,4-conjugate addition (Section 5) and Corey's bicycloheptane (Section 8) approaches. The A, B, C, D, and H-type prostaglandins and the prostacyclin and thromboxanes are treated individually in Sections 12 to 18. Readers interested in 11-deoxyprostaglandins, which have often served as models for synthesis of more complex prostaglandins and are interesting biological substances in their own right, are referred to a detailed review elsewhere[2].

3. ACYCLIC PRECURSORS

One of the earliest approaches to prostaglandins have involved closure of the cyclopentane ring from acyclic precursors, usually by an aldol condensation. In general, however, the cyclization approaches have been limited by the rather formidable problems involving stereochemical control of ring substituents and selective protection of functional groups. Successful syntheses of natural prostaglandins using the cyclization approach were achieved by Corey and his associates at Harvard, Miyano at G. D. Searle, and by Stork and his group at Columbia. The Harvard Syntheses represent the first total synthesis of optically active natural prostaglandins, while Miyano's work is appealing because of its direct simplicity, and may have considerable potential in the manufacture of analogs. The synthesis of $PGF_{2\alpha}$ from D-glucose, developed by the Columbia group is a noteworthy example of the approach to natural products using carbohydrate-derived "chiral templates."[12]

Operationally, formation of the cyclopentane ring may be accomplished in a variety of ways: Miyano's approach involves formation of the C_8-C_{12} bond, while Corey forges the $C_{11}-C_{12}$ bond. Syntheses involving $C_{10}-C_{11}$ and C_9-C_{10} bond formation have also been described and noteworthy among these is the Columbia approach. No examples involving C_8-C_9 bond formation have been reported.

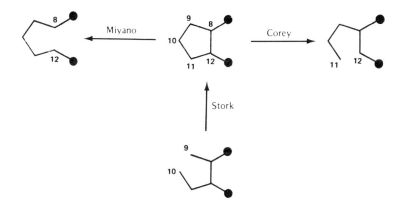

A. $C_{11}-C_{12}$ Bond Formation

Corey's syntheses, utilizing the cyclization approach, were reported in 1968. In these syntheses the five-membered ring was formed either by an aldol condensation, or a vinylogous aldol from open-chain precursors (Chart 3.1),

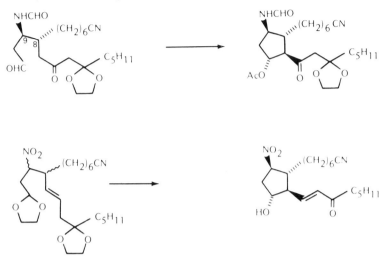

Chart 3.1. Corey's cyclization approaches.[13]   a.   b.

and have been reviewed earlier.[1,2,13] Although far from practical, Corey's syntheses were the first significant breakthroughs to natural prostaglandins and triggered the entry of several industrial laboratories into the search for therapeutically useful analogs. These syntheses were quickly supplanted by more practical approaches and are now only of historical interest.

B. C_8-C_{12} Bond Formation

Miyano's Syntheses

The synthesis of PGE_1 and PGF_1 developed by Miyano and co-workers at Searle involves formation of the C_8-C_{12} bond, thus proceeding through PGB-type intermediates (Chart 3.2). Although this approach has not been extended to prostaglandins of the two series, the relatively small number of steps involved, and good yields make this synthesis quite attractive for analog synthesis.

The synthesis, as originally disclosed by Miyano and Dorn,[14] was not entirely stereoselective and subsequently had to be modified.[15] The cyclopentane nucleus was formed by two aldol-type condensations starting with the β-ketoacid 1 and styrylglyoxal 2 readily available by selenous acid oxidation of benzalacetone. Cleavage of the styryl double bond in 4, using osmium tetroxide and periodate, afforded the unsaturated aldehyde 5. Prior to the key reduction step, which establishes the relative stereochemistry of C-8, C-11, and C-12 substituents, one face of the molecule was made less accessible to approach by the reducing species by anchoring the relatively bulky tetrahydropyranyl ether

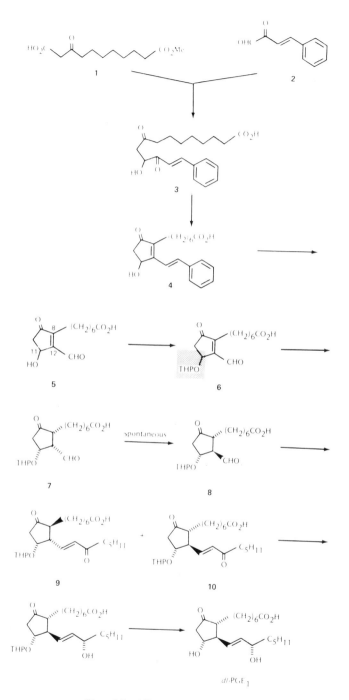

Chart 3.2. Miyano's synthesis.[15]

group to the C-11 hydroxyl. Reduction was accomplished using aqueous chromous sulfate solution and the resulting aldehyde 8 was subject to the Wittig reaction for introduction of the ω-chain, furnishing 10 as the major product, and with less than 20% contamination by the undesired 8,12-bisepiisomer (9). Isomerization of 7 to 8 appears to have occurred spontaneously during workup of the reduction mixture. The 15-dehydro-PGE$_1$ THP ether (10) was obtained in 25-30% overall yield from the unsaturated aldehyde 5. Reduction of 10 with lithium thexyl tetrahydrolimonyl borohydride preferentially introduced the 15(S)-hydroxyl (4:1 ratio), and was followed by deprotection to furnish $d\ell$-PGE$_1$. It is noteworthy that the key styryl intermediate 4 has been resolved microbiologically[16] and chemically via the (R)-(-)-a methoxyphenylacetic esters[17]. The (-)-4(R)-isomer was converted as described above into (-)-PGE$_1$.[15]

Kojima-Sakai Syntheses

Another example of the C$_8$–C$_{12}$ bond formation approach is the general route to prostaglandins developed by Kojima and Sakai at Sankyo Co., Japan.[18,19] The synthesis (Chart 3.3) involves as its key step the base-catalyzed cyclization of a diketoester 2 to afford the functionalized cyclopentenone 3. Introduction of three of the asymmetric centers (C-8, C-9, and C-12) on the cyclopentane ring is achieved simply by hydrogenating 3 over palladium on charcoal to give the *trans-cis*-cyclopentanone 4, while the fourth center (C-11) is introduced under influence of the already established C-8 and C-9 centers by reduction with sodium borohydride (4→5). Stereochemistry of the key intermediate 4 has been rigorously established by chemical and spectral means.[20] The synthesis has been utilized for synthesis of PGF$_{1\alpha}$[18] as well as PGE$_1$.[19] This approach is rather lengthy, and the overall yield is poor. While the yield of 4 from the starting ketoester is quite good (37%), the remaining steps from 4 to PGF$_{1\alpha}$ proceed in a total yield of only 6.7%.

Johnson's Synthesis

The utility of PGB-type intermediates generated by forging the C$_8$–C$_{12}$ bond is further illustrated by Johnson's chiral synthesis of the key prostaglandin intermediate 9, starting from (S)-(-)-malic acid (Chart 3.4).[21] The lactone 9 is a key intermediate in Corey's bicycloheptane approach and has been transformed into a variety of prostaglandins.

Treatment of (S)-(-)-malic acid with acetyl chloride gave S-(-)-2-acetoxysuccinic anhydride 1, which was converted to the corresponding succinyl choride 2 with dichloromethyl ether in presence of zinc chloride. Bishomologation using the dianion of methyl hydrogen malonate furnished the unstable dimethyl (S)-(-)-4-acetoxy-3,6-dioxosuberate (3), which was cyclized to a mixture of cyclopentenones 4 and 5. The fact that 5 was obtained predominantly is probably due

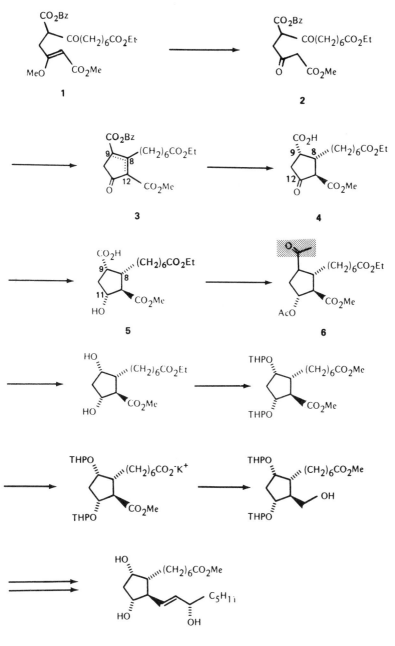

Chart 3.3. Synthesis of PGF$_{1\alpha}$ by Kojima and Sakai.

363

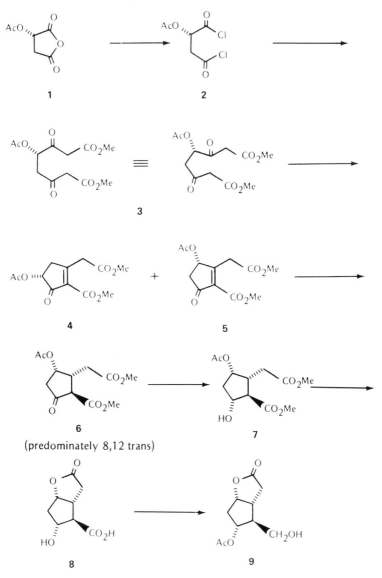

Chart 3.4. Johnson's synthesis of optically active Corey lactone.[21]

to the fact that the enolate ion (1) leading to 4 is destabilized with respect to (2) because of an interaction between the acetoxy group with the enolate oxygen atom. In any event pure 5 was obtained by direct crystallization from the reaction mixture, and the overall yield of 5 from 2 was 50%.

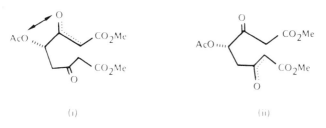

(i)   (ii)

Catalytic reduction of **5** afforded predominantly the cyclopentanone **6**, containing the thermodynamically stable *trans*-stereochemistry. Reduction of **6** with sodium borohydride in buffer followed by hydrolysis of the crude reduction product with KOH in methanol, and subsequent acidification led to the carboxylactone **8**. Reduction of the carboxyl group in **8** was accomplished, after protection of the hydroxyl group as the acetate, by converting it to the acid chloride and treatment with borohydride. The overall yield of optically active **9** from (S)-(−)-malic acid is impressive, exceeding 30%.

Stork's Synthesis

An interesting example of the C_8-C_{12} bond formation approach, using an intramolecular thermal ene reaction (**3→4**) to construct the cyclopentane ring has been reported by Stork and Kraus (Chart 3.5).[22] The resulting enone **7**, which was also synthesized by a different procedure has been converted to $PGF_{2\alpha}$ (Section 5.6).

C. C_9-C_{10} Bond Formation

Undoubtedly the most successful and elegant resolution of problems of the cyclization approach to prostaglandins was accomplished by Stork and co-workers[23] in their synthesis of $PGF_{2\alpha}$ from an acyclic chiral precursor (Chart 3.6). The chiral building block in this synthesis is D-glucose, which already contains the future C-11 and C-15 hydroxyl groups in the correct absolute stereoarray. The remaining centers of asymmetry in the molecule are systematically introduced under influence of existing chirality (**1→2**). Establishment of the

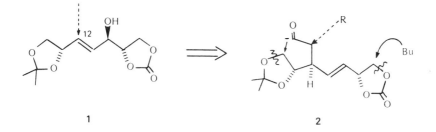

1   2

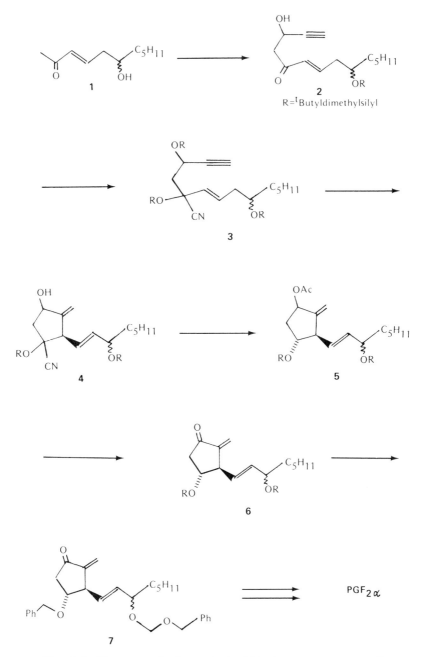

Chart 3.5.   The ene reaction for synthesis of 3-hydroxycyclopentanones.[22]

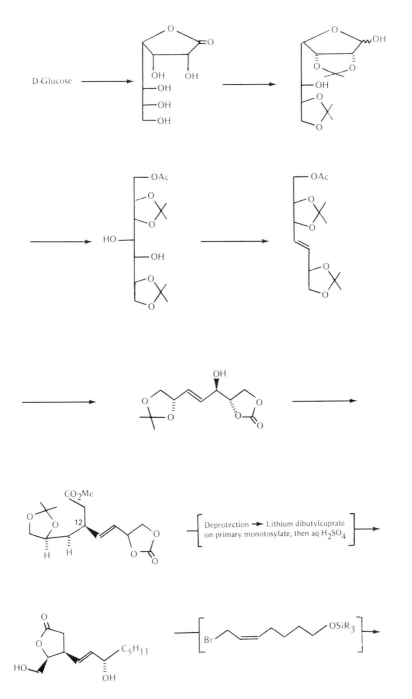

Chart 3.6.   Chiral synthesis of PGF$_2\alpha$ from D-glucose.[23]

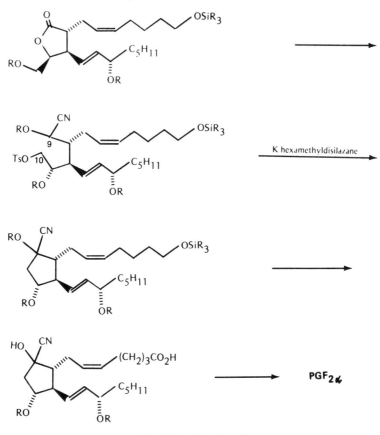

Chart 3.6. (Continued)

proper chirality at the eventual C-12 center was effected by the orthoester Claisen method. Construction of the cyclopentane ring involves formation of the C_9-C_{10} bond and was carried out using the protected cyanohydrin method.[24]

A synthesis of $d\ell$-PGE$_1$, involving formation of the C_9-C_{10} bond of the cyclopentane nucleus was reported by Finch and coworkers at Ciba-Geigy.[25] Although the synthesis, commencing from readily available starting materials, is characterized by good overall stereochemical control, it suffers from the lack of a suitable protecting group for the C-9 carbonyl, which is generated very early in the synthesis.

D. $C_{10}-C_{11}$ Bond Formation

This approach was utilized by Strike and Smith[26] of Wyeth laboratories for synthesis of a stereoisomeric mixture of 13,14-dihydro-PGE$_1$s. In this approach the cyclopentane ring is formed at a relatively late stage in the synthesis by a thermodynamically controlled intramolecular aldolization to give an A-type prostaglandin. Epoxidation and catalytic hydrogenation introduces the 11-hydroxyl group to furnish 13,14-dihydro-PGE$_1$ as a difficulty separable mixture containing all of the four possible C-11 and C-15 epimers.

4. CYCLOPENTANE PRECURSORS

Several promising and convenient approaches to prostaglandins commence with a preformed cyclopentane nucleus bearing requisite functionality for elaboration of at least one of the side chains. The second chain may be attached by a variety of methods, which are summarized in this section. However, by far the most successful variant of this general approach involves the conjugate addition of elements of the ω-chain to a cyclopentenone precursor, as discussed separately in the next section.

A. Epoxide Opening – Corey's Approach

The synthesis of PGF$_{2\alpha}$ in which the ω-chain is attached to the five-membered ring by epoxide displacement (1→2), using the latent β-formylvinyl anion bis(methylthio)allyllithium, with simultaneous generation of the C-11 hydroxyl, was described by Corey and Noyori in 1970.[27] Although this route to

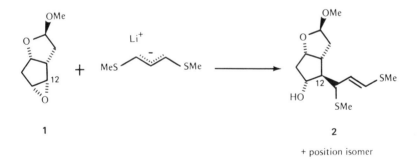

1	2

+ position isomer

PGF$_{2\alpha}$ was attractive by virtue of its directness, the low regiospecificity exhibited during the epoxide opening step make it impractical. Later work by Corey et al.[28] showed that the problem of lack of regiospecificity observed during opening of the epoxide could be overcome by the use of divinylcopper-lithium (vinyl Gilman reagent). Thus reaction of 3 with vinyl Gilman reagent gave a 94% yield of 4 as the major product, together with only a minor amount of the undesired position-isomer resulting from attack at C-11 (Chart 4.1). The

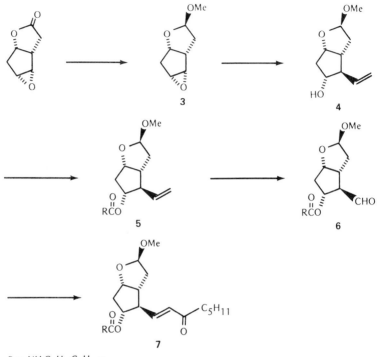

$R = -NH-C_6H_4-C_6H_5-p$

Chart 4.1.   Regiospecific opening of the epoxide.[28]

aldehyde **6** has been converted, via the enone **7**, to various prostaglandins by well-established pathways.

B. Epoxide Opening – Fried's Approach

A general synthesis of prostaglandins involving epoxide displacement with dial-kylalkynylalanes was developed by Fried and co-workers at the University of Chicago.[29] In this synthesis the fully functionalized lower side chain, in optically active form, is introduced on the oxidocyclopentane as an ethynylalane reagent. Like Corey's initial approach, described in the last section, this synthesis too suffers from a lack of stereospecificity during the key epoxide opening step. Subsequent work by Fried and co-workers[30] showed the regiospecificity of the alane-epoxide reaction is dependent on (1) composition of the alane reagent, and (2) the presence of a suitably placed carbinol function elsewhere in the mole-cule capable of forming a covalent aluminum-oxygen bond to ensure intra-molecular alkynylation at C-12 (cf. 1). Maximum regiospecificity was achieved

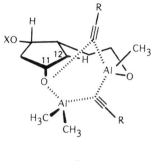

1

with the octynylmethoxymethylalane generated by the use of methoxy-methylchloroalane (MMCA).

The final improved synthesis of E and F prostaglandins using the regiospecific alane-epoxide reaction is outlined in Chart 4.2. Reaction of **4** with the alanate **8**, prepared from (S)-3-*t*-butyloxy-1-octyne and MMCA, gave exclusively the triol **5a**. The latter was debutylated with trifluoroacetic acid to give **5b** and

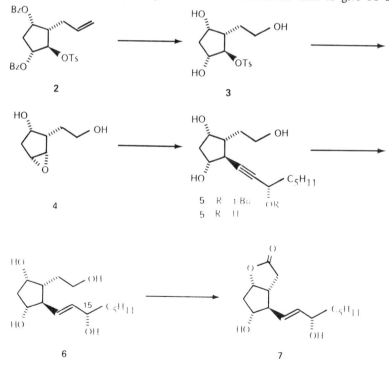

Chart 4.2.   Fried's regiospecific opening of the epoxide.[30,31]

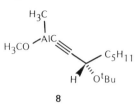

8

reduced with lithium aluminum hydride to **6** and converted to the Corey lactone **7** in several steps[31] or directly, using a novel selective dehydrogenation of **6** with platinum and oxygen.[30]

C. Asymmetric Synthesis of Prostaglandin Intermediates

Partridge et al. at Hoffman-La Roche[32] developed an ingenious asymmetric synthesis of the epoxy diol **6**, an important intermediate in Fried's general prostaglandin synthesis. The key to this approach is an asymmetric hydroboration, which proceeds in reasonable yields to furnish the product in high optical purity, without having to resort to a chemical resolution (Chart 4.3).

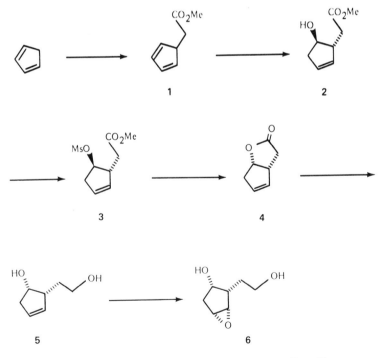

Chart 4.3.   Chiral synthesis of prostaglandin intermediates.[32]

The diene **1** was generated *in situ* by treatment of the sodium salt of cyclopentadiene with methyl bromoacetate in tetrahydrofuran. Asymmetric hydroboration with (+)-di-3-pinanylborane followed by alkaline hydrogen peroxide oxidation yielded the optically active hydroxy ester **2**, in 45% yield. The product thus obtained was at least of 96% optical purity. The hydroxy ester **2** was converted into the mesylate **3**, which afforded the optically pure lactone **4** upon treatment with base. The overall yield of **4** from cyclopentadiene was 40%. The lactone **4** was reduced with lithium aluminum hydride and converted to the desired epoxydiol **6**. Alternatively, the lactone **4** may be converted to the *cis*-epoxylactone **7**, and reduced directly to **6** with lithium aluminum hydride. This short asymmetrically induced synthesis also provides access to the epoxyacetal **8**, which has been converted into $PGF_{2\alpha}$ by Corey et al.[28]

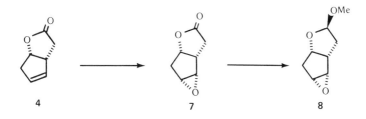

D. **Stork's Approach**

A novel general approach to prostaglandins via methylene cyclopentanone intermediates already bearing the ω-chain has been developed by the Columbia group.[33] Methylenecyclopentanones (cf. **1**) are readily prepared using a procedure developed by Stork and D'Angelo,[34] involving trapping of regiospecifi-

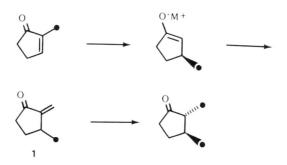

cally generated enolates with formaldehyde. Introduction of the remaining elements of the top chain in **1** is accomplished through conjugate addition to the a,β-unsaturated ketone. The synthesis is outlined in Chart 4.4.

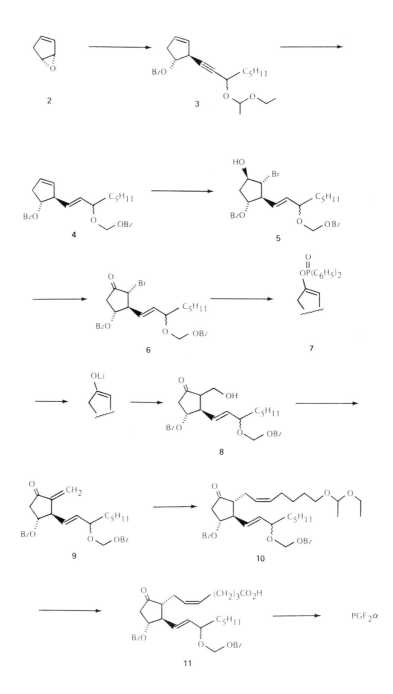

Chart 4.4. Stork's methylenecyclopentane approach to PGF$_2\alpha$.[33]

Introduction of the ω-chain on cyclopentadiene oxide (2) by epoxide opening with the lithium salt of 2-ethoxyethyl ether of 1-octyn-3-ol, followed by reduction to *trans*-vinylcarbinol and protection gave **4**. The choice of protecting groups in **4** was critical to the success of the synthesis. The groups had to be stable to acid conditions, especially during deprotection of the primary alcohol in **10**. The protecting group at C-15 had to be as electron-withdrawing as possible, while the reverse had to be true for the C-11 hydroxyl to prevent its elimination by base during the conversion of **8** to **9**. Finally, both protecting groups had to be removable without harm to the C-15 allylic alcohol function.

Addition of the elements of hypobromous acid to **4** proceeded regiospecifically to give the bromohydrin **5**. The observed regiospecificity was a consequence of the kinetic α-bromonium ion being displaced at the less hindered site, away from the side chain. Oxidation of the bromohydrin **5** with Jones reagent gave the bromoketone **6**, which was reduced with methyl diphenylphosphinite to the enol phosphinate **7**. Cleavage of the enol phosphinate with *t*-butyllithium smoothly afforded the corresponding lithium enolate, which was trapped with formaldehyde to give the hydroxymethylcyclopentanone **8** in excellent yield. Dehydration of **8** via the corresponding mesylate furnished the key methylenecyclopentanone **9** in 60% overall yield from **5**.

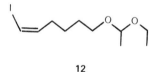

12

For addition of the top chain, the divinylcuprate prepared from the vinyl iodide **12** was reacted with **9** to provide **10**. Removal of the protecting group and oxidation gave the cyclopentanone acid **11**, which was converted to PGF$_{2\alpha}$ and its C-15 epimer. The two epimers, as the corresponding methyl esters, were separated by high pressure liquid chromatography.

E. Syntheses via Organoboranes

Several groups of workers have attempted boron-mediated cross-coupling reactions[35] for the addition of a carbon fragment or a fully functionalized ω-chain to cyclopentene precursors for generation of prostanoid structures. However, since only one of the carbon ligands attached to the boron is utilized during the cross-coupling process, such an approach is wasteful of one of the reaction components. The use of "mixed" dialkylboranes where one of the boron-bound carbon ligands (e.g., thexyl) has a low migratory aptitude is not entirely satisfactory, because migratory aptitude of ligands is both reaction-specific and substrate-dependent, and therefore not entirely predictable. Thus, while initial

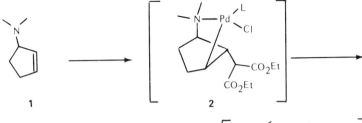

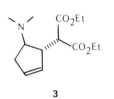

1

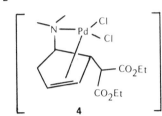

2

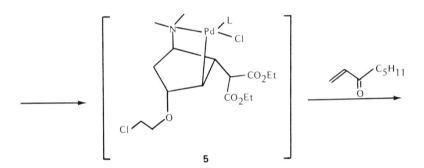

3

4

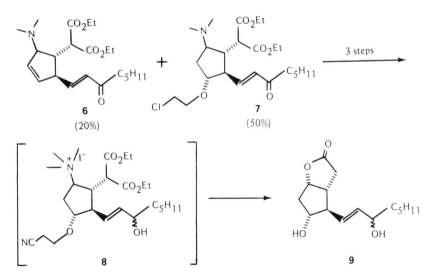

5

6
(20%)

+

7
(50%)

3 steps

8

9

Chart 4.5. Holton's carbopalladation approach to PGF$_2\alpha$.[38]

studies with model systems have established the feasibility of this approach,[36] it has not been extended to the synthesis of a natural prostaglandin.

Evans et al.[37] have explored the use of boronic esters for the synthesis of prostanoids. Since oxygen ligands do not compete with carbon in the rearrangement of boronate complexes, boronic esters offer a promising alternative to "mixed" dialkylboranes as a means to conserve the carbon ligands attached to boron during the cross-coupling process.

F. Syntheses via Carbopalladation

A novel approach to prostaglandins via carbopalladation has been described by Holton (Chart 4.5).[38] The synthesis involves direct attachment of the two side-chain fragments to the unactivated double bond of an appropriately substituted cyclopentene. In Holton's synthesis the four contiguous stereocenters about the cyclopentane ring are established from a single amine directing functionality. Following the dehydrocarbopalladation of **1**, the homoallylic alkoxypalladation-retrovinylation of **3** occurs *trans* to the sterically bulky diethyl malonate group. Although the production of **7** from cyclopentadiene is achieved in three chemical steps and in 44% overall yield, the synthesis is handicapped by formation of 20% of the unwanted enone **6** along with the desired enone **7**. The alcohol **9** is the Corey lactone diol, which has been converted to $PGF_{2\alpha}$.

5. CYCLOPENTANE PRECURSORS: CONJUGATE ADDITION APPROACHES

The conjugate addition of organometallic derivatives to α-substituted cyclopentenones provides an attractive, convergent approach to prostaglandins (**i**). This concept has been successfully applied, notably by Sih and co-workers at Wisconsin and a group at Syntex, to develop synthetic routes to PGE_1, PGE_2, 11-deoxyprostaglandins, and related compounds.

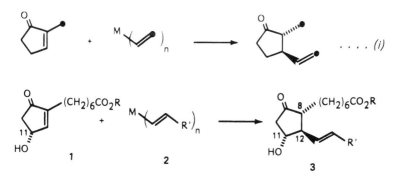

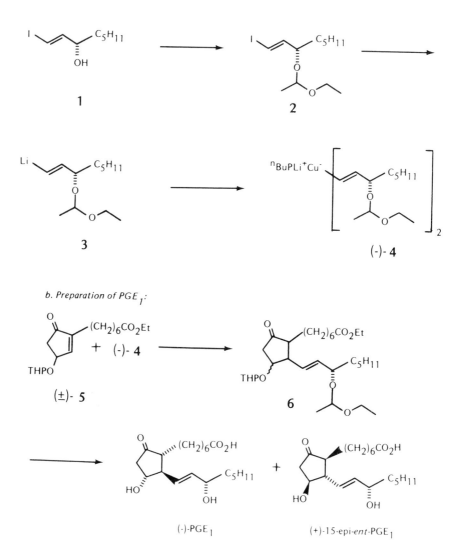

a. Preparation of the organocuprate 4:

1

2

3

(-)- **4**

b. Preparation of PGE₁:

(±)- **5**

6

(-)-PGE₁

(+)-15-epi-*ent*-PGE₁

Chart 5.1. Synthesis of PGE₁ by Sih et al.

Ample precedent exists for the efficient and stereospecific transfer of a vinyl group from vinyl-ate complexes to the β-carbon atom of conjugated enones to provide γ,δ-unsaturated ketones.[39] Moreover, addition of the vinyl species 2 is expected to proceed from the less-hindered side of the "C-11" hydroxylated cyclopentenone 1, giving rise to a vinylated cyclopentanone (e.g., 3) in which the two side chains are in the thermodynamically more stable *trans* relationship. The cyclopentanone 3 thus has the desired prostaglandin configuration at all three of its asymmetric centers. An intriguing feature of this approach is that once configuration of the single asymmetric center at C-11 in 1 has been established, it dictates the eventual configurational outcome at C-8 and C-12, thus making possible a highly stereospecific and concise synthesis of prostaglandins.

A. Conjugate Addition of Vinylcopper Reagents

The conjugate addition of a vinyl copper reagent to an α-alkylated cyclopentenone for the synthesis of a prostanoid, 15-deoxy-PGE$_1$, was first reported in 1972 by Sih and co-workers[40]. This was followed shortly afterward by a synthesis of PGE$_1$ (Chart 5.1) involving the addition of fully functionalized cuprate 4, complexed with tri-*n*-butylphosphine, to 2-(6-carbethoxyhexyl)-4-(2-tetrahydropyranyloxy)cyclopent-2-en-1-one (5).[41,42]

When the optically active cyclopentenone (-)-5 was used for the coupling, only (-)-PGE$_1$ methyl ester (65-70% yield based on 5) was isolated from the reaction mixture, thus providing a highly stereospecific synthesis of (-)-PGE$_1$.[43,44] Similarly, using the optically active cyclopentenone 7, which incorporates the future 5,6-double bond in its side chain, Sih et al. accomplished a completely stereospecific synthesis of (-)-PGE$_2$.[44,45]

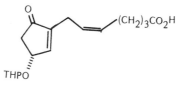

(-)-7

An almost identical approach to PGE$_1$ was developed independently by a Syntex group.[46] However, the yield obtained by these workers, in contrast to the Wisconsin group, was very low, and has been attributed to choice of trimethyl phosphite as the complexing ligand. A variant of this approach, also developed at Syntex,[47] involves conjugate addition of divinylcopperlithium instead of the fully functionalized side chain. The vinyl group introduced in

this manner serves as a latent aldehyde and is used as a handle for elaborating the lower side chain.

B. Alanate Additions to Cyclopentenones

A group at Lederle Laboratories[48, 49] explored the utility of conjugate addition of lithium *trans*-alkenyltrialkylalanate reagents to cyclopentenones for prostaglandin synthesis. *trans*-Alkenylalanates undergo selective transfer of the alkenyl ligand with retention of configuration,[50] 1,2-addition of the alkenyl group to the unsaturated carbonyl is observed, and there is no reaction at the ester function. Furthermore, the alkyl ligands do not appear to add to carbonyl function of the cyclopentenone **4**. The approach has been utilized for a synthesis of dl-PGE$_1$ (Chart 5.2).[49] The alanate required for the synthesis was initially prepared from a propargylic ether via hydroalumination, but later an improved

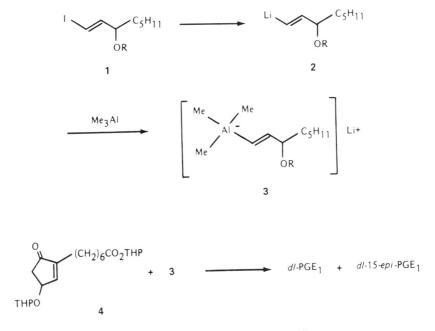

Chart 5.2.   Lederle synthesis of prostaglandins.

procedure was developed. The process involved preparation of the 3-oxy-*trans*-1-alkenyllithium by an exchange reaction, followed by treatment of the alkenyl-lithium with trialkylaluminum to generate the ate complex $(2 \rightarrow 3)$.[51]

C. Copper(1) Catalyzed Conjugate Addition of Grignard Reagent

Copper(1) catalyzed conjugate addition of the Grignard reagent prepared from 3-trityloxy-*trans*-1-bromide (**1**) to an appropriate cyclopentenone precursor has also been used for synthesis of PGE$_1$ (Chart 5.3).[52] Like the alanate reagents, the Grignard conjugate addition process also makes efficient use of the β-chain precursor (**1**), unlike the corresponding cuprate procedures in which half of the precursor cannot be utilized.

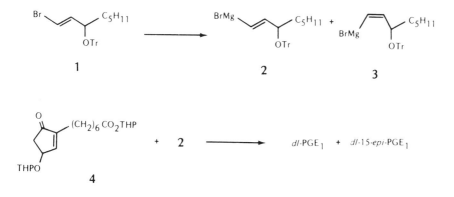

Chart 5.3. Synthesis of PGE$_1$ via conjugate addition of a Grignard reagent.[52]

D. The *cis*-Vinyl Route to PGE$_1$

Extending their studies on the conjugate addition approach to prostaglandins, the Syntex group also examined the addition of *cis*-divinylcuprates to cyclo-pentenones.[53] In marked contrast to the corresponding *trans*-divinylcuprates, the condensation of *cis*-divinylcuprates proceeded with high stereoselectivity and in excellent yields to furnish 9-oxo-13-*cis*-15β-prostenoic acids. Subsequently, Miller et al.[54] found that the 13-*cis*-prostenoid could be efficiently converted to the corresponding 13-*trans*-prostenoic acids via a highly stereo-specific sulfenate-sulfoxide rearrangement, thus completing an ingenious new total synthesis of prostaglandins (Chart 5.4).

When the 11-THP ether **8**, derived from **7** and **6**, was treated with *p*-toluene-sulfenyl chloride in presence of triethylamine it gave a sulfenate ester (**9**), which

a. Preparation of cis-divinylcuprate 6:

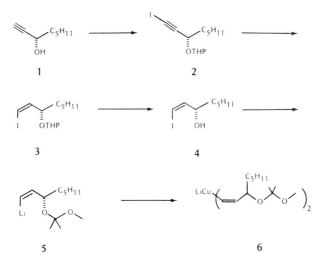

1   2

3   4

5   6

b. Preparation of PGE₁ methyl ester:

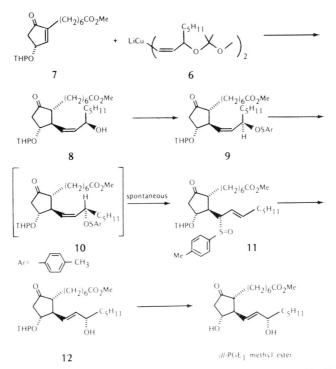

7   6

8   9

10   11

Ar = ⬡—CH₃

12   dl-PGE₁ methyl ester

Chart 5.4. Synthesis of prostaglandins via stereospecific sulfenate-sulfoxide transformations.

spontaneously underwent [2,3] sigmatropic rearrangement to the sulfoxide 11 in 81% yield. It should be noted that rearrangement of the sulfenate ester proceeds through the thermodynamically more stable transition state, which resembles the conformer 10 more than 9. Treatment of the sulfoxide 11 with trimethylphosphite gave the ketol 12, which was subjected to acid hydrolysis to furnish $dℓ$-PGE$_1$ methyl ester. The approach has also been extended for the preparation of natural PGE$_1$,[55] and for the synthesis of $dℓ$-19-hydroxy-PGE$_1$.[56]

E. Preparation of the Cyclopentyl Synthon

Several practical syntheses of the key α-alkylated cyclopentenone, 2-(6-carbomethoxyhexyl)-4-hydroxycyclopent-2-en-1-one (5), the PGE$_1$ precursor synthon, and the corresponding precursor for PGE$_2$ synthesis have been developed.

PGE$_1$ Precursor Synthon

One approach to the PGE$_1$ precursor synthon is based on the well-known[1,2,44] 1,2,4-cyclopentane trione acid 1. Conversion of the triketone 1 into 5 (Chart 5.5) has been accomplished by Sih, using the sequence: (1) asymmetric reduc-

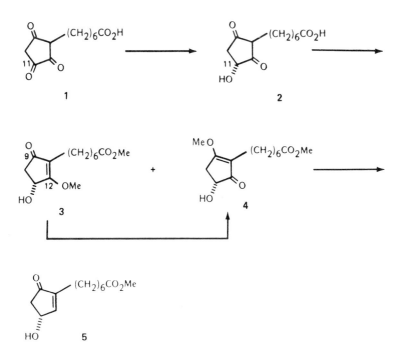

Chart 5.5. Synthesis of PGE$_1$ precursor synthon by Sih et al.

tion of the C-11 carbonyl to yield the (*R*)-alcohol **2**; (2) conversion of **2** into the enolate **3**; and (3) reduction of the C-9 carbonyl followed by allylic rearrangement to yield **5**.

Monoreduction of **1** to give racemic material is readily achieved chemically by hydrogenation over palladium catalyst, but asymmetric reduction is best accomplished by a wide variety of microorganisms.[44] Although not entirely satisfactory, partial asymmetric catalytic hydrogenation of **1** has also been carried out using a soluble rhodium catalyst in the presence of chiral phosphine ligands. The diketone **2** was converted to a mixture of methyl enol ethers **3** and **4**. Since the enolate at C-9 is sterically less crowded than at C-12, treatment of **3** with a trace of acid gave **4**, which was reduced to **5**. Pappo et al.[57] have reported a novel resolution of racemic **5**. The hydroxycyclopentenone **5** has also been prepared by either allylic[46,49] or microbial[58] hydroxylation of the readily available cyclopentenone **6**. Sih and co-workers[40,42] also developed an ingenious synthesis of the cyclopentenone **5** utilizing the 1,4-cycloaddition

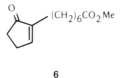

6

of singlet oxygen to an appropriately substituted cyclopentadiene (Chart 5.6). The chiral synthesis of the enone **5** from D-glyceraldehyde by Stork and Takahashi is also noteworthy (Chart 5.7).[55]

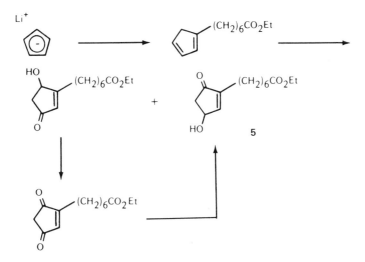

Chart 5.6.   The 1,4-cycloaddition approach to PGE_1 precursor synthon.

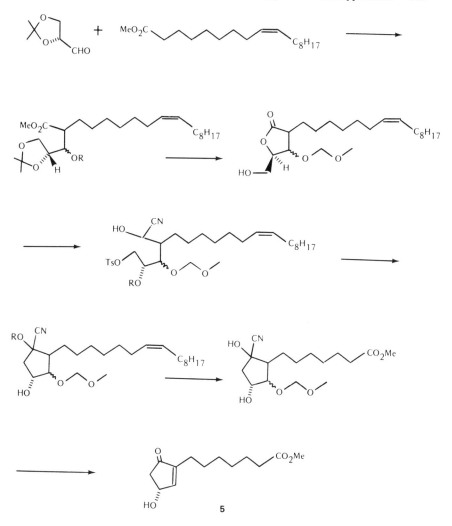

Chart 5.7. Synthesis of PGE$_1$-precursor synthon from D-glyceraldehyde.[55]

PGE$_2$ Precursor Synthon

Synthesis of the PGE$_2$ precursor, 2-(6-carbomethoxy-*cis*-2-hexenyl)-4(*R*)-hydroxy-2-cyclopenten-1-one (**1**), reported by Sih et al.,[44,45] is patterned after their earlier general approach to the corresponding PGE$_1$ precursor. The overall synthesis, however, is rather lengthy and has been superseded by other, more convenient, approaches.

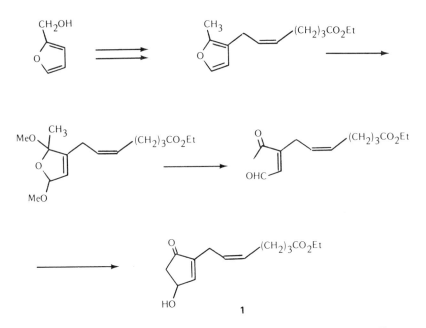

Chart 5.8. Synthesis of hydroxycyclopentenones from furfuryl alcohol.[59]

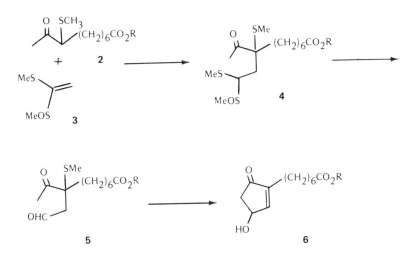

Chart 5.9. Cyclization approach to hydroxycyclopentenones.[60]

Two recent noteworthy cyclization approaches to the hydroxycyclopentenones involve intramolecular aldol condensation of open-chain precursors
generated from the dihydrofuran 1 (Chart 5.8)[59] or constructed by the addition
of the enolonium ion equivalent 2 to the ketone enolate 3 (Chart 5.9).[60]
Several practical approaches to the PGE_2 precursor synthon commence from
the readily accessible cyclopent-3-ene-1-ones (7). Stork et al. and a Lederle
group have shown that epoxyketones of the type 8, obtained from 7, are readily
isomerized to the desired 4-hydroxycyclopentenones 9, while Gruber et al.
obtained the same products from the isopropylidene derivative of the diol
11. The epoxyketone 8 can a priori undergo base-catalyzed elimination in two

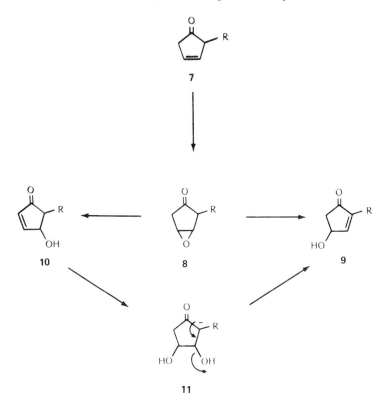

possible directions to give either 9 or 10. However, treatment of 8 with base
gives rise to a mixture of 9 and 10 in which the latter, the product of kinetically
formed (less substituted) enolate, predominates. Isomerization of 10 to the more
stable 9 requires rapid hydration of 10 to the diol 11, which then undergoes
dehydration. Stork et al.[61] found that making the hydration step effectively
intramolecular, by the addition of chloral during treatment of 10 with base,

rapidly and efficiently produced isomerization of **10** to **9**, presumably via the acetal intermediate **12**. Thus the key epoxyketone **13**, obtained by Stork[61]

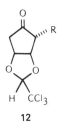

12

using the route outlined in Chart 5.10, upon treatment with triethylamine followed by chloral, furnished a 69% yield of the desired enone **14**. Equilibration of a mixture of the corresponding α- and β-epoxyketones in aqueous sodium

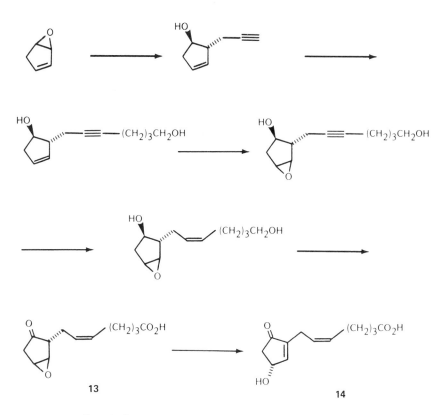

Chart 5.10.   Stork's synthesis of PGE$_2$ precursor synthon.

carbonate over 24 h, also furnished the enone **14**, but in 40% yield.[62] An alternative synthesis of the epoxyketone has been reported.[63]

The synthesis of PGE_2 precursor **16** by Gruber et al.[64] is outlined in Chart 5.11. This approach appears to be of considerable practical importance. An interesting feature of this synthesis is that it makes use of the unwanted antipode **15** obtained from the resolution of racemic **15**, itself an important intermediate in Corey's prostaglandin synthesis.

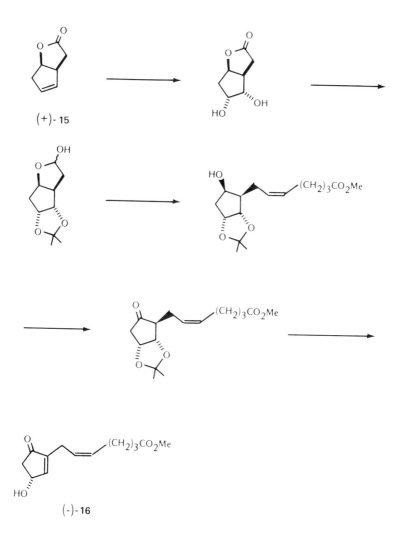

Chart 5.11.   Synthesis of PGE_2 precursor synthon by Gruber et al.[64]

Rearrangement of cyclopentenols of the type **10** to **9** also provides a convenient access to the PGE$_2$ precursor synthon. The requisite cyclopentenol **17** may be obtained from furyl precursors via a dimethoxyfuran[65] or directly by an acid-catalyzed rearrangement of a furyl carbinol.[66]

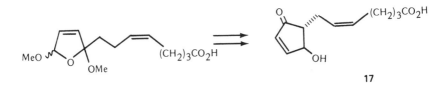

F. Conjugate Addition – Enolate Trapping Approach

A converging approach to prostaglandins would be available if the regiospecific cyclopentane enolate **2**, generated by the conjugate addition of organocopper reagent, can be trapped by alkylation with the fully functionalized top chain, or

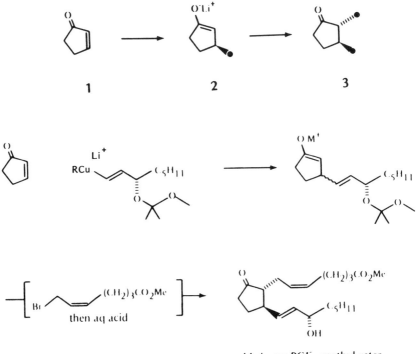

11-deoxy PGE$_2$ methyl ester

Chart 5.12.   Conjugate addition-enolate trapping approach to prostaglandins.

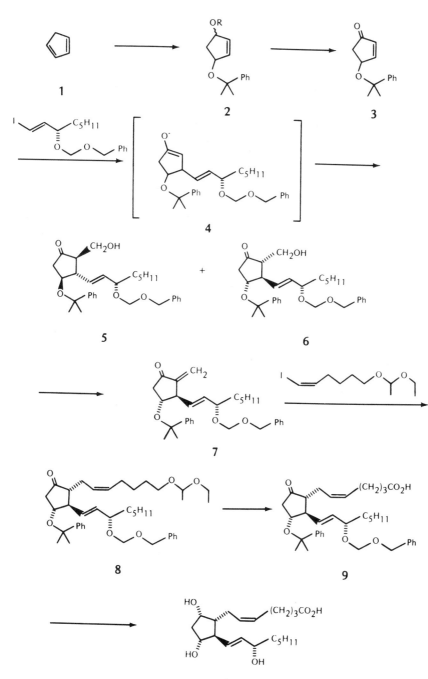

1

2

3

4

5   +   6

7

8   9

PGF$_{2\alpha}$

Chart 5.13. Methylenecyclopentanone approach to prostaglandins.[69]

a group that would provide a handle for its attachment. Patterson and Fried[67] reported just such a synthesis of 11-deoxyprostaglandins from cyclopenten-2-one (Chart 5.12), but were unable to extend it to E prostaglandins. Model studies, using this approach have also been described by Posner.[68]

Stork and Isobe[69] successfully applied this approach for the synthesis of natural prostaglandins (Chart 5.13). The regiospecific enolate **4** was efficiently trapped by monomeric formaldehyde to give a hydroxymethylcyclopentanone **6**, which provides the necessary functionality for further elaboration of the top side chain. The crucial 1,4-addition formaldehyde-trapping sequence reproducibly afforded yields of 50-60% for the conversion of **4** to **6**. Conversion of the hydroxymethylcyclopentanone **6** to a mesylate followed by elimination gave the key methylenecyclopentanone **7**, to which the remainder of the top chain was grafted by the addition of a divinylcuprate.[70] It is noteworthy that this remarkably short synthesis provides pure (+)-PGF$_{2\alpha}$ in only eight steps from 4-cumyloxy-2-cyclopentenone (**3**) in approximately 9% overall yield.

6. CYCLOHEXANE PRECURSORS

An important approach to prostaglandins involves contraction of six-membered carbocyclic precursors for production of the appropriately functionalized cyclopentane nucleus. The strategy has proved quite effective, owing to the large variety of efficient ring contraction procedures available to the organic chemist. Both Woodward and Corey used pinacolic deaminations to effect the key ring contraction step in their syntheses. In another approach, which has proven exceptionally successful for synthesis of 11-deoxyprostaglandins, Corey utilized a thallium(III) promoted semipinacolic ring contraction to establish the five-membered ring.[73]

A. The Woodward Synthesis

The stereospecific synthesis of PGF$_{2\alpha}$ developed at the Woodward Research Institute, Basel,[71, 72] is noteworthy for its ingenious use of functional groups as internal protecting agents and their applications to exert stereochemical control during synthetic operations. As pointed out earlier, the key step in this synthesis

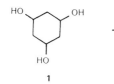

1

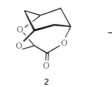

2

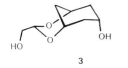

3

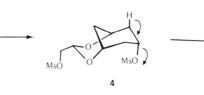

4

5

6

7

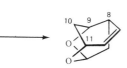

8

9
(minor)

+

10
(major)

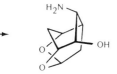

11

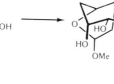

12

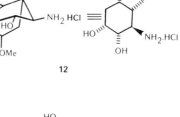

13

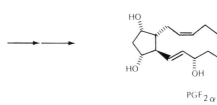

$PGF_{2\alpha}$

Chart 6.1. Woodward's synthesis of $PGF_2\alpha$.[71,72]

is the pinacolic deamination-ring contraction of the cyclohexyl aminohydrin **12**, which efficiently establishes the prostanoid cyclopentane nucleus (Chart 6.1).

The synthesis commences with *cis*-cyclohexane-1,3,5-triol (**1**), which was converted to the crystalline tricyclic lactone **2** with glyoxalic acid, and subjected to sodium borohydride reduction for production of the diol **3**. The dimesylate **4** was treated with base to yield a bicyclic olefin mesylate **5**, which furnished the crystalline tricyclic carbinol **6** after solvolysis. Mesylation of **6** followed by treatment with refluxing isopropanolic potassium hydroxide solution gave the key tricyclic olefin **8**, which contains the 11-hydroxyl group and the γ-lactol moiety of **13** in a mutually and internally protected form. In addition, the double bond in **8** also provides a handle for effecting the 6→5 ring contraction sequence with simultaneous introduction of the C-12 aldehyde function. Epoxidation of **8** gave a 62% yield of **10**, the product of preferential attack on the double bond from the concave side of the molecule, along with a 25% yield of the undesired epoxide **9**. After separation from **9**, ammonolysis of **10** furnished the aminoalcohol **11**, which was then converted to the dihydroxyaminoacetal hydrochloride **12**. In contrast to the rigid tricyclic system of **11**, which forces the amino group to orient axially, **12** has a flexible skeleton that allows it to undergo a conformational change, such that the amino function destined to serve as a leaving group is equatorially positioned. The key ring contraction step was accomplished by diazotization in aqueous acetic acid and followed by neutralization with mild base to give the desired hydroxyaldehyde acetal. The synthesis is completed by attachment of the lower side chain to the free aldehyde group in **13**, reduction of the C-15 carbonyl, and attachment of the top chain to the aldehyde function masked as the acetal in **13**, to give PGF$_{2\alpha}$.

Corey and Snider[74] used a similar pinacolic deamination for the key ring contraction step in their synthesis (Chart 6.2). The starting cyclohexylaminohydrin **15** was obtained from the readily available lactone **14**, by allylic functionalizing via the "ene" reaction with *N*-phenyltriazolinedione followed by cleavage to give the allylic amine.

B. The Hoffmann-La Roche Synthesis

The synthesis of prostaglandins developed by Rosen and co-workers at Hoffmann-La Roche[75] utilizes an interesting Favorskii-type ring contraction of 2-chloro-2-alkylcyclohexane-1,3-diones to establish the cyclopentane nucleus. The ring contraction sequence was originally developed by Buchi and Egger[76] in connection with their synthesis of methyl jasmonate. Another noteworthy feature of this synthesis is the use of a novel carboxy-inversion reaction[77] for generation of the C-11 hydroxy group (Chart 6.3).

Acid-catalyzed reaction of 5-carboxy-1,3-cyclohexanedione with allyl alcohol gave **2**, which rearranged to the enol acetate **3**. Treatment of the corresponding

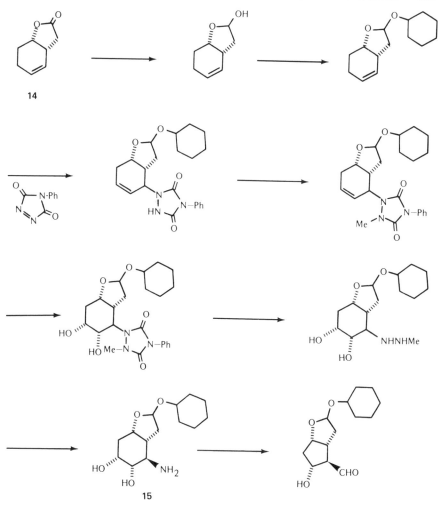

Chart 6.2. Corey's ring contraction approach.[74]

dione **4** with *tert*-butyl hypochlorite gave the chloro derivative **5**, which rearranged to the ring-contracted cyclopentenone **6**. Conjugate addition of the anion from nitromethane to **6** gave **7**, which was oxidized to the keto acid **8**. Reduction of the C-9 carbonyl followed by lactonization afforded the *cis*-fused lactone **9**. After transformation of the nitro group in **9** to an aldehyde, introduction of the lower chain, the carboxy inversion procedure and deprotection afforded the known lactone diol **13**, which has previously been converted to E and F prostaglandins by Corey.

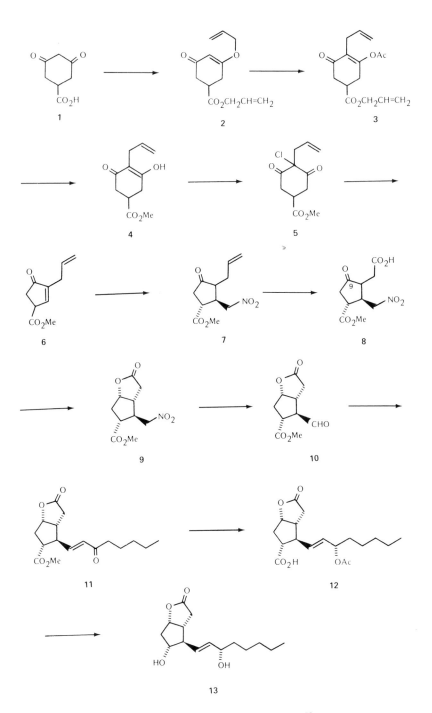

Chart 6.3. Hoffman-La Roche approach.[75]

C. The Merck Synthesis

A synthesis of PGE_1 developed at Merck[78] utilizes oxidative cleavage of a cyclohexene intermediate followed by cyclization of the resulting diacid to generate the desired functionalized cyclopentanone nucleus. The synthesis is stereoselective, but nonetheless, quite lengthy.

7. BICYCLOHEXANE PRECURSORS

A. Bicyclo[3.1.0]hexane Approach

The use of bicyclo[3.1.0]hexane precursors for prostanoid synthesis has been reviewed in an earlier volume.[1] Essentially, the approach is based on the expectation that the prostanoid C-11 and C-15 hydroxyl functions, along with the *trans*-13,14-double bond, could be simultaneously generated by solvolysis of an appropriate cyclopropylcarbinyl cation **1**, for instance, derived from an epoxide. Attack by water on the cation **1** is expected to occur at C-11 because of steric interference at C-12 by elements of the top-chain attached at C-8. Formation of the *trans*-13,14-olefinic linkage is expected from a consideration of the most favorable disposition of substituents during the ring-opening sequence.

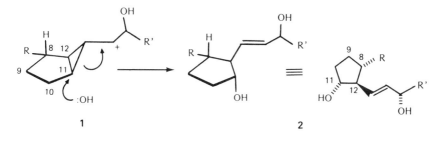

1   2

B. Combined Bicyclohexane-Bicycloheptane Approach

In 1970 Corey described a useful variant of the bicyclo[3.1.0]hexane approach in which he combined elements of the bicyclo[3.2.0]heptane approach for introduction of 8,9-substituents on the cyclopentane nucleus (Chart 7.1).[79] This synthesis, however, was not very efficient because the key solvolysis step proceeded in rather low yield. The strategy was later adapted by Kelly and Van Rheenen (Chart 7.2),[80] who combined it with the Upjohn solvolysis approach and used it to develop an efficient synthesis of the key, optically active, prostaglandin lactone diol **17**, which has previously been converted to natural prostaglandin by Corey. Resolution of the ketone **11** was effected using a novel

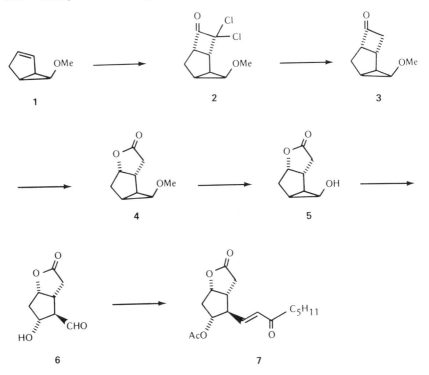

Chart 7.1. Corey's combined bicyclohexane-bicycloheptane approach.[79]

procedure involving the oxazolidine **19**.[81] Oxidation of the olefin **14** may be accomplished via the epoxide, which is subject to several cycles of solvolysis with formic acid to produce the desired 15α-diol **17** in 45% overall yield based on the epoxide **15**. Alternatively, the olefin **14** may be directly converted to the *cis*-glycol **16** using an improved catalytic osmium tetroxide oxidation with tertiary amine oxides as the oxidant.[82] Solvolysis of the cyclopropyl carbinyl system to furnish **17** is best accomplished via the orthoester **20** (Chart 7.3).[83] In another approach, the aldehyde **13** was converted to a masked mixed acyloin

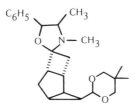

19

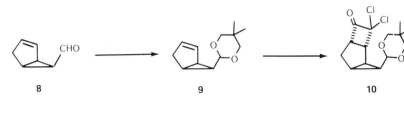

8　　　　　　　　　**9**　　　　　　　　　**10**

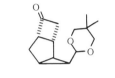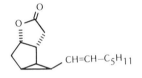

11　　　　　　　　　　　　　　**12**

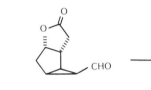

13　　　　　　　　　　　**14**

15

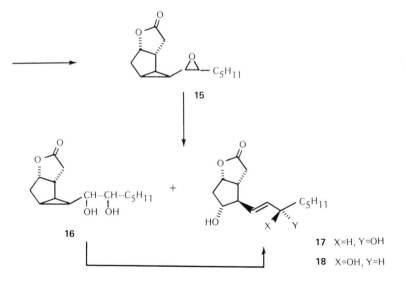

16

17 X=H, Y=OH

18 X=OH, Y=H

Chart 7.2. Kelly's synthesis of bicyclohexane intermediates.

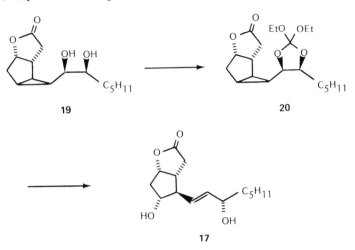

Chart 7.3.   Improved opening of bicyclohexane intermediates.[83]

21 which could be solvolyzed to a masked enone (as the cyanohydrin) (Chart 7.4). Unmasking and reduction of the enone **23** gave **17**.[84] The diol **17** was identical with material prepared by Corey et al. and has been converted to natural prostaglandins.

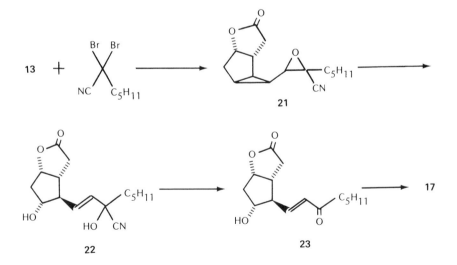

Chart 7.4.   From Ref. 84.

C. Miscellaneous Bicyclohexane Approaches

Cleavage of bicyclohexanes by organocopper reagents also provides a promising approach to prostanoids.[85] This reaction is of potential interest for direct production of 3 by the homoconjugate addition of the (S)-vinylcopper reagent 2 to optically active 1.

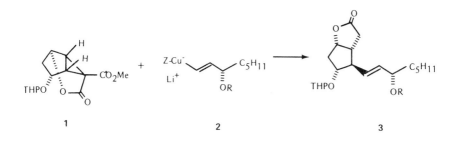

1   2   3

8. BICYCLOHEPTANE PRECURSORS: COREY'S SYNTHESIS

A versatile approach to prostaglandins, based on the bicyclic lactone aldehyde C, was developed by Corey and his associates at Harvard during 1969. This synthesis, generally known as the bicycloheptane approach, derives its name from the fact that the cyclopentane nucleus, along with its latent functionalities and key stereochemical centers is embedded in a bicyclo[2.2.1]heptane ring system. Scission of one of the bicycloheptane rings generates the latent functionalities, which are then elaborated into the ring appendages in a stepwise fashion (A→B):

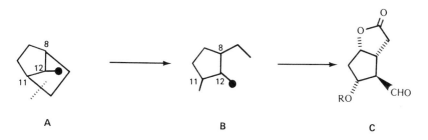

A   B   C

The synthesis is stereospecific and has been used for preparation of E and F prostaglandins of the one, two, and three series. It has also been modified to yield A, C, and D prostaglandins as well as numerous prostaglandin analogs. By changing the phosphonate and/or phosphorane Wittig reagents used for introduction of the side chains, it has been possible to prepare modified prostaglandins, which display greater metabolic stability and selectivity of biological effects (section 19). Although the synthesis is quite lengthy, the overall yield for the 17-step sequence from cyclopentadiene to racemic PGE_2 and $PGF_{2\alpha}$ is quite acceptable. The synthesis competes favorably with biosynthetic procedures, and in addition has the distinct merit of offering much greater flexibility in analog synthesis. Since optical resolution can be carried out at an early stage and the reactions are amenable to scale-up, the synthesis has been varied and extended, not only at Harvard and other academic laboraties, but also in several industrial laboratories where it has been adapted for pilot plant production.

Corey's synthesis has been outlined in an earlier volume,[1] and has been discussed in detail elsewhere.[2] The overall synthetic plan is summarized in Chart 8.1 and is discussed briefly below. Corey's own account of the early Harvard work is summarized in a lecture delivered to the National Academy of Sciences in 1971.[86] A noteworthy feature of the Corey synthesis is the use of a Diels-Alder reaction for construction of the key bicyclic intermediate 4 in which four

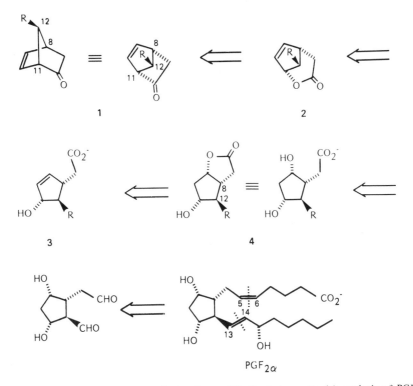

Chart 8.1. Corey's overall synthetic plan—antithetic (retrosynthetic) analysis of $PGF_2\alpha$.

of the ring appendages and chiral centers of the basic prostaglandin nucleus are established efficiently and with complete stereospecificity. The Diels-Alder addition assures the correct *trans-trans* relationship at C_8–C_{12} in 1, while introduction of the ring hydroxyls is smoothly accomplished by a Baeyer-Villiger oxidation of the ketone functionality to a lactone (1→2). This is followed by saponification and iodolactonization for placement of the fourth contiguous asymmetric center on the cyclopentane ring (3→4). This use of a neighboring group not only serves to establish the *cis* relationship between the 9-hydroxyl and the C-8 substituent but also facilitates the selective protection of the carbonyl functions required for sequential introduction of C-12 and C-8 side chains. The *cis*-Δ^5 and *trans*-Δ^{13} linkages are established using modified Wittig reactions to complete this versatile synthesis. The choice of tetrahydropyranyl ether-protecting groups, which are labile to mild acid hydrolysis, overcomes the difficulties associated with masking and unmasking of the sensitive β-ketol system of the E prostaglandins. Alternative protecting groups have since been developed and have found application in newer prostaglandin syntheses. Minor modifications of the synthesis lead to the synthesis of other primary prostaglandins.

A. Modifications of the Corey Synthesis

Several improvements in the Corey synthesis have been reported by the Harvard team and others.[2] The synthesis as originally described involved the use of a number of toxic and hazardous reagents, not suited for large-scale preparations. Another problem was isomerization of the diene observed during the initial Diels-Alder cycloaddition, which resulted in the desired bicycloheptene being contaminated with its isomer, requiring chromatographic purification. The use of thallous cyclopentadiene gives a product less prone to isomerization,[87] but offers special problems on a large scale owing to the highly toxic nature of thallium.

2-Chloroacrylonitrile, which served as the ketene equivalent during the cycloaddition in the early syntheses, has been replaced by other dienophiles. α-Chloroacrylyl chloride has been used,[88] and reacts with 5-benzyloxymethyl cyclopentadiene without prototropic rearrangement, but requires the use of sodium azide for unmasking the carbonyl function (Chart 8.2). The use of thallium or catalysts during the cycloaddition, and base-catalyzed isomerization of the intermediate 5-monoalkylated cyclopentadiene is best avoided by using acrylic acid as the dienophile (Chart 8.3). Conversion of the resulting bicyclic acid 2 to the ketone 3 is smoothly accomplished using a novel oxidative decarboxylation procedure developed by Trost and Tamura.[89] Wasserman and Lipshutz[90] have also reported an efficient oxidative decarboxylation, which appears suited for effecting the conversion 2→3. Synthesis of the optically active

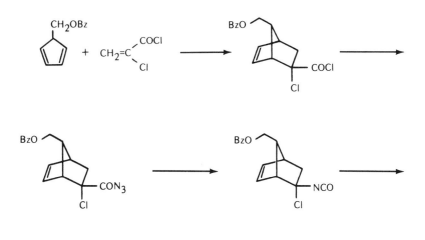

Chart 8.2. Improved synthesis of the bicyclic ketone.[88]

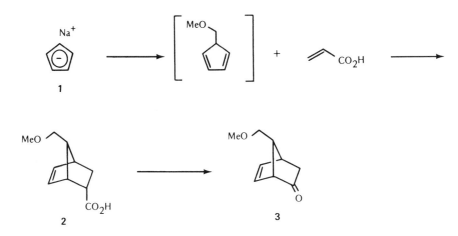

Chart 8.3. Oxidative decarboxylation approach to Corey's bicyclic ketone.

bicyclic ketone has been accomplished by asymmetric induction using the chiral acrylate **4**, prepared from (S)-(−)-pulegone.[91]

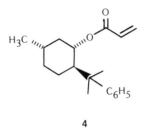

4

Another promising approach to the bicyclic ketone, developed by Ranganathan et al.,[92] utilizes nitroethylene as the ketene equivalent (Chart 8.4). The highly reactive nitroethylene makes it possible to replace the undesirable thallium cyclopentadienide with the corresponding sodium derivative for preparation of the nitronorbornenone **5**. The nitronorbornene → norbornenone transformation (**5→3**) utilizes the mild titanium-catalyzed nitro to ketone conversion developed by McMurry and Melton.[93] The nitroethylene approach has been further extended by Bartlett and co-workers.[94]

Several operational improvements have been reported in various steps of the Corey syntheses, and have been reviewed in detail elsewhere.[2] Particularly noteworthy is the use of p-phenylphenylcarbamoyl-protecting group and lithium trialkylborohydrides for stereospecific generation of the 15S configuration.[95]

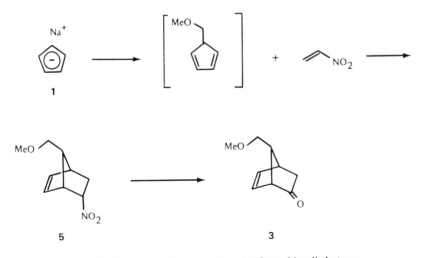

Chart 8.4.   Ranganathan's approach to the Corey bicyclic ketone.

A promising, short synthesis of the Corey aldehyde, involving a regiospecific Prins reaction on the readily available lactone 1, has been described by a Hungarian team (Chart 8.5).[96] Addition of formaldehyde in glacial acetic acid gave a 75-85% yield of the diacetate 2. While selective unmasking of the primary hydroxyl group could not be successfully achieved in 2, the diacetate was readily hydrolyzed to the known lactone diol 3, and then selectively oxidized to the hydroxyaldehyde 4 with the thioanisoe-chlorine[97] or more conveniently by the Pfitzner-Moffat reagent. Attachment of the ω-chain to the crude unstable hydroxyaldehyde, followed by p-phenylbenzoylation, gave the known bicyclic enone 5 in good overall yield.

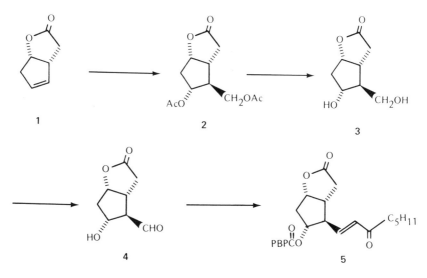

Chart 8.5.   From Ref. 96.

B.   The Nortricyclane Approach to Corey Intermediates

An alternative approach to the Corey alcohol 9, starting with Prins reaction of norbornadiene, was independently developed at Pfizer[98] and by Peel and Sutherland at the University of Manchester.[99] The Pfizer synthesis is outlined in Chart 8.6.

Reaction of norbornadiene (1) with paraformaldehyde in formic acid containing a catalytic amount of sulfuric acid gave the nortricyclane diformate 2 in good yield. Oxidation of 2 with Jones reagent led directly to the keto acid 3, which was readily secured in the correct absolute configuration required for prostaglandin synthesis by resolution with L-(−)-α-methylbenzylamine. Reaction of the (+)-acid 3 with aqueous hydrochloric acid cleanly produced the chloro

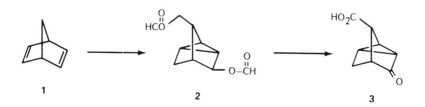

1   **2**   **3**

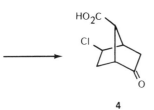 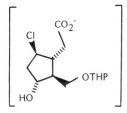 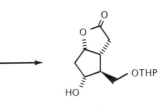

4   **5**

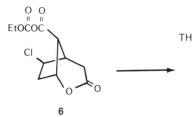 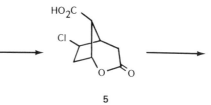

6   **7**

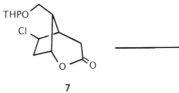

8

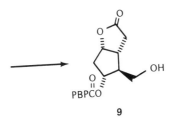

9

Chart 8.6.   Pfizer's nortricyclane approach.[98]

acid **4** which underwent Bayer-Villiger oxidation to give the chlorolactone acid **5**. Reduction of the carboxyl function, followed by protection of the resulting primary alcohol function as the tetrahydropyranyl ether gave **7**. The key displacement reaction establishing the γ-lactone proceeded in aqueous base buffered with 30% hydrogen peroxide to give **8**. Acylation with p-phenylbenzoyl chloride followed by acid cleavage of the tetrahydropyranyl group gave the known lactone alcohol **9**. Conversion of the racemic **3** to **9** was accomplished by the Manchester group by a slightly different route.[99]

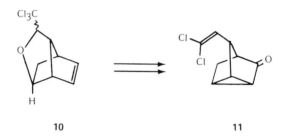

                    10   **11**

The tricyclic compound **10**, prepared from norbornadiene and chloral in the presence of aluminum chloride, provides another entry to the Corey aldehyde via nortricyclane intermediates.[100]

C. The Acetoxyfulvene Approach

ICI workers have developed a short and practical synthesis of the Corey aldehyde.[101] The synthesis avoids toxic thallium reagents and has been successfully operated on a large scale for the manufacture of prostaglandin analogs, cloprostenol and fluprostenol. The synthesis commences with 6-acetoxyfulvene in which isomerization of the 1,3-diene system is precluded by presence of an exocyclic enol acetate group. Another advantage of this approach is that the formyl group at C-12, required for attachment of the lower side chain, is introduced at the correct oxidation level at the very outset, viz., the enol acetate grouping in 6-acetoxyfulvene (**1**) (Chart 8.7).

Acid-catalyzed hydrolysis of the enol acetate **3** gave the *anti*-aldehyde **4** (product of kinetic control), which is isomerized to the more stable *syn*-aldehyde **5** upon prolonged treatment with acid at 84°C or via the Schiff's base with p-chloroaniline. The aldehyde **5**, which now contains the appropriate stereochemistry at C-12 for synthesis of **11**, is readily protected as the dimethylacetal prior to conversion of the chloronitrile moiety to give the ketone **7**. The ketone **7** may be resolved at this stage via d-amphetamine salt of hemiphthalate ester of the corresponding *endo*-alcohol. The latter is reoxidized to the ketone after the resolution. Conversion of the acetal ketone **7** to aldehyde **11** proceeds

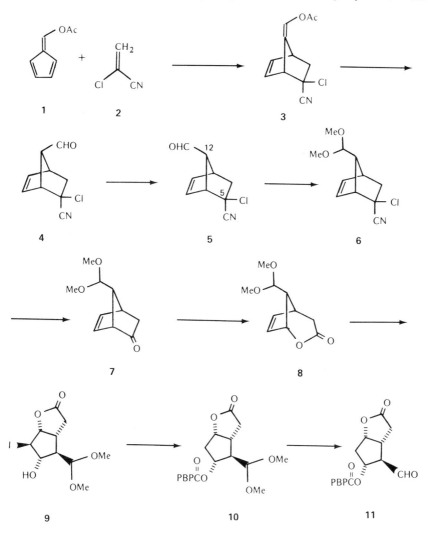

Chart 8.7. Acetoxyfulvene approach to Corey's aldehyde.[101]

smoothly, using essentially the sequence developed earlier by Corey. The synthesis has been modified for preparation of PGE_2.[102]

In a variant of this approach, early attachment of the lower side chain avoids the need for protecting and deprotecting the aldehyde. This strategy has been employed by Brown and Lilley to develop a short efficient synthesis of $PGF_{2\alpha}$.[103] Reaction of aldehyde 5 with the lithio derivative of dimethyl-2-

oxoheptylphosphonate furnishes the *trans*-enone **12**, which, after reduction and hydrolysis of the chloro-cyano group, gives the bicyclic ketone **13** in 56% overall yield from **5**. PGF$_{2\alpha}$ is obtained in five additional steps from **13** (Chart 8.8).

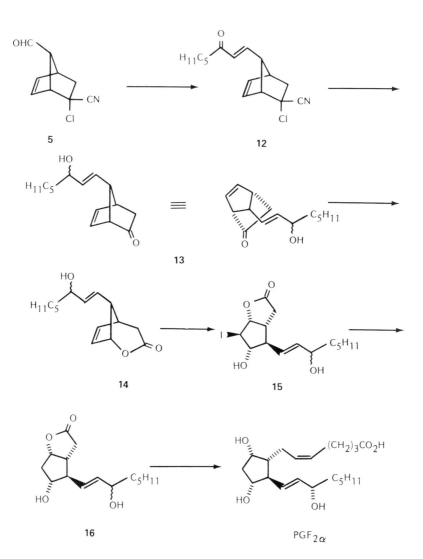

Chart 8.8.   ICI synthesis of PGF$_2\alpha$.[103]

D. Bicycloheptane Intermediates from a Tricyclo[3.2.0.0^{2,7}]-heptanone

Roberts in collaboration with an Allen and Hanburys team[104] developed a facile synthesis of bicycloheptane intermediates, suitable for preparation of prostanoids, involving the homoconjugate addition of organocuprates to the readily available tricyclo[3.2.0.0^{2,7}]heptanone **4**. Reaction of the known bicycloheptenone **1** with *N*-bromoacetamide in aqueous acetone proceeds stereospecifically

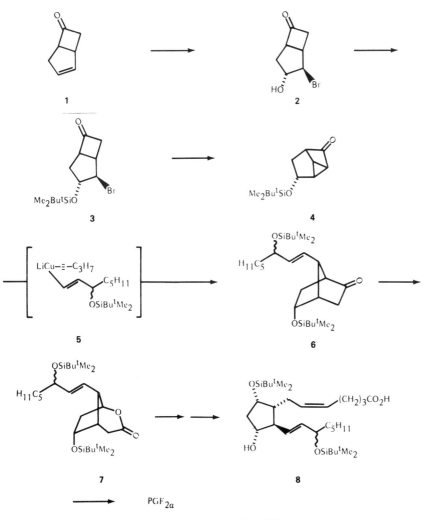

Chart 8.9. From Ref. 104.

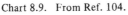

to yield the crystalline bromohydrin **2**. Protection as the *t*-butyldimethylsilyl derivative followed by treatment with base gave the key tricycloheptanone **4**. Opening of the strained cyclopropyl ketone in **4** proceeded smoothly with a mixed organocuprate (**5**) to furnish the norbornenone **6**. The latter was converted to $PGF_{2\alpha}$ along lines of the ICI approach (Chart 8.9).

Acid-catalyzed rearrangement of the δ-lactone **9** to the γ-lactone **10** provides an entry into the PGE_2 series.[105]

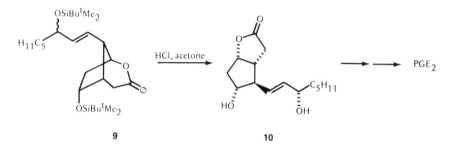

9   **10**

E. The Inverse Approach

In the original Corey synthesis of prostaglandins the eight-carbon lower chain is attached first, followed by elaboration of the carboxyl-containing top chain. A modification of this approach consists of reversing the order in which the two chains are attached to the cyclopentane ring. Since in this approach the carboxylic acid side chain is added first, opportunity is presented for reduction of the 5,6-double bond prior to introduction of the lower side chain, thus allowing the preparation of prostaglandins of the one-series. Schaaf and Corey[106] used this approach for synthesis of PGE_1 and $PGF_{1\alpha}$. Doria et al.[107] developed a similar approach to PGE_1 and $PGF_{1\alpha}$, differing only in the choice and timing of removal of the protecting groups at C-9 and C-11.

9. BICYCLOHEPTANE PRECURSORS: MISCELLANEOUS APPROACHES

A. Norbornane Precursors

A second bicycloheptane approach to prostanoid synthons utilizes oxidative cleavage of substituted norbornene derivatives for introduction of the *cis*-9,11-dihydroxy groups. Thus the double bond, readily incorporated into the norbornene system (**1**) by means of a Diels-Alder addition reaction with cyclopentadiene, serves as a latent source of two carbonyl functions, which in turn provide the 9,11-oxygen functions by a C—O inversion procedure.

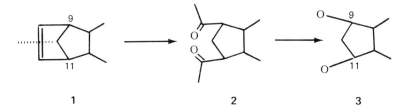

1   **2**   **3**

This strategy was first developed by Katsube who utilized it for a nonstereo-specific synthesis of the PGF_1 skeleton,[108] and later adapted it for a nonstereo-specific synthesis of the Corey lactone.[109]

The ICI Syntheses

In a variant of the oxidative cleavage approach developed at ICI, Jones et al.[110,111] commenced with an unsymmetrically substituted norbornene deriva-tive (2) bearing two differentially protected one-carbon pendants destined to become the C-8 and C-12 substituents. After oxidative cleavage and generation of the *cis*-9,11-dihydroxy groups, the C-8 pendant was selectively unmasked and homologated to a two-carbon residue (4→5). However, the homologation step,

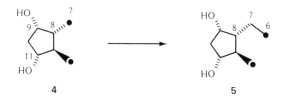

4   **5**

as originally planned, was complicated by a competing 1,3-acetate shift involving C-9 and the newly unmasked C-8 functionality. The synthesis was subsequently modified,[111] commencing with a symmetrically substituted norbornene deri-vative 6 (Chart 9.1). After cleavage and the C—O inversion sequence, which gave the diacetate 8, the benzyl groups were removed by catalytic hydrogenation to furnish, after equilibration, a mixture (3:2) of the primary acetate (10) and the secondary acetate (9). Thus the previously troublesome 1,3-diace-tate rearrangement was used to an advantage as a means of distinguishing between the two primary alcoholic functions on the α- and β-faces of the cyclo-pentane ring. A sequence of reactions, following addition of the C-6 one-carbon unit as a nitrile (12→15), afforded the lactone diol 16, which has earlier been converted to natural prostaglandins by the Harvard group.

The Hoffmann-La Roche Synthesis

A practical, but nonetheless lengthy synthesis of Corey's bicyclic lactone 18 in optically active form has been developed at Hoffmann-La Roche, Basel, using

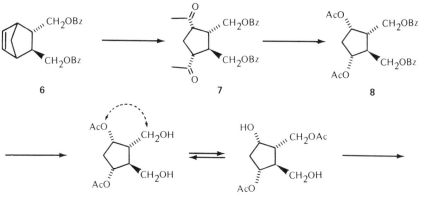

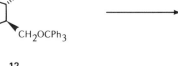

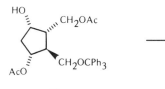

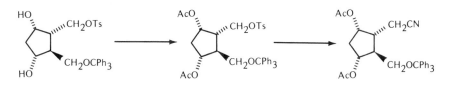

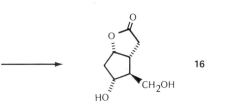

Chart 9.1. The ICI oxidative-cleavage approach.[111]

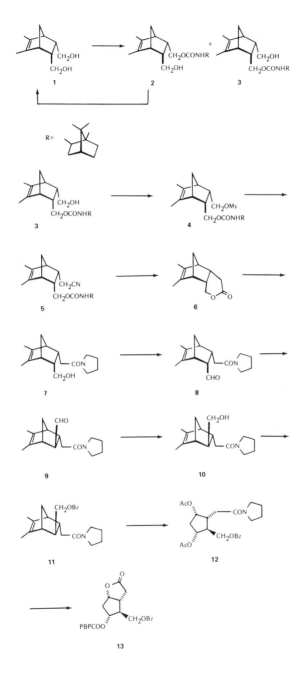

Chart 9.2. From Ref. 112.

the oxidative cleavage approach (Chart 9.2).[112] The synthesis commences with a symmetrically substituted norbornene derivative (1), but unlike the ICI synthesis utilizing almost identical starting material, cleavage of the double bond has been deferred until after elaboration of the two side-chain precursor pendants. Resolution is carried out early in the synthesis by fractional crystallization of the *meso*-intermediate (1), which is asymmetrically substituted with a chiral moiety. Since the undesired enantiomer is readily recycled, the chiral efficiency of this synthesis is very high. Moreover, it is possible to prepare both enantiomeric configurations of prostaglandins simply by interchanging the sequence of reactions by which the pendant chain is elongated and the 8,12-*trans* relationship is established.

The *meso*-alcohol 1 is reacted with phosgene to give a seven-membered cyclic carbonate, which is readily attacked by isobornylamine to furnish a mixture of diasteromers 2 and 3 separable by fractional crystallization. The undesired enantiomer is hydrolyzed back to 1 and recycled.

The hydroxyurethane 3 was homologated to the cyanourethane 5, and saponified to a hydroxy acid, which spontaneously lactonized to 6. Ring opening of the tricyclic lactone 6 with pyrrolidine furnished a hydroxy amide 7 in which the primary alcoholic function was transformed to an aldehyde using the Pfitzner-Moffatt oxidation procedure. Isomerization of the resulting *endo*-aldehyde 8 to the *exo*-aldehyde 9 established the desired *trans* relationship between the two side chains. After reprotection of the aldehyde function by reduction and benzylation, the tetrasubstituted double bond in 11 was ozonized and the intermediate ozonide converted with dimethyl sulfide to a labile bismethyl ketone. The latter was subjected, without purification, to Baeyer-Villiger oxidation to give the diacetoxyamide 12, which has the desired *cis-trans-trans* orientation of substituents on the cyclopentane ring. Saponification, lactonization, and reaction of the intermediate hydroxylactone with *p*-phenyl-benzoyl chloride furnished the desired bicyclic lactone 13.

B. Bicyclo[3.2.0]heptane Precursors

The oxidative cleavage of bicyclo[3.2.0]heptanes provides an efficient way for simultaneous generation of the C-9 oxygen function on the cyclopentane

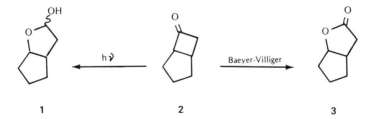

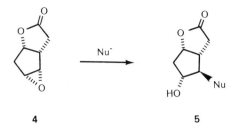

4   5

nucleus along with the two-carbon precursor of the carboxylic chain at C-8 via Bayer Villiger ring expansion[113] or photolytic conversion of the cyclobutanone to a γ-lactol.[114]

Some of the early routes to prostaglandins were based on nucleophilic opening of the bicyclic epoxide **4**, readily available from bicyclo[3.2.0]heptane precursors. This approach, however, suffered from a lack of selectivity during the crucial ring-opening step. Newton and Roberts recently discovered that the S_N2 attack by carbanions on the protected epoxybicycloheptane **11** proceeds in excellent yield and with a high degree of regioselectivity.[115] This strategy was applied for preparation of the bis(methylthio)propenyl derivative **6**, which was converted to the known prostaglandin precursor **7**.

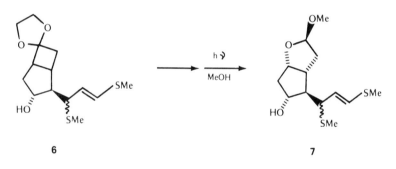

6   7

Regiospecific addition of the cuprate **15** to **11** furnishes, in 65% yield, the hydroxyketal **12**, which may be converted to PGE_2,[115] or photolytically transformed to the known γ-lactol **14**, which has been converted to $PGF_{2\alpha}$ (Chart 9.3).[116]

10. BICYCLOOCTANE PRECURSORS

Cleavage of one of the rings in a suitably functionalized bicyclo[3.3.0]octane precursor (2) may be used to generate the prostanoid cyclopentane nucleus bearing substituents at C-8 and C-12, suitable for further elaboration of the α

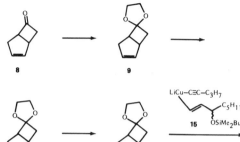

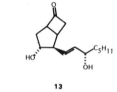

10

11

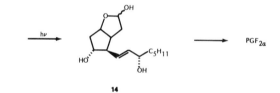

12

13

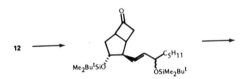

14

PGF$_{2\alpha}$

12

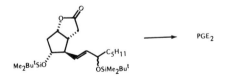

PGE$_2$

Chart 9.3. From Ref. 116.

418

and ω side chains. This approach was utilized by Turner et al.,[117] who prepared the key aldehydic synthon 3 from the bicyclo[3.3.0]octane 2 in a relatively short number of steps and converted it to $PGF_{2\alpha}$. Even though the latent C-12 functionality in 2 is initially in the wrong configuration, it can be readily cor-

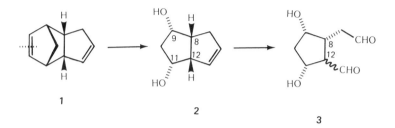

1   2   3

rected by epimerization adjacent to the C-12 aldehyde group once it is generated. The 9,11-*cis*-disposed hydroxy groups on the cyclopentane nucleus have been obtained by oxidative cleavage of a norbornene ethylene bridge, a strategy discussed in the preceding section. Turner's synthesis[117] commences with endo-dicyclopentadiene 1 readily obtained by crystallization from inexpensive commercial dicyclopentadiene. Cleavage of the norbornene double bond of dicyclopentadiene proceeds selectively, since it is the more reactive of the two double bonds present in the molecule. However, the overall yield of $PGF_{2\alpha}$ by this synthesis, starting from dicyclopentadiene is less than 0.1%.

11. BICYCLONONANE PRECURSORS

Two syntheses of PGE_1 involving cleavage of a hydrindane precursor for generating the ring appendages were developed by workers at Merck Sharp & Dohme. In both these syntheses oxidative cleavage of the key *cis*-hydrindenone 2 furnishes elements of the lower side chain and the C-11 hydroxyl group. Epimerization of the resultant acetyl function in 3 serves to establish the proper *trans, trans* geometrical disposition of substituents about the cyclopentanone ring (cf. 4).[118] Subsequently, the Merck workers developed an alternative route to the key hydrindenone intermediate (1) from the Diels-Alder adduct of *trans*-piperylene and maleic anhydride.[119] An improved two-carbon degradation sequence was also developed for shortening of the C-12 three-carbon side chain in 4. Although these syntheses exhibit a fair amount of stereocontrol, they are quite lengthy (over 25 steps), involving a large number of protection and deprotection steps (Chart 11.1).

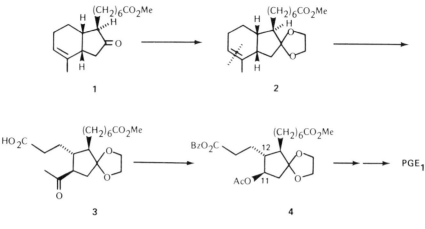

Chart 11.1

12. A PROSTAGLANDINS

Corey's bicyclo[2.2.1]heptane route to primary prostaglandins is readily modified to provide a convenient access to A prostaglandins. Cyclization approaches also appear to be particularly suited for preparation of PGAs, and a number of successful pathways have been described.

A. Syntheses from Bicyclo[2.2.1]heptane Precursors

The unsaturated bicyclic lactone 2 is a key starting point for synthesis of A prostaglandins modeled after Corey's bicycloheptane route.[120,121] The lactone is readily available from the known iodolactone 1,[120,122] or directly by a cationic rearrangement cyclization sequence from the lactone 3.[123] Alternatively, the bicyclic lactone may be prepared by an abnormal Nef reaction of the nitro norbornene 4, readily available from Diels-Alder addition of nitroethylene to 5-methoxymethylcyclopentadiene.[124]

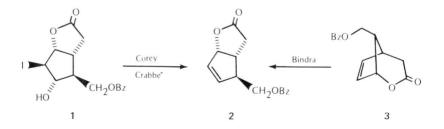

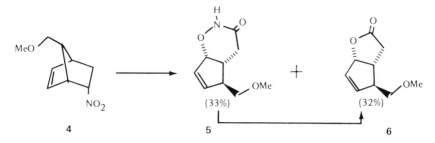

A modification of Corey's bicyclo[2.2.1]heptane route for synthesis of A prostaglandins has been reported by Corey and Moinet.[125] The synthesis incorporates several of the improvements and reagents developed in the primary approach, but differs from it essentially in avoiding the lactonization step until after introduction of the lower chain. Two noteworthy features of this synthesis are the use of the biphenylurethano group to facilitate the key lactonization step (7→8) and the unusually facile resolution step, which is accomplished early in the sequence. The synthesis is outlined in Chart 12.1.

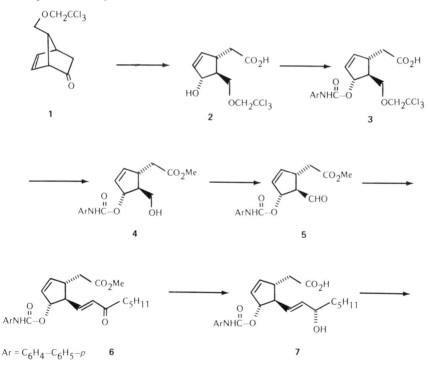

Chart 12.1. Synthesis of PGA$_2$ by Corey and Moinet.[125]

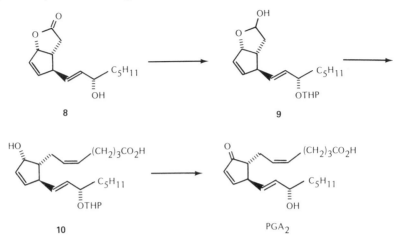

The racemic carbamate carboxylic acid **3** was prepared essentially as described in the general bicyclo[2.2.1]heptane approach. Resolution of **3** with (+)-amphetamine afforded (−)-**3**, which was converted to the hydroxyurethane **4** in two steps, viz., esterification of the carboxyl group in **3** followed by removal of the trichloroethyl group with zinc-copper couple. Collins oxidation of the primary hydroxyl group in **4** gave the aldehyde **5**, which was converted to the enone **6** by Wadsworth-Emmons condensation with sodiodimethyl 2-oxoheptyl-phosphonate. Selective reduction to the 15(*S*)-alcohol (91%, together with 9% of 15(*R*)-isomer) followed by saponification of the methyl ester with base gave the acid **7**. Heating a solution of **7** in aqueous dimethoxyethane buffered to pH 7 produced the hydroxylactone **8**. The synthesis was then completed by protection of the secondary alcohol function as a tetrahydropyranyl ether, reduction to the lactol **9**, followed by Wittig reaction to give the oily hydroxy acid **10**. Collins oxidation of **10** and cleavage of the protecting group cleanly afforded (+)-PGA$_2$.

An alternative synthesis of the alcohol **8**, developed by Corey and Mann[126] (Chart 12.2), involves the attachment of ω-chain to the cyclopentane nucleus

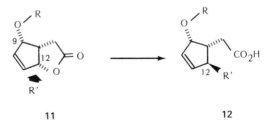

11   12

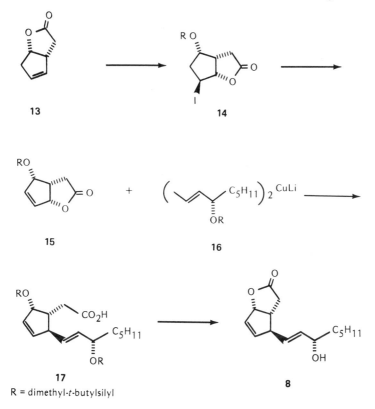

13

14

15   +   16

17

R = dimethyl-*t*-butylsilyl

8

Chart 12.2. From Ref. 126.

by cross-coupling of a vinylic copper reagent with the allylic electrophilic site at C-12 (prostanoic acid numbering) as shown in **11**. The second possible site of attachment on the allylic substrate (C-9) is effectively screened by a bulky dimethyl-*t*-butylsilyl group during the coupling.

Another noteworthy approach to PGA$_2$ has recently been described by Roberts and Newton.[127] This synthesis, commencing from the bicyclic lactone **18**, involves addition of the heterocuprate **22** to an allylic epoxide **21** as the key step (Chart 12.3).

B. The Roussel-Uclaf Approach

The stereoselective total synthesis of *dl*-PGA$_2$ described by the Roussel-Uclaf group[128] is outlined in Chart 12.4. The synthesis involves as its key step the cyclization of the γ,δ-epoxyolefin keto ester **2** to furnish the cyclopentanone

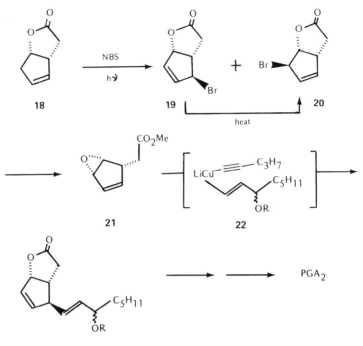

18   19   +   20

heat

21   22

PGA₂

48% + 14% position isomer

Chart 12.3.   From Ref. 127.

3, with simultaneous generation of the desired relative stereochemistry at C-12 (R) and C-15(S). An examination of various transition states leading to the cyclized product suggests that the latter is obtained only if the precursor 2 has either the *cis-trans* or *trans-cis* configuration. These workers chose the *cis-trans* configuration for 2, electing to generate the *cis*-olefin by partial hydrogenation of an acetylenic linkage, and the *trans*-epoxide using Cornforth's epoxide synthesis.[129] Following cyclization and protection of the C-15 hydroxyl, introduction of the top chain is accomplished by alkylation of the enolate of 3 with 7-bromohept-5-enoate to give the prostanoid 4. Treatment of 4 with sodium ethoxide causes *retro*-Dieckmann and recyclization of 5. Introduction of the 10,11-double bond is then carried out by bromination of the enol ether 6 in methanol to give a bromoacetal 7, which dehydrobrominates to PGA₂.

C. Biogenetically Patterned Synthesis of PGA₂

An interesting biogenetically patterned synthesis of the key PGA precursor 6 (Chart 12.5) has been described by Corey et al.[130] The PGA ring has been con-

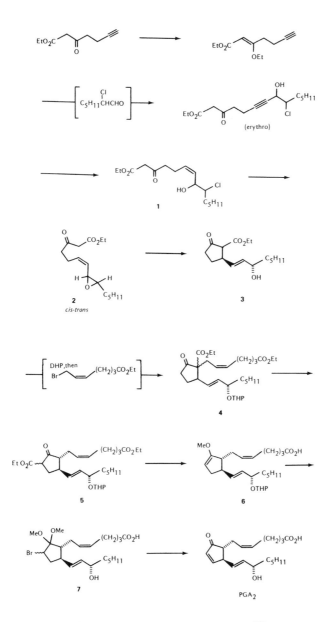

Chart 12.4. Roussel-Uclaf synthesis of PGA$_2$.[128]

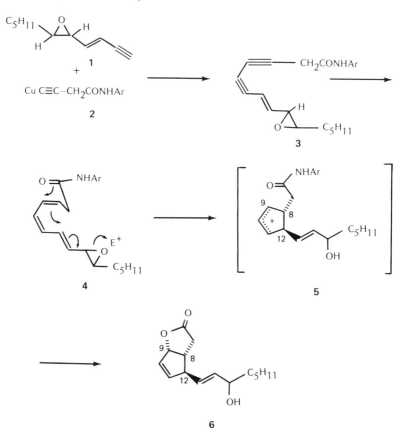

Chart 12.5.   Corey's biogenetically patterned synthesis of PGA$_2$.[130]

structed from the triene-amide-oxide **4** by a cationic cyclization process of the pentadienyl cation → cyclopentenyl cation type.[131] The acetyl amide chain in the intermediate allylic cation **5** is effectively situated to deliver an oxygen functionality specifically to C-9 with simultaneous establishment of the *cis* arrangement of C-9 and C-8 substituents. *trans* Orientation of substituents at C-8 and C-12 was expected on thermodynamic and kinetic grounds.

Coupling of the cuprous acetylide **2** with the lithium salt of the acetylene **1** gave the enediyne **3**, and catalytic hydrogenation furnished the desired *trans-trans-cis*-triene epoxyamide **4**. The cyclization was carried out in a mixture of 2-nitropropane and 1-nitropropane at −104° by treatment with boron trifluoride. After hydrolysis of this mixture the lactone **6** was obtained along with an equal amount of its C-15 epimer in about 15% yield. The lactone **6** has previously been converted into PGA$_2$ in five additional steps.

D. The Columbia Approach

The synthesis of PGA$_2$ by Stork and Raucher,[132] outlined in Chart 12.6, is an elegant example of chiral synthesis of prostaglandins from carbohydrates pioneered at Columbia. The synthesis commences from a readily available optically active sugar, L-rhamnose, already incorporating the C-15 asymmetric center. An interesting feature of the synthesis is construction of the key open-chain precursor (10) with the use of two Claisen rearrangements: one for production of the necessary *trans* geometry of a double bond (3→4), and the other as a means of transferring the chirality of the carbon-oxygen bond at C-14 to the carbon-carbon bond at C-12 (6→7).

2,3-Isopropylidene-L-erythrose (2), upon treatment with vinyl magnesium chloride, gives the vinyl carbinol 3. Protection of the primary alcohol in 3 as the methyl carbonate, followed by Claisen rearrangement gave the unsaturated ester 4. Hydrolysis of the acetonide group in 4 with aqueous acetic acid serves to unmask the allylic alcohol function required for the next Claisen rearrangement. Treatment of 5 with triethylamine gave the allylic alcohol cyclic carbonate 6, which was submitted to Claisen rearrangement with the orthoester 1 to give the trimethyl ester 7. The cyclic carbonate function in 7 was hydrolyzed with potassium carbonate in methanol and furnished the diol 8 in 59% overall yield from the acetonide 4. The crucial transfer of chirality during the transformation 6→7 occurs with the carbonate ring equatorial in the chair transition state 13. However, since the side chain R in 13 has no special steric preference, the product is a mixture of epimers at C-8, but the latter center is readily epimerized to the correct arrangement at the end.

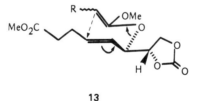

13

Partial hydrogenation of the triple bond in 8, selective tosylation, followed by protection of the secondary hydroxyl group, gave the monotosylate 9. The latter coupled with lithium di-*n*-butylcuprate to give the desired 10 in 67% yield. The feasibility of carbon-carbon σ-bond formation by coupling of monotosylates of vicinal diols with lithium dialkylcuprates was examined in model studies.[133] Dieckmann cyclization of 10, followed by hydrolysis and acidification, gave the 11-deoxy-PGE$_2$ derivative 11 in 77% yield (18% overall from 2). Final transformation to natural PGA$_2$ was carried out by selective introduction of the α-phenylseleno group with phenyl selenyl chloride on to the dianion of 11

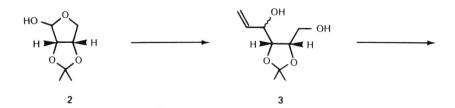

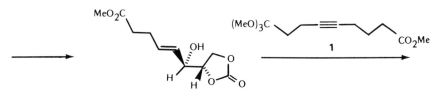

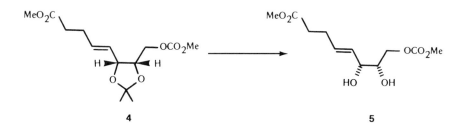

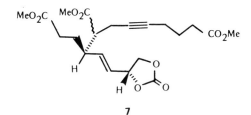

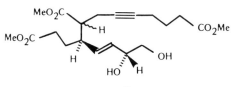

428

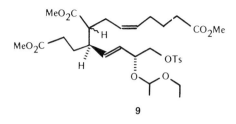

9

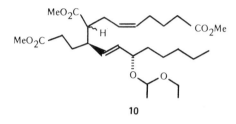

10

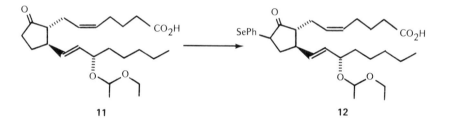

11   **12**

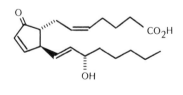

PGA₂

Chart 12.6.   Columbia synthesis of PGA₂.[132]

generated with lithium diisopropylamide. Oxidation of **12** with sodium *meta*-periodate and removal of the ethoxyethyl protecting group gave (+)-PGA$_2$ in 46% overall yield from **11** or 7.7% overall yield from **2**.

13. B PROSTAGLANDINS

The synthesis of PGBs attracted some attention briefly during the 1960s, but shortly afterward emphasis shifted to the synthesis of more challenging and biologically interesting prostaglandins of the E, F, and A series. The synthesis of B prostaglandins was discussed in Volume 1,[1] and no significant new developments have occurred since then. Interested readers may wish to consult recent reviews dealing with the synthesis of jasmone and related cyclopentenones, which are regarded as models for PGB compounds.[134,135]

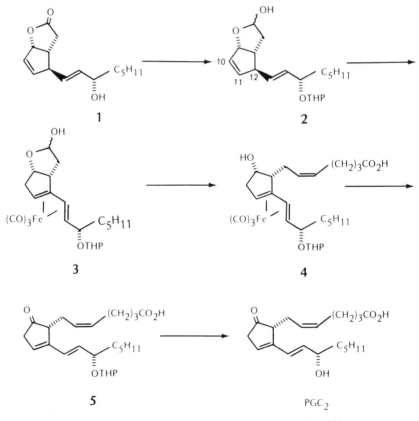

Chart 14.1. Synthesis of PGC$_2$ by Corey and Moinet.[137]

14. C PROSTAGLANDINS

Prostaglandins of the A type are deactivated in mammalian blood by enzymatic conversion to an 11,12-double bond isomer, PGC, which is subsequently isomerized by base to the inactive PGB.[136] The C prostaglandins have powerful blood pressure-lowering activity and are therefore of considerable biological interest.

A. Syntheses from Corey's Intermediates

The synthesis of PGC_2, described by Corey and Moinet,[137] is based on a controlled triiron carbonyl-catalyzed isomerization of the 10,11-double bond in the known lactol 2 to the 11,12-position (Chart 14.1). Since PGC_2 is extremely sensitive to traces of acid or base, an extremely mild method was required for the removal of iron from 4. This was accomplished in 80% yield by reaction of 4 with excess Collins reagent, with simultaneous oxidation of the C-9 hydroxyl, to give the 15-tetrahydropyranyl derivative of PGC_2 (5). Cleavage of the THP group followed by reverse-phase chromatography gave PGC_2.

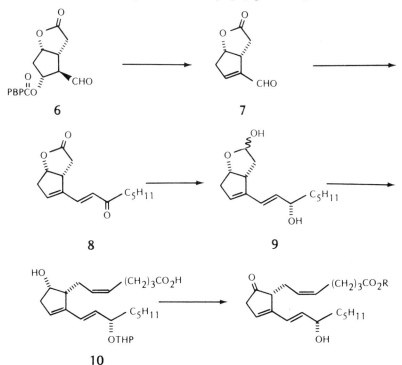

Chart 14.2. From Refs. 138 and 139.

Two almost identical syntheses of PGC$_2$ closely modeled after Corey's bicyclo-heptane approach and proceeding via the key lactol **9** have been described by Kelly et al.[138] and by Crabbe's group (Chart 14.2).[139] An efficient photosyn-thetic route to the lactol **12** has been described recently by Roberts and Newton.[140]

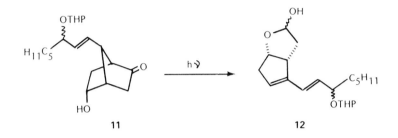

11   **12**

B.   Synthesis from PGA$_2$

In view of its highly sensitive nature, PGC$_2$ is probably most conveniently prepared from PGA$_2$ rather than by total synthesis. Corey and Cyr[141] have described a practical method for the conversion of PGA$_2$ to PGC$_2$. The process involves generation of a γ-extended enolate ion **13** by proton abstraction from C-12 in PGA$_2$, followed by α-protonation at C-10 under carefully controlled conditions to give PGC$_2$.

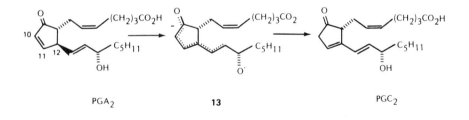

PGA$_2$ **13**   PGC$_2$

15.   D PROSTAGLANDINS

Two groups of workers have described total syntheses of PGD$_2$, commencing from known intermediates of Corey's bicycloheptane approach. The shorter and simpler of the two syntheses is by Hayashi and Tanouchi,[142] and affords a 2:1 mixture of 15-OTHP derivatives of PGD$_2$ and PGE$_2$, which are separated chromatographically (Chart 15.1). The synthesis by a Ciba-Geigy group in Basle[143] is based on selective unmasking and oxidation of the C-11 hydroxyl

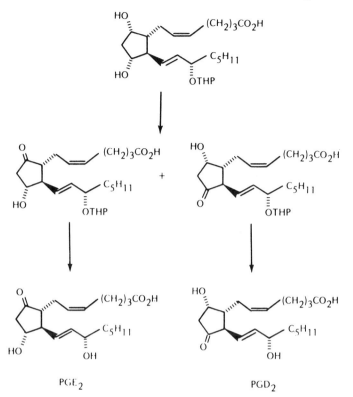

Chart 15.1. Synthesis of PGD$_2$ by Hayashi and Tanouchi.[142]

group in a differentially protected PGF intermediate and furnishes a 3:1 mixture of PGD$_2$ and PGE$_2$ (Chart 15.2).

By far the cleanest synthesis of PGD$_2$ has been reported by a Glaxo group,[144] who obtained the requisite differentially protected PGF derivative 2 from the lactone 1.

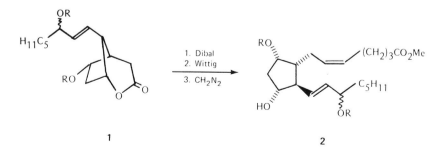

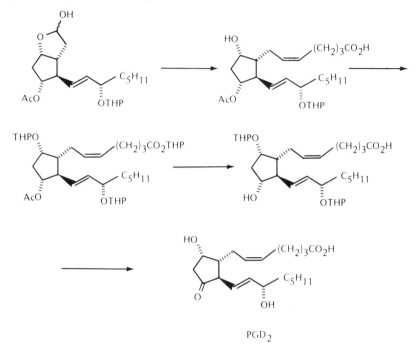

Chart 15.2.   Ciba-Geigy synthesis of PGD$_2$.[143]

16. H PROSTAGLANDINS

The prostaglandin endoperoxide (PGH$_2$) lies at a crucial biochemical branch point in the arachidonic acid cascade.[145] It may be converted to either thromboxane A$_2$ or prostacyclin, which exert diametrically opposite actions on platelet aggregation and vascular smooth muscle.[146]

An Upjohn group reported the synthesis of PGH$_2$ methyl ester in 3% yield from 9β,11β-dibromo-9,11-dideoxy prostaglandin F$_{2\alpha}$ methyl ester by S_N2 displacement of bromide by hydroperoxide anion.[147] Subsequently, based on model studies with cis-1,3-cyclopentane dibromide, Porter and co-workers[148] prepared PGH$_2$ methyl ester in 20-25% yield from the dibromide using silver salts and hydrogen peroxide. The method has been extended successfully for the preparation of PGH$_2$ in overall 2-3% yield from PGF$_{2\alpha}$ (Chart 16.1).[149] Porter's approach opens the way for the preparation of analogs of prostaglandin endoperoxide that have previously been difficult or impossible to obtain by biochemical routes. Analogs of PGH$_2$ in which the peroxide linkage has been replaced by other atoms have been extensively investigated.[149]

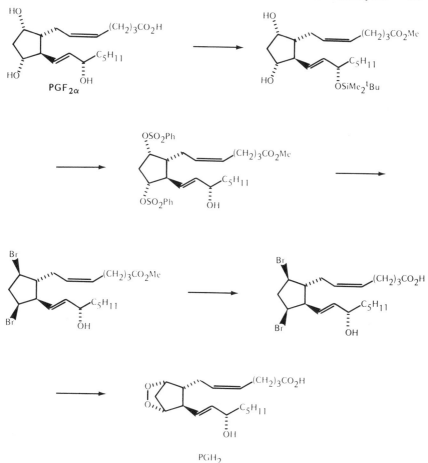

Chart 16.1.   Synthesis of PGH$_2$.[149]

17. PROSTACYCLIN

Prostaglandin endoperoxides, PGH$_2$ and PGG$_2$, are enzymatically transformed into prostacyclin (PGI$_2$), a substance with potent vasodilatory and antiaggregatory properties.[146, 150]

Several simple and efficient syntheses of prostacyclin have been reported, and involve addition of halogens to PGF$_{2\alpha}$ derivatives for formation of the 5-membered ring ether involving C-6 and the C-9 oxygen. Corey et al.[151] (Chart 17.1) reacted the 11,15-bis-THP derivative of PGF$_{2\alpha}$ with N-bromosuccinimide

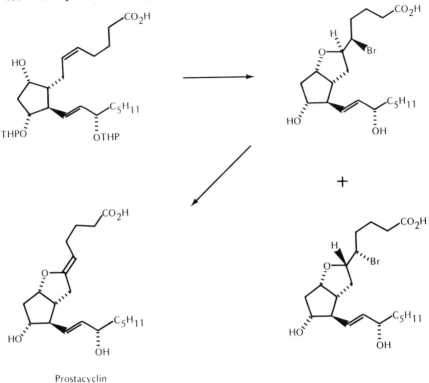

Prostacyclin

Chart 17.1. Synthesis of prostacyclin.

to obtain diastereomeric bromoethers, which were separated after depyranylation, and the major isomer treated with potassium *tert*-butoxide to furnish PGI₂. In the procedure developed by Nicolaou et al.[152,153] the methyl ester of PGF$_{2\alpha}$ was converted with iodine in methylene chloride in the presence of potassium carbonate to an iodoether, which was dehalogenated with sodium ethoxide for production of prostacyclin. Dehalogenation with DBU furnishes the prostacyclin methyl ester.[153] Using phenylselenium chloride as the electrophilic reagent furnishes a phenylselenoether, which eliminates to give the (4*E*)-isoprostacyclin isomer 1.[154] Similar methodology was employed by Upjohn workers in their synthesis of prostacyclin.[155]

Because of the striking biological activities of prostacyclin, which make it a potential drug for treatment of thrombosis, stroke, and heart attack, the synthesis of prostacyclin analogs has attracted considerable interest, and a large variety of analogs have been prepared. Interested readers should consult the excellent review by Nicolaou et al.[6]

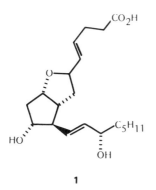

1

18. THROMBOXANES

The labile thromboxane A_2, formed from the endoperoxides PGG_2 and PGH_2, is a potent inducer of platelet aggregation and is a profound constrictor of vascular smooth muscle. Thromboxane A_2 is rapidly inactivated to thromboxane B_2. While the unstable thromboxane A_2 has not yet been synthesized, several syntheses of thromboxane B_2 have been reported.

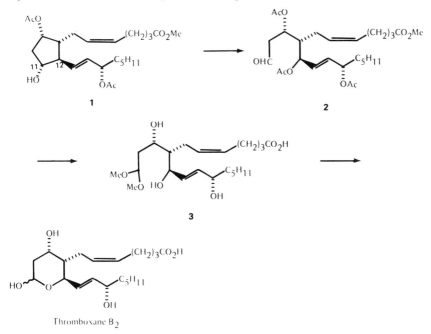

Chart 18.1. First synthesis of thromboxane B_2.[156]

One of the first synthesis reported by an Upjohn group[156] involves cleavage of the 11,12-bond in 9,15-diacetoxy PGF$_{2\alpha}$ (1) to provide a direct access to thromboxane B$_2$ (Chart 18.1). Rupture of the cyclopentane nucleus using lead tetraacetate is facilitated by the homoallylic alcohol functionality in 1. The unstable aldehyde obtained in this manner is protected as the dimethyl acetal. Removal of the acetate groups and hydrolysis of the acetal with aqueous phosphoric acid affords thromboxane B$_2$.

Two alternative routes to thromboxane B$_2$ developed at Upjohn are based on cleavage of the cyclopentane nucleus in known prostanoid intermediates. The

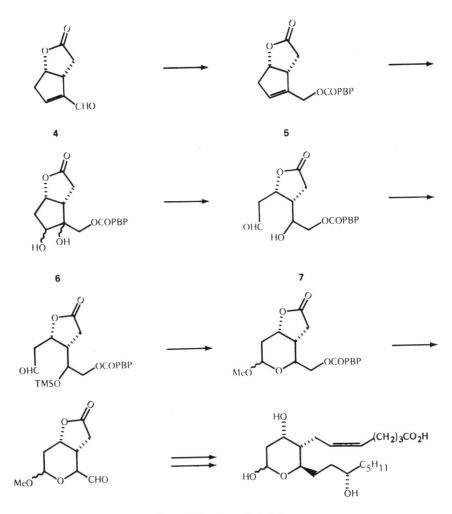

Chart 18.2.   From Ref. 157.

key step in the synthesis by Nelson and Jackson[157] is based on periodate cleavage of the 11,12-bond of **4** to give the unstable aldehyde **7**, which furnishes the desired pyran nucleus (Chart 18.2). A variant of this strategy involves cleavage of the 5-membered ring via a Bayer-Villiger oxidation of the cyclopentanone **8** (Chart 18.3).[158] Treatment of the resulting dilactone with base, followed by reduction with diisobutylaluminium hydride and protection of the hemiacetal, allows unmasking of the C-13 aldehyde functionality for attaching the ω-chain.

A particularly elegant approach to thromboxanes involves their total synthesis from optically active monosaccharides. Hanessian's synthesis[159] is based on the systematic and stereospecific introduction of substituents on the chiral framework of a 2-deoxysugar derivative readily available from D-glucose (Chart 18.4).

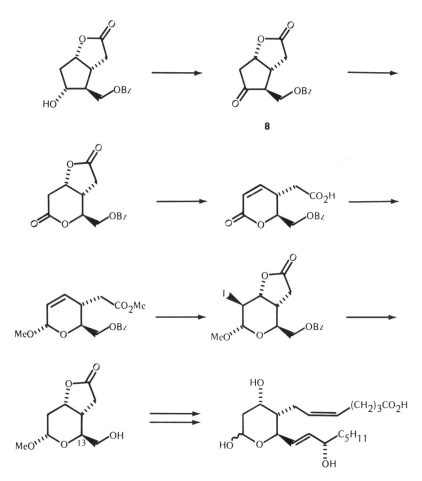

Chart 18.3. From Ref. 158.

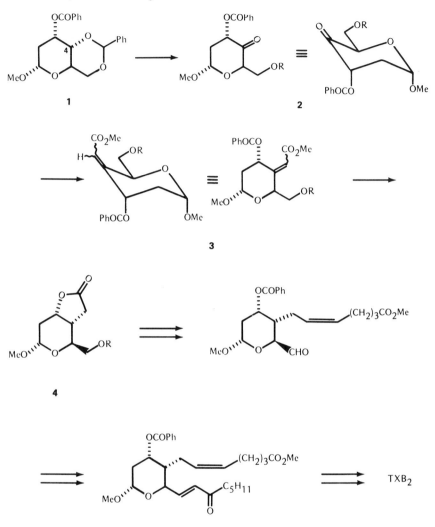

Chart 18.4. Hanessian synthesis of thromboxane B$_2$.[159]

Hydrogenolysis and selective protection of the primary alcohol in the sugar derivative **1**, followed by oxidation of the secondary alcohol functionality at C-4 provides for introduction of the acetic acid side chain on the pyran nucleus via a Wittig-Horner reaction. Catalytic hydrogenation of **3** furnishes the equatorially disposed side chain, which upon deprotection lactonizes to the key intermediate **4**. The remainder of the synthesis follows the standard prostaglandin methodology to give thromboxane B$_2$.

Corey's stereocontrolled synthesis of the key bicyclic lactone **4** (Chart 18.5)[160] is based on Claisen rearrangement of the unsaturated sugar derivative **6** with *N,N*-dimethylacetamide dimethylaminal to the dimethylamide **7**. Iodolactonization and deiodination furnishes the desired hydroxylactone **4**.

A total synthesis of thromboxane B_2 from a nonprostanoid, acyclic precursor was also developed by Corey et al.[161]

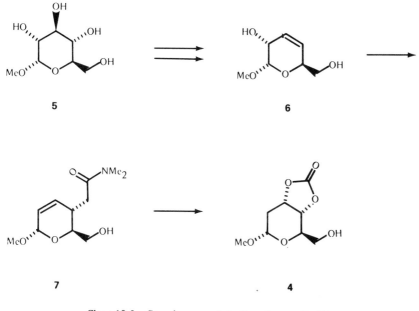

Chart 18.5.   Corey's approach to thromboxane B_2.[160]

19.   THERAPEUTIC PROSPECTS

Progress during the last ten years in the art of prostaglandin synthesis can only be described as phenomenal. All of the natural prostaglandins are now readily available by total synthesis, and the chemist interested in a specific analog can choose from any number of practical approaches to his target. Progress toward developing prostaglandin analogs with useful therapeutic properties has been disappointing, however. Breeding selectivity into these molecules and taming side effects has been altogether a difficult proposition. Only a few from among the handful of analogs described below are expected to find application in medical use during the near future.

Not unexpectedly, some of the earliest and most promising practical applications of prostaglandins have been in obstetrics and gynecology.[162] Natural

$PGF_{2\alpha}$ (dinoprost) and PGE_2 (dinoprostone) were introduced in the United States by Upjohn as early as 1973 for hospital use for the induction of labor at term, and on a limited basis for termination of second trimester pregnancy. The use of natural prostaglandins in the clinics, however, is severely limited by their short duration of action and a high incidence of side effects that are undoubtedly related to their relatively poor target organ specificity, and a number of analogs have appeared designed to alleviate these problems. Upjohn's 15-methyl-$PGF_{2\alpha}$ (carboprost) and 16,16-dimethyl-PGE_2, designed to enhance metabolic stability, exhibit greater potency and longer duration of action on the human uterus than the parent prostaglandins.

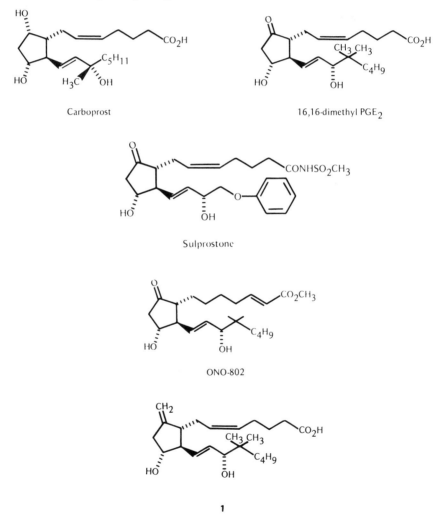

Carboprost

16,16-dimethyl PGE₂

Sulprostone

ONO-802

1

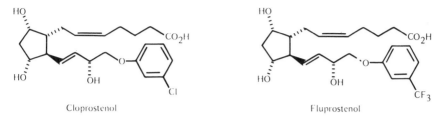

Cloprostenol   Fluprostenol

Sulprostone, which is being developed by Pfizer and Schering AG, was also designed to be metabolically more stable. Pharmacological studies indicate that sulprostone is one of the most selective prostaglandin uterine stimulants known. Clinical trials confirm that sulprostone combines high efficacy and low incidence of side effects when used for pregnancy termination.[163]

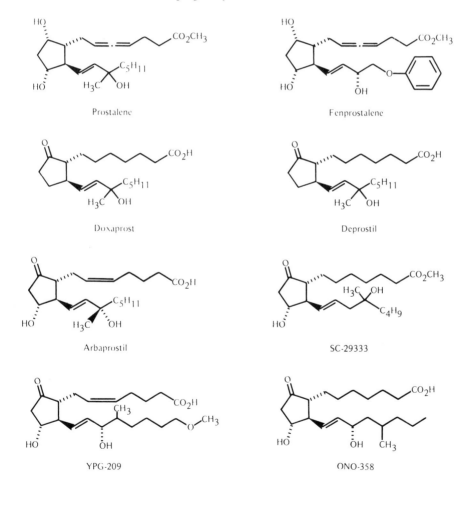

Prostalene   Fenprostalene

Doxaprost   Deprostil

Arbaprostil   SC-29333

YPG-209   ONO-358

Encouraging results have been reported in studies with ONO-802 and Upjohn's 9-deoxo-16,16-dimethyl-9-methylene-PGE$_2$ (1) for termination of first and second trimester pregnancies. Cloprostenol (Estrumate) and fluprostenol (Equimate) are phenoxyprostaglandins introduced into the United Kingdom in 1975 and 1976, respectively, by ICI as veterinary luteolytics for synchronization of estrus in animals and for treatment of infertility in horses. The allenic prostaglandin analogs, prostalene and fenprostalene, are under study as luteolysins by Syntex.

Doxaprost (15-methyl-11-deoxyprostaglandin E$_1$) and the corresponding 13,14-dihydro analog, deprostil, are under study by Ayerst as bronchodilator and gastric antisecretory agents. Other antisecretory and antiulcer analogs of interest include Upjohn's 15-methyl-PGE$_2$ analog arbaprostil and 16,16-dimethyl-PGE$_2$. Searle's SC-29333 (15-deoxy-16-hydroxy-16-methyl-PGE$_1$, antisecretory), Yamanouchi's YPG-209 (a 16-methyl-20-methoxy-PGE$_2$ analog, bronchodilator), and Ono's ONO-358 (coronary vasodilator) appear promising based on animal pharmacology tests. Clinical work with prostacylin and prostacyclin analogs is as yet at a preliminary stage.

REFERENCES

1. U. Axen. J. E. Pike, and W. P. Schneider, in *The Total Synthesis of Natural Products*, Vol .1, ApSimon, Ed., Wiley, New York, 1973, p. 81.

2. J. S. Bindra and R. Bindra, *Prostaglandin Synthesis*, Academic, New York, 1977.

3. P. Crabbé, Ed., *Prostaglandin Research*, Academic, New York, 1977.

4. S. M. Roberts and F. Scheinmann, Eds., *Chem. Biochem. Pharmacol. Activity of Prostanoids*, Pergamon, New York, 1979.

5. M. P. L. Caton and K. Crowshaw, *Progr. Med. Chem.*, **15**, 357 (1978).

6. K. C. Nicolaou, G. P. Gasic, and W. Barnette, *Angew. Chem. Int. Ed.*, **17**, 293 (1978).

7. R. Clarkson, *Prog. Org. Chem.*, **8**, 1 (1973), P. H. Bentley, *Chem. Soc. Rev.*, **2**, 29 (1973).

8. N. M. Weinshenker and N. H. Andersen, *The Prostaglandins*, **1**, 1 (1973).

9. G. Pattenden, *Aliphatic Chem.*, **2**, 258 (1974); **3**, 311 (1975).

10. T. K. Schaaf, *Ann. Rept. Med. Chem.*, **11**, 80 (1976).

11. N. A. Nelson, *J. Med. Chem.*, **17**, 911 (1974).

12. S. Hannesian, *Acct. Chem. Res.*, **12**, 159 (1979).

13. E. J. Corey, *Ann. N.Y. Acad. Sci.*, **180**, 24 (1971).

14. M. Miyano and C. R. Dorn, *Tetrahedron Lett.*, 1615 (1969); M. Miyano, C. R. Dorn and R. A. Mueller, *J. Org. Chem.*, **37**, 1810 (1972).

15. M. Miyano and M. A. Stealey, *Chem. Commun.*, 180 (1973); *J. Org. Chem.*, **40**, 1748 (1975).

16. W. J. Marsheck and M. Miyano, *Biochim. Biophs. Acta*, **316**, 363 (1973).

17. M. Miyano and C. R. Dorn, *J. Am. Chem. Soc.*, **95**, 2664 (1973).

18. K. Kojima and K. Sakai, *Tetrahedron Lett.*, 3333 (1972).

19. K. Kojima and K. Sakai, *Tetrahedron Lett.*, 2837 (1975).

20. K. Kojima and K. Sakai, *Tetrahedron Lett.*, 3337 (1972).

21. K. G. Paul, F. Johnson, and D. Favara, *J. Am. Chem. Soc.*, **98**, 1285 (1976).

22. G. Stork and G. Kraus, *J. Am. Chem. Soc.*, **98**, 6747 (1976).

23. G. Stork, T. Takahashi, I. Kawamoto, and T. Suzuki, *J. Am. Chem. Soc.*, **100**, 8272 (1978).

24. G. Stork and L. Maldonado, *J. Am. Chem. Soc.*, **92**, 5286 (1971).

25. N. Finch and J. J. Fitt, *Tetrahedron Lett.*, 4639 (1969); N. Finch, L. Della Vecchia, J. J. Fitt, R. Stephani, and I. Vlattas, *J. Org. Chem.*, **38**, 4412 (1973).

26. D. P. Strike and H. Smith, *Tetrahedron Lett.*, 4393 (1970); *Ann. N.Y. Acad. Sci.*, **180**, 91 (1971).

27. E. J. Corey and R. Noyori, *Tetrahedron Lett.*, 311 (1970).

28. E. J. Corey, K. C. Nicolaou, and D. J. Beams, *Tetrahedron Lett.*, 2439 (1974).

29. J. Fried, C. H. Lin, J. C. Sih, P. Dalven, and G. F. Cooper, *J. Am. Chem. Soc.*, **94**, 4342 (1972).

30. J. Fried and J. C. Sih, *Tetrahedron Lett*, 3899 (1973).

31. J. Fried, J. C. Sih, C. H. Lin, and P. Dalven, *J. Am. Chem. Soc.*, **94**, 4343 (1972).

32. J. J. Partridge, N. K. Chadha, and M. R. Uskokovic, *J. Am. Chem. Soc.*, **95**, 7171 (1973); U. S. Patent 3,933,892 (1976).

33. G. Stork and M. Isobe, *J. Am. Chem. Soc.*, **97**, 4745 (1975).

34. G. Stork and J. D'Angelo, *J. Am. Chem. Soc.*, **96**, 7114 (1974).

35. G. Zweifel, R. P. Fisher, J. T. Snow, and C. C. Whitney, *J. Am. Chem. Soc.*, **93**, 6309 (1971); **94**, 6560 (1972).

36. E. J. Corey and T. Ravindranathan, *J. Am. Chem. Soc.*, **94**, 4013 (1972).

37. D. A. Evans, R. C. Thomas, and J. A. Walker, *Tetrahedron Lett.*, 1427 (1976); D. A. Evans, T. C. Crawford, R. C. Thomas, and J. A. Walker, *J. Org. Chem.*, **41**, 3947 (1976).

38. R. A. Holton, *J. Am. Chem. Soc.*, **99**, 8083 (1977).

39. J. Hooz and R. B. Layton, *Can. J. Chem.*, **48**, 1626 (1970). See also G. H. Posner, *Org. React.*, **19**, 1 (1972).

40. C. J. Sih, R. G. Salomon, P. Price, G. Peruzzoti, and R. Sood, *Chem. Commun.*, 240 (1972).

41. C. J. Sih, P. Price, R. Sood, R. G. Salomon, G. Peruzzotti, and M. Casey, *J. Am. Chem. Soc.*, **94**, 3643 (1972).

42. C. J. Sih, R. G. Salomon, P. Price, R. Sood, and G. Peruzzotti, *J. Am. Chem. Soc.*, **97**, 857 (1975).

43. C. J. Sih, J. B. Heather, G. Peruzzotti, P. Price, R. Sood, L. F. Hsu Lee, *J. Am. Chem. Soc.*, **95**, 1676 (1973).

44. C. J. Sih, J. B. Heather, R. Sood, P. Price, G. Peruzzotti, L. F. Hsu Lee, and S. S. Lee, *J. Am. Chem. Soc.*, **97**, 865 (1975).

45. J. B. Heather, R. Sood, P. Price, G. P. Peruzzotti, S. S. Lee, L. F. Hsu Lee, and C. J. Sih, *Tetrahedron Lett.*, 2313 (1973).

46. A. F. Kluge, K. G. Untch, and J. H. Fried, *J. Am. Chem. Soc.*, **94**, 7827 (1972).

47. F. S. Alvarez, D. Wren, and A. Prince, *J. Am. Chem. Soc.*, 94, 7823 (1972).

48. K. F. Bernady and M. J. Weiss, *Tetrahedron Lett.*, 4083 (1972).

49. M. B. Floyd and M. J. Weiss, *Prostaglandins*, 3, 921 (1973).

50. G. Zweifel and R. B. Steele, *J. Am. Chem. Soc.*, 89, 2754 (1967).

51. K. F. Bernady, J. F. Poletto, and M. J. Weiss, *Tetrahedron Lett.*, 765 (1975).

52. K. F. Bernady and M. J. Weiss, *Prostaglandins*, 3, 505 (1973).

53. A. F. Kluge, K. G. Untch, and J. H. Fried, *J. Am. Chem. Soc.*, 94, 9256 (1972).

54. J. G. Miller, W. Kurz, and K. G. Untch, *J. Am. Chem. Soc.*, 96, 6774 (1974).

55. G. Stork and T. Takahashi, *J. Am. Chem. Soc.*, 99, 1275 (1977).

56. C. Luthy, P. Konstantine, and K. G. Untch, *J. Am. Chem. Soc.*, 100, 6211 (1978).

57. R. Pappo, P. Collins, and C. Jung, *Tetrahedron Lett.*, 943 (1973).

58. S. Kurozumi, T. Toru, and S. Ishimoto, *Tetrahedron Lett.*, 4959 (1973).

59. T.-j. Lee, *Tetrahedron Lett.*, 2297 (1979).

60. G. R. Kieczykowski, C. S. Pogonowski, J. E. Richman, and R. H. Schlessinger, *J. Org. Chem.*, 42, 175 (1977).

61. G. Stork, C. Kowalski, and G. Garcia, *J. Am. Chem. Soc.*, 97, 3258 (1975).

62. M. B. Floyd, *Synth. Commun.*, 4, 317 (1974).

63. M. Kobayashi, S. Kurozumi, T. Toru, and S. Ishimoto, *Chem. Lett.*, 1341 (1976).

64. L. Gruber, I. Tomoskozi, E. Major, and G. Kovacs, *Tetrahedron Lett.*, 3729 (1974).

65. M. B. Floyd, *J. Org. Chem.*, 43, 1641 (1978).

66. G. Piancatelli and A. Scettri, *Tetrahedron Lett.*, 1131 (1977).

67. J. W. Patterson and J. H. Fried, *J. Org. Chem.*, 39, 2506 (1974).

68. E. H. Posner, J. J. Sterling, C. E. Whitten, C. M. Lentz, and D. J. Brunelle, *J. Am. Chem. Soc.*, 97, 107 (1975); *Tetrahedron Lett.*, 2591 (1974).

69. G. Stork and M. Isobe, *J. Am. Chem. Soc.*, 97, 6260 (1975).

70. G. Stork and M. Isobe, *J. Am. Chem. Soc.*, 97, 4745 (1975).

71. R. B. Woodward, J. Gosteli, I. Ernest, R. J. Friary, G. Nestler, H. Raman, R. Sitrin, Ch. Suter, and J. K. Whitesell, *J. Am. Chem. Soc.*, 95, 6853 (1973).

72. I. Ernest, *Angew, Chem. Int. Ed.*, 15, 207 (1976).

73. E. J. Corey and T. Ravindranathan, *Tetrahedron Lett.*, 4753 (1971).

74. E. J. Corey and B. B. Snider, *Tetrahedron Lett.*, 3091 (1973).

75. F. Kienzle, G. W. Holland, J. L. Jernow, S. Kwoh, and P. Rosen, *J. Org. Chem.*, 38, 3440 (1973).

76. G. Buchi and B. Egger, *J. Org. Chem.*, 36, 2021 (1971).

77. D. B. Denney and N. Sherman, *J. Org. Chem.*, 30, 3760 (1965); T. Kashiwagi, S. Kozuka, and S. Oae, *Tetrahedron*, 26, 3619 (1970).

78. C. H. Kuo, D. Taub, and N. L. Wendler, *Tetrahedron Lett.*, 5317 (1972).

79. E. J. Corey, Z. Arnold, and J. Hutton, *Tetrahedron Lett.*, 307 (1970).

80. R. C. Kelly, V. VanRheenen, I. Schletter, and M. D. Pillai, *J. Am. Chem. Soc.*, 95, 2746 (1973).

81. R. C. Kelly and V. VanRheenen, *Tetrahedron Lett.*, 1709 (1973).

82. V. VanRheenen, R. C. Kelly, and D. Y. Cha, *Tetrahedron Lett.*, 1973 (1976).

83. R. C. Kelly and V. VanRheenen, *Tetrahedron Lett.*, 1067 (1976).

84. D. R. White, *Tetrahedron Lett.*, 1753 (1976).

85. E. J. Corey and P. L. Fuchs, *J. Am. Chem. Soc.*, **94**, 4014 (1972).

86. E. J. Corey, *Ann. N.Y. Acad. Sci.*, **180**, 24 (1971); E. J. Corey, N. M. Weinshenker, T. K. Schaaf, and W. Huber, *J. Am. Chem. Soc.*, **91**, 5675 (1969).

87. E. J. Corey, S. M. Albonico, U. Koelliker, T. K. Schaaf, and R. Kumar, *J. Am. Chem. Soc.*, **93**, 1491 (1971).

88. E. J. Corey, T. Ravindranathan, and S. Terashima, *J. Am. Chem. Soc.*, **93**, 4326 (1971).

89. B. M. Trost and Y. Tamura, *J. Am. Chem. Soc.*, **97**, 3528 (1975).

90. H. Wasserman and B. H. Lipshutz, *Tetrahedron Lett.*, 4611 (1975).

91. E. J. Corey and H. E. Ensley, *J. Am. Chem. Soc.*, **97**, 6908 (1975).

92. S. Ranganathan, D. Ranganathan, and A. K. Mehrotra, *J. Am. Chem. Soc.*, **96**, 5261 (1974).

93. J. E. McMurry and J. Melton, *J. Org. Chem.*, **38**, 4367 (1973).

94. P. A. Bartlett, F. R. Green and T. R. Webb, *Tetrahedron Lett.*, 331 (1977).

95. E. J. Corey, K. B. Becker, and R. K. Verma, *J. Am. Chem. Soc.*, **94**, 8616 (1972).

96. I. Tomoskozi, L. Gruber, G. Kovacs, I. Szekely, and V. Simonidesz, *Tetrahedron Lett.*, 4639 (1976).

97. E. J. Corey and C. U. Kim, *J. Am. Chem. Soc.*, **94**, 7586 (1972); *J. Org. Chem.*, **38**, 1233 (1973).

98. J. S. Bindra, A. Grodski, T. K. Schaaf, and E. J. Corey, *J. Am. Chem. Soc.*, **95**, 7522 (1973).

99. R. Peel and J. K. Sutherland, *Chem. Commun.*, 151 (1974).

100. S. Takano, N. Kubodera, and K. Ogasawara, *J. Org. Chem.*, **42**, 786 (1977).

101. E. D. Brown, R. Clarkson, T. J. Leeney, and G. E. Robinson, *Chem. Commun.*, 642 (1974); *J. C. S. Perkin I*, 1507 (1978).

102. K. B. Mallion and E. R. H. Walker, *Synth. Commun.*, **5**, 221 (1975).

103. E. D. Brown and T. J. Lilley, *Chem. Commun.*, 39 (1975).

104. M. J. Dimsdale, R. F. Newton, D. K. Rainey, C. F. Webb, T. V. Lee, and S. M. Roberts, *Chem. Commun.*, 716 (1977); *J. C. S. Perkin I*, 1176 (1978).

105. N. M. Crossland, S. M. Roberts, R. F. Newton, and C. F. Webb, *Chem. Commun.*, 660 (1978); R. F. Newton, D. P. Reynolds, C. F. Webb, S. N. Young, Z. Grudzinski, and S. M. Roberts, *J. C. S. Perkin I*, 2789 (1979).

106. T. K. Schaaf and E. J. Corey, *J. Org. Chem.*, **37**, 2922 (1972).

107. C. Doria, P. Gaio and C. Gandolfi, *Tetrahedron Lett.*, 4307 (1972).

108. J. Katsube, H. Shimomura, and M. Matsui, *Agr. Biol. Chem.*, **35**, 1828 (1971).

109. H. Shimomura, J. Katsube, and M. Matsui, *Agr. Biol. Chem.*, **39**, 657 (1975).

110. G. Jones, R. A. Raphael, and S. Wright, *Chem. Commun.*, 609 (1972).

111. G. Jones, R. A. Raphael, and S. Wright, *J. Chem. Soc. Perkin I*, 1676 (1974).

112. A. Fischli, M. Klaus, H. Mayer, P. Schonholzer, and R. Ruegg, *Helvetia*, **58**, 564 (1975).

113. E. J. Corey, Z. Arnold, and J. Hutton, *Tetrahedron Lett.*, 307 (1970).

114. N. M. Crossland, D. R. Kelly, S. M. Roberts, D. P. Reynolds, and R. F. Newton, *Chem. Commun.*, 681 (1979).

115. R. F. Newton, C. C. Howard, D. P. Reynolds, A. H. Wadsworth, N. M. Crossland, and S. M. Roberts, *Chem. Commun.*, 662 (1978).

116. R. F. Newton, D. P. Reynolds, N. M. Crossland, D. R. Kelly, and S. M. Roberts, *Chem. Commun.*, 683 (1979).

117. D. Brewster, M. Myers, J. Ormerod, M. E. Spinner, S. Turner, and A. C. B. Smith, *Chem. Commun.*, 1235 (1972); D. Brewster, M. Myers, J. Ormerod, P. Otter, A. C. B. Smith, M. E. Spinner, and S. Turner. *J. Chem. Soc. Perkin I*, 2796 (1973).

118. D. Taub, R. D. Hoffsommer, C. H. Kuo, H. L. Slates, Z. S. Zelawski, and N. L. Wendler, *Chem. Commun.*, 1258 (1970); *Tetrahedron*, 29, 1447 (1973); D. Taub, *Ann. N.Y. Acad. Sci.*, 180, 101 (1971).

119. H. L. Slates, Z. S. Zelawski, D. Taub, and N. L. Wendler, *Chem. Commun.*, 304 (1972); *Tetrahedron*, 30, 819 (1974).

120. E. J. Corey and P. A. Grieco, *Tetrahedron Lett.*, 107 (1972).

121. P. Crabbe and A. Guzman, *Tetrahedron Lett.*, 115 (1972).

122. A. Guzman, P. Ortiz de Montellano, and P. Crabbé, *J. C. S. Perkin I*, 91 (1973).

123. J. S. Bindra and A. Grodski, *J. Org. Chem.*, 43, 3240 (1978).

124. S. Ranganathan, D. Ranganathan, and A. K. Mehrotra, *J. Am. Chem. Soc.*, 96, 5261 (1974); S. Ranganathan, D. Ranganathan and R. Iyengar, *Tetrahedron,* 32, 961 (1976).

125. E. J. Corey and G. Moinet, *J. Am. Chem. Soc.*, 95, 6831 (1973).

126. E. J. Corey and J. Mann, *J. Am. Chem. Soc.*, 95, 6832 (1973).

127. M. A. W. Finch, T. V. Lee, S. M. Roberts, and R. F. Newton, *Chem. Commun.*, 677 (1979).

128. J. Martel, E. Toromanoff, J. Mathieu, and G. Nomine, *Tetrahedron Lett.*, 1491 (1972).

129. J. W. Cornforth, R. H. Cornforth, and K. K. Mathew, *J. Chem. Soc.*, 112 (1959).

130. E. J. Corey, G. W. J. Fleet, and M. Kato, *Tetrahedron Lett.*, 3963 (1973).

131. I. N. Nazarov, I. I. Zaretskaya, and T. I. Sorkina, *J. Gen. Chem. USSR*, 30, 765 (1960); E. A. Braude and J. A. Coles, *J. Chem. Soc.*, 1432 (1952); N. C. Deno, C. U. Pittman, and J. O. Turner, *J. Am. Chem. Soc.*, 87, 2153 (1965).

132. G. Stork and S. Raucher, *J. Am. Chem. Soc.*, 98, 1583 (1976).

133. S. Raucher, *Tetrahedron Lett.*, 1161 (1976).

134. R. A. Ellison, *Synthesis*, 397 (1973).

135. T. -L. Ho, *Synth. Commun.*, 4, 265 (1974).

136. R. L. Jones, *J. Lipid Res.*, 13, 511 (1972); R. L. Jones and S. Cammock, *Adv. Biosci.*, 9, 61 (1973).

137. E. J. Corey and G. Moinet, *J. Am. Chem. Soc.*, 95, 7185 (1973).

138. R. C. Kelly, I. Schletter, and R. L. Jones, *Prostaglandins*, 4, 653 (1973).

139. P. Crabbé, A. Guzman, and M. Vera, *Tetrahedron Lett.*, 3021, 4730 (1973); P. Crabbé and A. Cervantes, *Ibid.*, 1319 (1973).

140. N. M. Crossland, S. M. Roberts, and R. F. Newton, *J. C. S. Perkin I*, 2397 (1979); N. M. Crossland, S. M. Roberts, R. F. Newton, and C. F. Webb, *Chem. Commun.*, 660 (1978).

141. E. J. Corey and C. R. Cyr, *Tetrahedron Lett.*, 1761 (1974).

142. M. Hayashi and T. Tanouchi, *J. Org. Chem.*, 38, 2115 (1973).

143. E. F. Jenny, P. Schaublin, H. Fritz, and H. Fuhrer, *Tetrahedron Lett.*, 2235 (1974).

144. S. M. Ali, M. A. W. Finch, S. M. Roberts, and R. F. Newton, *Chem. Commun.*, 679 (1979).

145. B. Samuelsson, G. Folco, E. Granstrom, H. Kindahl, and C. Malmsten, *Adv. Prostaglandin Thromboxane Res.*, **4**, 1 (1978).

146. S. Moncada, R. Gryglewski, S. Bunting, and J. R. Vane, *Nature*, **263**, 663 (1976).

147. R. A. Johnson, E. G. Nidy, L. Baczynskyj, and R. R. Gorman, *J. Am. Chem. Soc.*, **99**, 7738 (1977).

148. N. A. Porter, J. D. Byers, R. C. Mebane, D. W. Gilmore, and J. R. Nixon, *J. Org. Chem.*, **43**, 2088 (1978).

149. N. A. Porter, J. D. Byers, K. M. Holden, and D. B. Menzel, *J. Am. Chem. Soc.*, **101**, 4319 (1979).

150. S. Moncada, R. J. Gryglewski, S. Bunting, and J. R. Vane, *Prostaglandins*, **12**, 685, 715 (1976).

151. E. J. Corey, G. E. Keck, and I. Szekely, *J. Am. Chem. Soc.*, **99**, 2006 (1977).

152. K. C. Nicolaou et al., *Lancet*, **I**, 1058 (1977).

153. K. C. Nicoulaou, W. E. Barnette, G. P. Gasic, R. L. Magolda, and W. J. Sipio, *Chem. Commun.*, 630 (1977).

154. K. C. Nicolaou and W. E. Barnette, *Chem. Commun.*, 331 (1977).

155. R. A. Johnson, F. H. Lincoln, J. L. Thompson, E. G. Nidy, S. A. Mizsak, and U. Axen, *J. Am. Chem. Soc.*, **99**, 4182 (1977).

156. W. P. Schneider and R. A. Morge, *Tetrahedron Lett.*, 3283 (1976).

157. N. A. Nelson and R. W. Jackson, *Tetrahedron Lett.*, 3275 (1976).

158. R. C. Kelly, I. Schletter, and S. J. Stein, *Tetrahedron Lett.*, 3276 (1976).

159. S. Hanessian and P. Lavallee, *Can. J. Chem.*, **55**, 562 (1977).

160. E. J. Corey, M. Shibasaki, and J. Knolle, *Tetrahedron Lett.*, 1625 (1977).

161. E. J. Corey, M. Shibasaki, J. Knolle, and T. Sugahara, *Tetrahedron Lett.*, 785 (1977).

162. *Population Reports, Series G*, No. 8, March 1980, The Johns Hopkins University, Baltimore, Md.

163. K. Friebel, A. Schneider and H. Würfel, Eds., *International Sulprostone Symposium, Vienna, November 1978*, Schering AG Berlin, 1979.

The Synthesis of Monoterpenes, 1971—1979

ALAN F. THOMAS

Research Laboratory,
Firmenich SA,
Geneva, Switzerland

and YVONNE BESSIERE,

Institute of Organic Chemistry,
University of Lausanne,
Lausanne, Switzerland

1. INTRODUCTION

Perhaps the best beginning to a second edition of the synthesis of monoterpenes would be a list of errata from the first edition! This list is at the end of this section.

The first edition (Vol. 2) has been described by Hoffmann as "wissenschaftlich."[1] If this means "theoretical" as opposed to "industrial," it is not because the industrial syntheses were ignored in the first edition – on the contrary, they are nearly all there (but see citral, p. 471 in this edition) – but rather because the proportion of published work is small. Much of it is in the patent literature, about which some academics still take the attitude that it is not "publication" in the true sense of the word. (To illustrate how wrong this idea is, one might read the reviews of Hoffmann[1] or Nürrenbach.[2]) Since the proportion of published industrial to academic synthesis has possibly even decreased recently, this edition may appear even more "theoretical." Nevertheless, the supply situation of natural products is not improving, and industry keeps a "weather eye" on academic synthesis.

In this chapter an attempt is made to present those syntheses that introduce some novel aspect in preparing the natural products, even though they may not be economically practical. Multistage syntheses of common products are mentioned, but not generally discussed, since these are usually of interest only in cases where labeling is required.

There are, in fact, two quite different types of synthesis associated with monoterpenoids, and these different types are linked with specific classes of monoterpenoids. Generally speaking, the syntheses of dimethyloctane derivatives fall into "industrial" if not "classical" patterns, as do syntheses associated with menthanes. The other extreme is represented by the iridoids. Syntheses are long and highly complex, and it is not difficult to see that one must always attach a higher price label to compounds like this, which are characterized by an increased number of asymmetric centers, and furthermore have few readily available members of the class from which others may be made. In this they differ from most of the bicyclic monoterpenes, which, apart from being derivable from menthanes at least on paper, nearly all have one or more members available in large amounts, with the possible exception of the thujanes.

Phenolic monoterpenoid synthesis has not been discussed in this chapter.

Concerning the occurrence in nature of monoterpenoids, we do not pretend that this chapter gives a complete list of natural products, although an attempt has been made to be precise about whether a given synthesis leads to a natural product or not. A useful list of monoterpenoids (and other materials) occurring in natural aromas is given by Ohloff.[3]

Literature is complete up to approximately October 1979, with occasional later references.

Errata to Chapter 1, Volume 2

p. 15, line 5	For "Ref. 44" read "Ref. 55".
p. 64	A methyl group is missing from C(7) in iridomyrmecin.
p. 88	The formula **225** depicts $(-)$-(S)-limonene, not the $(+)$-(R)-isomer.
p. 96	Scheme 34. The first two formulas should have isopropyl not isopropenyl side chains.
p.108	In Scheme 41, the solvent for transformation of **225** to **282** should be CH_2Cl_2.
p. 116	The reference to **332**, evodone, should be **366**.
pp. 156, 157	The reaction of N-bromosuccinimide with α-pinene does not yield the 7-brominated pinene derivative to any extent, myrtenyl bromide (**449**) being one of the main products.

2. TERPENE SYNTHESIS FROM ISOPRENE

Of the many synthons suitable for monoterpene synthesis that are prepared from isoprene, a selection has been made in the Table below. Other more specific syntheses from isoprene will be discussed under the sections devoted to the compounds prepared.

Conditions	Products	Literature
$CH_3CO_2NO_2$, room temp.		4
C_6H_5SH / O_2		5
1) HBr, $-10°$; 2) $(C_6H_5)_3P$		6
$ClCH_2OMe/SnCl_4/C_2H_5OH$	1	7
Peracid	2	8^a
Epoxide 2, Pd(acac), $(C_6H_5)_3P$	3	12^b

*Isoprene epoxide (2) has been made in poor yield by perbenzoic acid[9] or peracetic acid[10] oxidation, as well as via the bromohydrin using N-bromosuccinimide.[11] The method quoted in the table[8] is the most efficient.

†For another efficient synthesis of tiglic aldehyde (3) not from isoprene, see Loewenthal.[13]

Some C_5 acetals, 4, 5, have been prepared in stereochemically pure form (though not from isoprene), and these could be very useful for terpene synthesis.[14]

The telomerization of isoprene produces a very large literature, but few of the papers report highly specific synthesis of natural products. Use of sodium and secondary amines can lead to fairly efficient syntheses of myrcene (6)[15], while palladium catalysts with alcohols can be made to give either tail-to-tail dimers,[16] or head-to-tail dimers,[17] one system [PdCl$_2$-C$_6$H$_5$CN + (C$_6$H$_5$)$_3$P and NaOR] leading mainly to 7.[18] Dimerization with Pd-phosphine complexes and CO$_2$ also gives mostly tail-to-tail isomers.[19] Stannic chloride telomerization can produce, in addition to tail-to-tail isomers, up to 61% of (E)- and (Z)-isomers of geranyl chloride (8),[20] the (Z)-isomer cyclizing under these conditions to the 8-chloro-menthene (9).[21] The dimerization of isoprene using lithium naphthalene has also been studied.[22]

Use of these mixtures has been made to synthesize monoterpenes, for example, treatment of the mixture of dimers obtained from isoprene and formic acid in the presence of triethylamine and various palladium catalysts with hydrochloric acid yields a chloride 10, which can be separated from other dimers. t-Butyl peracetate oxidation of 10 in the presence of cuprous bromide, followed by lithium aluminum hydride reduction and heating yields linalool (11).[23] A few further papers on the subject might be quoted: head-to-tail dimeri-

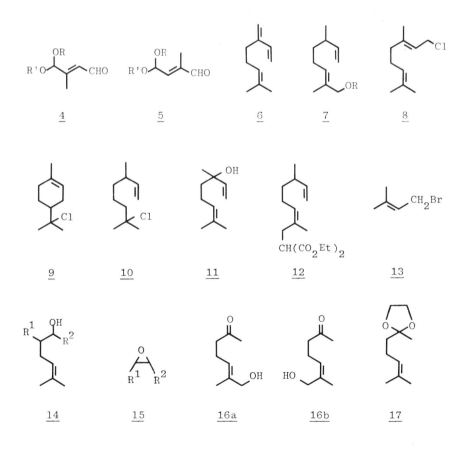

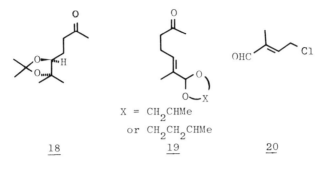

X = CH₂CHMe
or CH₂CH₂CHMe

<u>18</u>   <u>19</u>   <u>20</u>

zation can be favored in the presence of nickel catalysts,[24] and Baker has used this fact to prepare 12.[25]

Prenyl bromide (13) (and other prenyl halides) are excellent functionalized isoprenes; further examples of their use will be found later, but a few general reactions are given here. With $CrCl_3$ and $LiAlH_4$ in tetrahydrofuran, 13 mostly yields tail-to-tail dimers.[26] Treatment with the anion of a functionalized dithiane enables the bromine atom to be replaced by an alkyl group,[27] and the Grignard reagent, well known to react in the allylically rearranged form, reacts in the unrearranged form (to give 14) with epoxides 15 in the presence of 10% of cuprous iodide.[28]

The anodic oxidation of isoprene has been studied, but yields (in methanol at –25°) are very low, with ca. 10% head-to-tail dimers.[29]

3. 6-METHYL-5-HEPTEN-2-ONE

In Vol. 2 of this series (p. 7), it was briefly mentioned that this vital intermediate for terpene synthesis could be made from prenyl chloride and acetone. Improvements in the method consist in carrying out the substitution using solid alkali in the presence of an organic amine and/or ammonia,[30] a reaction which also forms higher ketones. The latter can be cracked by water at 230-310°, resulting in greatly increased yields of methylheptenone.[31]

The preparation of hydroxylated methylheptenones, specifically (E)- (16a) and (Z)- (16b), using selenium dioxide oxidation of the acetal 17 for 16a,[32, 33] and Wittig reaction (the "*cis*-modification") of the aldehydoacetal obtained by ozonolysis of 17 for 16b has also been described.[33]

Attention is drawn to the preparation of a methylheptenone in which the double bond is protected by a chiral group, 18.[34] This was prepared from (S)-phenylalanine, and can in principle be used to synthesize chiral terpenes.

A functionalized methylheptenone 19 has been made from 20 (one method for the preparation of which is described in Ref. 8) by acetoacetate ketone synthesis on the acetal of 20.[35]

4. 2,6-DIMETHYLOCTANE DERIVATIVES

A. Hydrocarbons

There is very little new work in the synthesis of naturally occurring hydrocarbons of this series. What is practically a one-step synthesis of myrcene (**6**) by addition of 1-buten-3-yne to the organocuprate **21**,* is reported,[36] but myrcene is still readily available from pinene (Vol. 2, p. 8) by pyrolysis. It has been reported that the problem of the isomerization of (*E*)-ocimene (**22**) to alloocimene (**23**) occurring extensively under these conditions can be circumvented by rapid cooling of the pyrolysate.[37] Vig et al. have synthesized myrcene from ethyl acetoacetate[38] and from ethyl 5-methyl-4-hexenoate,[39] and α-ocimene (**24**) from **25**.[40] All these syntheses are straightforward, and all involve Wittig reactions.

"Achillene" is not the compound **26** synthesized by Schulte-Elte (Vol. 2, p. 12)[41], but is identical with santolinatriene (**27**).[42] Although Schulte-Elte was unable to obtain from the Russian authors a direct comparison of his synthetic product, this has not stopped Dembitskii et al. from correcting their work without reference to either Schulte-Elte, or to the original publication[43] of the NMR spectrum of their "achillene" (**26**).[42] The name should now be deleted from the literature.

Nerol (**28**) has been converted into myrcene (**6**) by Yasuda et al. It is first epoxidized with *t*-butyl hydroperoxide and vanadium(IV),[44] and the trimethylsilyl ether of the epoxide **29** is treated with 2,2,6,6-tetramethylpiperidine diethylaluminum. This reacts stereospecifically to yield the diol **30**, which was converted to myrcene (**6**) by cuprous bromide and phosphorus tribromide at $-78°$ followed by zinc. Following the same route, geraniol gives (*E*)-ocimene (**22**).[45]

A review of the marine terpenoids, a large number of which are halogenated myrcenes, but very few of which have been synthesized, has been published by Faulkner.[46]

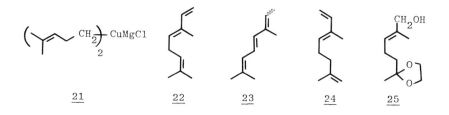

21   22   23   24   25

*The stereospecific addition of organocopper compounds to acetylenes was developed by Normant et al., and a representative synthesis of citral acetal is mentioned on p. 000.

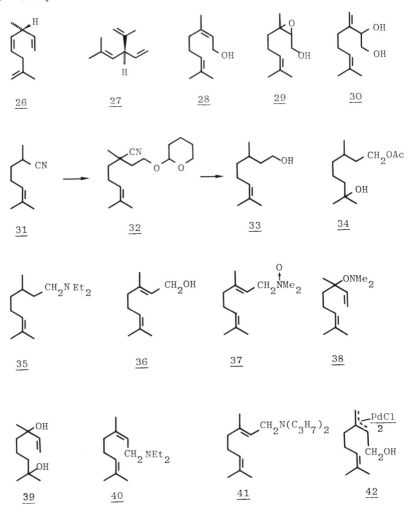

B. Alcohols

Citronellol

The synthesis of natural (−)-citronellol is still as reported in Vol. 2 (p. 14), but a new synthesis of the racemate has been achieved by reaction of 4-methyl-3-pentenyl bromide with the anion of propionitrile in hexamethylphosphoramide. The resulting nitrile **31** is allowed to react in the form of its anion with 2-chloroethanol tetrahydropyranyl ether, and the product **32** is reduced with potassium and *t*-butanol in hexamethylphosphoramide to citronellol (**33**).[47]

Treatment of 3,7-dimethyloctyl acetate with ozone adsorbed on silica gel at $-78°$ (the technique of Mazur[48]), leads to a mixture of oxygenated substances, 65% of which is the hydroxylated citronellyl acetate (34).[49] The diethylamine 35 corresponding to citronellol has been isolated from Réunion geranium oil and synthesized from citronellal oxime.[50]

Linalool (11), Nerol (28), Geraniol (36)

The wide occurrence of these alcohols and their esters precludes a list of natural products, but we cannot resist mentioning that neryl formate is an alarm pheromone of the cheese mite *Tyrophagus putrescentiae*,[51] and wondering, in view of the fact that it is used in perfumery,[52] what the effect of a perfumed hand over a dish of doubtful cheese would be!

Because geranic esters are readily converted to 28 and 36 their synthesis is included in this section. The novel "reverse" allylic rearrangement from the dimethylamine oxide 37 to the hydroxylamine 38 occurs on heating, and Rautenstrauch has used this fact to make very pure linalool (11) by reduction of 38 with zinc in acetic acid.[53] Takabe et al. have repeated this work[54] using diethylgeranylamine prepared from isoprene,[15] also making the diol 39 by the Rautenstrauch route after hydroxylating the amine with aqueous sulfuric acid.[54] Rautenstrauch also found that if the linalyl hydroxylamine 38 was heated, it reverted to the geranyl hydroxylamine, zinc-acetic acid reduction of which yielded geraniol and nerol (2:1).[53] This has also been repeated by Takabe et al.[55] The reverse transformation can also be carried out by reduction of the mesylate of geraniol epoxide with sodium (or calcium) in liquid ammonia.[56] Using the nerylamine 40 prepared by the action of lithium or butyllithium on isoprene and diethylamine, Takabe et al. prepared nerol via the chloride (obtained with ethyl chloroformate and 40) and neryl acetate.[57] They have also shown that when myrcene (6) is treated with lithium and dipropylamine, the geranylamine 41 is obtained, which can be converted to geraniol (36) by the same route.[57]

Of the various publications concerning the complexes of myrcene with palladium,[58,59] one is particularly interesting. The complex in question 42 is formed from myrcene (6) and $(CH_3CN)_2PdCl_2$ in hexamethylphosphoramide and water (10:1). When it is treated with sodium methoxide in methanol it yields 45% of nerol (28), 36% of citral (below), but no geraniol.[60]

One type of synthesis of dimethyloctane alcohols employs activated anions (cf. the early work, Vol. 2, p. 54, in a synthesis of chrysanthemic acid[61]). Two groups adopted this approach: Evans et al. used the 4-carbon unit phenyl sulfoxide anion 43, adding isohexenyl iodide to obtain the sulfoxide 44, readily convertible to a 9:1 mixture of geraniol (36) and nerol (28).[62] Julia used the phenyl sulfone (45) anion,[63] which, together with prenyl bromide (13), yields the sulfones 46. Reduction of this mixture gave esters having the geranic acid

(**47**) and lavandulic acid (**48**) skeletons. The condensation step can be carried out under phase-transfer conditions, which leads to a considerable improvement in the accessibility of geraniol by this route.[64] Julia has since developed the principle further by using the anion of a prenyl phenyl sulfone, which, with

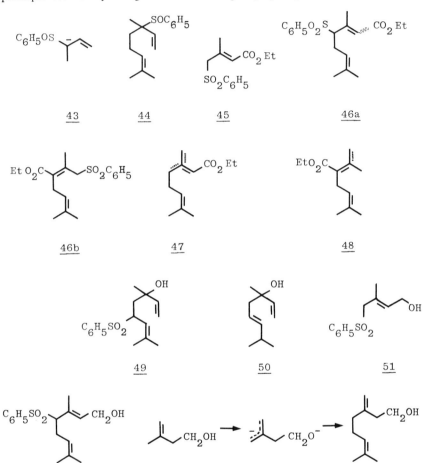

43   44   45   46a

46b   47   48

49   50   51

52   53   54

55   56   57   58

isoprene epoxide (2), gives the sulfone 49 derived from linalool (11). Reduction of 49 by either sodium amalgam or lithium in ethylamine leads to a mixture of linalool (11) and its unnatural double-bond isomer 50.[65] One of the synthons mentioned in Table 1[5] is closely related to a sulfone 51 that Julia et al. made from isoprene. The dianion of this sulfone 51 can be condensed with a prenyl halide (they used the chloride corresponding to 13) to yield a sulfone 52 derived from geraniol. Reduction of the latter sulfone with sodium amalgam leads to (Z)- and (E)-ocimene (22), but with lithium in ethylamine, geraniol is obtained.[66] Prenyl bromide (13) has also been condensed with the dianion of 3-methyl-3-butenol (53), the isobutylene-formaldehyde Prins product (see a new industrial citral synthesis below). This reaction leads to the methylene isomer (54) of geraniol,* which has been converted to a mixture of the geometrical isomers of geranic acid.[67]

A group of $C_5 + C_5$ syntheses of geraniol and nerol has dealt with the selective formation of a (Z)- or (E)-double bond. One approach consists in the addition of prenyl bromide (13) to the dianion of 3-methyl-2-butenoic acid (55, "senecioic acid"), followed by methyl iodide. Esters of geranic acids (E-56, and Z-56) and the ester corresponding to lavandulol (57, R = OH) are obtained, the presence of a cuprous salt, added before the halides, markedly increasing the proportion of the products of γ-addition 56.[68] Katzenellenbogen and Crumrine also found a 9:1 ratio of γ- to α-regioselectivity (with 45:55 (Z):(E)-isomers of 43 esters) from the reaction of the ethyl ester of 55 and prenyl bromide (13) in the presence of cuprous iodide.[69] Another $C_5 + C_5$ synthesis is the copper-catalyzed Wurtz-type reaction between the Grignard reagent of a prenyl halide (e.g., 13 − care is needed to avoid allyl rearrangement) and the chloro-alcohol 58.[70]

Related to these syntheses are some $C_5 + C_4 + C_1$ types. One, leading to highly stereoselective (Z)-56, consists in prenylating (with 13) the dianion of 2-butynoic acid (made from the acid with lithium diisopropylamide). This leads to the acetylenic ester 59, to which lithium dimethylcuprate adds very stereoselectively, yielding the (Z)-ester of 56 which can be reduced to nerol (28) with lithium aluminum hydride.[71] An alternative route from 59 to ethyl geranate consists in treating the acetylenic ester (actually the ethyl ester was employed) with thiophenol and sodium ethoxide. The resulting thioester 60 is methylated (99% stereospecific) with methylmagnesium iodide and cuprous iodide.[72]

The dianion of methyl acetoacetate[73] can also be prenylated, forming the ketoester 61. Isopropenyl acetate gives an enol acetate 62 with the latter, and 62 can be converted into geranic ester with lithium dimethylcuprate, reduction

*Sometimes called γ-geraniol; see p. 536 for the relationship of 54 with the dimethylethyli-denecyclohexanes.

yielding geraniol (**36**).[74] This reaction becomes stereoselective if advantage is taken of the fact that when the formation of the enol acetate **62** is carried out using isopropenyl acetate in the presence of *p*-toluenesulfonic acid it is mainly the (*Z*)-isomer (as written in **62**) that is formed (20:1 ratio), while with acetyl chloride and triethylamine in hexamethylphosphoramide, the (*E*)-isomer of **62** is formed (1:12 ratio). The stereoselectivity is even higher in the case of the enol benzoates.[75] Isoprene reacts with Grignard reagents in the presence of diethyltitanium dichloride. The use of methallyl chloride as starting material gives the Grignard reagent **63**, which can be treated with formaldehyde to give α-geraniol (**64**), or with carbon dioxide to give the corresponding acid (erroneously called "geranic acid" in the publication).[76]

A further method of joining isobutenyl bromide to isoprene is to use morpholine in the presence of catalytic amounts of palladium acetate. Treatment of the reaction mixture with methyl chloroformate replaces the morpholine in the adduct thus obtained **65** (as the major adduct, 54%) by chlorine, and conversion to isogeranic ester **66** was effected by carbonylation with carbon monoxide and ethanol over a palladium chloride catalyst, but in poor yield.[77]

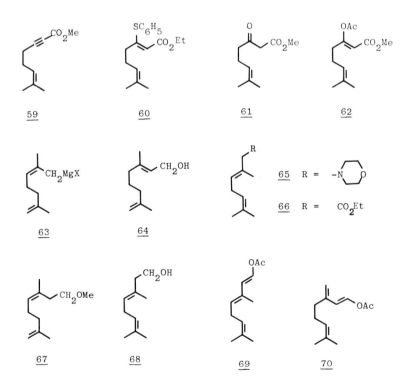

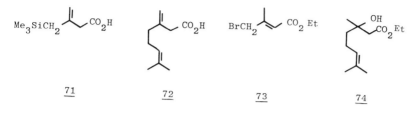

<u>71</u>   <u>72</u>   <u>73</u>   <u>74</u>

Use of the C_6 synthon (1) in Table 1[7] only means that the units have been joined together in a different way ($C_5 + C_1 + C_4$), and the Grignard reagent from methallyl halide reacts with this synthon 1 to give α-isogeranyl methyl ether (67).[78] The same synthon 1 reacts with isoprene in the presence of titanium tetrachloride, dechlorination of the product this time giving the α- and β-isomers of isogeranyl methyl ether.[79]

The same authors made (E)-β-isogeraniol (68) stereospecifically by reducing citral enol acetate (69) with sodium borohydride, when the other citral enol acetate (70) remained unchanged.[79]

The organometallic isohexenyl derivative (21, in the form of its lithium complex) adds to the dimethylacetal of 2-butynal to yield the dimethylacetal of citral in high yield.[80] The magnesium bromide complex (21, Br) will also add to the trimethylsilyl ether of 3-butynol; after aqueous workup 54 is obtained (for which the authors of the method employ the unacceptable name "myrcenol").[81]

Reaction of diketene with Me_3SiCH_2MgCl in the presence of nickel chloride in tetrahydrofuran gives the functionalized isoprene 71. In the form of its dicopper dienolate (made with lithium diisopropylamide and cuprous iodide in tetrahydrofuran), this can be prenylated with prenyl bromide, and after removal of the silicon, the product is the methylene isomer 72 of geranic acid. The latter is obtained on treatment of 72 with strong base.[82] With nickel carbonyl, prenyl bromide (13) yields a nickel complex, and the latter reacts with the bromoester 73 to form ethyl geranate.[83]

The lithium enolate of ethyl acetate adds to 6-methylhept-5-en-2-one, thus constituting a new $C_8 + C_2$ route, but the hydroxyester 74 obtained did not undergo dehydration with aluminum diethoxychloride and lithium diisopropylamide so stereoselectively as had been hoped (the hope having been based on studies with model compounds).[84]

6-Methylhept-5-enyne, made from prenyl bromide (13) and 2-propynylmagnesium bromide can be converted to geranic esters via addition of ethyl chloroformate to the anion (giving 59) and following the routes given above.[71,72] Alternatively, treatment with the complex $Me_3Al-Cl_2Zr(C_5H_5)_2$ in dichloroethylene yields an aluminum complex which gives geraniol after adding butyllithium and formaldehyde, or ethyl geranate after adding ethyl chloroformate.[85]

A mixture of prenyl pyrophosphate and isopentenyl pyrophosphate gives nerol (28) pyrophosphate in the presence of farnesyl pyrophosphate synthetase,[86] such biosynthetic routes being commonly supposed to pass through a carbenium ion equivalent. When prenyl acetate and isopentenyl acetate (excess) are heated with lithium perchlorate in acetic acid (solvent), a reaction resembling the biosynthesis occurs, but the products obtained are only minor amounts of neryl and geranyl acetates, the major products being the acetates of 54 and the two geometrical isomers of 68.[87] Isopentenyl acetate (53-acetate) with excess 2-methyl-3-buten-2-ol in the presence of trifluoroacetic acid yields the acetate 75 by a similar reaction.[88]

Perhaps the fact that the epoxyketone 76 undergoes a unique ring opening to the ketoester 77 on treatment with sodium carbonate in methanol, and that 77 is readily converted to methyl geranate[89] can hardly count as an economic approach to geraniol (36), but it certainly involves the most original route published in recent years.

Among the labeled geraniols that have been synthesized, leading references can be found in papers referring to the preparation of [7-^{14}C]-,[90, 91] [7′,8-^{14}C]-,[91] [7′,8-^{3}H]-,[91] and [3′-^{14}C]-geraniol.[92] Geraniols and linalools labeled with fluorine have been prepared,[93, 94] and 2-fluorogeraniol was converted into its pyrophosphate for studies on the biogenesis of farnesyl pyrophosphate.[94]

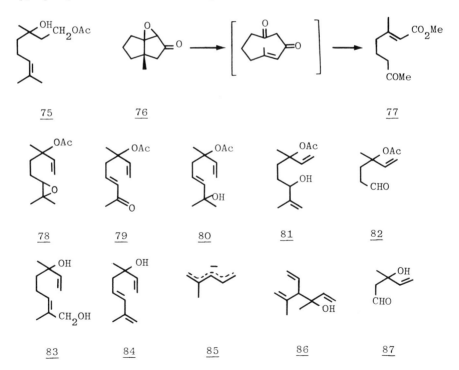

| 75 | 76 | 77 |

| 78 | 79 | 80 | 81 | 82 |

| 83 | 84 | 85 | 86 | 87 |

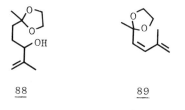

88   89

Oxidized Linalools, Hotrienol, Dehydrogenated Nerol, and Geraniol

A number of oxidized linalyl acetates, 78-82, occur in lavendin oil, possibly as artifacts; these can be made by the photooxygenation of linalyl acetate.[95] The photooxygenation of linalool (11) has also been reexamined.[96] Compound 81 occurs in lavender and lavendin with the acetoxy and hydroxy groups interchanged.[97] The diol 83,* found in Greek tobacco, was synthesized from linalool by selenium oxide oxidation,[99] and the glucosides of the two possible geometrical isomers of 83 have been found in *Betula alba* bark and the fruits of *Chaenomeles japonica* (Rosaceae),[100] although no synthesis of the (Z)-isomer has been reported.

The acetate of the aldehyde corresponding to 83 has been identified in lavendin oil.[95] More recent reports of hotrienol (84) in grape and wine flavor[101] and in nutmeg oil have appeared.[102]

Pentadienyl anions can be made from pentadienes and butyllithium in tetrahydrofuran.[103] The 2-methylpentadienyl anion (85) exists in a mixture of conformations,[103] but with methyl vinyl ketone, it yields about 35% of hotrienol (84) in one step, accompanied by the alcohol 86, also in about 35%.[104] This synthesis of Wilson's has also been adapted to prepare another natural monoterpenoid alcohol (below).

The aldehyde 87 has been prepared from the acetal of acetylacetaldehyde (3-oxobutanal), and Wittig reaction leads to the (±)-trienol 84.[105] Another synthesis involves potassium permanganate fission of the double bond of the acetal (17) of methylheptenone,[106] but the most interesting of this series by Vig et al. is his claim to have made the (Z)-isomer of hotrienol by dehydration of the photooxygenation product 88 of methylheptenone acetal (17). Using phosphorus oxychloride in pyridine, this is said to yield the (Z)-diene 89, from which (Z)-hotrienol is readily made by deacetalization and Grignard reaction.[107] Julia's synthesis of the (±)-(E)-isomer compares favorably with the earlier ones and depends on the formation of the sulfoxide 90; this can again form a carbanion from which isoprene epoxide (1) yields the hotrienol skeleton, the phenylsulfoxide group being removed in refluxing toluene (Scheme 1).[108]

*This diol is also formed by microbial hydroxylation of *Nicotiana tabacum*.[98]

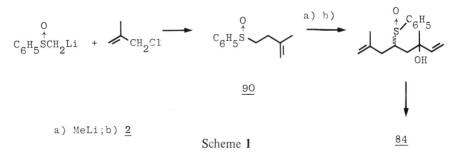

90

a) MeLi;b) **2**

Scheme 1   84

The diol **91** occurs in the hairpencils of the Monarch butterfly, and has been synthesized by Kossanyi et al. The key step of the synthesis is a Norrish Type I photolysis of the cyclopentanone **92** to yield the aldehydes **93**. The latter were easily converted to the diol **91**.[109]

The isovalerate of dehydronerol (**94**) was isolated from *Anthemis montana* (Compositae),[110] and synthesized from 2-methylhepta-2,4-dien-6-one by the Horner-Wittig reaction to the esters (2*Z*, 4*E*)- and (2*E*, 4*E*) corresponding to **94**. After separation of these (thin-layer chromatography), the natural product was obtained by reduction of the (*Z*)-isomer with lithium aluminum hydride and esterification.[111] Bohlmann and Zdero have also isolated the most oxidized dimethyloctane monoterpenoid alcohol **95** in the form of various esters from another Compositae, *Schkuhria senecioides*,[112] but the synthesis is not yet reported.

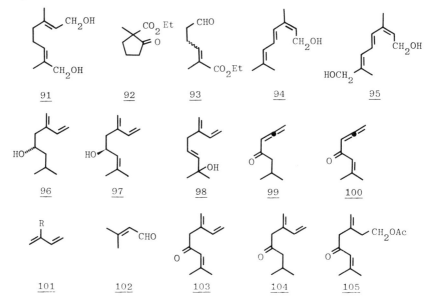

Ipsenol (96), Ipsdienol (97), 2-Methyl-6-methylen-3,7-octadien-2-ol (98)

The sex attractants of *Ips confusus* (Vol. 2, p. 26) and other beetles[113] have received considerable synthetic attention. The natural products **96**, **97** have the (S) configuration, that is, have opposite stereochemistries,[114,115] and are usually accompanied by the allyl isomer **98** of **97**, which also occurs in *Pinus ponderosa*[116] and *Ledum palustre*.[117] Ipsdienol (**97**) is the biosynthetic precursor of ipsenol (**96**), the transformation being effected only by male *Ips* beetles.[118]

If the starting material were readily available, a simple synthesis of both **96** and **97** starts from metallated allene. (Metallated allenes can be made either by direct lithiation of allenes with butyllithium in tetrahydrofuran, or by metal exchange of bromoallene.) The metallated allene is treated with an amide of 3-methylbutyric acid (for **96**), or 3-methyl-2-butenoic acid (for **97**). The products (respectively **99** and **100**) can be vinylated with vinyl cuprate in ether, yielding ipsenol (**96**) or ipsdienol (**97**) as racemates.[119] A number of other syntheses also involve organometallic reactions. The straightforward Grignard reaction of the butadienylmagnesium chloride (**101**, R = MgCl) on 4-methyl-1-pentene epoxide gives a 72% yield of ipsenol (**96**).[120] Similarly straightforward is the reaction of the bromoisoprene (**101**, R = CH_2Br) with 3-methyl-2-butenal (dimethylacrolein or senecia aldehyde, **102**) and zinc,[121] although the bromo compound is not available at a low price. Another functionalized isoprene used in a similar way for making the ipsenol series is the trimethylsilyl derivative (**101**, R = Me_3SiCH_2), made by the reaction of the Grignard reagent from chloromethyltrimethylsilane and 2-chlorobutadiene. In the presence of a Lewis acid ($AlCl_3$ or $TiCl_4$), this silane will react with 3-methylbutanal to give (±)-ipsenol (**96**), or with the acid chloride of 3-methyl-2-butenoic acid (**55**) to give the ketone **103** corresponding to ipsdienol (**97**) to which it can be reduced by metal hydrides. Alternatively, the reagent **101** (R = Me_3SiCH_2), can react with 3-methylbutyryl chloride to give the ketone **104** corresponding to ipsenol (**96**), into which it is converted with dibutylaluminum hydride.[122] A similar reaction has been used to prepare the tagetenones (see below). The ketone **103** has also been prepared by Garbers and Scott by treatment of the anhydride of senecioic acid (**55**) and isopentenyl acetate (the acetate of alcohol **53**) with boron trifluoride in dichloromethane. The major product of the mixture of double bond isomers thus obtained has an *exo*-methylene group **105**, and pyrolysis of this gives mainly **104**, together with a small amount of cyclized (menthane) material.[123]

Katzenellenbogen and Lenox have shown that ipsenol can be obtained by treatment of 3-methylbutanal and the mesitoate of alcohol **106** with zinc in tetrahydrofuran.[124]

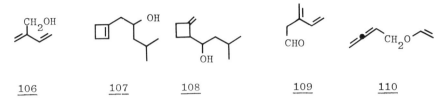

106   107   108   109   110

Methylenecyclobutane can be lithiated with butyllithium in tetramethyl-ethylenediamine, and the anion reacts with 3-methylbutanal to give the cyclobutene alcohol **107** together with **108**. Pyrolysis of **107** at 150° yields **96**.[125] Wilson has recently improved this method by using the trimethylsilane obtained from the methylcyclobutenyl anion. With 3-methylbutanal and titanium tetrachloride, **107** is obtained without contamination by **108**.[126] However, the price of methylenecyclobutane at present places these syntheses out of economic feasibility. Care must be taken in any method employing pyrolysis in the last step, since at high temperatures, ipsenol (**96**) undergoes fission to isoprene and methylbutanal, as Haslouin and Rouessac found in the last step of their somewhat lengthy synthesis from itaconic anhydride (Scheme 2).[127] Ipsdienol is also unstable to temperatures above 100°, when it is converted into the allylically rearranged isomer **98**.[121]

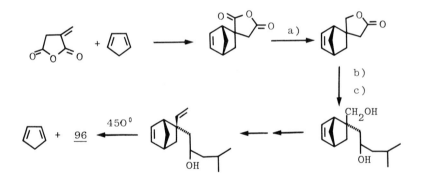

a) NaBH$_4$; b) Bu$_2$AlH; c) i-BuMgBr

Scheme 2

Perhaps the most straightforward Grignard synthesis of all is the reaction of an isobutyl (C$_4$) unit on 3-methylene-4-pentenal (**109**). The problem here is the preparation of the aldehyde, but, based on the Cope reaction of 1,2,6-heptatriene,[128] Skattebøl showed that the vinyl ether (**110**) of 2,3-butadien-1-ol undergoes a Claisen reaction to yield **109**,[129] **110** being available in two steps from 2-butyne-1,4-diol.[130] This reaction has been carried out again with minor variations (and without acknowledgment of Skatteb∮l) by Bertrand and Viala.[131] Clearly 96 and 97 may be directly prepared from 109.

Three methods are available for the oxidation of myrcene (6) that lead to the ipsenol group of alcohols. Myrcene epoxide (111) is generally opened in the base-catalyzed reaction with attack on the proton of the methyl group (leading to 112),[132, 133] but Sharpless has developed a method consisting in the addition of diphenyl diselenide to 111, followed by sodium borohydride and hydrogen peroxide (Scheme 3).[133] Alternatively, photooxygenation of myrcene (6) yields, after reduction of the hydroperoxides, two alcohols 112 and 98; 112 is, however, the major product, and is used as an intermediate in the synthesis of certain furanoid monoterpenes (see below).[134] Mori has converted the alcohol 98 into the allylically rearranged, more valuable pheromone 97 by treatment of the two

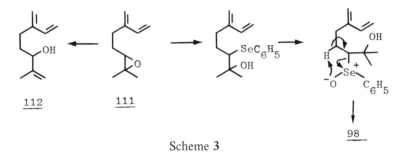

Scheme 3

corresponding chlorides (obtained by the action of thionyl chloride on 98 with silver oxide in acetic acid). The proportion, however, was 3:1 in favor of the acetate of 98, and the alcohols obtained after metal hydride reduction had to be separated by reacetylation, only the secondary alcohol 98 reacting with acetic anhydride in pyridine.[135] Myrcene (6) also adds benzene sulfenyl chloride specifically to the isopropylidene terminus, and the product 113 has been converted to ipsdienol (97) in 25% yield.[136]

Although Skatteb∮l induced some optical asymmetry into his synthesis by using an equimolar amount of (2S, 3S)-(+)-N,N,N',N'-tetramethyl-1,4-diamino-2,3-dimethoxybutane during the Grignard step,[129] the first synthesis of (−)-ipsenol was carried out starting from L-(+)-leucine (114) by Mori (Scheme 4).[137] This was followed by a synthesis, similar in the later stages, of (+)-ipsdienol from 115 (Scheme 5). While these syntheses were very valuable for confirming the absolute configurations of the molecules, it cannot be pretended that they are applicable for large-scale preparation. More likely from the latter aspect, is the preparation (although not a synthesis really, since it starts ultimately from pinene) of both isomers of ipsdienol (97) by Ohloff and Giersch from (+)- and (−)-verbenone (116).[138] Deconjugation of the double bond with sodium

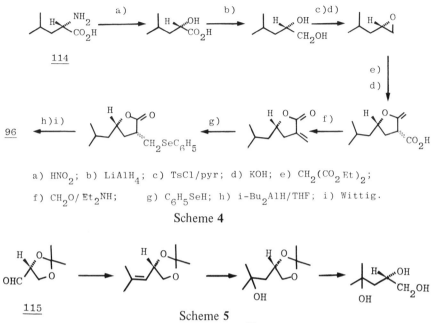

a) HNO_2; b) $LiAlH_4$; c) $TsCl/pyr$; d) KOH; e) $CH_2(CO_2Et)_2$;

f) CH_2O/Et_2NH;   g) C_6H_5SeH; h) i-Bu_2AlH/THF; i) Wittig.

Scheme 4

Scheme 5

borohydride and quenching with boric acid[139] gave the ketones 117, which were reduced to the corresponding alcohols; after flash pyrolysis, these yielded the ipsdienols (98). Each ketone 117 gave two alcohols (*cis* and *trans*), and each of the optically active ipsdienols was thus available from an optically active verbenone (116), thus providing a double proof of configuration.[138]

2,6-Dimethyl-1,5,7-octatrien-3-ol (118)

This alcohol was isolated from *Ledum palustre* without precision of the stereochemistry,[117] although the published data would seem to fit (*E*) stereochemistry better. The alcohol was synthesized by Wilson et al. by treating the anion of 3-methylpentadiene (119), which has been shown by ^{13}C NMR spectrometry to exist almost exclusively in the W-conformation,[140] with methylacrolein, yielding 65% of 118. It would be desirable to clear up the double-bond geometry.

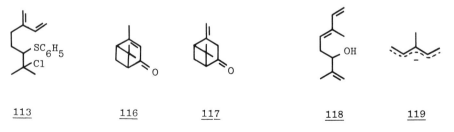

113   116   117   118   119

C. Aldehydes and Ketones

Citral (120)

The conversion of dehydrolinalool (**121**) to citral (**120**) is an important commercial process (Vol. 2 p. 28), and is now carried out using vanadium catalysts,[141] improvements being achieved with catalysts incorporating silicon,[142] a particularly effective one and easy to handle in air, being a polymer formed from sodium orthovanadate and diphenyldichlorosilane.[143] The mechanism of the reaction is believed to be as shown in Scheme 6.[144,145]

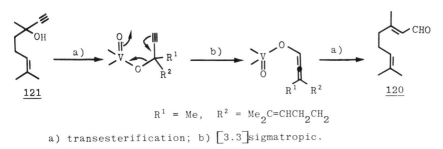

$$R^1 = Me, \quad R^2 = Me_2C=CHCH_2CH_2$$

a) transesterification; b) [3.3] sigmatropic.

Scheme 6

Both prenol (**122**) and 3-methyl-2-butenal (**102**) are available from the reaction between isobutylene and formaldehyde (leading initially to **53**). Their condensation under various conditions, usually in presence of acid,[146,147] leads directly to citral, as does the condensation of the acetal diacetate of dimethyl-acrolein and prenol.[148] The most important industrial preparation of citral (**120**), however, is probably by dehydrogenation of geraniol (**36**), usually over copper catalysts.[149]

On a less industrial level, linalool (**11**) can be converted to citral (**120**) by forming the thiocarbamate **123** using butyllithium followed by thiocarbamyl chloride. Mercuric chloride and calcium carbonate in tetrahydrofuran then give citral in 65% yield by this route, with a majority of the (*E*)-isomer.[150] Takabe's geranylamines[15] undergo conjugation of the double bonds (to **124**) with powerful bases (NaH, KNH$_2$, etc. in ethylenediamine), and in the presence of acid, the latter are hydrolyzed to dihydrocitral (**125**).[151]

Addition of a one-carbon unit to the epoxide **126** can be carried out using the anion of dithiane. This leads to the dithiane **127**, from which citral is readily obtained. The compound **127** was actually made to synthesize pyrans (see below).[152]

8-Hydroxy-2,6-dimethyl-2,6-octadienal

This compound occurs as its 7-hydroxycoumarin ether **128** in *Capnophyllum peregrinum*, and has been synthesized by oxidizing the tetrahydropyranyl ether of linalool (11) with selenium dioxide, and effecting the allyl rearrangement on the product with phosphorus tribromide. The resulting bromide 129 is then readily etherified with 7-hydroxycoumarin.[153]

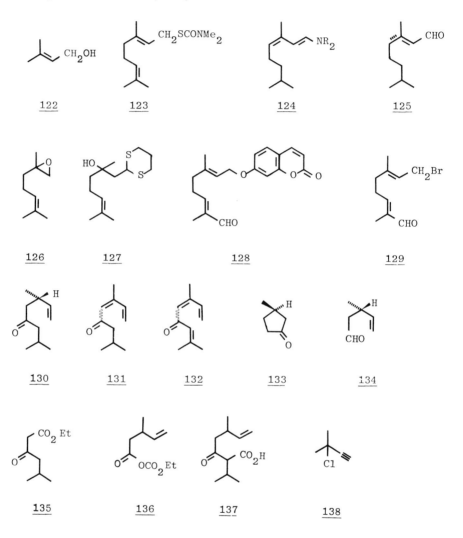

Dihydrotagetone (130), Tagetones (131), Ocimenones (132), and Related Compounds

Apart from the isolation from numerous *Tagetes* sp.[113] (+)-dihydrotagetone (130) constitutes 97% of the essential oil of *Phebalium glandulosum*,[154] an Australian Rutaceae, in which it occurs with high optical purity, compared with that reported from the *Tagetes* sp. This question of optical purity was examined in the course of a synthesis from (+)-3-methylcyclopentanone (133). Irradiation of this gives 20% of 3-methyl-4-pentenal (134) with some racemization, from which (+)-dihydrotagetone (130) was obtained in 83% optical purity by Grignard reaction and oxidation,[155] about the same as the optical purity of the Australian natural product.

(±)-Dihydrotagetone has also been made by a Carroll reaction of ethyl 3-oxo-5-methylhexanoate (135) and 2-butenol.[156] Another method involves treatment of the trimethylsilyl ester of 3-methylbutyric acid with lithium diisopropylamide, a proton is abstracted, and the resulting lithium derivative reacts with the mixed anhydride 136 made from 3-methyl-4-pentenoic acid. The resulting acid 137 is very readily decarboxylated on heating, giving 66% of racemic dihydro-tagetone (130).[157]

Both the (Z)- and (E)-tagetones (131) and ocimenones (=tagetenones, 132) are naturally occurring. A recent synthesis of the former is that of S. Julia, which is actually a $C_5 + C_5$ type.[158] Starting from 3-chloro-3-methylbutyne (138), addition of benzenethiol under phase-transfer conditions gives the allenic sulfide 139, readily isomerized to the thioenol ether, peracid oxidation of which yields the corresponding sulfoxide 140. The latter was allowed to react with the protected cyanohydrin anion (141)[159] derived from 3-methylbutanal, and the resulting derivative 142 yields a mixture of (Z)- and (E)-tagetones (131) after 1 hour in contact with sodium hydroxide. Shorter periods remove the ketone protecting groups, but not the phenylsulfoxide.[158]

In 1971 de Villiers et al. showed that isoprene and the chloride of senecioic acid (55) react in the presence of antimony pentachloride, when a mixture of (Z)- and (E)-ocimenones (132) is obtained.[160] When the reaction was carried out with stannic chloride at –78°, Adams et al. isolated the intermediate chloroketone 143, and they also synthesized the two tagetones (131) by substituting the acid chloride of 3-methylbutyric acid for that of senecioic acid.[161] The latter work is important also for the synthesis of filifolone (144) (see below) from ocimenone. Indeed, the relationship between chrysanthenone (145), filifolone (144), and ocimenone (132) is well known, since it has been shown in 1958 that 132 is obtained from chrysanthenone (145) by the action of water at 150°.[162]

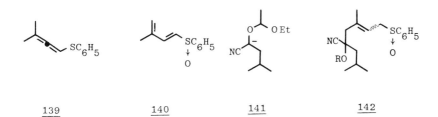

139   140   141   142

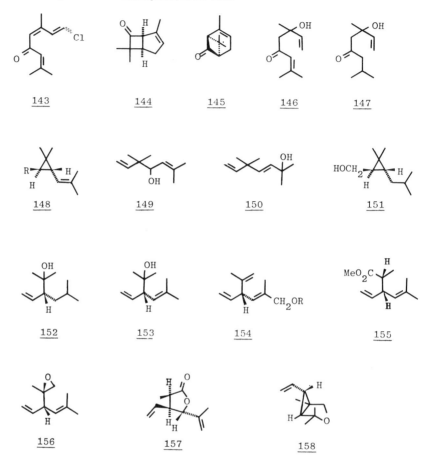

143   144   145   146   147

148   149   150   151

152   153   154   155

156   157   158

Another $C_5 + C_5$ synthesis arose from the intermediate **105** of Garbers and Scott's synthesis of ipsdienol. When this intermediate acetate was treated first with triethylamine in tetrahydrofuran, then with strong base, the main products were the ocimenones (**132**).[123]

A $C_6 + C_4$ route consists in the addition of methyl vinyl ketone to the anion of mesityl oxide (prepared with lithium diisopropylamide in tetrahydrofuran at $-78°$). The resulting hydroxyocimenone (**146**) can be dehydrated with acid to a mixture of the ocimenones.[163] Tagetone (**131**) was made by the same route but starting from 4-methyl-2-pentanone instead of mesityl oxide.[163] Vig et al. have also prepared the tagetonol (**147**) from 4-methyl-2-pentanone by a somewhat longer route.[164] This tagetonol (the (+)-enantiomer) is a constituent of *Cinnamomum camphora*.[165]

5. SUBSTANCES RELATED TO CHRYSANTHEMIC ACID

The relationship between natural *trans*-chrysanthemic acid (148 R = CO_2H) and the santolinyl, artemisyl, and lavandulyl skeletons was described in Vol. 2, p. 34, and the biogenetic implications have been reviewed by Epstein and Poulter[166] (see also the first comment that the lavandulyl skeleton does not necessarily have to arise after a chrysanthemyl precursor, while the other two must pass through chrysanthemyl[167]). The solvolysis of the chrysanthemyl carbenium ion leads mainly to artemisia alcohol (149) or its allyl isomer yomogi alcohol (150), depending on the conditions, with only very little of the substances derived from the santolina skeleton.[168] Dihydrochrysanthemyl alcohol (151) can, however, be converted to a dihydrosantolinyl alcohol (152),[169] a conversion that established the relationship of the absolute configuration between chrysanthemic acid (148) and santolina alcohol (153), isolated from *Ormenis multicaulis*.[170] Other conversions of chrysanthemyl derivatives to lavandulyl[171] and santolinyl[172] compounds are published, but most of these cannot really be classed as "syntheses," with the exception of a recent synthesis of natural (*S*)-lyratol (154, R = H) from (-)-148 (R = CO_2H) (see below).

A. The Santolinyl Skeleton

New natural products in this group have been isolated from sagebrush, *Artemisia tridentata*, methyl santolinate (155)[173] and the epoxide 156, in which the (3*R*) chirality is contrary to that expected if (1*R*, 3*R*)-chrysanthemyl pyrophosphate (151-pyrophosphate) is precursor, although from another location a diastereoisomeric mixture was isolated.[174] Four stereoisomers of the lactone (santolinolides, 157) have also been found in *A. tridentata*.[175] The epoxide 156 is converted by formic acid in 15 minutes at room temperature to the ether arthole* (158), also a constituent of *Artemisia tridentata*.[176] (Only one of the epoxides 156 gives arthole with Lewis acids; the other diastereoisomer rearranges to the corresponding aldehyde.) "Achillene," isolated from *Achillea filipendulina*,[41] (see comments, p. 457) is the same as santolinatriene (27). The full paper of

---

*The history of this compound is confused. It seems to have been first isolated in 1973,[177] and again isolated, with an incorrect structure assignment, in 1977,[178] although the correct structure was published (with the name arthole and a mass spectrum, but no other commentary) in 1976[179] and appears in *Chemical Abstracts* under this name and with the correct structure in 1976. In 1977 the same group that had used the name arthole published the reasoning that led to the correct structure 158, but proposing the name "artemeseole," without even referring to their earlier name.[176] Meanwhile, Banthorpe and Christon had synthesized one of the earlier incorrect structures, and supported the structure 158.[180] It would be preferable to keep the original name, arthole, particularly as this has penetrated to sciences beyond the strictly chemical context.[181]

syntheses in the artemisyl-santolinyl group described in Vol. 2, p. 38 has appeared.[182]

Methyl santolinate has been synthesized from 5-methyl-2,4-hexadienol (159, R = H) readily available from senecia aldehyde (102). The propionate 159, (R = COEt) was converted to its lithium enolate with lithium isopropylcyclohexylamide, then to the enol silyl ether 159, (R = MeCH=C−OSiMe₂−t-Bu). The latter rearranged after 3 hours at 65° to a mixture of the silyl esters 160a and 160b, which were converted by hydrolysis with acetic acid and methylation with diazomethane to the natural sagebrush ester 155 and its isomer 161.[183]

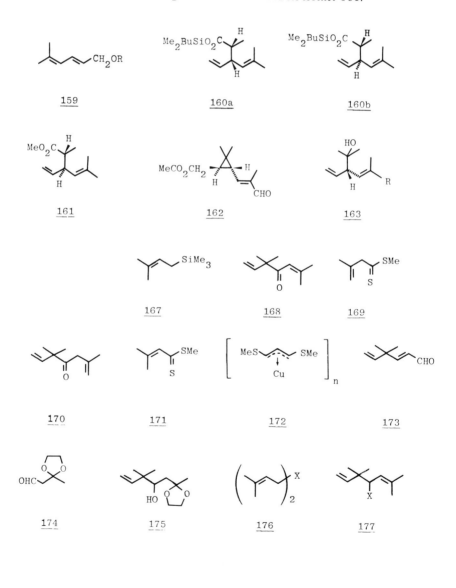

Sucrow's original synthesis of lyratol (**154**, R = H, Vol. 2, p. 39) produced the racemate, now Gaughan and Poulter have synthesized the natural (*S*)-isomer from chrysanthemic acid. The latter was reduced to the (1*R*, 3*R*)-alcohol **148** (R = CH_2OH), and the acetate was oxidized with selenium dioxide to **162**. Treatment of the mesylate corresponding to the acetate **162** with sodium bicarbonate in aqueous acetone yielded the ring-opened unstable hydroxy-aldehyde **163** (R = CHO), which was immediately reduced with sodium borohydride to the diol **163** (R = CH_2OH). Dehydration of the acetate **163**, (R = CH_2OAc) with thionyl chloride in pyridine gave lyratyl acetate (**154**, R = Ac).[184]

Some years ago, a lactone **164** with the santolinyl skeleton was isolated from *Chrysanthemum flosculosum*,[185] and this has now been synthesized (Scheme 7).[186]

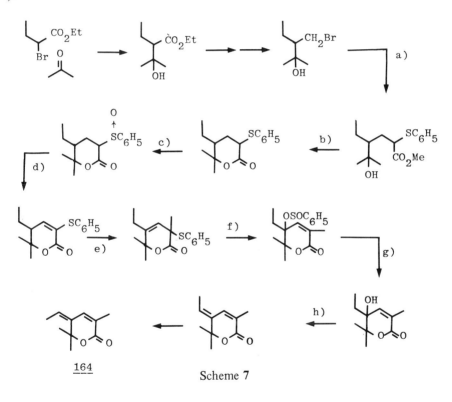

<u>164</u>   Scheme 7

a) $C_6H_5\overset{-}{S}CHCO_2MeNa^+$; b) TsOH; c) $NaIO_4$; d) $(CF_3CO)_2$);
e) t-BuO$^-$/MeI; f) m-Cl-perbenzoic acid; |2,3|sigmatropic, then oxidation; g) NaOMe; h) dehydration.

B. The Artemisyl Skeleton

The aldehyde 165 has been isolated from verbena[187] and neroli[188] oils. This aldehyde was synthesized by Corbier and Teisseire[188] by a modification of a reaction described in Vol. 2, p. 44, but before being identified as a natural product, it had already been prepared by Re and Schinz.[188a] Condensation of prenyl bromide (13) with methylacetoacetic ester gave a product 166, which already had the desired skeleton, and conventional steps led to the aldehyde 165.

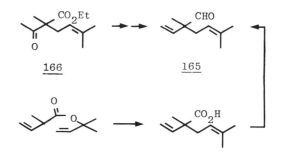

The most straightforward route to the skeleton is by coupling of two prenyl units (Vol. 2, p. 42). The Grignard reagent from prenyl bromide (13) generally reacts in the allylically rearranged form (Vol. 2; see also Ref. 189), but several improvements have been suggested to effect this reaction. The Grignard reagent can be converted with trimethylsilyl chloride to the silane 167, and the latter reacts with senecioic acid (55) chloride to yield artemisia ketone (168).[190] The prenyl Grignard reagent also reacts with the thioester 169 to yield the dithioketal of isoartemisia ketone (170), into which it is converted by silver nitrate solution and cadmium carbonate. Base treatment of isoartemisia ketone yields the conjugated ketone 168, which can be obtained by the same route (though in poorer yield) from the isomeric thioester 171.[191] Prenyl bromide and senecia aldehyde (102) give a 91% yield of artemisia alcohol (149) when passed over a heated zinc column,[192] an alternative way of carrying out this coupling is using chromous chloride.[193] It should not be forgotten, however, that artemisia alcohol (149, like yomogi alcohol, 150) is not a good intermediate for the ketone because the oxidation proceeds so badly. To obtain yomogi alcohol (150) from prenyl bromide (13) via an organometallic route is a little longer; addition to a three-carbon unit (in the form of the copper complex 172) leads to the aldehyde 173,[194] or addition (using chromous chloride) to the C_4 unit 174 leads to the protected ketone 175, which can be dehydrated and deacetalized with polyphosphoric acid. From this product[193] or the aldehyde 173, yomogi alcohol is obtainable by Grignard reaction (two stages in the case of 173).

New developments in the [2,3] sigmatropic approach to the skeleton (Vol. 2, p. 41) are the use of a chiral base to effect the rearrangement of the sulfonium salt **176** (X = $\overset{+}{S}$Me) to the thioether (**177**, X = SMe), when only 12% optical yield was achieved,[195] and the rearrangement of the ammonium salt **176** (X = $\overset{+}{N}$Me$_2$) to the amine **177** (X = NMe$_2$).[196] A series of papers by S. Julia's group has shown how a similar reaction may be carried out by forming what is probably the allenic zwitterion **178** from prenyl thioether (**179**, R = Me) and 3-chloro-3-methylbutyne (**138**). When **138** is employed as its lithium salt, the allene **180** is the product isolated, which can be converted into artemisia ketone (**168**) by hydrolysis (HgCl$_2$). Alternatively use of sodium hydroxide in a protonating solvent leads to the thioether **177** (X = SMe).[197] A route involving the use of the formaldehyde dithioacetal (**181**) has also been developed.[198] (These papers should be compared with Raphael's allenic carbene synthesis of chrysanthemyl alcohol described below.)

An original synthetic route by Franck-Neumann and Lohmann consists in building a prenyl group into a pyrazole by the reaction between the tosylthioacetylene (**182**) and dimethyldiazomethane. The product is a 41:1 mixture of two pyrazoles (**183a** and **183b**), irradiation of which in the presence of the prenyl thioether (**179**, R = Et) gave a mixture of thioacetals from which artemisia ketone (**168**) could be obtained by the action of silver nitrate. The mechanism conceivably involves reaction of a vinyl carbene, Me$_2$C=C$\overset{\cdot\cdot}{C}$–STs, with the prenyl thioether.[199]

Using the ester **184** (obtained by dehydration of the alcohol corresponding to ethyl dimethylacetoacetate), Vig et al. prepared the corresponding β-ketoester, which they converted to artemisia ketone (**168**) by means of a Grignard reaction and dehydration.[200]

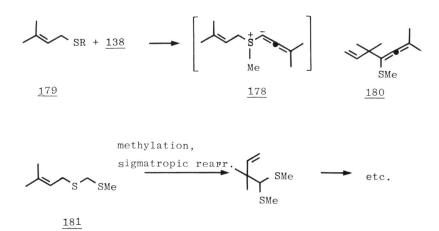

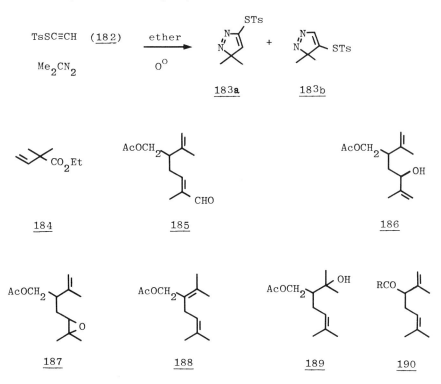

183a   183b

184   185   186

187   188   189   190

If one is fortunate enough to have a supply of 3,3-dimethylpent-1-en-4-yne,[201] it is a simple matter to convert it to yomogi alcohol (150) by addition of acetone to the lithium derivative, followed by partial reduction (lithium aluminum hydride) of the triple bond.[202]

C. The Lavandulyl Skeleton

The aldehyde 185 has been identified in lavendin,[203] but not yet synthesized. Photooxygenation of lavandulyl acetate (57, R = OAc) leads to a hydroxy-acetate (186), which occurs in lavender,[204] together with a number of related compounds.[205] Epoxylavandulyl acetate (187) occurs only in lavendin, not lavender.[205]

The direct coupling of two molecules of dimethylallyl acetate (=prenyl acetate, 122-acetate) to give lavandulyl acetate (57, R = OAc) has been in the literature for some time (see, e.g., Ref. 206), together with related reactions. Recently M. Julia et al. have studied the reaction of dimethylallyl acetate and lithium perchlorate systematically, obtaining good yields of lavandulyl acetate (57, R = OAc) and isolavandulyl acetate (188)[207], and have drawn attention to the synthesis of lavandulyl acetate using dimethylallyl acetate and isoprene,[208] where they feel that the reaction does not depend on the presence of the latter. Julia et al. were also able to isolate a hydrated lavandulyl acetate **189** from the reaction between dimethylallyl acetate and 3-methyl-1-buten-3-ol in the presence of trifluoroacetic acid.[207] The enolates of the dialkylamides of senecioic acid (**55**) react with prenyl halides to yield the amides of lavandulic acid [**190**, R = N(alk)$_2$], which can be readily converted to lavandulol (**57**, R = OH).[209] This reaction is somewhat reminiscent of the addition of prenyl halides to the carbanion **191** from the *t*-butylimine of 3-methylbutanal, leading to the imine **192** (R = C=N−*t*-Bu) of dihydrolavandulal (**192**, R = CHO).[210] Takabe et al. have attempted to carry out Lewis acid-catalyzed condensations of thio analogs of dimethylallyl acetate, but the yields of thiolavandulyl and thioisolavandulyl acetates were low, the main product being the sulfide **176**. (X = S).[211]

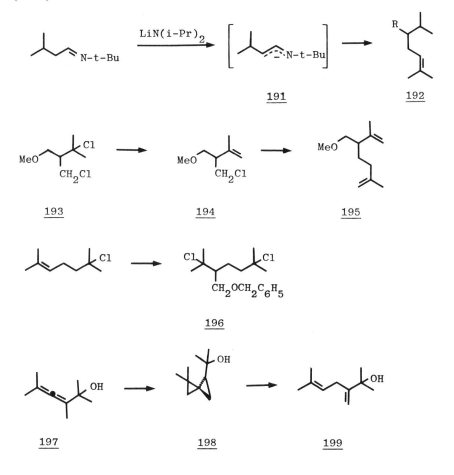

191   192

193   194   195

196

197   198   199

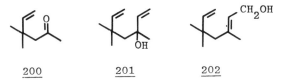

200   201   202

Addition of lithium dimethylcuprate to the allenic ester, ethyl 2,3-buta-dienoate gives the lithium enolate of ethyl 3-methyl-3-butenoate, which, on quenching with prenyl bromide in dimethoxyethane at $-30°$, gives the ethyl ester of lavandulic acid (190, R = OEt), but the yields reported are not above 50%.[212]

A C_5 + C_1 + C_4 route has been investigated. It consists in the addition of chloromethyl methyl ether to 3-chloro-3-methyl-1-butene (this is the compound first obtained by HCl addition to isoprene, which rearranges to prenyl chloride during workup unless precautions are taken) yielding the dichloride 193. Loss of HCl leaves a primary chloride 194 from which a Wurtz reaction using iso-butylene chloride gives the methyl ether 195.[213]

The Prins approach (C_9 + C_1, Vol. 2, p. 46) has been improved,[214] and another C_9 + C_1 route, starting from 2-chloro-2,6-dimethyl-5-heptene has been used by Garbers et al. Addition of benzyl chloromethyl ether yields the dichloro-ether 196, the benzyl group being replaceable by acetate with acetic anhydride and boron trifluoride etherate.[215] Very different is the C_9 + C_1 synthesis of Maurin and Bertrand. Addition of carbene to the allene 197 gave the spiro-alcohol 198, acid treatment of which led to the alcohol 199. This was converted to the known lavandulyl bromide (57, R = Br) with phosphorus tribromide, from which isolavandulyl acetate (188) is readily obtained.[216]

One of the by-products from the methylheptenone synthesis starting from acetone and acetylene (Vol. 2, p. 5-6) is 4,4-dimethyl-5-hexen-2-one (200). This can be treated with vinylmagnesium bromide to yield the divinyl alcohol 201, also obtainable from mesityl oxide by a copper-catalyzed double Grignard reaction. Rearrangement of 201 to the primary alcohol 202 or its acetate is straightforward, and both 202 and its acetate undergo a thermal Cope reaction to give, respectively, lavandulol (57, R = OH) or its acetate.[217]

D. Chrysanthemic Acid and Related Substances

It is not in the scope of this chapter to list the vast literature concerning the synthesis of putative insecticidal homologs and analogs of chrysanthemic acid. The original synthesis of chrysanthemic acid[218] by reaction of diazoacetates on 2,5-dimethyl-2,4-hexadiene has been improved by using asymmetric copper complexes as catalysts, when optically active chrysanthemates are obtained.[219] A related synthesis of chrysanthemolactone (203) from 204 in 34% optical yield using a chiral copper complex has been published.[220]

Replacement of Cu catalysis by Rh(II) is reported to give a higher *cis:trans* ratio of cyclopropane carboxylic esters.[221] Poulter et al. have prepared a homolog (without the *gem*-dimethyl group on the cyclopropane ring) by the original diazoacetate route, but resolving the racemic acid with quinine, this lower acid being proposed as an intermediate in the biosynthesis.[222] Another method for functionalizing 2,5-dimethyl-2,4-hexadiene is by treatment with methyl acetoacetate in the presence of manganese(III) acetate and acetic anhydride. The product **205** (R = Me) yields the cyclopropane **206** on irradiation (the reaction works more smoothly when R = Cl), from which methyl chrysanthemates [(±)-**148**, R = CO_2Me] can be obtained (when R = Me) with potassium *t*-butoxide.[223]

The synthesis of Martel and Huynh[61] described in Vol. 2, p. 53 has been adapted for an undergraduate experiment.[224]

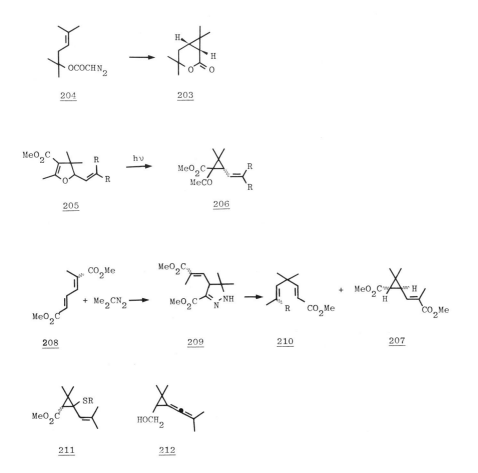

213   214   215

An older synthesis of chrysanthemum dicarboxylic acid methyl ester (=pyrethric acid, 207) was not described in Vol. 2;[225] perhaps this is not too serious, since Scharf and Mattay have reexamined it, after the necessary starting material, dimethyl 2-methyl-2,4-hexadienedicarboxylate (208) became more readily available.[226] Reaction with diazoisopropane gives the pyrazoline 209 (different from the pyrazoline reported earlier[225]), thermolysis of which leads mainly to the esters 210 (R = CO₂Me), the chrysanthemyl skeleton forming only 11-17% of the products.[226]

It is possible to pass from 210 (R = Me) to chrysanthemic acid by irradiation,[227, 228] but there are several by-products, and the starting material must be made.[228] The (E)-isomer (but not the (Z)-isomer) of 210 (R = Me) gives methyl chrysanthemate ((±)-148, R = CO₂Me).[227] Despite the relatively modest yield, Franck-Neumann has extended his photochemical synthesis involving the pyrazole 183a to lead to the thiolated chrysanthemic esters 211, obtained when 183a is irradiated in the presence of methyl 3-methyl-2-butenoate (55, methyl ester). Raney nickel quantitatively removes the thiol group, and an interesting feature of the synthesis is that the cis-chrysanthemate always predominates, whichever isomer of 211 is used.[229] This synthesis probably involves a vinyl-carbene (see above), and carbene addition is also proposed in the synthesis of Raphael et al. of chrysanthemyl alcohol ((±)-148, R = CH₂OH) by treatment of chlorobutyne (138) and prenol (122) with potassium t-butoxide, when the allene 212 is obtained. This allene can be reduced with sodium in liquid ammonia to trans-chrysanthemyl alcohol ((±)-148, R = CH₂OH).[230]

One approach to the chrysanthemate skeleton discussed in Vol. 2 (p. 52) is by base treatment of compounds such as 213 (where X may be OCOR and Y, CN or CO₂R) and 214 (where X is frequently halogen). Most of these routes are actually attempts to prepare the starting material (i.e., 213 or 214) more easily, and can be illustrated by the synthesis of Ficini and d'Angelo. They reduced the ketone 215 (made by the Lewis acid-catalyzed addition of isobutyryl chloride to isobutylene then HCl elimination) to the corresponding alcohol, which reacts with ethyl orthoacetate in the presence of acid to give 70% yield of 214 (X = H, Y = CO₂Et). t-Butyl phenylperacetate then gives 214 (X = OCOC₆H₅, Y = CO₂Et), which cyclizes under the influence of lithium diisopropylamide to a 1:1 mixture of cis- and trans-ethyl chrysanthemates.[231] Very similar is the synthesis of Kondo et al., except that in order to introduce the function X they treat 214 (X = H, Y = CO₂R) with N-bromosuccinimide. Actually, this work was intended as a synthesis for analogs of chrysanthemates rather than the chrysanthemates themselves.[232] The syntheses passing through pyrocine (216, the pyrolysis product of chrysanthemic acid) also involve a 1,3-elimination by base. For example, Takeda et al. treated the lactone 217 (R = OC_6H_5) with 2-methyl-1-propenylmagnesium bromide in a copper-catalyzed addition, and from the product 218 (R = OC_6H_5), obtained pyrocine (216) by hydrolysis. This was then treated with thionyl chloride and ethanol to give the chloroester 219, from which ethyl *trans*-chrysanthemate (($\pm$)-148, R = CO_2Et) is obtained by treatment with lithium diisopropylamide in hexamethylphosphoramide.[233] Three other syntheses of pyrocine (216) are relatively straightforward, particularly that of De Vos and Krief. They treated the monoepoxide (220) of 2,5-dimethyl-2,4-hexadiene with diethylmalonate carbanion, obtaining 218 (R = OEt), later steps being as before.[234] From a mixture of isobutanal and dimethyl malonate with N-bromosuccinimide, the bromoester 221 can be obtained. Torii et al. prepared the methyl ester 217 (R = OMe) by pyrolysis of 221 (or its chloro analog), following the same path afterwards, although the hydrolysis of 218 (R = OMe) to 216 was carried out by the unusual method of sodium chloride in wet dimethyl sulfoxide at 170°,[235] 2,2,5,5-Tetramethyltetrahydrofuran-3-one has also been converted into pyrocine (216).[236] The Horner-Wittig reaction yields a mixture of two unsaturated esters 222, 223, reduction and hydrolysis of which lead to the acid 224. Action of boron trifluoride etherate then leads to pyrocine. Alternatively, thionyl chloride gives with the ester of 224 a dichloride (corresponding to 219 + HCl), the conversion of which to chrysanthemic acid is known.[236] A related paper deals with the asymmetric hydrogenation of 222 + 223.[237]

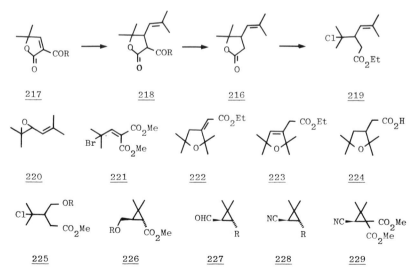

217   218   216   219

220   221   222   223   224

225   226   227   228   229

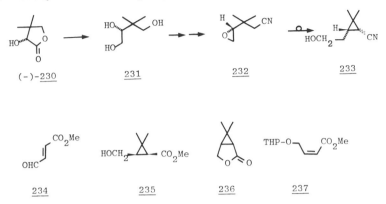

Garbers et al. constructed chloroesters **225**, corresponding to **219** but without the isopropylidene group, by reaction of methyl 4-methyl-3-pentenoate and chloromethyl alkyl ethers in the presence of a Lewis acid. The chloroesters **225** can be cyclized by base, and the product **226** was converted into chrysanthemic acid via the aldehyde **227** (R = CO_2Me).[238]

The ready availability of **221** induced De Vos and Krief to develop a one-pot synthesis of the nitrile **228** (R = CO_2H) from it using sodium cyanide in dimethyl sulfoxide.[239] The intermediate diester **229** was known to be decarboxylated by sodium cyanide in dimethyl sulfoxide, but the others of this work[240] seem to have the wrong configuration for the product **228** (R = CO_2H). Using borane-dimethyl sulfide, **228** (R = CO_2H) can be converted to the alcohol **228** (R = CH_2OH), then oxidized to the aldehyde **228** (R = CHO), from which chrysanthemic nitrile can be obtained by a Wittig reaction.[239] The optically active nitrile **228** (R = CH_2OH, opposite isomer to that shown) was used to make (+)-*trans*-chrysanthemic acid (+)-(**148**). The starting material was optically active (–)-hydroxylactone **230**. Lithium aluminum hydride reduction gave the triol **231**, conventional steps then leading to the epoxide **232**, from which the nitriles **233** (position carrying the CH_2OH group defined) were obtained. (+)-Chrysanthemic acid was then made by known steps from **233**.[241]

Sorbic esters can be selectively ozonized, and the aldehyde **234** thus obtained in yields reported to be over 80% reacts with 2.4 equivalents of isopropylidene-triphenylphosphorane to give methyl *trans*-chrysanthemate in over 60% yield. Alternatively, protection of the aldehyde group as an acetal enables preparation of **227** (R = CO_2Me) acetal, from which both chrysanthemate and pyrethric acid esters are accessible.[242] Another series of syntheses involve the *cis*-substituted cyclopropane **235**. This was prepared by reaction of isopropylidenediphenyl-sulfurane on various substrates: butenolide gives **236**, and **237** the tetrahydropyranyl ether of **235**. A second reaction of isopropylidenediphenylsulfurane on the aldehyde **239** leads to methyl *cis*-chrysanthemate.[243]

The original idea of adding both substituents to dimethylcyclopropene in a one-pot process was carried out by Lehmkuhl and Mehler by adding it to 2-methyl-2-propenylmagnesium bromide (Scheme 8), then carbonating the resulting Grignard compound. The ethyl *cis*-isochrysanthemate thereby obtained can be converted to chrysanthemolactone (**203**) by acid or to *cis*- or *trans*-chrysanthemates by other known methods.[244]

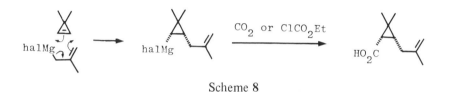

Scheme 8

The conversion of other monoterpenes to chrysanthemic acids has also received attention, particularly since this type of synthesis contains a built-in chiral center in most cases. One that does not contain such a center starts from eucarvone (**238**), methylation of which followed by treatment with osmium tetroxide and sodium chlorate yields the methylcarone (**239**). Ozonolysis of **239** gives the *cis*-chrysanthemic skeleton in the form of the acetal **240**, which can be converted to chrysanthemic acids by conventional techniques.[245] Apart from the cost of the steps in this synthesis, eucarvone is usually prepared from compounds having a chiral center (like carvone), and it is unsatisfactory to destroy this when the product is to be a chiral molecule. Improvements to the known methods are available from (+)-3-carene (**241**) (particularly those of Sukh Dev[246] to the method beginning with oxonolysis[247]), and a new method has been described by Cocker et al. In the course of an investigation on the Prins reaction of (+)-3-carene they prepared **242**,[248] which was converted in four steps to the ketoester **243**, from which methyl (−)-*cis*-chrysanthemate (**244**) is obtained by Grignard reaction and dehydration.[249]

Mitra and Khanra have carried out a synthesis of methyl *trans*-chrysanthemate from α-pinene (**245**). The first stages (ozonolysis, Baeyer-Villiger reaction and esterification with diazomethane) lead to the ester **246**, and after Grignard reaction, oxidation to the cyclobutanone, and bromination, the bromoketone **247** undergoes ring-contraction on treatment with sodium methoxide in dry ether, to **248**.[250]

Some optically active chrysanthemic acid derivatives have been synthesized from a pyranoside.[250a]

Although it does not lead to a natural chrysanthemic acid, attention is drawn to another Favorskii ring contraction, that of the chlorocyclobutanone **249** (made by a very ingenious and rapid method from acrylic acid), which yields exclusively the *cis*-acid **250** with potassium hydroxide at 90°.[251]

Tritiated *trans*-chrysanthemate has been synthesized from (−)-*cis*-chrysanthemate.[252]

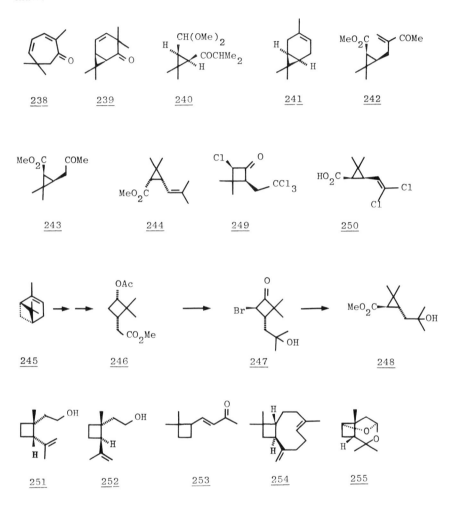

6. CYCLOBUTANE MONOTERPENES

Besides grandisol (251; see Vol. 2, p. 58) and its stereoisomer fragranol (252, isolated from *Artemisia fragrens*),[253] junionone (253) has been isolated from the oil of *Juniperus communis*.[254] Although junionone (253) has been prepared by degradation of caryophyllene (254),[254] there is no total synthesis so far recorded. An acetal, lineatin (255) has been isolated from the frass of the

Douglas fir beetle, *Trypodendron lineatum*,[255] and its synthesis (see below) has confirmed the structure.

Because of the commercial interest in grandisol as part of the sex hormone system of the male boll weevil (*Anthonomus grandis*), considerable effort has been devoted to its synthesis, and reviews have already appeared.[256]

The original synthesis[257] of grandisol (251) had as its key step the photochemical addition of ethylene to 4-methyldihydropyranone to yield 256[258], and this is also the principle of a more recent synthesis which used slightly different steps in the later stages.[259] Cargill and Wright also used a photoproduct 257 (R = Me) in order to form the cyclobutane (Scheme 9). Grignard reaction on the isopropylidene ketone then gave an alcohol 258, which was ozonolyzed to 259. This could then be transformed conventionally to grandisol (251).[260] The same approach was used by Mori to make both optically active forms of grandisol, except that he had to use an intermediate that would permit resolution. In

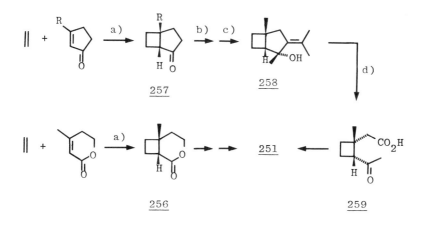

a) hν; b) acetone–base; c) MeMgI; d) O₃

Scheme 9

fact he used the ester (R = CO₂Et), resolving the acids after the photochemical stage and hydrolysis 257 (R = CO₂H).[261] With optically active 257 (R = CO₂H) he then converted the acids to optically active 257 (R = Me) and followed Cargill and Wright's route to make both (+)-grandisol, the naturally occurring one (251) and (–)-grandisol,[261] which also turned out to be biologically active.[262] Mori has also adapted this approach to synthesize lineatin (255), obtaining the isomer 260 at the same time (see Scheme 10),[263] which had originally been considered as an alternative structure to 255.[255] Full details have not been included in Scheme 10, but it should be noted that the final steps unfortunately gave poor yields.

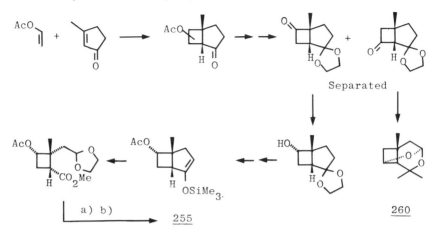

a) MeMgI; b) HCl

Scheme **10**

The photoisomer **261** of eucarvone (**238**) has the advantage over **257** (R = Me) that all the carbon atoms of grandisol are present in the correct position. Catalytic reduction, nitrosation, and Wolff-Kishner reduction give the oxime **262** (R = NOH), and PCl$_5$ in lutidine yields the nitrile **263** which is readily converted into grandisol.[264] Wenkert's synthesis also uses this ketone **262** (R = O) and its oxime; however, the route for the preparation of the ketone is not photochemical, but starts from **264**[265] as shown in Scheme 11.[266]

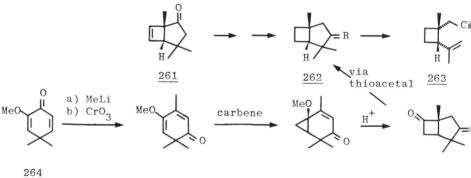

Scheme **11**

Three schemes for the synthesis of grandisol (**251**) start from isoprene. One is photochemical, and consists in irradiating a mixture of isoprene and methyl vinyl ketone. Two cyclobutanes **265a** and **265b** are obtained, and conventional chemistry is used to transform the *cis*-isomer **265a** into grandisol (**251**), and the *trans*-isomer **265b** into the other isomer, fragranol (**252**), unknown as a natural product at the time the work was carried out.[267] The other route from isoprene involves dimerization on a catalyst of bis(1,5-cyclooctadiene)nickel and tris(2-biphenylyl)phosphite, when **266** (R = CH=CH$_2$) is formed in 12-15% yield, together with many other products, and can be converted to grandisol (**251**) by hydroboration-oxidation.[268] Isoprene reacts with the acid chloride **267**

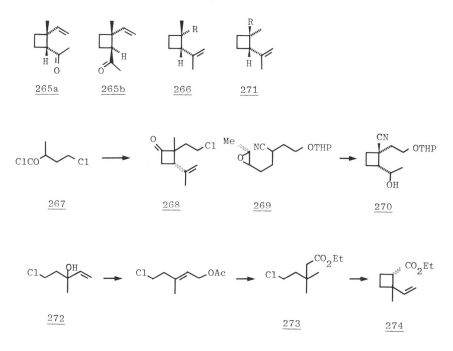

(obtained from 3-methylbutanolide and thionyl chloride) in triethylamine solution in an autoclave to yield the carbon skeleton of grandisol, but it was not found possible to convert the product **268** to the alcohol.[269]

If four- or five-membered rings are alternative cyclization products of epoxynitriles, Stork has found that the four-membered ring is always preferred, and synthesized grandisol from **269** by this principle. The cyclobutane thus obtained **270** (from **269** and hexamethyldisilazane) was converted to grandisol by conventional means.[270]

Cyclobutanecarboxylic acid is commercially available (although expensive), and, by a somewhat tedious route, involving bromination-dehydrobromination to cyclobutenecarboxylic acid; this is then alkylated, first with isopropenyl-magnesium bromide in a copper-catalyzed reaction at −70°, then with methyl iodide and lithium diisopropylamide to yield a mixture of the esters **266** and **271** (R = CO$_2$Me). The (desired) acid **266** (R = CO$_2$H) was converted via the diazoketone to the ester **266** (R = CH_2CO_2Me); the acid chloride **266** (R = COCl) cyclized to a lactone. From **266** (R = CH_2CO_2Me), grandisol (**251**) is easily prepared.[271]

1-Chloro-3-butanone, a methyl vinyl ketone equivalent, is readily obtainable, and reacts with vinylmagnesium chloride to yield the chloroalcohol **272**. Acetylation proceeds with allyl rearrangement under acid conditions, then hydrolysis and reaction of the alcohol under Claisen conditions with ethyl orthoacetate yields the key intermediate **273**. The latter compound cyclizes to a mixture of diastereomers **274** under basic conditions. The subsequent transformations to grandisol and fragranol are conventional, but still somewhat lengthy. A full paper on this synthesis, particularly the key cyclization step is promised.[272]

Trost has used his method of cyclobutanone formation (1-lithiocyclopropyl phenyl sulfide, then *p*-toluenesulfonic acid at reflux) to make a grandisol-fragranol mixture (**251** + **252** = 5:1, Scheme 12).[273]

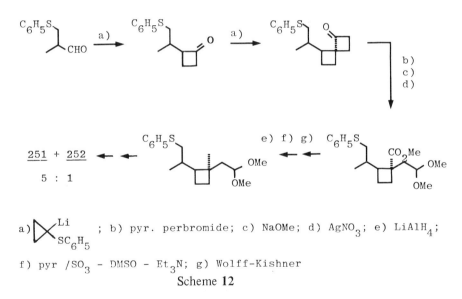

a) ⟨diagram⟩ SC_6H_5 ; b) pyr. perbromide; c) NaOMe; d) $AgNO_3$; e) $LiAlH_4$;

f) pyr $/SO_3$ – DMSO – Et_3N; g) Wolff-Kishner

Scheme 12

Starting from other monoterpenoids is always an attractive means of access to the more difficult ones; undoubtedly the quickest route on paper for the transformation of readily available monoterpenoids into grandisol and fragranol is that of Rautenstrauch, who found that treatment of the *t*-butyl sulfide corresponding to geraniol epoxide (i.e., **275**) with butyllithium in *N,N,N',N'*-tetramethylethylenediamine followed by hydrolysis led to the *t*-butyl sulfides **276**, related to grandisol and fragranol. Unfortunately this idea remained there, for Rautenstrauch found no method of converting **276** to the desired products.[274]

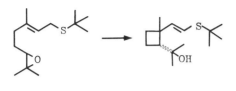

275   276

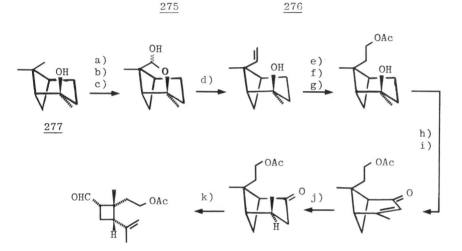

277

a) Nitrite ester, hν; b) Δ; c) Et$_2$O-acetone-2% HCl; d) Wittig;
e) hydroboration; f) H$_2$O$_2$; g) Ac$_2$O/pyr; h) POCl$_3$-pyr;
i) CrO$_3$-pyr-CH$_2$Cl$_2$; j) H$_2$/Pd-C; k) hν.

Scheme **13**

Magnus's synthesis of (+)-grandisol, despite its length, is of great importance, in that it was the first synthesis of the optically active substance from a clearly defined starting material (*trans*-pinan-2-ol, **277**, derived from β-pinene). It is shown in Scheme 13,[275] and it is noteworthy that Mori's synthesis, described above,[261,263] arrived at chiral isomers with rotations comparable with the earlier ones of Magnus.

7. CYCLOPENTANE MONOTERPENES

Campholenic derivatives (**278**, R = CO$_2$Me, R = CH$_2$CO$_2$Me, **279**) have been found in *Cistus ladaniferus*[276] and *Aden olibanum*,[277] but there is no novelty about the synthesis. There are a number of cyclopentane monoterpenoids occurring as minor components of tobacco, 280[278] and 281,[279] for example, and one, 282 (R = CO_2H),[280] that is of somewhat more interest in view of the use of the corresponding aldehyde 282 (R = CHO) in syntheses of higher molecular weight compounds. The synthesis consists in ring-opening the glycol 283 derived from menth-1-ene (the (-)-(S)-isomer (284) is shown), yielding the keto-aldehyde 285.[281] The unsaturated ketoaldehyde (derived similarly from limonene) had been shown earlier[282] to cyclize to a ketone with potassium hydroxide; alternatively, using piperidine acetate[283] an unsaturated aldehyde was obtained. Now 282 has been prepared in both (R)-[284] and (S)-[281] forms, and it can be oxidized (silver oxide) to the acid.[280]

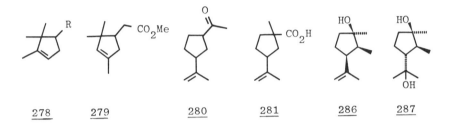

278   279   280   281   286   287

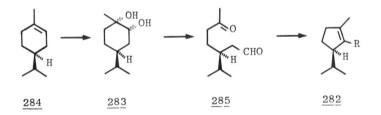

284   283   285   282

The thermal cyclization of linalool (11, Vol. 2, p. 61) yields plinol C (the cis, cis-isomer, 286) as the major product. Now it has been shown that the cyclization of linalool (11) with mercuric salts favors formation of the cis,trans-diol 287, corresponding to plinol A.[285]

Iridoids

It is no longer possible to give a list of the iridoids isolated from natural sources because of the enormous number described in recent years.[286] This section will be restricted to recent syntheses of simple iridoids, proceeding to loganin and some related secoiridoids.

The acetals tetrahydroanhydroaucubigenone (288) and tetrahydroanhydro-aucubigenin (the β-alcohol corresponding to 288), known for many years as being related to aucubenin (289), noriridoids that are widely distributed in the

Scrophulariaceae), have recently been synthesized as racemates for the first time.[287] The starting material **290** is obtained from Dieckmann condensation of diethyl tetrahydrofuran-*cis*-3,4-dicarboxylate,[288] and the key step is the irradiation of the nitroso compound **291** (X = NO) obtained from the alcohol **291** (X = H), which exists in equilibrium with its cyclic acetal **291a**.[287] Metal hydride reduction of **288** was already known,[289] although in their total synthesis Obara et al. employed sodium in moist ether to effect this reduction.[287]

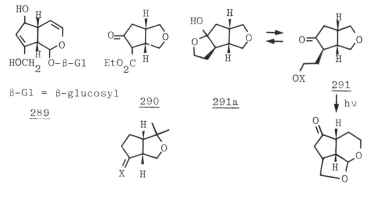

β-Gl = β-glucosyl

289

290

291a

291

↓ hν

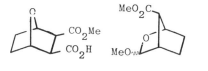

292

288

293

294

295

296

297

298

299

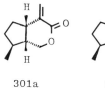

300a

300b

300c

301a

301b

The length of this synthesis (about 10 steps) is characteristic of syntheses of quite simple iridoid systems. For example, hop ether (292, X = CH$_2$) (occurring in Japanese hop oil) is one of the simplest iridoids derivable directly from a geraniol precursor, but its first synthesis dates from 1979.[290] The starting material is the monoester 293, accessible by well-known Diels-Alder methods from furan. A key step is the anodic oxidation of this material,[291] which leads to what is actually a protected 2-substituted 3-formylcyclopentanol 294. Lithium aluminum hydride reduction, then acetylation followed by m-chloro-perbenzoic acid oxidation lead to the lactone 295, which is converted to the ketone 292 (X = O) by Grignard addition and oxidation (Collins reagent).[290] The methylene group 292 (X = CH$_2$) was added using the CH$_2$Cl$_2$–Zn–TiCl$_4$ reagent.[292]

A novel approach to the bicyclic ether skeleton of 292 by Yamada et al. involves the cyclization of geraniol (36) with Tl(III) salts, leading to the stereo-isomeric mixture 292 (X = Me, OH). By conventional means, they then prepared the ketone 296 (X = O)[293] which was converted in turn to 296 (X = H$_2$), then with acetic anhydride in pyridine-HCl to the acetate 297,[294] from which iridomyrmecin (298) was available by a route described by Wolinsky.[295] Both stereoisomers (at C(8)) of cis,cis-298 have been made from all-cis-nepetalactone (299) with lithium aluminum hydride followed by oxidation, and the structures determined by X-ray analysis.[296]

The dolichodials (300a, 300b, Vol. 2, p. 78) have been isolated from *Teucrium marum* (L. *abiatae*), together with the related lactones 301a, 301b, which were suggested as possibly arising from biogenetic Cannizzaro reactions,[297] and the *trans, trans*-isomer 300c has been reported from *Iridomyrmex humilis*.[298] The well-known synthesis of these aldehydes (Vol. 2, p. 78) has recently been described again,[299] but otherwise there has been little synthetic work.

Isomeric with these aldehydes are chrysomelidial (302a, present, together with plagiolactone (303), in the defense secretion of *Plagiodera versicolara*[300] and alone from *Gastrophysa cyanea*[301]) and dehydroiridodial (302b), isolated from *Actinidia polygama*.[302] The synthesis of 302a was effected from (+)-limonene (304) (Scheme 14),[303] a similar route from (−)-limonene leading to 302b.[302] While it will be remarked that this scheme is based on that of Wolinsky's earlier method for nepetalactone[282] (Vol. 2, p. 73), it includes the important modifica-tion that C(9) of the limonene skeleton is functionalized before ozonolysis.

A method from monoterpenoids not involving ozonolysis (further ozonolysis methods are discussed below) involves treatment of carvone epoxide (305) with sodium ethoxide. This yields the cyclopentanecarboxylic acid 306 directly, and Asaka et al. converted the latter into the tosylate 307 by standard tech-niques. They had been hoping to convert this substance into the diiridoid jasminin (308) by condensing it with the secoiridoid 309, which they obtained from jasminin (308). Although this partial synthesis was not successful, since

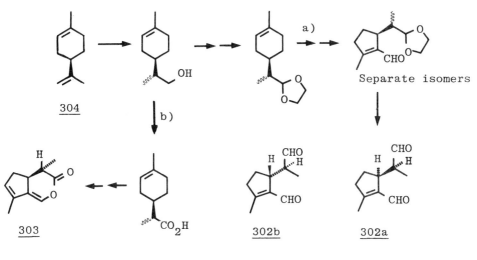

304

303   302b   302a

a) O$_3$ then piperidine; b) CrO$_3$

Scheme 14

it led to **310**, the authors were able to prove the structure of jasminin by their results.[304]

A fairly short route to isodihydronepetalactone (**311**), together with some of its isomers, was developed by Ficini using her ynamine reaction. The first step

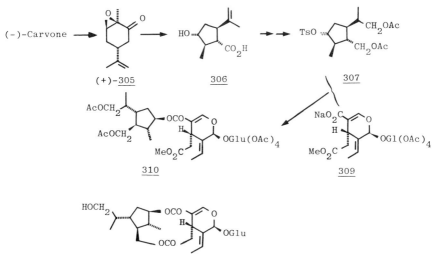

(−)-Carvone

(+)-305   306   307

310   309

308

is the creation of a bicyclo[3.2.0]heptane system **312** from 5-methyl-1-cyclo-pentenecarbonitrile (**313**). Reaction of borohydride on **312** gives the alcohol **314** when hydrolysis yields finally a mixture of lactones, 50% of which is **312**. (Scheme 15).[305] Au-Yeung and Fleming also pass through a bicyclo[3.2.0]-heptene (Scheme 16) to make loganin aglucone acetate (**315**), but this is in order

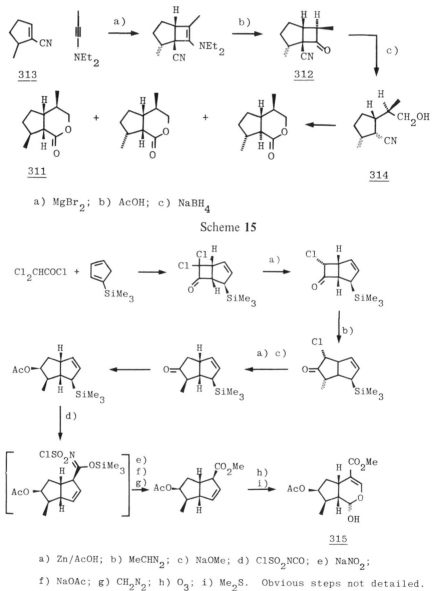

a) MgBr$_2$; b) AcOH; c) NaBH$_4$

Scheme 15

a) Zn/AcOH; b) MeCHN$_2$; c) NaOMe; d) ClSO$_2$NCO; e) NaNO$_2$;

f) NaOAc; g) CH$_2$N$_2$; h) O$_3$; i) Me$_2$S.   Obvious steps not detailed.

Scheme 16

to make a bicyclo[3.3.0]octene, which, when suitably substituted, can be ozonized to prepare iridoids,[306] a method extensively studied by Whitesell (see next paragraph).

The syntheses of Whitesell et al. all depend on this ozonolysis of a suitable bicyclo[3.3.0]octene; a typical example being the ozonolysis, followed by borohydride and acidic methanol, of the ester **316**, which leads to the simple iridoid model **317**, as a mixture of anomers[307] previously prepared by Tietze[308] as an extension of the loganin syntheses discussed later. In the case of his synthesis of iridomyrmecin (**298**), Whitesell made the bicyclic ring system from 1,5-dimethylcycloocta-1,5-diene (**318**) by a transannular solvolysis shown in Scheme 17.[309]

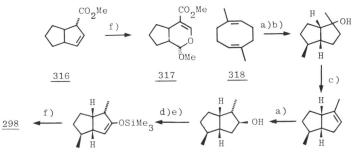

a) Hydroboration-oxidation; b) Solvolysis of sulfonate; c) dehydration; d) Jones oxidation; e) LiN(i-Pr)$_2$ + Me$_3$SiCl; f) O$_3$ etc.

Scheme **17**

Following similar lines, several unsuccessful attempts were made to synthesize genipic acid (**319**).* For example, ozonolysis of the epoxide **320**, followed by zinc/acetic acid reduction gave the acetal **321**, and this could be converted to **322**, but no way was found to carry out the last stages to **319**.[310] A report of the total synthesis of the secoiridoid xylomollin (**323**), isolated from an East African tree, and reported to have antifeedant activity,[311] depends on the ozonolysis of the alcohol **324**,[312] but full information is not available at time of writing. The structure of xylomollin (**323**) was revised, and it is the first natural iridoid with a *trans*-ring junction to be described, although the epimerization was suggested to be possibly due to the method of isolation.[313] A partial synthesis of the *cis*-fused isomer of **323** from loganin (**325**) has also been carried out.[313]

One of the finest achievements of the bicyclo[3.3.0]octene approach is Whitesell's synthesis of sarracenin (**326**).[314] (The numbering of secoiridoids

*There is now doubt as to the correctness of the structure of genipic acid.[309a]

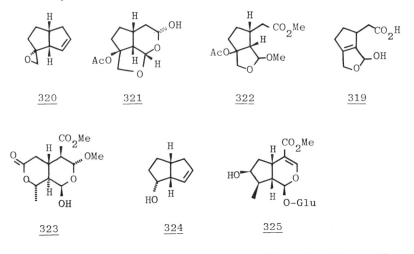

always presents a problem. Iridoids are generally numbered as on formula **298**, following the menthanes and other cyclic monoterpenoids. Secoiridoids thus have a broken C(6)–C(1) bond. The Ring Index numbering follows **327** and **328**, a system used by some authors, see Ref. 247, for example. Nakane et al.[313] prefer a different system, **329**, which, coincidentally, leads to C(8) being the same atom as C(8) in the regular bicyclo[3.3.0]octane numbering used by Whitesell, and apparently extended to sarracenin.[314] Most authors avoid numbering secoiridoids, and Nakanishi et al. seem to have no system at all![315].)

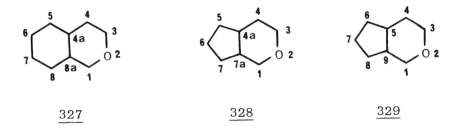

Sarracenin (**326**) was isolated from *Sarracenin flavia*, an insectivorous plant growing in the Okefenokee swamp. It has been suggested that it may represent a route to the indole alkaloids that is complementary or alternative to secologanin (**330**),[316] especially since the absolute configuration at C(1) (monoterpenoid numbering) is the same as that of the corresponding carbon atom of ajmalicine and reserpine, and which is not chiral in secologanin (**330**, R = β-glucosyl).

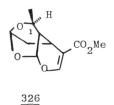

326

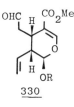

330

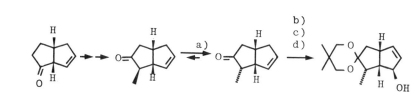

331

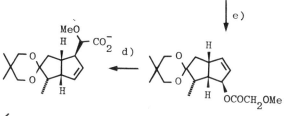

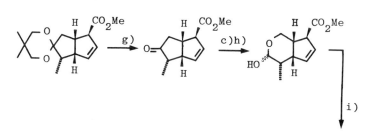

326 ←

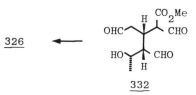

332

a) NaOMe; b) $Me_2C(CH_2OH)_2$; m-Cl-perbenzoic acid; d) $LiNEt_2$

e) $MeOCH_2COCl$; f) O_2; g) H^+/H_2O; h) Dibal; i) O_3 etc.

Scheme **18**

The functionalization of the starting material, bicyclo[3.3.0]oct-7-en-2-one (331) is carried out first, on the carbonyl-containing ring, then an ester is introduced on the cyclopentene side, prior to Claisen rearrangement to bring it into the correct position (Scheme 18). Cleavage of the cyclopentanone is effected by the Baeyer-Villiger reaction, the other ring then being ozonolyzed, yielding the precursor 332 of sarracenin (326).[314]

In their synthesis of the dimethyl ester of forsythide (333)[317] Furuichi and Miwa open a bicyclo[3.3.0]octene system with osmium tetroxide followed by lead tetraacetate. The bicyclic system was here constructed from the more readily available bicyclo[2.2.1]heptanes (Scheme 19).[318] The same authors have also reported the synthesis of sweroside aglucone O-methyl ether (334) from a

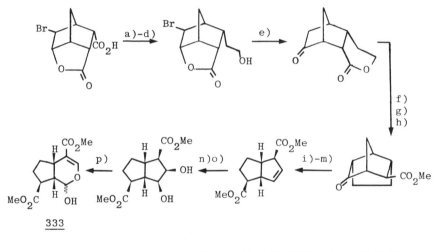

a) $SOCl_2/C_6H_6$; b) CH_2N_2; c) hν; d) B_2H_6/THF; e) 1.5N NaOH;

f) NaOMe; g) TsCl/pyr/CH_2Cl_2; h) t-BuOK; i) hν; j) 0.1N aq.

NaOH/MeOH; k) CH_2N_2; l) Jones oxidation; m) CH_2N_2; n) OsO_4;

o) H_2S; p) $Pb(OAc)_4/C_6H_6/K_2CO_3$.

Scheme 19

bicyclo[2.2.1]heptane system (Scheme 20), in which the key step is the Norrish Type 1 fission of the substituted norcamphor 335.[319]

It will be recalled that Büchi's loganin (325) synthesis (Vol. 2, p. 82, full details of which were published in 1973[320]) depended on the photochemical coupling of a protected cyclopentene and the ester enol 336. Partridge's improvement consists in placing a methyl group of the correct absolute configuration on the cyclopentene side of the molecule 337 before the photo-

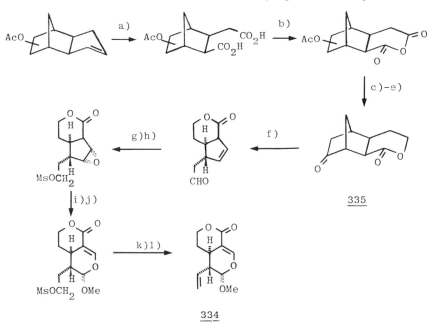

a) Lemieux oxidation; b) Ac_2O/Δ; c) MeOH; d) Na/NH_3; e) Jones oxidation; f) $h\nu/MeCN$; g) $NaBH_4$; h) MsCl then $C_6H_5CO_3H$; i) HIO_4; j) $MeOH/H^+$; k) $NaI/acetone$; l) 1,9-diazabicyclo[5.4.0]undec-7-ene.

Scheme 20

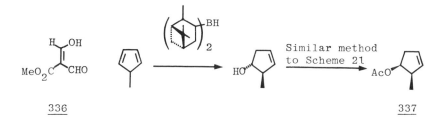

chemical step.[321] Tietze achieved only low yields of iridanes from cyclopentene-carbaldehyde and ethyl 3-ethoxyacrylate, but realized a "biogenetic type" synthesis of secologanin aglucone methyl ether (330, R = Me) and sweroside aglucone methyl ether (334) from one of the intermediates 338 of the Büchi synthesis as shown in Scheme 21.[322,323] Recently a very simple two-step conversion of 338 into secologanin (330) has been described.[324] It was not possible to cleave the ring of the *cis*-loganin series 339 to give the seco compounds, although **339** was glucosylated to hydroxyloganin.[323] The difficulties of glucosylation have also been discussed.[325]

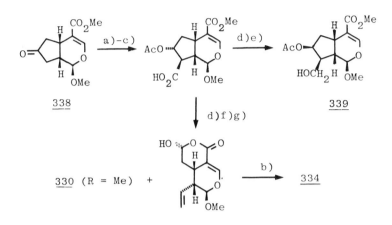

338   339

330 (R = Me) +   b)   →   334

a) $MeMgCO_3$; b) $NaBH_4$; c) Ac_2O/pyr; d) B_2H_6; e) hydrolyse, then treat dimesylate with $AcONEt_4$; f) $TsCl/pyr$; g) $NaCH_2SOMe$.

Scheme 21

The photochemical reaction involving ester **336** does not have to have a cyclopentene as the other component. Reaction with 1,4-cyclohexadiene under the influence of light yields 62% of the *cis*-fused bicyclic compound **340** (with only 5% of the *trans*-fused isomer). After methylation of the hydroxyl group and fission at the double bond, the glycol **341** can be converted into sweroside aglucone methyl ether (**334**), or secologanin aglucone methyl ether (**330**, R = Me) straightforwardly.[326] Another approach to the bicyclic type of compound like **340** involves a Diels-Alder reaction of butadiene with **342**, the product being converted to **343** before ring opening.[327] So far, these routes give only racemates.

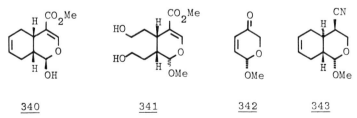

340   341   342   343

The hydrolysis product, elenolic acid (**344**), from oleuropein (**345**)[328] has been synthesized following Scheme 22 by Kelley and Schetter.[329]

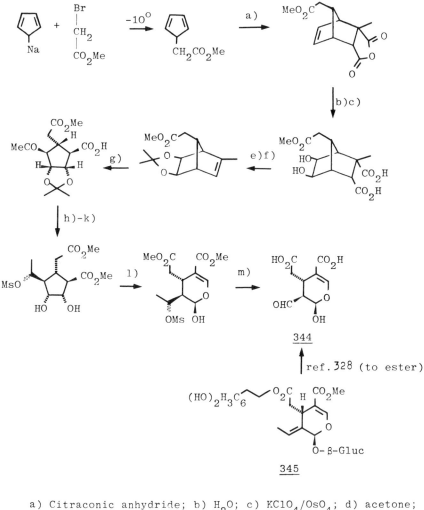

a) Citraconic anhydride; b) H_2O; c) $KClO_4/OsO_4$; d) acetone;

e) electrolytic decarboxylation; f) $KMnO_4/KIO_4$; g) CH_2N_2;

h) $NaBH_4$; i) $MsCl/pyr$; j) $60\% HCO_2H$; k) HIO_4; l) pyr/H_2O.

Scheme 22

Elenolide (**346**) was isolated from *Olea europea* at the same time as secologanin (**330**, R = β-glucosyl),[330] and Tietze has now shown that a connection is possible by converting secologanin into elenolide (which thus constitutes a synthesis of the latter).[331] Enzymatic cleavage of the glycoside bond of secologanin gives **347**, which yields the acetal **348** on treatment with HCl in ether.

Pyridinium chlorochromate oxidation in the presence of sodium acetate then gives elenolide (346).

It has recently been found that one of the products of heating the butenyl ester 349 at 135° for 200 hours is a lactone 350 closely related to elenolic acid (344), and experiments are said to be in progress for the conversion of 350 to 344.[332]

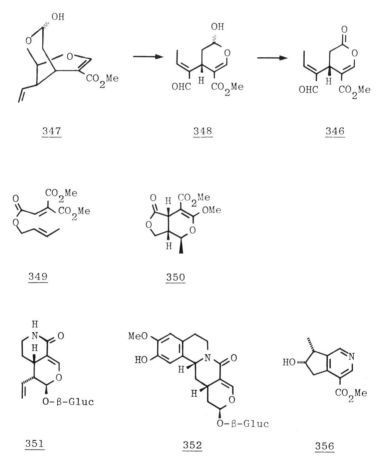

The relationship between secologanin (330, R = β-glucosyl) and the indole alkaloids is well known [see discussions,[316, 333] and a more recent conversion of secologanin to corynanthé alkaloids via elenolic acid (344)[334]] but there is a small group of alkaloids that are much more clearly monoterpenoid in nature. One of the most directly related is bakankosin (351), isolated from *Strychnos vacacoua*,[335] and for which the structure proposed[336] was not confirmed before its synthesis from secologanin (**330**, R = β-glucosyl). The synthesis was effected by Tietze[337] in fairly low yield by direct reductive amination (ammonium acetate and sodium cyanoborohydride in methanol[338]); secologanin tetraacetate has also been converted to bakankosin tetraacetate.[339] Another alkaloid directly derived from secologanin, alangiside (**352**),[340] has been synthesized as its methyl ether from secologanin.[341] Secologanin has been converted into useful synthons for alkaloid synthesis using similar reductive amination to that of Lane[338] as the key step.[342] Related to loganin (**325**) rather than the seco compound are the alkaloids of the actinidine series. (+)-Actinidine [**353**, the natural compound is the (−)-isomer] has been synthesized from (+)-pulegone (**354**) via pulegenic acid (**355**, Scheme 23 and Vol. 2, p. 72).[343] Cantleyine (**356**) may be formed from

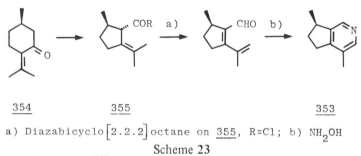

354	355	353

a) Diazabicyclo[2.2.2]octane on <u>355</u>, R=Cl; b) NH₂OH

Scheme 23

loganin and ammonia,[344] but there is not yet a total synthesis. Valerianine (**357**), isolated from *Valeriana officinalis*, has been synthesized by a route in which the key step is a Diels-Alder reaction between 5-methyl-1-cyclopentene-carbaldehyde and **358**.[345]

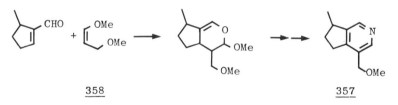

358	357

8.   THE *p*-MENTHANES

A.   Hydrocarbons

A careful study of the (mostly classical) methods of preparation of the more common menthadienes, prepared in high purity and of which the NMR spectra are given, has been published.[346] The phellandrenes (**359** and **360**) were not included in this study.

Methods for the synthesis of α- (359) and β-phellandrene (360) are interesting because there is only a limited supply available from natural sources. A conventional, though effective way of preparing a mixture of phellandrenes is by Singaram and Verghese, employing bromination and dehydrobromination of 1-menthene (284). The debromination was effected by silver nitrate in dimethyl sulfoxide, and separation of the phellandrenes depended on the fact that only α-phellandrene (359) reacts with maleic anhydride.[347] A similar method was employed for making terpinolene (361).[348] β-Phellandrene (360) is obtained in 89% yield on pyrolysis of sodium 1-menthene-7-sulfonate.[349]

A multi-step synthesis of terpinolene (361) from 1-(4-methylphenyl)-1-methylethylmagnesium halide involves a Birch-type reduction of trimethylsilyl derivative,[350] and, while it may be useful for preparing other compounds, it must be recalled that terpinolene is one of the cheapest of monoterpenes, and no synthesis of more than one step is economically justifiable.

Introduction of an isopropyl or isopropenyl group into a suitably substituted cyclohexane remains one of the most common ways of synthesizing menthanes. Hydroboration of 1-methyl-1,4-cyclohexadiene is, however, not stereospecific. Using borabicyclo[3.3.1]nonane followed by isopropenyllithium and iodine, the ratio of 1,4- to 1,3-menthadienes was 43:57.[351] Birch found that the iron tricarbonyl complexes of cyclohexadienes can be alkylated with zinc or cadmium alkyls. Thus from the ion 362, derived from 2-methyl-1,3-cyclohexadiene, the α-phellandrene iron tricarbonyl complexes (363) can be obtained.[352] These are rearranged to 364 by acid,[353] but can be converted to α-phellandrene (359) by cupric ion.[354]

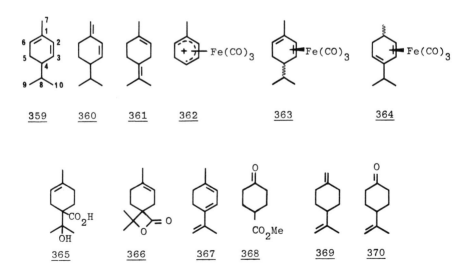

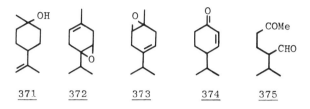

371   372   373   374   375

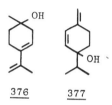

376   377

Addition of acetone to the dilithium derivative of 4-methyl-3-cyclohexene-carboxylic acid leads to a hydroxy-acid **365**, which lactonizes with benzenesulfonyl chloride, the lactone **366** giving terpinolene (**361**) on heating.[355]

When 1-methyl-1,3-cyclohexadiene is treated with isopropenyl bromide and piperidine in the presence of palladium acetate and triphenylphosphine, 39% of 1,3,8-menthatriene (**367**) and 13% of another (uncharacterized) triene are obtained.[356] While **367** and the 1,4,8-isomer[357] (Vol. 2, p. 91) are well known and characterized, menthatrienes are occasionally reported in the literature without adequate characterization. For example, 1,5,8-menthatriene, for which only scanty NMR data are published[358] is reportedly available from carveol[359], or carene oxides,[360] or α-pinene epoxide,[361] and there are two reports of 1(7),2,4(8)-menthatriene[362] for which we are unaware of any NMR data.

An interesting new application of the Wittig reaction has been discovered by Uijttewaal, Jonkers, and van der Gen. In dimethyl sulfoxide, esters react with excess methylenetriphenylphosphorane to yield, after hydrolysis, the corresponding isopropenyl compounds.[363] Thus treatment of the ketoester **368** with excess of the phosphorane yielded 78% of 1(7),8-menthadiene (**369**), and by protecting the ketone group as an acetal before the reaction then removing the protecting group, 4-isopropenylcyclohexanone (**370**) could be obtained—a convenient route to β-terpineol (**371**),[364] a compound belonging to the following section.

9-^{14}C-Limonene has been synthesized.[365]

One of the epoxides (**372**) of γ-terpinene was reported to occur in *Origanum heracleoticum*, this same epoxide being said to be the sole product from mono-perphthalic acid epoxidation of the hydrocarbon.[366] Other work has shown that γ-terpinene gives a mixture of monoepoxides, **372** and **373**, on oxidation with peracetic acid[367] or peroxybenzimidic acid,[368] diepoxides being formed on further reaction.[369] Furthermore, the properties of **372** do not agree with those of the *Origanum* substance.[370,367]

B. Oxygenated Derivatives of *p*-Menthane

The 1-, 4-, and 8-Oxygenated Menthanes

Cryptone (**374**, Vol. 2, p. 97) has been prepared very simply from methylheptenone, carbonylation of which yields the precursor **375**, which cyclizes readily.[371]

In addition to the compounds listed as naturally occurring in Vol. 2, the menthadienols **376** and **377** have been isolated from pepper oil,[372] but no synthesis has been reported to our knowledge. *p*-3-Menthene-8-ol (**378**) is a component of the essential oil of *Mentha gentilis*,[373] and has been synthesized from 4-methylcyclohexanone by a novel route.[374] This involves reaction of the ketone with the anion of chloroethyl phenyl sulfoxide, leading to the epoxysulfoxide **379**. Pyrolysis of the latter gives the known[375] ketone **380**, from which 3-menthene-8-ol (**378**) is obtained with methyllithium.[374]

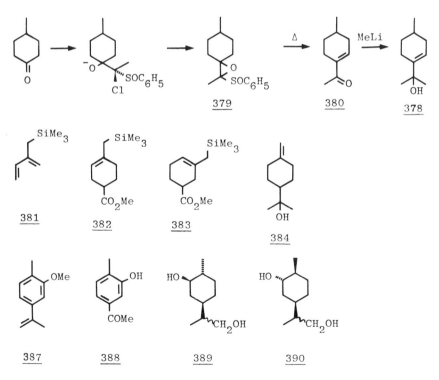

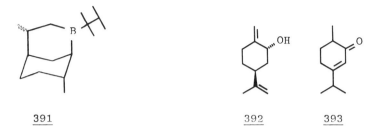

391   392   393

The radicals from the 2-menthenes with *t*-butyl perbenzoate in the presence of cupric octanoate react with lithium aluminum hydride to give mixtures containing the *t*-alcohols,[376] but the method does not seem to be readily adaptable for synthetic use.

Pyrolysis of the silane obtained from methylenecyclobutane (discussed in a previous section)[126] leads to a substituted isoprene 381, which, in a Diels-Alder reaction with methyl acrylate yields an 8:2 mixture of the adducts 382 and 383. The former can be converted to δ-terpineol (384) with methyllithium and acid, the last two reactions being carried out in either order.[126]

In Vol. 2, p. 100, the preparation of 1,8-menthadien-4-ol (385) was discussed. It was not mentioned, however, that the easiest route is probably from terpinolene (361), the epoxide of which is converted to 385 directly by *p*-toluenesulfonic acid at 25-30°.[377] Another very simple route is the Grignard reaction of 4-methyl-3-cyclohexenone,[378] which has been repeated without acknowledgment.[379] Sharpless has carefully reinvestigated the selenium dioxide oxidation of limonene (304),[380] because, although this should give a racemate, at least one report[381] claimed optically active products. There is no doubt that the compound obtained from optically active limonene in this reaction is racemic, any optical activity detected being due to impurities.[380] Although the naturally occurring 1,8-menthadien-4-ols are still of unknown absolute configuration, Delay and Ohloff have prepared both optical isomers from (+)-(R)-limonene (R-304) by the routes shown in Scheme 24.[382] In this connection, the ozonolysis of optically active limonene epoxides and the subsequent reduction (Zn-NaI) to optically active 386 has been thoroughly described.[383]

1-Menthanols labeled with deuterium at C(2) have been synthesized.[384]

The 2-Oxygenated Menthanes

Two syntheses of 2-methyl-5-isopropenylanisole (387, Vol. 2, p. 103) have been reported. The first of these includes the bad first stage (20% yield) of converting 2-amino-4-chlorotoluene to 2-hydroxy-4-chlorotoluene, then methylating the phenol and inserting the isopropenyl group with lithium diisopropenylcopper.[385] The other synthesis relies on compound 388, an intermediate in the original

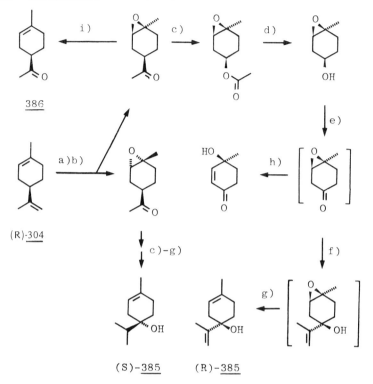

386

(R)-304

(S)-385   (R)-385

a) AcO$_2$H; b) O$_3$; c) m-chloroperbenzoic acid; d) MeONa/MeOH;

e) (PyrH)$_2$Cr$_2$O$_7$; f) i-PrLi; g) NaI/Zn/NaOAc/HOAc;

h) H$^+$ or OH$^-$; i) ref. 383.

Scheme 24

synthesis, using Grignard addition and dehydration to put in the methylene group. This method necessitates care in the final step to avoid dimerization.[386]

One of the most economical ways of approaching C(2) oxygenated menthanes is from limonene, as we have noted in Vol. 2, p. 108. Direct hydroboration of limonene gives mixtures,[387] but Brown and Pfaffenberger have developed a method for producing a single configuration about the ring. Thus oxidative hydroboration of (+)–(R)-limonene produces a 70:30 mixture of the diols 389 and 390 (they were unable to separate the *threo* and *erythro* isomers at C(8), which are called *"cis"* and *"trans"* in the paper) but under certain conditions, particularly distillation of the product obtained with thexylborane, a cyclic borane **391** was obtained, from which a 83% yield of the pair of compounds (−)-(1*R*,2*R*,4*R*)-carvomenthol (**389**) was obtained.[388]

Oxidation of limonene generally is less specific, anodic oxidation giving, for example, mixtures of stereoisomers of 2-oxygenated 1(7),8- and 5,8-menthadienes, together with a small amount of 7-oxygenated limonene.[389] Ring-opening of limonene epoxide (Vol. 2, p. 108) also gives mixtures, but one of us has examined the various conditions leading to either 2-substituted **392** or 7-substituted compounds very carefully.[390] Ring-opening of the *cis*- and *trans*-menthene-1,2-epoxides over various catalysts has been examined by Arata et al., and gives mixtures of 2-oxygenated menthanes with ring-contracted products.[391]

Dehydrolinalool (**121**) acetate can be cyclized with zinc chloride to a mixture containing carvenone (**393**), its enol acetate, a 2-carone enol acetate and a pyran derivative.[392]

Concerning the syntheses of specific natural 2-oxygenated menthanes, new natural products derived from 1(7),5-menthadien-2-ol have been found. The two alcohols **394a**, **394b** (R = H) were reported in *Cinnamomum japonicum*,[393] and one isomer (not determined) in pepper oil.[372] They were suspected to be present in angelica (*Angelica archangelica*) root oil,[394] and when Escher et al. showed clearly that both alcohols and one of the acetates **394b** (R = Ac) were present in this oil, they synthesized both isomers stereospecifically from α-phellandrene (**359**). Direct epoxidation yielded the *trans*-epoxide **395b**, the *cis*-epoxide **395a** being prepared by the action of *N*-bromosuccinimide (to the bromohydrin) and sodium hydroxide. The alcohols were then obtained by the action of lithium diisopropylamide on the corresponding epoxides.[395]

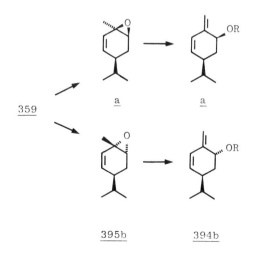

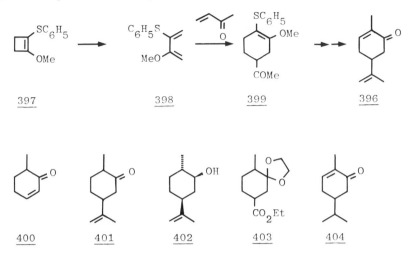

A review of the synthesis of carvone has been published.[396]

A synthesis of carvone (396) was the vehicle Trost used to demonstrate how cyclobutanone might be used. Thiophenylation and conversion to the enol ether gave 397, which was pyrolyzed to the diene 398. In a Diels-Alder reaction, the latter yielded the cyclohexenyl ketone 399, which was converted conventionally to carvone (396).[397]

2-Methylcyclohexanone has been converted to carvone (396) in an overall yield of only 28%, but by a very attractive route on paper. The trimethylsilyl enol ether of 2-methylcyclohexanone is first oxidized to the unsaturated ketone 400 with dichlorodicyanobenzoquinone and collidine, and the isopropenyl side chain is introduced using a copper-catalyzed Grignard addition. The dihydrocarvone (401) thus obtained was converted to carvone by the same oxidation as before, with DDQ/collidine on the trimethylsilyl ether. Piperitone (see below) has been made by a similar oxidation of menthone trimethylsilyl enol ether.[398]

Vig has synthesized dihydrocarveol (402) conventionally from 4-methyl-3-cyclohexenyl methyl ketone,[399] and dihydrocarvone (401) from the ester 403.[400] It should be noted that carvotanacetone (404), the synthesis of which is described in *Organic Syntheses*,[401] is wrongly listed there as "dihydrocarvone."

Various deuterium-labeled dihydrocarvones, dihydrocarveols, and carvomenthols have been prepared.[384,402]

The scheme used by Rao (Vol. 2, p. 107, and Ref. 403) to convert one optical antipode of carvone to the other has been applied to the examination of the very different odors of the two naturally occurring antipodes.[404] Yasuda et al. have used the metal-ammonia reduction of the mesylate of *cis*-carveol epoxide (derived from the (2R,4R)-isomer, 405) to prepare the (2S,4S)-isomer of 405.[405]

The 3-Oxygenated Menthanes, Menthofuran, and Other 3,7-Oxygenated Derivatives

The cyclization of open-chain monoterpenoids to 3-oxygenated menthanes is well known (see Vol. 2, pp. 116, 120), and the literature is large; we shall confine this aspect to a few selected recent developments. The use of tris-(triphenylphosphine)chlororhodium or stannic chloride in the cyclization of citronellal leads to different amounts of the isopulegol isomers.[406] A pulegol ester can be obtained by heating 3,7-dimethyloctane-1,7-diol above 200° in the presence of an acid anhydride.[407] Simultaneous cyclization and reduction can lead directly from citronellal (**406**) to menthone (**407a**) and isomenthone (**407b**).[408] Cyclization is most easily carried out using zinc bromide in benzene (toluene?) when the optical activity of the citronellal (**406**) is retained in the isopulegol (**408**) obtained (in 70% yield with 96% optical retention).[409] This same cyclization can be carried out with retention of optical activity by treatment of citronellol (**33**) with pyridinium chlorochromate (obtaining isopulegone **409**).[410] Cyclization of the acid chloride of (*Z*)-geranic acid (*Z*-**56**) in the presence of a Lewis acid yields a mixture of piperitenone (**410**) and its hydrochloride (**411**). Lithium chloride in dimethylformamide converts **411** to **410**, and treatment of the mixture with lithium in ethylenediamine yields thymol (**412**).[411]

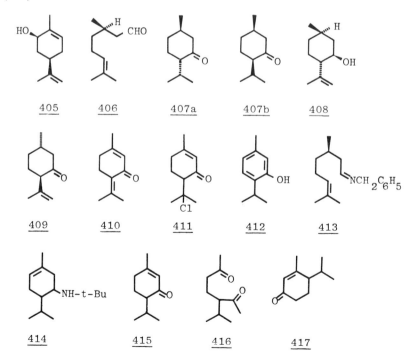

| 405 | 406 | 407a | 407b | 408 |

| 409 | 410 | 411 | 412 | 413 |

| 414 | 415 | 416 | 417 |

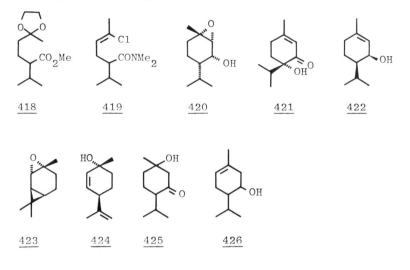

The oxygenated thymols are discussed in the section on 9-oxygenated menthanes.

Imines of open-chain compounds have been used in two ways to synthesize menthanes. (+)-(R)-citronellal benzylimine **413** is cyclized at 25° in benzene to a mixture of 8-menthenylamines with retention of configuration at C(1); catalytic reduction gives a mixture of 3-menthylamines.[412] These were not well characterized in the preliminary note, and it is to be hoped that the full paper makes the proportions of the stereoisomers quite clear. 3-Methylbutanal *t*-butylimine reacts with isoprene in the presence of sodium in an autoclave at 40° to give the menthenylamine **414**.[413]

Piperitone (**415**) is accessible by cyclization of the diketone **416** (Vol. 2, p. 119), but some of the alternative *o*-menthenone (**417**) is formed at the same time.[414] Actually, piperitone "unzips" even during dienamine formation, and dienamines corresponding to **417** and its double bond isomers are always found among the products.[415] Two methods have been described for making **416**; one involves the protected ketoacid **418**, converting the ester group to a methyl ketone using the dimethyl sulfoxide anion and reduction with aluminum amalgam,[414] the other consists in making the chloroamide **419** (from the enol of *N*-dimethyl-3-methylbutyramide and $ClCH_2CH=CClMe$), forming the methyl ketone with methyllithium, then hydrolyzing the vinyl chloride with acid.[416]

A certain number of piperitone derivatives are present in various mint species (cf. Vol. 2, p. 136), and more recently, *trans*-piperitol epoxide (**420**)[417] and the 4-hydroxy compound **421**[418] have been identified in the wild Japanese mint, *Mentha gentilis*. The former must be accessible from *cis*-piperitol (**422**), but we have not been able to find a clear synthetic description of either. The piperitols themselves have been made from 2-carene epoxide (423) by Prasad and Sukh Dev (who inexplicably say that they were not described earlier—relevant references are in Vol. 2, p. 120, and include a detailed description of the NMR spectra[419]). The route they used consists in ring opening with metatitanic acid to the menthadienol 424, rearrangement (acetic acid) and reduction (tristriphenylrhodium chloride).*[421]

Still has shown how the trimethylstannyllithium reagent can be added to cryptone (374) and the product oxidized to yield 1-hydroxy-3-menthone (425).[422] Unfortunately he does not mention the stereochemistry of the product, although both isomers of 425 are well described.[423] An acetate of 425 occurs in *Mentha gentilis*[424] (which was also not compared with the known 425).

The isomer of piperitol, 1-menthen-5-ol 426 has been found in pepper oil.[372] The authors do not publish a stereochemistry, but it was probably *trans*.[425] Once again, although these compounds 426 have been in the literature for some time, it is difficult to pin down a good synthetic description. The first description of one of the isomers was by Read and Swann in 1937, who prepared it as a by-product of the catayltic reduction of piperitone (415) epoxide.[426] Among the products of the thermolysis of *trans*-verbenol (427) is 30% of a mixture of *cis*- and *trans*-1,8-menthadien-5-ols (428). Separation of these and catalytic reduction leads to the isomers of 426.[427] Marshall and Babler have mentioned the mesylate of *trans*-426 without comment on its preparation.[428] The unconjugated ketone, 1-menthen-5-one has also been described,[415,427,429] indeed, Read and Swann comment on the difficulty of oxidation of the alcohol 425,[426] but it has not yet been reported in nature.

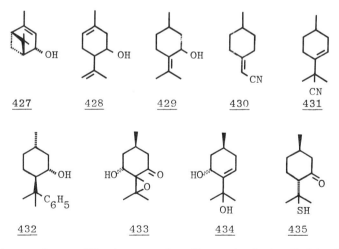

| 427 | 428 | 429 | 430 | 431 |

| 432 | 433 | 434 | 435 |

*The ring opening to 424 using peracetic acid was already described by Ohloff and Giersch,[420] but is contested by Prasad and Sukh Dev.

The pulegols (429) have been synthesized by Marshall et al. from 4-methyl-cyclohexanone. Condensation with diethyl sodiocyanomethyl phosphate yields 430, which can be methylated to 431 with lithium diisopropylamide and methyl iodide in hexamethylphosphoramide. Epoxidation (peracid) and treatment with sodium in ammonia yield a 2:1 ratio of *trans:cis*-429. The oxidation of the pulegols was studied at the same time. Manganese dioxide oxidation of *cis*-429 yields pulegone [(±)-354] in 80% yield, but the other isomer leads to the epoxide. The pulegone epoxides are also discussed.[430]

The compound 432 is a good chiral director, and a synthesis of it now exists from the cheaper (+)-(R)-pulegone (354).[431] Before this, however, it was necessary to start from (–)-(S)-pulegone, which is not of easy access. The conversion of the (R)- to the (S)-antipode was carried out by Ensley and Carr. Photooxygenation (with reduction) followed by peroxidation of the α,β-unsaturated ketone leads in 70% yield to the epoxides 433. Treatment of the latter with hydrazine gives the diols 434; catalytic reduction of the double bond, oxidation of the carbinol group, and dehydration with iodine then give (–)-(S)-pulegone [(–)-354].[432] This series of reactions is reminiscent of an earlier conversion of (–)-(1R,4S)-menthone (407a) to the antipode, required for the synthesis of optically pure citronellols.[433] (–)-(S)-Pulegone can also be prepared by the classical cyclization of (S)-citronellal, recent improvements being the use of (S)-citronellol (generally of better optical purity than the aldehyde) and pyridinium chlorochromate, which yields optically active (S)-isopulegol directly, it remaining only to isomerize the double bond with base.[434]

Hydrogenation of piperitenone (410) with chiral homogeneous rhodium catalysts gives pulegone with up to 38% optical purity.[435]

A synthesis of isopulegone (409) from ethyl 2-keto-4-methylcyclohexanecarboxylate has been described.[400]

The thiol 435 is an interesting pulegone derivative; it is made by treatment of pulegone with hydrogen sulfide in triethylamine, and it occurs naturally in Buchu oil and black currants (*Ribes nigrum*).[436] The corresponding thioacetate also occurs in Buchu oil.[437]

Owing to the economic importance of menthone (407a) and the menthols there is a great deal of literature. Few recent syntheses, however, can be considered economically justified. (See Leffingwell's review[438] on the synthesis of optically active menthol.) The following are, however, worthy of mention. A fairly simple approach starts from the enol acetate of 4-methylcyclohexanone. Treatment with methanol under phase-transfer conditions (with tetraethyl-ammonium *p*-toluenesulfonate) gives the methoxyketone 436, Grignard reaction on which gives menthol (437) directly via a 1,2-carbonyl transposition.[439] By a double alkylation of ethyl acetoacetate, first with 4-bromobutene then with isopropyl iodide, Conia et al. prepared the β-ketoester 438 (R = CO$_2$Et). This may either be hydrolyzed and decarboxylated (NaCN and dimethyl sulfoxide) to the ketone **438** (R = H), pyrolysis of which leads to menthone and isomenthone (**407**), or pyrolyzed directly. At 300° the cyclization occurs without decarboxylation, but after 2 hours at 350°, menthones are formed.[440]

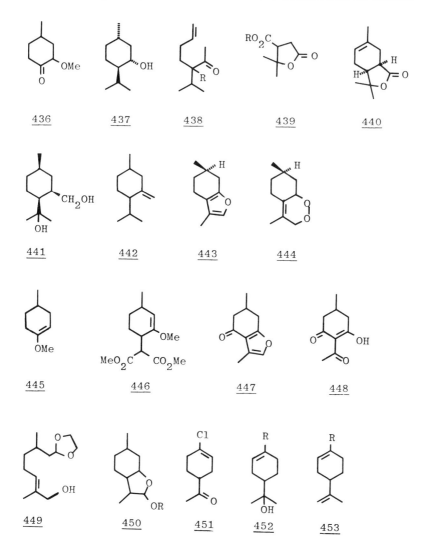

436   437   438   439   440

441   442   443   444

445   446   447   448

449   450   451   452   453

One method which appears superficially of only academic interest has nevertheless given rise to a patent. This is based on the fact that α-butenolides carboxylated in the β-position, **439**, are dienophiles, and the acid **439** (R = H) reacts with isoprene to give 91% of the *cis*-adduct **440**. After catalytic reduction, ring-opening with lithium aluminum hydride leads to the diol **441**. Conventional reactions lead to either *cis*- or *trans*-**442**, from which isomenthone (**407b**) or menthone (**407a**) are obtained by ozonolysis.[441]

The synthesis of menthol glycosides has been investigated.[442]

A simple synthesis of menthofuran (**443**) involves treatment of the cyclic peroxide **444**, obtained from 3,8-menthadiene and singlet oxygen, with lithium diisopropylamide at −70° and *p*-toluenesulfonyl chloride.[443] (Surely an improvement would be to use the ferrous reagent.[444]) Other, more complicated syntheses start from ethyl 2-oxo-4-methylcyclohexanecarboxylate. Vig uses the carboxylate group to activate the position which is to receive the side chain,[400] but Wenkert introduces it in order to prepare specifically the enol ether **445** reaction of which with dimethyl diazomalonate in the presence of copper phosphite gives 39% of **446**. Menthofuran (**443**) is prepared conventionally from **446** in 3 steps.[445] This (and another equally long synthesis in the same paper) is clearly only the vehicle for showing how the addition of the diazomalonate occurs to an enol ether, for not only is it of prohibitive length, but the addition of a carboxyl group, which is later removed in order to add a C_3 unit, is not esthetically satisfying. An older process involving peracid oxidation of isopulegone enol acetate (see Vol. 2, p. 122) has recently been improved.[445a]

A synthesis of evodone (**447**, Vol. 2, p. 123) has been effected by the reaction of diazomethane with the dihydroresorcinol **448**.[446]

Acid cyclization of the hydroxycitronellal acetal **449** leads to an acetal **450** related to a hydrogenated menthofuran.[447]

The 7-Oxygenated Menthanes

The Diels-Alder reaction between chloroprene (2-chloro-1,3-butadiene) and methyl vinyl ketone leads to the chloroketone **451**, and this is a useful intermediate for the preparation of 7-oxygenated menthanes. For example, oleuropic acid (**452**, R = CO_2H) has been made by protecting the ketone group as an acetal, introducing the aldehyde group with lithium and dimethylformamide, removal of the protecting group on the ketone and subsequent protection of the aldehyde, then Grignard addition and oxidation to **452** (R = CO_2H).[448] Alternatively, a Grignard reaction on **451** gives an alcohol that can be dehydrated to **453** (R = Cl), from which perilla aldehyde (**453**, R = CHO) is obtained with lithium and dimethylformamide.[449]

For approaches to perilla alcohol (**453**, R = CH_2OH) from **392**, which is a photooxidation product of limonene, see Ref. 312. The corresponding aldehyde **453** (R = CHO) can be reduced with sodium in the two-phase system of benzene-aqueous ammonia to a mixture of *cis*- and *trans*-shisool (**454**),[450] two alcohols isolated from *Perilla acuta* var. *viridis*,[451] the *trans*-isomer also having been found in the oil of *Juniperus communis*.[452] More recent work has shown how, with different conditions, the lithium in ammonia-isopropanol reduction of perilla aldehyde (**453**, R = CHO) can lead to other products of different oxidation levels.[453]

One of the aldehydes **455** occurring in cumin seeds has been synthesized by Birch and Dastur using Birch reduction of the protected cuminaldehyde **456**. Removal of the protecting group after reduction yielded the aldehyde **455**.[454] Reduction of cuminic acid (**457**) with sodium in liquid ammonia was already known.[455]

Besides occurring as part of the bitter principle of the olive tree (Vol. 2, p. 135), oleuropic acid (**452**, R = CO$_2$H) and other 7-carboxylic acids are metabolic products of *Pseudomonas* sp., **452** (R = CO$_2$H) being produced from α-terpineol (**452**, R = Me),[456] and limonene giving perillic acid (**453**, R = CO$_2$H).[457]

The 9-Oxygenated Menthanes

In addition to the previously known 9-oxygenated menthanes, the enol acetate **458** has been found in tangerine oil.[458]

The access to 9-oxygenated menthanes through lead tetraacetate oxidation of limonene was discussed in Vol. 2 (p. 130), but a few modifications have been published more recently,[459] particularly with the end of synthesizing the aldehyde **459**, isolated from rose oil,[460] cotton seed oil,[461] and Buchu leaf oil,[462] (the structure of which was determined some time ago[460]) and the widely distributed unsaturated alcohol **460**.[459] Vig has also synthesized the isomer **461** (not yet reported naturally occurring), but the stereochemistry was not established.[463]

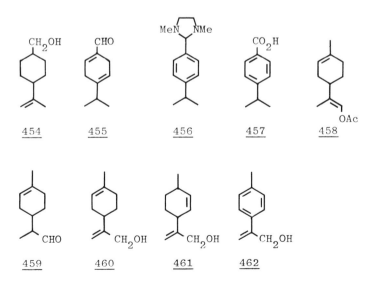

454   455   456   457   458

459   460   461   462

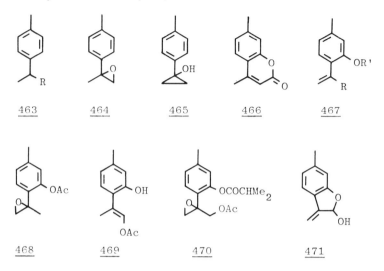

463   464   465   466   467

468   469   470   471

Of the aromatic 9-hydroxylated *p*-cymenes, **462** is not yet reported naturally occurring, but **463** (R = CH$_2$OH) and the related acid **463**, (R = CO$_2$H) are products of the metabolism of *p*-cymene (**463**, R = Me) by a soil pseudomonad,[464] and **463** (R = CH$_2$OH) also occurs in the oil of *Juniperus communis*,[465] and has been known for a long time. Recent preparations of **462** include the preparation of the epoxide **464** from 4-methylacetophenone and ring opening with lithium diisopropylamide[466] [acid-catalyzed ring opening gives mainly the aldehyde **463** (R = CHO)[467]] , and solvolysis of the *p*-nitrobenzoate of **465**, prepared by the reaction of dichloroacetone with 4-tolylmagnesium bromide.[468] Another entry to 9-functionalized α,4-dimethylstyrenes is via the reaction of α,4-dimethylstyrene with *N*-bromosuccinimide.[469] Apart from the catalytic reduction of **462** or related substances, **463** (R = CH$_2$OH) can also be prepared directly by the action of propylene oxide and triethylaluminum on toluene. After hydrolysis, the alcohol is obtained in 61% yield.[470]

Related to these synthetic methods are the routes used to prepare some of the many thymol (**467**, R = Me, R' = H) derivatives substituted at C(9) that Bohlmann's group has isolated particularly from Compositae. Divakar and Rao, improving the known reaction of 4,7-dimethylcoumarin (**466**) with base, obtained a good yield of the unsaturated thymol **467** (R = Me, R' = H). After acetylation and perbenzoic acid oxidation, they obtained the epoxide **468**, which they pyrolyzed to the enol acetate **469**;[471] this could be useful in the synthesis of enol acylates that occur naturally.[472] Divakar et al. have also synthesized the esters **467** (R = CH$_2$OCO-*i*-Pr, R' = CO-*i*-Pr) and **467** (R = CH$_2$OCO-*i*-Pr, R' = Me).[473] In order to synthesize the compound **470** which they had isolated from *Wedelia forsteriana*,[474] Bohlmann et al. prepared the ester **467** (R = Me, R' = CO-i-Pr) by the conventional route (Grignard addition, dehydration and esterification) from methyl 4-methylsalicylate. Introduction of an oxygen function at C(9) was effected using selenium dioxide in acetic anhydride, and the product **467** (R = CH$_2$OAc, R' = CO-*i*-Pr) was epoxidized to **470**.[475] Divakar et al. have also always introduced the oxygen function at C(9) with selenium dioxide, but using dimethyl sulfoxide as solvent in order to make the aldehydes **467** (R = CHO). Another compound made by them following similar lines is the naturally occurring benzofuranol **471**.[476]

An ingenious route by Ficini and Touzin does not lead specifically to a natural product, but is a useful entry to racemates of stereochemically established configurations about C(8) of the menthane-9-carboxylic acids. A [2 + 2] cycloaddition of an ynamine to 2-methylcyclohex-5-enone leads to the *cis*-substituted bicyclo[4.2.0]octenone **472a**, converted in neutral or basic solution into the isomer **473** (Scheme 25). Decomposition of these substances with acetic acid yields two of the possible 2-oxomenthane-9-carboxylic acids (**474a**, **474b**), while treatment of the enamines **472**, **473** with dry hydrogen chloride gives two other enamines **475a**, **475b**, which can likewise be decomposed with acetic acid to the enantiomers **476a** and **476b** of **474a** and **474b**.[477]

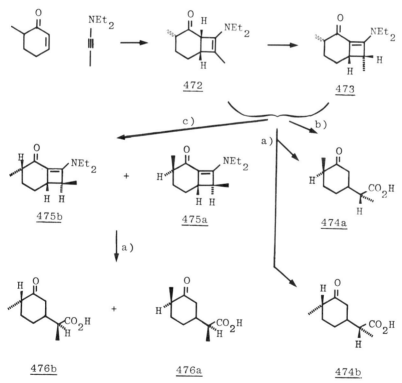

a) HOAc; b) HCl–H$_2$O; c) dry HCl, then Na$_2$CO$_3$

Scheme 25

Ethers, Ascaridole, Polyoxygenated Menthanes

A new synthesis of 1,4-cineole (**477**, X = H, Vol. 2, p. 133) gives 79% overall yield of a cleaner product than earlier methods. 1-Menthen-4-ol (**478**) is irradiated in the presence of mercuric oxide and iodine, and the resulting iodide **477** (X = I) reduced with lithium aluminum hydride to 1,4-cineole (**477**, X = H).[478]

1,8-Cineole (=eucalyptole, **479**) has been synthesized from limonene (**304**). When the latter is allowed to react with phenylselenic acid (C_6H_5SeOH, formed *in situ* from a mixture of phenylseleninic acid and diphenyl selenide[479]) the adduct **480** is formed, and this can be reduced with tri-*n*-butyltin hydride to give 80% of 1,8-cineole (**479**).[480] The dehydro analog **481** occurs in *Laurus nobilis*, and has been synthesized in low yield by the reaction of butyllithium on the *p*-tosylhydrazone of **482** (starting from the same material, this paper also describes the synthesis of the non-naturally occurring pinol and isopinol; see also Vol. 2, p. 131).[481]

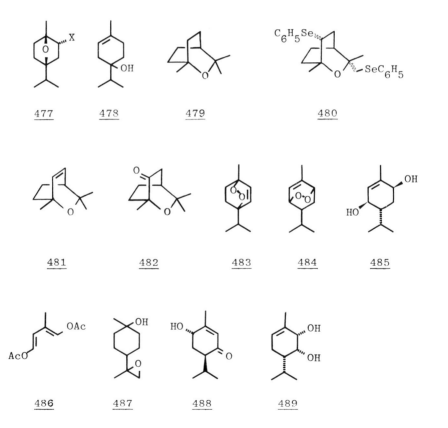

477   478   479   480

481   482   483   484   485

486   487   488   489

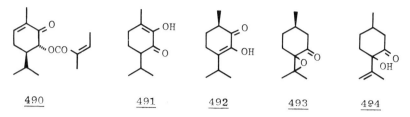

490   491   492   493   494

The formation of ascaridole (483) by the addition of oxygen to α-terpinene (Vol. 2, p. 133) is catalyzed by the triphenylmethylium ion.[482] A similar effect occurs in the presence of Lewis acid in methylene chloride at −90° or in SO_2 solution at −70° without Lewis acids.[483]

Two peroxides, *cis*- and *trans*-484 are accessible from α-phellandrene (359) in the same way as ascaridole (483) is from α-terpinene.[484] Ring opening of 484 has been known for some time[485,444] one of the products being the diol 485, isolated from *Eupatorium macrocephalum* by González et al.[486] The diacetate of 485 has also been made by the Diels-Alder reaction between 486 and 3-methyl-1-butene.[487]

Polyoxygenated menthanes are frequently found among the products of microbiological oxidation of common terpenoids. For instance, besides the 7-oxygenated compounds mentioned above, α-terpineol (452, R = Me) gives one of the epoxides 487,[488] and piperitone (415) gives various hydroxylated piperitones, particularly 488,[489] accessible from 484.[444,485]

2,3-Dioxygenated menthanes occur naturally, 489 in *Mentha gentilis*,[490] for example, and an ester 490 of the corresponding ketone in the roots of *Pluchea odorata*.[491] Diosphenol (491) and its isomer, ψ-diosphenol (492), have long been known as constituents of Buchu leaf oil (various species of *Barosma*; for other related constituents see Ref. 492),[493] but 492 was apparently unknown to Shibata and Shimizu,[494] who prepared it as a by-product of diosphenol (491) when repeating the original ferric chloride oxidation of menthone (407).[495] A reexamination of the acid-catalyzed ring opening of pulegone epoxide (493) has shown that the hydroxy-8-menthenone (494) obtained originally by Reusch et al. using *p*-toluenesulfonic acid[496] is converted by 20% aqueous sulfuric acid into a 3:7 mixture of diosphenol (491) and ψ-diosphenol (492), the same mixture as is obtained by direct treatment of 493 with sulfuric acid. In this route to 492, it should be noted that the chirality at C(1) is retained.[497] A "building-block"-type synthesis of diosphenol (491) and its isomer 492 has been effected by Ohashi et al., during which they found that the ester 495 could not be decarboxylated. This induced them to oxidize 2-isopropylcyclohexanone directly, adding the extra methyl group after.[498]

The limonene 8,9-epoxides (496), prepared from limonene by the action of peroxybenzimidic acid,[499] can be separated by spinning-band distillation[500] (as can also the 1,2-epoxides[501]). Kergomard and Verschambre have correlated the uroterpinols (**497**), obtained on acid-catalyzed ring opening of **496**, with the products from limonene metabolism.[500] They assigned the stereochemistry of the glycols using europium shift NMR techniques, which, in the case of polyfunctional compounds, are not always easy to interpret. This could account for a discrepancy that has been noted between the stereochemistry of bisabolols (**498**) prepared from the epoxides **496**[500] and by other means,[502] and until this is resolved,* the configuration of the uroterpinols (**497**) about C(8) must also be in doubt. Uroterpinol (**497**) and its monoacetate are products of the lead tetraacetate oxidation of limonene mentioned above.

The synthesis of both stereoisomers of 1(7)-menthen-2,8-diol (**499**) has been carried out by Vig;[503] they are not yet reported in nature. The dimethyl ether of **499** has been made by thallation of myrcene (**6**) in methanol.[504]

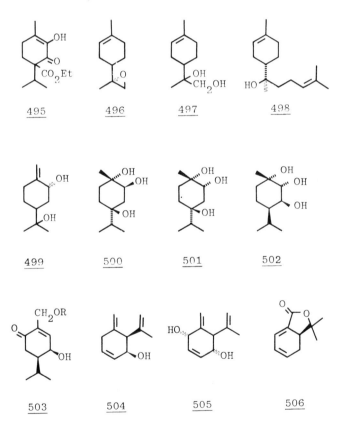

*After this text was written, an X-ray study of α-bisabolol[502a] has shown that the stereochemistry Kergomard attributed to **497** must be modified.

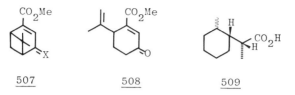

507   508   509

The triol **500** was isolated from *Zanthoxylum budrunga*. It was not synthesized, but the structure was determined by comparison with the isomeric triol **501**, made by the reaction of osmium tetroxide on 1-menthen-4-ol.[505] The 1,2,3-triol **502** is the first monoterpenoid to be isolated from a microorganism, *Fusicoccum amygdali*,[506] and it has been synthesized from piperitone (**415**), together with six other stereoisomers.[506a]

Bohlmann et al. have found the angelate ester (**503**, R = angeloyl) in *Blumea wightiana* (Compositae).[507]

9. THE o-MENTHANES

The first discovered member of this small group of monoterpenoids, carquejol (**504**) and its acetate[508,509] have not yet been synthesized.* Some derivatives of related diols **505** have been isolated from other *Baccharis* and *Piqueria* species,[511] but again no synthetic work is reported.

The only successful preparation so far of a natural o-menthane is that of the lactone **506**, isolated from the urine of the koala bear (*Phascolarctos cinereus*).[512] Lander and Mechoulam[513] started with methyl myrtenate (**507**, X = H$_2$), which can readily be prepared by oxidation of α-pinene (**245**).[514] Oxidation with sodium chromate yields the verbenone derivative (**507**, X = O), and the latter gives the o-menthene **508** on treatment with p-toluenesulfonic acid in acetic acid and acetic anhydride. After metal hydride reduction, **508** can be converted to the lactone **506** with p-toluenesulfonic acid. This paper also contains much useful information about the opening of pinane systems to o- rather than to p-menthanes.[513] This is likely to be a similar route to the biogenetic one; Lander and Mechoulam even suggest that the lactone **506** might be an artifact arising from the acid hydrolysis of the urine employed in its workup.

It is clear that a small modification of Ficini's ynamine synthesis (Scheme 25, p. 96) will lead to an entry into substituted o-menthanes, and she has made the acids **509** in this way from 4-methyl-2-cyclohexenone, subsequently correlating them with o-menthan-4-ones prepared from the copper-catalyzed addition of isopropylmagnesium bromide to 4-methyl-2-cyclohexenone.[515]

---

*The structure of carquejol is frequently incorrect in the literature.[510] The original structure[508] was later modified.[509]

10. THE *m*-MENTHANES

As stated in Vol. 2 (p. 137), most naturally occurring *m*-menthane derivatives reported are probably to be ascribed to artifacts from carenes, notably the "sylveterpineols" (**510a, 510b**) of Wallach,[516] one of which **510a** has recently been isolated from *Laggera aurita*.[517] Perhaps the 6,8-*m*-menthadien-3-ol (**511**) isolated from *Cannabis sativa* resin,[518] but not yet synthesized, falls into this category too. Probably more authentically "natural" are isothymol (**512**), synthesized many decades ago, and now isolated from *Neurolaena oaxacana*,[519] its esters from *Eupatorium* species,[520] and, together with an ether, from *Senecio* sp.[521]

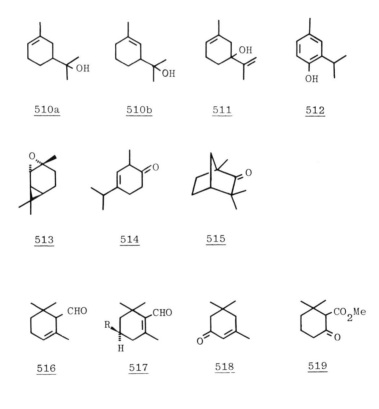

510a   510b   511   512

513   514   515

516   517   518   519

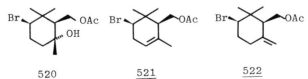

520   521   522

Among the convertions of carenes to *m*-menthenes, a recently reported reaction of Lewis acids with carene 2,3-epoxide (513) is very interesting, because the main product, reported to be 514,[522] does not have the same spectral properties as another compound said to possess this structure, and made by the action of trifluoromethanesulfonic anhydride on fenchone ((±)-515).[523] At first sight, the evidence presented in Ref. 522 appears to be more reliable. Ficini's ynamine synthesis has also been adapted to the *m*-menthane group.[524]

11. THE TETRAMETHYLCYCLOHEXANES

A. 1,1,2,3-Substituted Cyclohexanes

The naturally occurring compounds of this group arise formally by cyclization of the dimethyloctane monoterpenoids, to form, for example, two of the most common members of the group, α- (516) and β-cyclocitral (517, R = H). The latter is widespread, and occurs in oxygenated forms. Saffron (*Crocus sativa*) for example, contains a number of cyclocitrals and related substances oxygenated at C(4) and C(5),[525] particularly picrocrocin, the full stereochemistry of which has been shown to be 517, R = β-glucosyl.[526] The corresponding acid to picrocrocin occurs in *Gardenia jasminoides grandiflora*.[527] A list of related compounds detected in virginia tobacco has been given,[528] and β-cyclocitral is also the only monoterpenoid detected in the blue-green algae *Microcystis wesenbergii* and *M. aeruginosa*.[529]

Some of the compounds of the series lack a carbon atom at C(1), and these can readily be made by various oxidations of isophorone (518).[525,530]

The cyclization of geraniol (36) and related monoterpenoids has been known for a long time. Recent developments include the cyclization of geranyl acetate by thiyl radicals generated by photolysis of thiols (for instance), after Raney nickel desulfurization, the product is fully hydrogenated 516.[531] A mixture of thioethers corresponding to 516 and 517 is obtained on acid treatment of geranyl phenyl sulfide.[532] Acid treatment of the addition product of geranyl acetate epoxide and C_6H_5Se yields a cyclized product that was then converted to safranal (2,6,6-trimethylcyclohexa-1,3-dienecarbaldehyde).[532a] The ketoester 61 has been cyclized to 519, a versatile intermediate (although it was only used to make the readily accessible methyl β-cyclogeranate in this paper).[533] Geranyl acetate is cyclized by bromine in the presence of stannic bromide at 0°, when 16% of 520 is formed. The latter can be dehydrated to a mixture of bromine-substituted α-(521) and γ-cyclogeranyl acetates (522).[534] The interest in this synthesis lies in the possibility that it might be a model for the biosynthesis of some of the halogen-containing monoterpenoids found in marine organisms (see Ref. 46), syntheses of which are very rare. Although not strictly speaking the synthesis of a naturally occurring monoterpene, the recent work by Rouessac and Zamarlik is an interesting approach to the synthesis of 6-chloro-1,1,2,3-substituted cyclohexanes. After placing a function in the 2-position by a sigmatropic reaction of, say, the vinyl ether of **523**, thus leading to **524** (several methods besides this were used, with products having different functions for R), they found that by refluxing the acid chloride **524** (R = COCl) in benzene, the chloroester **525** was obtained.[535] This type of intramolecular exchange was first described by Prelog in 1941.[536]

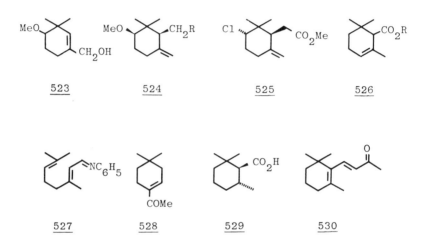

| 523 | 524 | 525 | 526 |

| 527 | 528 | 529 | 530 |

Geranyl cyanide yields 15% of cyclized material with Lewis acid and 2,4,4,6-tetrabromo-2,5-cyclohexadienone,[537] and graphite bisulfate is said to favor the formation of the α-acid **526** (R = H),[538] though full details are not available. Methyl geranate is cyclized by liquid sulfur dioxide at −70°[539] or by mercuric trifluoroacetate[540] very efficiently to **526** (R = Me). Breslow's mechanism for the cyclization of geranyl acetate by dibenzoyl peroxide and copper salts (cf. Vol. 2, p. 139 and Ref. 541) has been criticized by Beckwith et al., who point out that a pure radical mechanism should give a 5-membered rather than a 6-membered ring.[542] An old method for the preparation of the cyclocitrals was the treatment of citral anil (**527**) with sulfuric acid at low temperature; the main problem being the separation of the β-isomer **517** (R = H) from an accompanying by-product **528**.[543] A practical method for overcoming the difficulty has been given, consisting in avoiding a steam distillation during workup.[544] The situation during this synthesis is reminiscent of similar difficulties encountered during certain reactions of piperitone (**415**, see p. 000 and Ref. 415). Derived from this synthesis is the treatment of citral pyrrolidine enamine with 90% sulfuric acid at 0°. After hydrolysis, α-cyclocitral (**516**) is obtained.[545] Using an optically active amine to make the enamine leads to optically active α-cyclocitral, L-proline diethylamide, for example, gives (R)-α-cyclocitral.[546] Optically active α-cyclocitral has also been prepared from α-cyclogeranic acid (526, R = H), which is readily resolved.[547] Reduction to the alcohol is straightforward, but the Oppenauer oxidation to the optically active aldehyde 516 requires careful conditions so that the product is rapidly removed from the reaction medium.[548] The optically active 516 thus obtained was converted to the acid 529, present in Californian crude oil.[549]

Aside from the cyclization methods of dimethyloctane monoterpenoids, the best method of formation of β-cyclocitral is by ozonolysis of β-ionone (530).[550] The same method using α-ionone to obtain α-cyclocitral is less efficient because both double bonds of the ionone are attacked simultaneously.

Hunt and Lythgoe have described an intriguing method for the preparation of γ-cyclocitral (531), β-cyclocitral being obtained from this by base-catalyzed conjugation of the double bond. The bromide 532 is condensed with dithiane, and the sulfonium salt thus obtained is rearranged to the thioacetal of γ-cyclocitral by butyllithium (Scheme 26).[551]

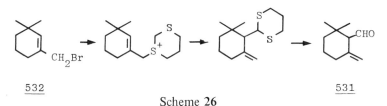

<u>532</u>   <u>531</u>

Scheme 26

Another type of oxidation of tetrasubstituted cyclohexanes is illustrated by esters of 533 (R^1 = Me, R^2 = CHO), the angelate of which has been isolated from *Bupleurum gibraltaricum*[552] and other Umbelliferae.[553] Esters of the compound 533 (R^1 = CHO, R^2 = Me) have also been found in *Piqueria trinervia* (Compositae).[554] Access to this type of substance is by oxidation of the anion of ethyl β-safranate (534) which occurs in the C(5) position.[555] Photooxidation of ethyl α-safranate (535) has also been examined, and while the main products obtained after reduction of the initially formed peroxide 536 are oxidized at C(3), 22% of a C(5) alcohol is obtained.[556]

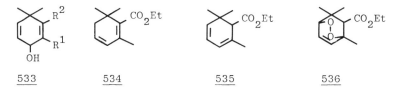

<u>533</u>   <u>534</u>   <u>535</u>   <u>536</u>

Recently, Mukaiyama et al. have described a seven-step synthesis of karahana ether (537) from the amide 538,[557] the latter having been made by a organometallically accelerated intramolecular Diels-Alder reaction the same authors described earlier.[558] This synthesis is long, and does not seem to offer any improvement over the earlier synthesis of Coates and Melvin[559] (which also involves the *cis*-diol **539** as a key intermediate; cf. Vol. 2, p. 139, and Refs. 431 and 432). On the other hand, nerol (**28**) is cyclized directly to a 15:45 mixture of karahana ether (**537**) and its *endo*-unsaturated isomer with thallium(III) perchlorate.[560]

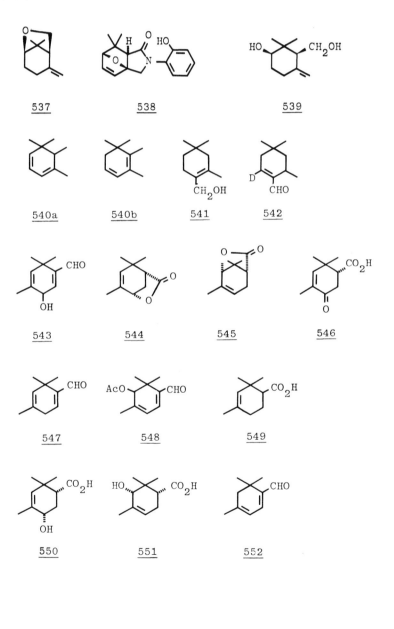

537

538

539

540a

540b

541

542

543

544

545

546

547

548

549

550

551

552

Before leaving this group, mention should be made of a possible synthetic approach from the pyronenes (**540a, 540b**), pyrolysis products of α-pinene (**245**). Various oxidations of α-pyronene (**540a**) were initiated by Cocker et al. in 1973,[561] then De Pascual Teresa et al. studied the oxidation of both pyronenes at a lower temperature and in a two-phase system which enabled further products to be isolated,[562] and finally Cocker completed the earlier work with an examination of β-pyronene (**540b**).[563] These oxidations did not, however, lead to natural products.

B. 1,1,3,4-Substituted Cyclohexanes

This is the cyclolavandulol skeleton (Vol. 2, p. 48), cyclolavandulol (**541**) itself being found as an ether with furocoumarins in *Seseli iliense* (Umbelliferae).[564] The acid has been isolated again from another species, *Carum roxburghianum*,[565] but there is nothing new concerning the synthesis.

A deuterium-labeled double-bond isomer **542** of cyclolavandulal has been synthesized.[566]

C. 1,1,2,5-Substituted Cyclohexanes

Esters of the hydroxyaldehyde **543** were isolated from *Ferula hispanica* in 1969,[567] and Bohlmann suggested the name "ferulol" for this alcohol, further esters of which were isolated later from other umbelliferae.[552,553,568] Two lactones of the group, (−)-filifolide A (**544**) and (+)-filifolide B (**545**), together with isophorone (**518**) and the ketoacid (**546**) occur in the compositae *Artemisia filifolia* (the skeleton being exaggeratedly called "unique" by the authors).[569] 1,1,2,5-Tetramethyl-5-cyclohexen-4-one has been found in iris rhizomes.[570]

Bohlmann's synthesis of ferulol (**543**) consists in oxidizing the Diels-Alder product **547** of propynal and 2,4-dimethyl-1,3-pentadiene with lead tetraacetate. Two products were obtained, **548** and the acetate of ferulol, from which the alcohol **543** was easily obtained.[571]

Allylic bromination (*N*-bromosuccinimide) of 2,2,4-trimethyl-3-cyclohexene-carboxylic acid (**549**) yielded a bromide, which, after losing hydrogen bromide, gave the lactone **544**.[569] More interesting is the conversion of (−)-chrysanthenone (**145**) to the same lactone. Epoxidation of chrysanthenone was known to give an epoxide,[572] (the stereochemistry of which had definitely been established as *trans*[573]), and this epoxide yielded the hydroxyacid **550** with hydroxide ion.[572] When either chrysanthenone epoxide or **550** are heated, they are converted to (−)-filifolide A (**544**), thereby proving the absolute stereochemistry of the natural product.[569] An interesting relationship also exists between **544** and **545**. They are interconverted by gas chromatography, but not by steam distillation, whereas the corresponding hydroxyacids, **550** and **551**, are interconverted by steam distillation. These facts enabled the absolute configuration of **545** also to be established.[569]

The formation of the isomer **552** of **547** by base-catalyzed condensation of two molecules of 3-methyl-2-butenal has been recently reinvestigated, and several dihydro compounds of **552** were prepared.[574]

12. THE DIMETHYLETHYLCYCLOHEXANES

This small group of compounds comprises four carbon skeletons. Violacene (**533**, revised structure[575]), isolated from the red alga *Plocamium violaceum*,[576] is a 1,3-dimethylethylcyclohexane. Plocamene B, (**554**) isolated from the same alga,[577] is a 2,4-dimethylethylcyclohexane. The 3,3-dimethylethylcyclohexane skeleton of the boll weevil pheromone constituents **555** (R = CHO or CH$_2$OH) is the same as that of a brominated terpenoid **556** isolated from the seaweed *Desmia (Chondrococcus) hornemanni*;[578] other marine terpenoids with this skeleton are known[579] but not synthesized. Finally the ketone **557**, isolated from *Juniperus communis*,[580] is a 1,4-dimethylethylcyclohexane.

The first two compounds, **553, 554** have not yet been synthesized. The ketone **557** is readily made by the stannic chloride-promoted Diels-Alder reaction between isoprene and 3-methyl-3-buten-2-one,[581] a six-step synthesis from diethyl 4-acetyl-4-methylheptane-1,7-dicarboxylate also having been described.[582] The bromide **556** has been synthesized by the brominative cyclization of myrcene (**6**) using 2,4,4,6-tetrabromo-2,5-cyclohexadienone, when **558** and **559** are the major by-products.[583]

The major synthetic effort in this group is thus concentrated on the boll weevil pheromones. Many routes require 3,3-dimethylcyclohexanone as a key intermediate. Pelletier and Mody prepared this substance by the copper-catalyzed Grignard addition to 3-methyl-2-cyclohexenone. Subsequent addition of acetylene and acetylation yielded **560**, from which the two stereoisomeric aldehydes **555** (R = CHO) were obtained using silver carbonate in acetic acid then hydrolysis. The alcohols **555** (R = CH$_2$OH) were obtained by borohydride reduction of the aldehydes.[584] A variant of the route consists in using the zinc chloride-catalyzed addition of ethyl vinyl ether to the acetal of 3,3-dimethyl-

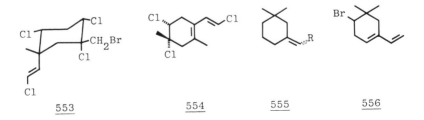

553   554   555   556

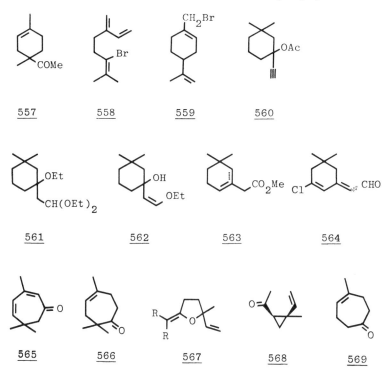

557   558   559   560

561   562   563   564

565   566   567   568   569

cyclohexanone. This yields **561**, from which the aldehydes **555** (R = CHO) are obtained by treatment with sodium acetate in acetic acid and pyrolysis.[585]

Several other methods involving 3,3-dimethylcyclohexanone have been published. The original synthesis used a Reformatsky reaction which led to the corresponding esters **555** (R = CO_2Et). These were then reduced with lithium aluminum hydride to the alcohols **555** (R = CH_2OH), from which the aldehydes are accessible with manganese dioxide.[586] Vig et al. used a modified Wittig reaction (ethyl trimethoxyphosphonoacetate),[587] as did Babler and Mortell (diethylcyanomethylphosphonate; these authors also made their 3,3-dimethyl-cyclohexanone from *m*-methylanisole)[588]. Less convenient reagents are the dihydro-1,3-oxazine reagent of Meyers et al. (which additionally gave an *endo*-double bond isomer of **555**, R = CHO)[589] and the acetaldehyde equivalent, (*Z*)-2-ethoxyvinyllithium, of Wollenberg and Pereis. The product **562** of this reaction is converted to **555** (R = CHO) by silica gel.[590] Alkylation of 3,3-dimethylcyclohexanone with *O*-ethyl *S*-ethoxycarbonylmethyl dithiocarbonate leads stereoselectively to the (*E*)-isomer of **555** (R = CO_2Et).[591] Apart from being economically more attractive, the synthesis of Bedoukian and Wolinsky is interesting in that it follows what is believed to be a biogenetic route. γ-

Geraniol (54) may be the precursor of all the boll weevil pheromones since, depending on the mode of cyclization, either grandisol (251) or 555 (R = CHO or CH_2OH) can be obtained. In this synthesis, methyl γ-geranate (the ester of 72) made by deconjugation of geranic acid (56) is cyclized with polyphosphoric acid to yield the *endo*-unsaturated esters 563. The double bond was brought into conjugation with the ester group by brominating (HBr) and dehydrobrominating (triethylamine), when the alcohols 555 (R = CH_2OH) were made as before.[592] Another cyclization, this time of 3,7-dimethyl-1,6-octadiene, leading to the 3,3-dimethylethylcyclohexane skeleton, but not, however, to the natural products, has been described.[593] The most straightforward synthesis is probably that of Hoffmann and Müller, who carried out a Vilsmeier reaction (formamide and phosphorus oxychloride) on isophorone (518) to obtain the imine of 564. This was converted to the aldehydes 555 (R = CHO) with sodium perchlorate or sodium iodide.[594] This approach has also been used by Traas, Boelens, and Takken, who found that the chloroaldehyde 564 can be reduced directly to the aldehydes 555 (R = CHO) over a poisoned palladium on charcoal catalyst.[595] These authors have also used an alternative route from isophorone (518), which involves reduction with sodium dihydrobis (methoxyethoxy)aluminate and dehydration before the Vilsmeier formylation.[596]

13. THE CYCLOHEPTANES

A. Trimethylcycloheptanes

In addition to the compounds described in Vol. 2 (p. 140), the doubly unsaturated ketone 565 has been isolated from pepper oil[372] and, together with karahanaenone (566) and the alcohol corresponding to 566, from *Cupressus sempervirens*.[597]

Thermolytic rearrangement of 2-methyl-2-vinyl-5-isopropylidenetetrahydrofuran (567, R = Me) leads directly to karahanaenone (Vol. 2, p. 142) and reactions that might lead to some 567 (R = Me) usually have 566 present in the products.[598] *cis*-1-Acetyl-2-methyl-2-vinylcyclopropane (568) rearranges thermally to 4-methyl-4-cycloheptenone (569), and so does 2,5-dimethyl-2-vinyl-2,3-dihydrofuran (570). Rhoads has shown that the two reactions are independent, 570 passing through an intermediate 567 (R = H) homologous to that of the karahanaenone synthesis.[599] The possibility that the acetylvinylcyclopropane synthesis passes through an enol intermediate was discussed by Rhoads,[599] and this approach has in fact been carried out by Wender and Filosa, who prepared the trimethylsilyl enol ether 571 conventionally from the alcohol 572 of the Grignard reaction between 2-methyl-2-vinyl-cyclopropyllithium and

isobutanal. Karahanaenone (**566**) was obtained by thermolysis of **571** followed
by desilylation (butyllithium in tetrahydrofuran at 25°).[600]

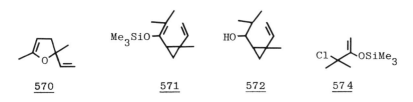

570   571   572   574

Cycloheptene carbocations can be constructed by the cycloaddition of an allyl
carbocation to a diene (see the review by H. M. R. Hoffmann,[601] for example),
and Hoffmann has synthesized karahanaenone (**566**) in this way from a
carbocation prepared by the action of a zinc-copper couple on 3-bromo-1-iodo-
3-methyl-2-butanone and isoprene under various conditions. The yields were
just over 50%, and considerable amounts of two other by-products (one, **573**,
the product of inverse addition of isoprene) were formed (Scheme 27).[602]
The dihaloketones can also be activated with $Fe_2(CO)_9$, and examples of
syntheses of this nature are described in the next section (thujaplicins). A similar
approach was followed by Shimizu and Tsuno, who made the trimethylsilyl
enol ether (**574**) of 2-chloro-2-methyl ˆ ' utanone, forming the carbocation
with silver perchlorate. In this case, the ˌoportion of karahanaenone (**566**) to
573 was 2.5:1.[603]

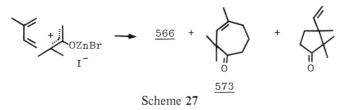

573

Scheme 27

The direct cyclization of a nerol derivative was effected by Hashimoto et al.;
this consisted in treating the trimethylsilyl ether **575** with methylaluminum
bis(trifluoroacetate). The ether **575** was made by opening neryl acetate epoxide
(**576**) with hydrogen bromide to give the bromohydrin, then oxidation and
treatment of the bromoketone with zinc in trimethylsilyl chloride.[604] A recent
modification using titanium complexes has led to a synthesis of nezukone (**577**,
see below).

It is clear that none of these methods really represents a synthesis for its own
sake; they are used to illustrate a principle of cyclization—but are possibly more
use for the preparation of sesquiterpenes.

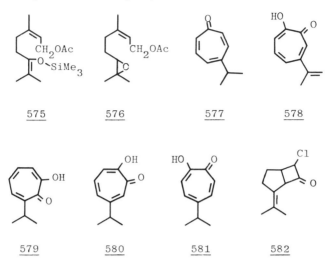

575   576   577   578

579   580   581   582

B. Isopropyl- and Isopropenylcycloheptanes

In addition to nezukone (**577**) and the thujaplicins, described in Vol. 2, p. 143, a member of the group that was not mentioned is β-dolabrin (**578**) isolated from *Thujopsis dolobrata*.[605]

This group of compounds was not given extensive treatment in Vol. 2, so a brief summary of early syntheses will be given. The first syntheses of α- (**579**), β- (**580** = "hinokitiol"), and γ-thujaplicin (**581**) consisted in the bromination of cycloheptane-1,2-diones, followed by dehydrobromination,[606,607] or oxidation of isopropylcycloheptatrienes, the latter being made by the ring expansion of benzenes (diazomethane and irradiation).[608] Later, Asao et al. attempted to synthesize β-dolabrin by the action of potassium acetate or tetramethyl-ammonium acetate on the product **582** of the [2 + 2] addition of chloroketene to dimethylfulvene, but unexpectedly, the main product was α-dolabrin (**583**), although small amounts of the β-isomer were obtained when the product derived from dichloroketene was used. By reduction of **583**, they were able to synthesize α-thujaplicin (**579**).[609] The [2 + 2] cycloaddition of dichloroketene to

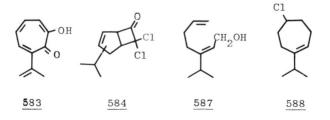

583   **584**   **587**   **588**

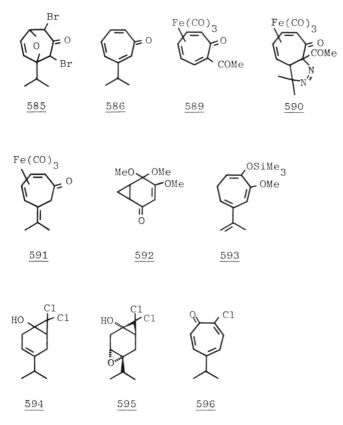

585   586   589   590

591   592   593

594   595   596

isopropylcyclopentadienes (mixture of isomers) gives a mixture of isomers **584**, the isopropyl group being attached at C(1), C(2), and C(3), treatment of which with potassium acetate in acetic acid yields a mixture of β- (**580**) and γ-thujaplicins (**581**).[610]

The [4 + 3] cycloadditions mentioned in the previous section were used by Noyori et al. to make isopropylcycloheptanes of this series.[611] The method required reaction of tetrabromoacetone with 2-isopropylfuran, for example, in the presence of iron nonacarbonyl, $Fe_2(CO)_9$, when the dibrominated cyclo-heptene oxide **585** is formed. Hydrogenation removes the bromine atoms, and boron trifluoride etherate removes the ether. Bromination (N-bromosuccini-mide) and dehydrobromination (lithium in dimethylformamide) give the isomer **586** of nezukone, which was converted to β-thujaplicin (**580**) by known techniques (hydrazine hydrate followed by potassium hydroxide).[612] By a similar route from 3-isopropylfuran (though with some improvements in the later stages), they prepared nezukone (**577**).[613] The synthesis of α-thujaplicin

(579) starting from 1,1,1,3-tetrabromo-4-methyl-2-pentanone and furan has also been described (together with further details about the synthesis of the other products of the series).[614]

Cyclization of the allyl alcohol 587 using a complex formed from methylaniline and titanium tetrachloride gave a chloride 588 that was converted to nezukone (577).[615]

In one of Noyori's papers[612] it is pointed out that there was at that time no way of introducing the isopropyl group into tropone site-specifically. Now Franck-Neumann et al. have supplied this deficiency with syntheses of β-thujaplicin (580) and dolabrin (578). The key step in these syntheses is the addition of 2-diazopropane to the complex 589, readily prepared from tropone. The resulting pyrazoline 590 undergoes loss of nitrogen and β-ketone hydrolysis in refluxing ethanol containing potassium carbonate, and the resulting complex 591 forms the material for synthesizing both 580 and 578. Removal of the iron (trimethylamine oxide) and conjugation of the double bond with potassium carbonate gives the ketone 586, which has already been converted to β-thujaplicin (580). Alternatively, oxidation of 591 (MnO$_2$) before removal of the iron gives the ketone 586, but having an isopropenyl side chain, from which β-dolabrin (578) can be obtained by the hydrazine hydrate method.[616]

Evans has developed a synthesis of β-dolabrin (578) that depends on ring expansion of a cyclohexenone. The action of oxosulfonium methylide on 3,4,4-trimethoxy-2,5-cyclohexadien-1-one gives the cyclopropane 592, the key intermediate to a variety of monocyclic tropolones. The side chain is introduced by a Grignard reaction, then potassium hydride in tetrahydrofuran, followed by trimethylchlorosilane gives the cycloheptatriene 593, from which β-dolabrin is obtained after oxidation (chloranil) and treatment with boron tribromide.[617] Another method involving a somewhat similar type of ring enlargement is that of Macdonald. Dichlorocarbene is added to the Birch reduction product of the triethylsilyl ether of 4-isopropylphenol. After desilylation, the product 594 is epoxidized, and the epoxide 595 treated with a trace of p-toluenesulfonic acid in refluxing benzene. The chlorocycloheptatrienone 596 thus obtained gives γ-thujaplicin (581) with aqueous phosphoric acid in refluxing acetic acid. By a similar route from the trimethylsilyl ether of 3-isopropylphenol it is possible to synthesize β-thujaplicin.[618]

Betains were suggested as sources of tropolones by Katritzky et al.,[619] and Tamura et al. have synthesized β-thujaplicin from the betain 597. Addition of methyl methacrylate to 597 gives the bicyclic compound 598, the methiodide of which is cleaved by sodium carbonate to 599. The hydrochloride of the latter gives the corresponding tropolone with potassium hydroxide, and decarboxylation to 580 is effected by copper chromite in quinoline.[620]

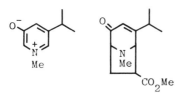

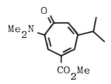

597   598   599

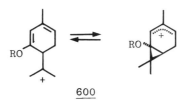

600

601a   601b   602   603

604   605   606   607   608

14. BICYCLO[3.2.0]HEPTANES

Ocimenone (132) is converted to filifolone (144) by the action of aluminum chloride,[161] and it was suggested that the route involves the ions 600, already proposed as intermediates in the formation of filifolone from chrysanthenone (145).[621] In their exhaustive examination of the possible routes to filifolone that such rearrangements might take, Erman et al. excluded a possibility that passed through a cycloheptene intermediate. In this connection, it is interesting that Capellini et al. found no substances with the bicyclo[3.2.0]heptane skeleton when they examined the solvolysis of 2,2,5-trimethyl-4-cycloheptenyl esters, but only menthane derivatives.[622]

15. BICYCLO[3.1.0]HEPTANES, THE THUJANES

A problem of nomenclature exists in this group. Some years ago, Brown et al. proposed that, as for comparable menthane derivatives, the prefix "iso-" be reserved for compounds having a *cis*-orientation of the methyl and isopropyl groups.[623] Most authors adhere to the principle now, but not all, and notably the review by Whittaker and Banthorpe[624] about the group employs the opposite convention (as these authors have continued to do in their publications). We use Brown's logical suggestion in this chapter, as we did in Vol. 2.

The group is widely distributed, and the thujones (601a, and isothujone 601b) are often believed to be associated with pharmacological activity,[625] but syntheses are very rare. Large amounts of isothujone (601b) are present in *Thuja plicata*, and it is possible to convert this to thujone (601a).[626] Sabinene (602) and the two isomers of sabinene hydrate (603) are also widely distributed, generally in small amounts.

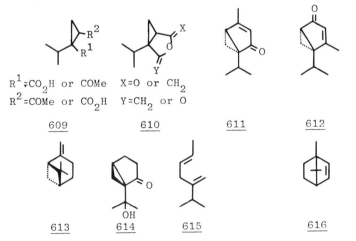

R^1 = CO_2H or COMe   X = O or CH_2
R^2 = COMe or CO_2H Y = CH_2 or O

609   610   611   612

613   614   615   616

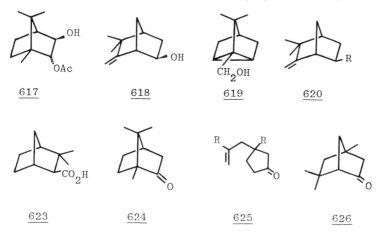

617   618   619   620

623   624   625   626

Alexandre and Rouessac prepared the cyclopentanone **604** by a fairly long route from methyl 1-isopropyl-3-cyclohexenylcarboxylate, then cyclized **604** to the bicyclic ketone **605** (R = H) with dicyclohexylcarbodiimide.[627] A more rapid synthesis is that of Gaoni, who found that the epoxide (**606**) of 4-isopropylidenecyclohexanone was converted to **605** (R = OH) with ethanolic sodium hydroxide, although the preparation of the isopropylidenecyclohexanone requires further stages.[628] 3-Isopropyl-2-cyclopentenone (the cyclization product of 6-methylheptane-2,5-dione), on treatment with excess sulfoxonium methylide yields a mixture of the two sabinene epoxides (**607**), making the sabinene hydrates (**603**) available by this route.[629]

These methods all depend on the formation of the cyclopropane ring from a cyclopentane or cyclohexane; another method starts with the cyclopropane in the form of the anhydride (**608**) of 1-isopropylcyclopropane-1,2-dicarboxylic acid. Reaction with lithium dimethylcuprate now yields a mixture of two isomeric ketoacids **609**, which cyclize with acid, to **610**. These substances are then allowed to react with the anion of diethyl methylphosphonate, to yield the natural product umbellulone (**611**) and its isomer **612**.[630]

There is one interesting conversion of the pinane skeleton to a "northujane." During a conventional oxidation of β-pinene (**613**) using the von Rudloff reagent, Jefford et al. observed the formation of the ketol **614**. They were able to dehydrate this to the corresponding isopropenyl derivative with thionyl chloride in pyridine.[631] If this method could be made more reliable and to work in higher yield, it would truly be a breakthrough so far as access to the bicyclo-[3.1.0]hexane system is concerned.

Salvan (**615**), the photolysis product from thujone (**601a**) has been synthesized by Sharma and Aggarwal; the synthesis is conventional except for the use of the nonconjugated 3-pentenal in a Grignard reaction with isopropylmagnesium iodide. Oxidation of the product and Wittig reaction to introduce the methylene group yielded salvan (615).[632]

16. BICYCLO[2.2.1]HEPTANES

Apart from many esters of borneol, the following bornane derivatives have been announced recently as natural products. Bornylene (616) has been found in *Abies alba* needles.[633] It has been known for many years, and the best synthesis is probably by the action of butyllithium on camphor *p*-toluenesulfonylhydrazone.[634] The monoacetate (617) of *trans*-bornanediol, isolated from *Artemisia vulgaris* has been named vulgarol.[635] The bornanediols are also well known, but a recent description of the physical properties includes adequate references for the synthesis.[636] The hydroxylated camphene nojigiku alcohol (618) has been isolated from *Chrysanthemum japonese*.[637]

The racemate of 618 had been known at least since it was prepared by Lipp in 1947 by the reaction of phosphorus tribromide on tricyclol (619), followed by alkaline treatment of the resulting bromide 620 (R = Br).[638] Other subsequent syntheses of the racemate followed a closely parallelled route, Sukh Dev using the action of *N*-bromosuccinimide on tricyclene to arrive at the same bromide 620 (R = Br),[639] While the chlorination of camphene (620, R = H) followed by the action of acetate gave an acetate, which, after several incorrect structures had been discussed, finally turned out to be 620 (R = Ac).[640] The first synthesis of the optically active material was by Julia et al., although they did not claim it to be a very efficient route for large amounts. This synthesis was interesting for a second reason, since it depended on the remote oxidation of (+)-camphene (620, R = H) using *t*-butyl perbenzoate in the presence of copper salts. Since the product, after treatment with potassium hydroxide, was optically active (+)-nojigiku alcohol (618), it must have been formed directly and not via tricyclol (619).[641] More detailed work on the remote oxidation of bornyl acetate (621, R = Ac) by chemical and microbiological means (using *Helminthosporium sativum* and *Fusarium culmorum*) was carried out by Money et al., who isolated various 3- and 5-oxygenated compounds,[642] following this up with a synthesis of (+)-nojigiku alcohol (618) from isobornyl acetate (622, R = Ac, Scheme 28).[643]

There was a slight confusion in the literature owing to an incorrect interpretation of the relationship between optically active isocamphanic acid (623) and camphene (620, R = H).[644] This error has been carefully corrected, (+)-isocamphanic acid (623) having been synthesized from the Diels-Alder product of cyclopentadiene and mesityl oxide (Vol. 2, p. 151), resolved,[645] and after its structure was confirmed by X-ray crystallographic work, converted by a Cope reaction of the corresponding amine to (–)-camphene (the enantiomer of 620,

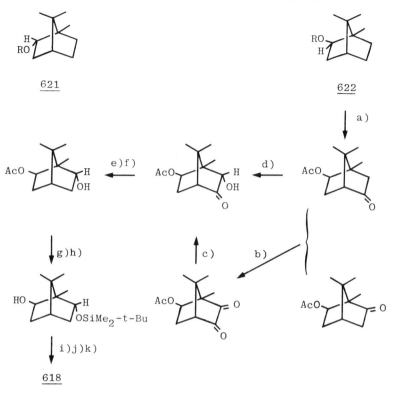

a) $CrO_3/Ac_2O/HOAc$; b) SeO_2; c) $Zn/HOAc$; d) $MoO_5/HMPA/pyr$;

e) $(CH_2SH)_2/BF_3 \cdot Et_2O$; f) Raney Ni; g) $t\text{-}BuMe_2SiCl/imidazole$;

h) $KOH/EtOH$; i) $MsCl/pyr/\Delta$; j) Bu_4NF/THF; k) Ac_2O/pyr

Scheme 28

R = H).[646,647] This work also produced the (−)-camphene with the highest recorded rotation.[647]

The full paper about Money's synthesis of racemic camphor (Vol. 2, p. 152) by treatment of dihydrocarvone (**401**) enol acetate with boron trifluoride has appeared;[648] soon after, Lange and Conia showed that simple pyrolysis of optically pure dihydrocarvone (**401**) yields camphor (**624**) which is 80% optically pure.[649] More recently, Wolff and Agosta heated the cyclopentanone **625** (R = Me) for 6 hours at 375° and obtained 71% of (±)-epiisofenchone (**626**), the simpler compound **625** (R = H) yielding 6-endo-methyl-2-norbornanone.[650]

The old principle of formation of a bicyclo[2.2.1]heptane by solvolysis of a 3-cyclopentenylethyl tosylate[651] was the basis of a synthesis of norcamphor from the tosylate **627**, which cyclizes via the enol to the extent of 3% in acetic acid, but 74% if urea is added.[652] This is also the principle of Cocker's synthesis of (+)-fenchone (**515**), although the starting material was (+)-α-pinene epoxide (**628**), the first step being the pinane-bornane rearrangement to **629**. Cleavage of the latter with silver acetate led to the aldehyde **630**. By conventional means, this was converted to the chloroketone **631**, which cyclized with ethoxide

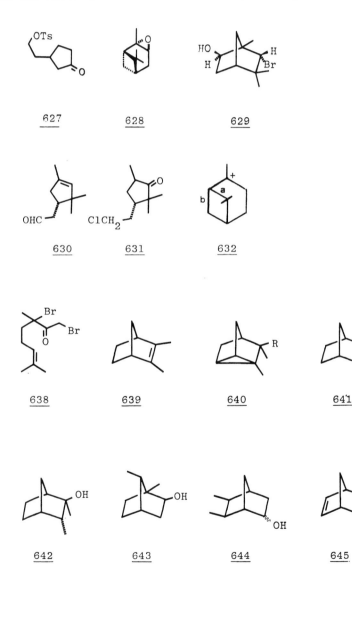

627	628	629
630	631	632

| 638 | 639 | 640 | 641 |

| 642 | 643 | 644 | 645 |

ion.[653] This is scarcely a "total" synthesis except indirectly.* Another interconversion from another monoterpenoid by ring opening to a suitable cyclopentene and ring closing was employed by Gream et al. to make high optical purity (−)-camphene from borneol (621, R = H); in the course of this work (related to bornyl carbonium ions) they also examined the conversion of isoborneol (622, R = H) to (+)- camphene (620, R = H).[654]

The classical rearrangements of pinyl ions (632) to bornanes (migration of bond a) or fenchanes (migration of bond b) are well-known entries to the two latter groups. One of us has evaluated the steric conditions necessary for migration of a or b,[655] and the alternatives between the two migrations and ring opening in the case of ions generated by the action of p-toluenesulfonic acid or zinc bromide on pinane 2-trans,3-trans-diol have been described by Pascual Teresa.[656] The situation has been extensively discussed by Whittaker et al.[657] More unusual is the rearrangement of the oxide 633 (prepared by Gibson and Erman[658]). This reaction has been studied by Bosworth and Magnus[659] and Grison and one of us,[660] in both instances the aim was to produce substances suited for elaboration to sesquiterpenes, although both described the synthesis of α-fenchol (634). A limited amount of acetic anhydride in the presence of boron trifluoride causes the oxide 633 to rearrange, giving the primary alcohol 635 as the major product, and lithium aluminum hydride reduction of the tosylate of 635 yields α-fenchol (634). The alcohol 635 is, however, unstable on prolonged treatment with acid (e.g., excess acetic anhydride and boron trifluoride, or reflux with carbon tetrachloride), and is readily converted to the secondary alcohol 636, acetolysis of which yields the acetate 637 by a second rearrangement (Scheme 29).

Camphor (624) has been synthesized in one step from the dibromide 638 using $Fe_2(CO)_9$ at 100-110°.[661]

A number of compounds related to the long-known santene (639)[662] have been described. Demole et al. have isolated several of these, including santalenone (640, R = COMe), 640 (R = CHO), and 641, from the top fraction of sandalwood oil (Santalum album).[663] The aldehyde 640 (R = CHO) was made from the well-known teresantalol (640, R = CH_2OH)[662] by oxidation. Photooxygenation of santene (639) followed by reduction of the hydroperoxide yielded the alcohol 641. The latter had probably already been isolated by Müller in 1900,[664] but not recognized.

An interesting compound identified by Demole was 642. This is the Wagner-Meerwein precursor of santenol (643), traditionally believed to occur in sandalwood oil, but which Demole did not find. 642 was made by hydroboration of santene (639), and there was no doubt about which isomer it was, all isomers

---

*Racemic 631 was also prepared from 2,2,4-trimethyl-3-oxocyclopentanecarboxylic acid.[653]

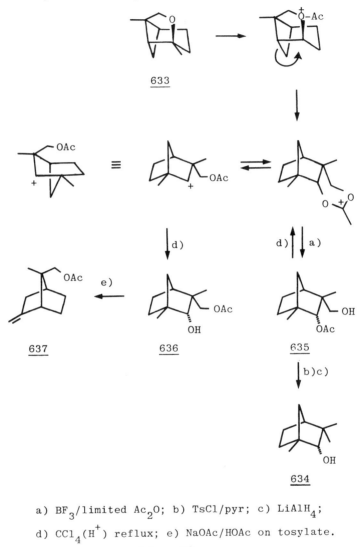

a) BF_3/limited Ac_2O; b) TsCl/pyr; c) $LiAlH_4$;

d) $CCl_4(H^+)$ reflux; e) NaOAc/HOAc on tosylate.

Scheme 29

having been described by Beckmann and Geiger in 1961.[665] Another hydrated santene **644** was suspected in *Cymbopogon jawarancusa* (a type of lemon grass), but not completely identified.[666] A synthesis of teresantalol (**640**, R = CH_2OH) uses the classical Diels-Alder way of making the bicycloheptane unit, but with the modification of having the allene, 2-methyl-2,3-butadienoic acid, as the dienophile. The product of the reaction with cyclopentadiene is a mixture of the *exo*- and *endo*-acids **645**. The *endo*-acid was separated by iodolactonization and regenerated with zinc, then formic acid yielded the tricyclene lactone **646**. Reduction of the ester of the acid **647** obtained by the action of sodium thiophenate on **646**, first with lithium aluminum hydride, then with Raney nickel, yielded teresantalol (**640**, R = CH$_2$OH).[667]

4-Chloro-2-methylbutenal is an intermediate in one industrial route to vitamin A and other polyenes,[668] and it reacts readily in a Diels-Alder addition both thermally and by Lewis acid catalysis with cyclopentadiene. The product **648** was hydrogenated and the dimethyl acetal of the saturated chloride dehydrochlorinated (potassium *t*-butoxide in dimethyl sulfoxide) to yield the aldehyde **649**,[669] which might also be a component of sandalwood oil.[663]

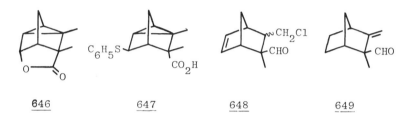

646   647   648   649

Teresantalic acid (**640**, R = CO$_2$H) has also been known as a constituent of sandalwood oil for a very long time, but syntheses are extremely rare. Now Monti and Larsen have made the racemate from the ketone **650**, the compound obtained by cyclization of 4-methyl-3-cyclohexenylacetic acid with trifluoroacetic anhydride. This synthesis is illustrated in Scheme 30,[670] the same authors have used the same starting material to synthesize racemic α-pinene (**245**) (below).

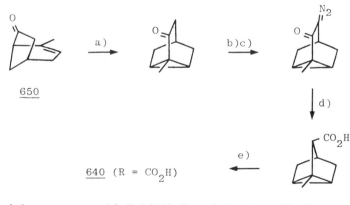

650

640 (R = CO$_2$H)

a) hν, acetone; b) NaH/HCO$_2$Et; c) tosyl azide/Et$_3$N;

d) hν, NaHCO$_3$/THF–H$_2$O; e) LiN–iPr$_2$, then BuLi/MeI.

Scheme 30

exo-Isosantene (**651**), not known to be naturally occurring, is difficult to synthesize. Now Castanet and Petit have found that treatment of 2-methylene-norbornane with palladium chloride yields 90% of the *exo*-methylated product **651**.[671]

Albene, now known to have the structure **652** was first isolated from plants of the genera *Petasites* and *Adenostyles*,[672] and in 1972 was attributed the formula **653**,[673] apparently confirmed by the preparation of a ketone which could also be obtained from albene.[674] In 1977 Kreiser et al. showed that something was wrong with this structure, and, after many difficulties, synthesized **653**, which was not identical with the natural product, and which they named isoalbene.[675] The synthesis started with the (difficult) Diels-Alder reaction between cyclopentadiene and dimethylmaleic anhydride, yielding **654**. This was converted via the dinitrile **655** ($R^1 = CH_2CN$, $R^2 = Me$) to the ketone **656**. Later

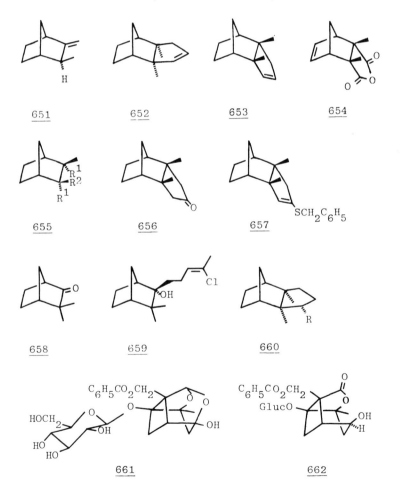

the ketone **656** was made by methylation of the more readily available ester **655** (R^1 = CO_2Me, R^2 = H)[676]. The benzyl thioenol ether **657** (following a method of Barton[677]) had the stereochemistry confirmed by X-ray crystallography, and it was found that the hydrocarbon obtained after desulfurization with Raney nickel was not albene, but isoalbene (**653**).[678] Kreiser then synthesized the correct structure **652** from camphenilone (**658**), following in fact a route that was fairly close to that described by Lansbury[674] in the earlier work. Grignard reaction made available the chloride **659**, and treatment of this with formic acid gave the ketone **660** (R = COMe)—this was the step where the rearrangement invoked by Lansbury had involved a different stereochemistry. Oxidation with trifluoroacetic anhydride and hydrogen peroxide gave a mixture of the acetate **660** (R = OAc) and the acid **660** (R = CO_2H), lead tetraacetate oxidation of which yielded albene (**652**).[679]

We cannot pretend to have listed the entire literature of labeled terpenoids of this series, which is voluminous, but the following will give key references. All the ^{14}C-methyl-labeled camphors have been made,[680] as has [14]C(8)-camphene[681]. Various deuterium-labeled camphors (in the C(8) methyl group)[682] and C(8)-deuterated camphene have been described.[683] Exchange reactions of various ketones of the series have been further examined,[684] and a series of deuterated alcohols (borneols and fenchols) has been made from α-pinene.[685] A synthesis of 3,3-difluorocamphor is also reported.[686]

17. BICYCLO[3.1.1]HEPTANES

The United States production of turpentine is around 95,500 tonnes annually (sulfate) and 9000 tonnes (wood), making the pinene skeleton by far the most commonly available from natural sources. Somewhat over half of this is converted to "synthetic" pine oil (mostly α-terpineol) and camphene. The former is used in soaps and the latter as a starting material for insecticides. Much of the remainder is used in the manufacture of the so-called "fine" terpene chemicals. linalool, geraniol, citral, etc.* It is thus not too surprising that there are still only very few syntheses of the pinane skeleton.

Recent novel terpenoids belonging to the group are the glycoside paeoniflorin (**661**) from Chinese paeony root (*Paeonia albiflora*),[687] where it is accompanied by albiflorin (**662**).[688] *cis*-Chrysanthenyl glucoside (**663**, R^1 = *O*-glucose, R^2 = H) has been found in *Dicoria canescens*,[689] and the corresponding acetate **663** (R^1 = OAc, R^2 = H) in the Australian Compositae *Centipeda cunninghamii*.[690] *trans*-Chrysanthenyl acetate (**663**, R^1 = H, R^2 = OAc) occurs in two species of *Chrysanthemum*.[691] Two hydroxylated *cis*-chrysanthenyl acetates

*We are most grateful to Dr. B. J. Kane, SCM Organic Chemicals, Jacksonville, Florida, for information about the U. S. turpentine industry.

(**664a** and **664b**), together with chrysanthenol and the corresponding ketone have been isolated from *Diotis maritima*.[692] A homoterpenoid, **665**, has been reported in *Artemisia annua*,[693] but the NMR data are somewhat sketchy and no stereochemistry was given; it was stated, however, that the NMR spectrum resembled the data given for the *trans*-isomer (**665**, as shown), which one of us had already synthesized from *trans*-pinocamphone **666** by making, first, the hydroxymethylene derivative, then reducing with formaldehyde.[694]

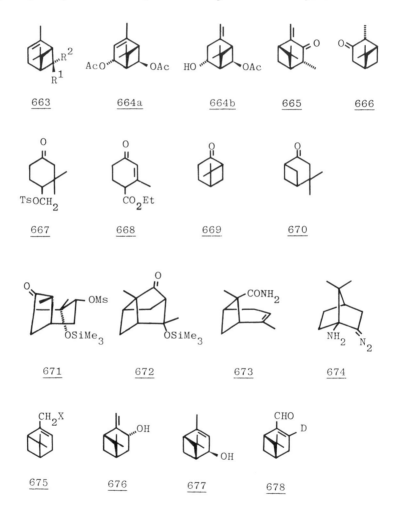

The first direct total synthesis of the pinenes was described by M. T. Thomas and Fallis in 1973.[695] It consists in cyclizing the tosylate **667**, prepared from Hagemann's ester (**668**) after protection of the keto group as a thioketal, lithium aluminum hydride reduction, additon of the extra methyl group by copper-catalyzed Grignard addition, and deprotection of the keto group using mercuric chloride and cadmium carbonate in acetonitrile. The cyclization of **667** (carried out using sodium hydride in dimethoxyethane) furnished nopinone (**669**) together with an isomer **670**, to avoid which it was necessary to block the C(6) position of the cyclohexanone **667** with benzaldehyde, removing the blocking group afterwards, best with potassium hydroxide and crown ether (see Ref. 696). Conversion of nopinone (**669**) to other pinane derivatives is well known.[697]

Larsen and Monti have also synthesized (±)-α-pinene (**245**) from the same material **650**[670] they used for the synthesis of teresantalic acid (**640**, R = CO$_2$H) (above).[698] Methylation of the ketone **650**, and protection of the glycol obtained by opening the corresponding epoxide gives the ester **671**, cyclization of which gives the pinane derivative **672**, that, after cleavage to the amide **673** opens the route to α-pinene (**245**).[698]

Kirmse's conversion of (+)-camphor (**624**) into optically pure (+)-β-pinene (**613**) also represents a total (though indirect) synthesis of the latter.[699] Generally ions of the type **632** rearrange to the bicyclo[2.2.1]heptane skeleton, but by placing a suitable leaving group at C(10) of camphor (i.e., on the bridgehead methyl group), Kirmse showed that rearrangement in the other direction could predominate (Scheme 31). Indeed, rearrangement of 1-amino-2-diazo-7,7-di-methylnorbornane (**674**) yields 91% of nopinone (**669**).[700] Another rearrangement of the Kirmse type using phosphorus pentachloride on 3-bromo-borneol is reported, but we have been unable to assess the experimental detail.[701]

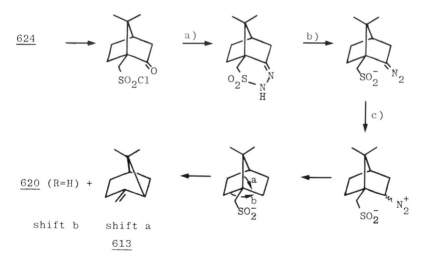

a) N$_2$H$_4$; b) hν; c) MeOH

Scheme **31**

Concerning the interrelation of different pinane derivatives, attention is drawn to the error in Vol. 2, pp. 156-157 concerning the reaction of N-bromosuccinimide with α-pinene, the major product from which is myrtenyl bromide (675, R = Br).[702] The best conditions for obtaining myrtenol (675, R = OH) from the readily accessible trans-pinocarveol (676) consist of treatment of the latter with phosphorus tribromide, when a good yield of the rearranged bromide (675, R = Br) is obtained.[703] Generally, conversion of β- to α-pinene is easy, while the reverse reaction, to obtain the more expensive (and frequently more useful) β-pinene (613) is more difficult. It has recently been reported that certain palladium on alumina catalysts are efficient for this reverse reaction.[704] Another important "synthesis" is that of the optically pure cis-verbenols (677) from α-pinenes (245), as a result of which Mori was able to show that the (1S,4S,5S) isomer (as shown, 677) has a positive rotation in chloroform and a negative rotation in methanol.[705] Mori also made the trans-verbenols (427).[706] These compounds are important as metabolites of pinene by the bark beetle Ips paraconfusus, the gut of which contains a microorganism, Bacillus cereus, which effects the oxidation.[707] The rabbit also metabolizes α-pinene to verbenol.[708]

Stereospecifically deuterium-labeled β-pinene [trans-C(3)],[709] C(3)-deuterated myrtenal (678),[566] and deuterated verbenones[711] have been prepared. Other deuterated β-pinenes have also been prepared.[710]

18. BICYCLO[4.1.0]HEPTANES

This group of monoterpenoids is as lacking in synthetic approaches as pinanes and thujanes. Once again, very large amounts of the hydrocarbon (+)-3-carene (679) are available (see Vol. 2, p. 157), making synthetic approaches less attractive. Suprisingly few oxidized carenes are naturally occurring, compared with other terpenoid groups. (–)-3-Caren-2-one (680, R = H) occurs in Zieria asplanthoides,[712] and the enedione (680, RR = O) in Asiasarum heteropoides,[713] but little else of novelty has been reported.

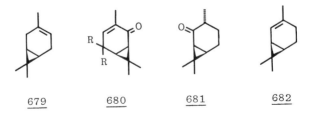

679   680   681   682

Some improvement in the synthesis of 2-carone (681) from dihydrocarvone (401), and conversion of 681 to 2-carene (682) have been described[714] (cf. Vol.

2, p. 157). Cyclization of 3-(2-chloro-2-propyl)cyclohexanones of this type is one of the best ways of making the carane skeleton, and forms the basis of the synthesis of chamic acid methyl ester (683) and chaminic acid (684) shown in Scheme 32.[715] Chamic acid and chaminic acid were isolated from *Chamaecyparis*

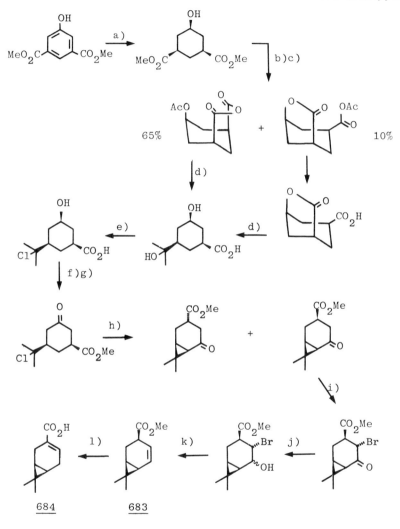

a) Rh/Al$_2$O$_3$-H$_2$; b) hydrolysis; c) AcCl; d) MeMgCl; e) HCl;

f) esterify; g) oxidize; h) KO-t-Bu; i) (PhNMe$_3$)Br$_3$;

j) NaBH$_4$; k) Zn/MeOH; l) alkali

Scheme 32

nootkatensis; their structure and stereochemistry was fully known,[716] and they had previously been obtained from other carane derivatives[717] (thereby having been synthesized indirectly), but this was the first direct synthesis.

The synthesis of optically active 2-carene (**682**) from the cyclohexenyl ester (**685**) of alanine has been effected by treatment of **685** with isopentyl nitrite in the presence of copper salts. This leads to a lactone **686** already containing the carane skeleton, and which is readily converted, first, to the alcohol **687**, then to 2-carene (**682**).[718] Cocker et al. have synthesized the carane skeleton by replacing the two bromine atoms of the norcarane **688** (obtained by dibromocarbene addition to the acetal of 3-cyclohexenone) with methyl groups using lithium dimethylcuprate for 4 days then adding methyl iodide. (Another approach using the ethylene thioketal gave a poorer yield.) The ketone **689** (X = O) obtained after deacetalization was converted to a mixture of 2-carene (**682**), 3-carene (**679**) and β-carene (**689**, X = CH_2) by Grignard reaction followed by dehydration.[719]

The conversion of pulegone (**354**) to carenes has been known for a long time (cf. Vol. 2, p. 157). Recent knowledge about the formation of hydroxycyclopropanes during zinc/hydrochloric acid reductions of unsaturated ketones[720] has led to a synthesis of the carane skeleton (as the hydroxy or acetoxy compound **690**, R = H or Ac) by the zinc/hydrochloric acid reduction of pulegone in the presence of acetic acid or acetic anhydride (the latter giving the acetate **690**, R = Ac).[721]

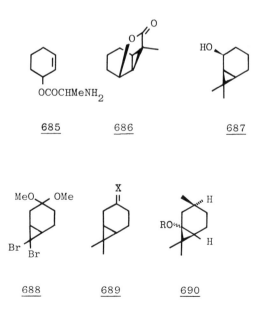

685   686   687

688   689   690

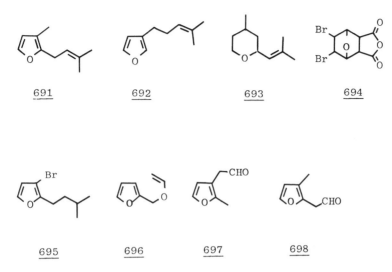

691   692   693   694

695   696   697   698

19. FURAN MONOTERPENOIDS

A. 3-Methyl-2-Substituted and 3-Substituted Furans

There has been a marked increase in the amount of synthetic work related to both this class of terpenoids and the pyrans (next section), possibly owing to greater facility in identification of these compounds from natural sources where they are fairly widespread, but nearly always in small amounts, and sometimes making an important contribution to the odor of the natural oil. A typical example is that of verbena oil, which contains small amounts of rose furan (691), perillene (692) and rose oxide (693).[722] Naves has reviewed rose furan and the related pyrans[723] (see next section).

The Büchi synthesis of rose furan (691, cf. Vol. 2, p. 161; Ref. 724) involved lithiation of 3-methylfuran, which involved replacement of a mercury salt. This awkward step has been avoided by Takeda et al. by making 2-bromo-3-methylfuran (known) from a Grignard reagent with magnesium and copper, and carrying out a Wurtz coupling with prenyl bromide (13).[725] Nguyên Dinh Ly and Schlosser have described a method which looks well on paper, but which from a technical point of view suffers from a number of disadvantages.[726] First, it requires 3-bromofuran, which generally involves a multi-step synthesis, for example, Diels-Alder addition of furan to maleic anhydride, bromination, and pyrolysis of the dibromo-adduct 694.[727] The 3-bromofuran thus obtained can be metallated with lithium diisopropylamide, but this requires a very low temperature, and reacts with prenyl bromide (13) in 66% yield. Metal exchange (butyllithium) and methylation (methyl iodide) of the bromide 695 then yields rose furan (691).[726] Another method involving introduction of the prenyl group to a preformed furan ring is that of Birch and Slobbe. 3-Methylfuran-2-carboxylic acid is reduced with lithium in liquid ammonia, and the resulting anion is quenched with prenyl bromide (13). It is then necessary to reoxidize and decarboxylate (using lead tetraacetate) to arrive at rose furan (691).[728] On a large scale, lead tetraacetate is an unsatisfactory reagent, making this too unacceptable.

In 1970 one of us reported that the vinyl ether (696) of 2-furfuryl alcohol on pyrolysis yielded the aldehyde 697.[729] By operating the other way round, that is, with the vinyl ether of 3-furfuryl alcohol, Vig et al. prepared the aldehyde 698, which they readily converted into rose furan by a Wittig reaction.[730] Unfortunately the yields of these steps are fairly low (apart from the use of mercury to make the vinyl ethers). Nevertheless, another rearrangement method has led to a technically feasible process for rose furan. This consists in treating the Grignard product 699 from furfural and 2-methyl-2-propenylmagnesium chloride with triethyl orthoacetate in the presence of pivalic acid, when a Claisen rearrangement occurs (Scheme 33) yielding the ester 700. Hydrolysis and decarboxylation, either preceded by or followed by isomerization (palladium on charcoal), then give rose furan (691).[731]

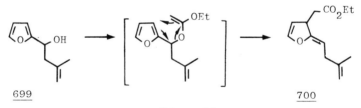

699   700

Scheme 33

Another technically feasible method has been described by Schulte-Elte, which involves construction of the furan ring by reaction with a mercury-impregnated ion exchange resin of the acetylene 701, obtained in turn from 2-methyl-2-propenyl chloride and the commercially available acetylenic alcohol 702.[732] The allylically rearranged isomers 703 of 701 also undergoes this cyclization;[732] the dangers of certain manipulations of the acetylenic precursors 702 and 704 should be noted.[733]

Elsholtzia ketone (705) has also been made by a "rearrangement" method. Cazes and Julia prepared the 3-furfuryl ether of 3-methylbutanal cyanohydrin (706) from 3-furfuryl bromide. This compound 706 undergoes a [2,3] sigmatropic rearrangement of the anion, made with lithium diisopropylamide, leading directly to 705.[159]

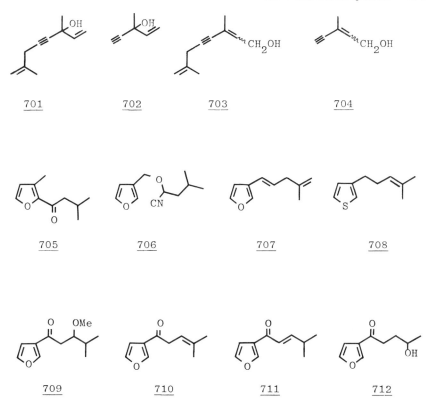

701   702   703   704

705   706   707   708

709   710   711   712

In addition to the list of naturally occurring 3-substituted furans given in Vol. 2 (p. 159), α-clausenane (707)[734] should be added.* We might also include the thio-analog 708 of perillene (692), which, together with some related substances containing additional sulfur atoms, has been isolated from hops,[735] but not yet synthesized. The ketol ether 709 has been reported in *Perilla frutescens*[736] — the same plant that contains egomaketone (710) and its isomer 711; such natural juxtapositions always make one wonder whether all of these compounds are really naturally occurring. Probably the highly toxic 4-ipomeanol (712) and related compounds, isolated from moldy sweet potatoes[737] are also isoprenoid.

All the older methods for preparing 3-substituted furans of this type involve attaching the side chain to a furan already substituted in the 3-position. A typical modification of this approach is that used in the synthesis of isoegoma-ketone (711) involving the addition of dimethylformamide dimethyl acetal to

---

*This compound (707) is also reported to occur in *Ledum palustre* and has been named lepalene,[734a] but the older name is preferable.

3-acetylfuran. The resulting enamine **713** reacts with isopropyllithium to give **711**, which the authors also reduced to perilla ketone (**714**). All the yields were fairly low, addition to the enamine **713** reaching only 37%.[738] Better yields were achieved by Hoppmann and Weyerstahl, who activated 3-furfural with a dithiane group, then added prenyl halide to the lithium derivative of the thioacetal **715** to obtain perillene (**692**).[739] A longer route of this type is that of Gosselin et al., who required 3-bromofuran[727] in order to prepare the thioester **716**. A Grignard reaction with prenyl bromide (and ethylmagnesium bromide) led to the thioacetal of egomaketone (**710**). The latter could then be obtained by dethioacetalization with cadmium carbonate and silver nitrate, or alternatively, sodium in liquid ammonia gave perillene (**692**).[740]

An ingenious approach starting from furan itself was published by Zamojski and Kozluk,[741] and immediately after by Kitamura et al.[742] This took advantage of the fact that when an aldehyde, RCHO, is irradiated in furan, a dioxabicyclo-[3.2.0]heptane **717** is formed. The yield is usually small (although the reaction works well when $R = CO_2Et$), but in view of the low cost, this is less important. p-Toluenesulfonic acid in carbon tetrachloride will open the 4-membered ring of **717** ($R = CH_2CH_2CHMe_2$), and the resulting alcohol is readily oxidized to perilla ketone (**714**).[741, 742]

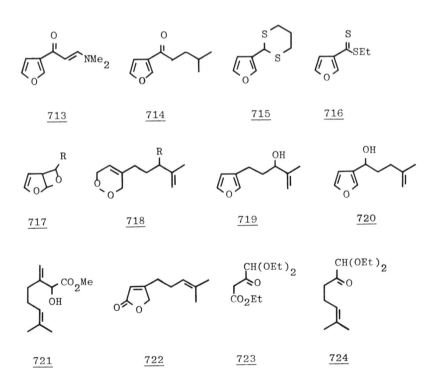

713   714   715   716

717   718   719   720

721   722   723   724

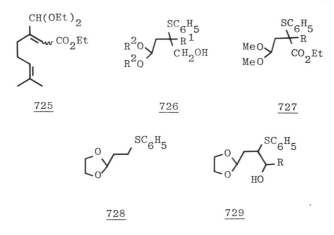

725   726   727

728   729

We have already spoken about the photooxygenation of myrcene (6) to two alcohols, 98 and 112.[134] Further addition of oxygen to 112 under the influence of light yields the endoperoxide 718 (R = OH), which can be converted to the corresponding furan 719* in various ways, Kondo having used lithium *t*-butoxide. Treatment of 719 with thionyl chloride, followed by lithium aluminum hydride reduction of the chloride thus obtained, gave perillene (692).[134] Several modifications of this scheme exist. For example, photo-oxidation of myrcene (6) to the hydroperoxide 718 (R = OOH) can be carried out in a single step,[743] and photooxygenation of ipsenol (96) can be used (following the same route) to prepare perilla ketone (714) or α-clausenane (707) via the alcohol 720.[744] An improved conversion of the peroxides like 718 to the corresponding furans using ferrous sulfate has also been described.[745]

Conversion of an ester of geranic acid (56) functionalized on the allylic methyl group to a γ-lactone is possible, the lactone being subsequently reduced with diisobutylaluminum hydride to a furan. There are examples of either direct functionalization of the geranic ester and of construction of such a molecule from smaller pieces. For the first example, Katsumura et al. oxygenated the anion of methyl geranate (obtained with lithium diisopropylamide in hexamethylphosphoramide), obtaining ester 721 as the main product (66%). The benzoate of 721 was cyclized to the lactone 722 with lithium iodide in dimethylformamide, and the lactone converted to perillene (692).[746] Takahashi's longer route to 722 consisted in addition of prenyl chloride to the substituted ethyl acetoacetate 723. After hydrolysis and decarboxylation of the product, the resulting substituted methylheptenone 724 must be converted to the corresponding geranic acid ester 725 with ethyl triethoxyphosphonoacetate. Cyclization to an ethoxybute-

*The alcohol 719 has been isolated from *Ledum palustre* essential oil and named lepalol.[734a]

nolide occurs on UV irradiation in the presence of acid, the reduction of this with sodium borohydride gives the butenolide 722.[747]

In the syntheses exploited by Inomata et al., the furan ring is constructed in the last step, by cyclization of the primary alcohol group and the aldehyde formed by the acid-catalyzed removal of the acetal of formula 726. Several ways have been used to make 726, which was not always isolated. One route involves condensation of bromoacetaldehyde dimethyl acetal with the anion of ethyl phenylthioacetate; the product 727 (R = H) can then form an anion which will react with an alkyl halide yielding the ester 727 (R = alkyl). Lithium aluminum hydride reduction then gives 726, cyclizable to the 3-substituted furan with acid.[748] Alternatively, the acetal 728 can be metallated (butyllithium) and condensed with an aldehyde, RCHO. This time the resulting alcohol 729 must be oxidized, then, if R = $CH_2CH_2CHMe_2$, the anion of the ketone obtained yields perilla ketone (714) after reaction with formaldehyde and acid workup.[749]

B. 2,5,5-Substituted Tetrahydrofurans

All four stereoisomers of lilac alcohol (730) are present in lilac flower oil (*Syringa vulgaris*),[750] and their synthesis has been accomplished from linalyl acetate (731, R = H_2). The first step, oxidation (selenium dioxide), yields an aldehyde 731 (R = O), which can be converted to the alcohol 732. The cyclization step to 733 can either be carried out directly on 731 (R = O)[751] or on 732,[752] but the yields are not very good in either case. Reduction of the aldehydes 733 to the lilac alcohols (730) is straightforward, and the various isomers of 730 can be separated by gas chromatography.[752] Lilac alcohols have also been synthesized by cyclization of the ester corresponding to 732, then lithium aluminum hydride reduction.[753, 109]

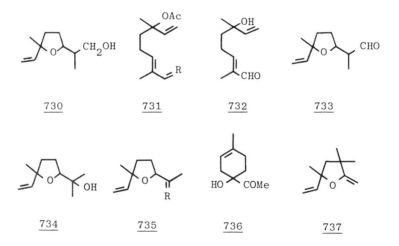

730   731   732   733

734   735   736   737

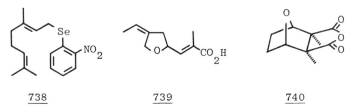

738   739   740

Linalool oxides (734) can be dehydrated to 735 (R = CH₂), and it might be thought that Grignard additions to the corresponding ketone 735 (R = O, obtained by ozonolysis of 735, R = CH₂ [754]) would provide an entry to other tetrahydrofurans of this type. This is not the case, and an attempt to synthesize linalool oxides (734) again from 735 (R = O) yielded 736, the product of a [2,3] sigmatropic reaction of the anion of 735 (R = O).[755] The other dehydration product 567 (R = Me) of linalool oxide rearranges to karahanaenone (566, see Vol. 2, p. 142), and it was synthesized by Chidgey and Hoffmann from isoprene and 1,3-dibromo-3-methyl-2-butanone with copper bronze and sodium iodide in acetonitrile. The isomer 737 was separated from the reaction mixture too, but yields were not very good.[756] A new synthesis of linalool oxides (734) from geraniol involves reaction with 2-nitrophenylselenocyanide in the presence of tributylphosphine. The selenide 738 thus obtained is oxidized to the linalool oxides (734) with 30% hydrogen peroxide in tetrahydrofuran.[757]

C. Other Furans

Menthofuran and 1,4-cineole occur under the section about menthanes.

The acid 739 appears to be a monoterpene, and occurs as an ester with a triterpene alcohol in *Acacia concinna*,[758] but it has not been synthesized.

It was once believed that cantharidin (740) was isoprenoid, especially since it is synthesized from farnesol and methyl farnesate by males of the "spanish fly" *Lytta vesicatoria*, more efficiently when stimulated by copulation;[759] later results, however, show that it is apparently not isoprenoid.[760]

20. PYRAN MONOTERPENOIDS

Attention is drawn to the review already mentioned.[723]

2,6,6-Trimethyl-2-vinyltetrahydropyran (741, R = H; Vol 2, p. 167) has been synthesized by Torii et al. (who incorrectly refer to it as "linaloyl oxide") from the epoxide 126 via the dithiane 127. The latter was cyclized with boron trifluoride, then the dithiane group removed with boron trifluoride/mercuric oxide. The resulting aldehyde was reduced to the alcohol, the xanthate of which was dehydrated to the pyran 741 (R = H).[152]

There is a close relation between the alcohols **741** (R = OH) and the linalool oxides (**734**); in *Lilium makinoi* they all have the same absolute configuration at C(2) as that of the congeneric linalool (**11**), which is presumed to be the biogenetic precursor[761] (see Vol. 2, p. 166). It was reported that the two acetates **741** (R = OAc) occurred in Indian linaloe (*Bursera delpechiana*),[762] but it transpired that they were not this structure, but were the acetates of linalool oxide (**734**, acetate).[763]

It is interesting to note that linalool (**11**) is cyclized by mercuric acetate to **741**, R = HgOAc, but borohydride reduction of this does not give simple demercuration to **741** (R = H) but mostly cyclopentane ethers.[764]

The antibiotic nectriapyrone (**743**) was isolated from the fungus *Gyrostroma missouriense*,[765] and synthesized from tiglic aldehyde [(*E*)-2-methyl-2-butenal (**3**)] and ethyl methylacetoacetate. This condensation yields the dihydropyrone **744**, where it is necessary to protect the double bond in the side chain (by addition of bromine) before dehydrogenating; removal of the two bromine atoms then yields the antibiotic.[766]

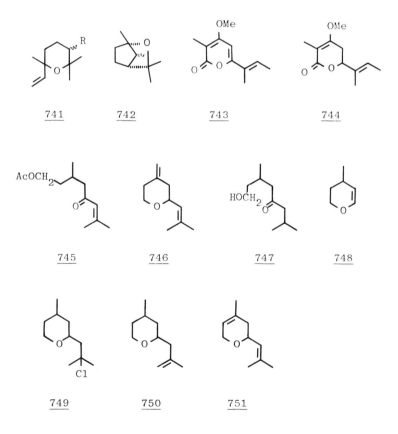

741	742	743	744

745	746	747	748

749	750	751

Most of the remaining synthetic work on pyran monoterpenoids concerns rose oxide (**693**). Many methods are based (like the earliest ones, Vol. 2, p. 168) on the oxidation of citronellol (**33**) or some closely related compound. Thus *N*-iodosuccinimide in carbon tetrachloride gives a 36% yield of rose oxide from citronellol without loss of the stereochemistry.[767] Citronellyl acetate (**33** acetate) is oxidized with *t*-butyl chromate to **745** and products of further degradation. Reduction of **745** to the diol and cyclization with sulfuric acid gives the two isomers of rose oxide (**693**).[768] The dehydro analog **105** of **745** after reduction with lithium aluminum hydride cyclized to **746**, and the latter can be hydrogenated, in particular with homogeneous rhodium catalysts, to the rose oxides.[123] Another similar scheme, this time for dihydrorose oxide, consists in synthesizing the precursor **747** from the trimethylsilyl enol ether of 4-methyl-2-pentanone and the dimethyl acetal of 2-butenal before metal hydride reduction and cyclization as usual.[769] Anodic oxidation of citronellol can be made to yield a 71:29 mixture of *cis:trans* rose oxides.[770]

The other "classical" route to the rose oxides starts from the dihydropyran **748**. One modification involves the reaction of **748** with hydrogen chloride, then reaction of the 2-chlorotetrahydropyran thus obtained with isobutylene in the presence of zinc chloride yields the chloride **749**. This can be dehydro-chlorinated to yield the two rose oxides, and their isomers **750**.[771] Alternatively, the dihydropyran **748** has been converted to the tetrahydropyranyl ether of 2,4-dichlorophenol. The latter reacts with isobutenylmagnesium bromide to give the rose oxides, or isobutylmagnesium bromide to give the dihydro analogs.[772]

Tyman and Willis succeeded in constructing the pyran ring at the same time as the side chain is introduced. 3-Methyl-2-butenal (**102**) reacts with 3-methyl-3-butenol (**53**) in the presence of acid to yield a mixture of nerol oxide (**751**) and **746**. Reduction then gives rose oxides (**693**). The authors examined a number of reduction systems, and found varying amounts of rose oxide stereo-isomers (they also prepared the dihydrorose oxides).[773] Recently Schulte-Elte has found that treatment of *trans*-rose oxide with a Lewis acid yields a high proportion of *cis*-oxide.[774]

Many pyran monoterpenoids are related to other monoterpenoid skeletons (as, in fact, are those that have been given a separate section here), but the relation is frequently so close that it has been more convenient to treat them in the sections relating to their parent compounds: see, for example, artemisia compounds and iridoids.

ACKNOWLEDGMENTS

The number of people we ought to thank for assistance in the preparation of this chapter is enormous. We must, however, single out Dr. F. Delay for a particularly searching scrutiny (indeed, we are tempted to blame him for any errors remaining!). Dr. G. Ohloff of the Firmenich Laboratories, and Dr. W. Hoffmann of BASF also read the entire manuscript and made many valuable suggestions. Drs. K. H. Schulte-Elte, Sina Escher, and A. P. Uijttewaal supplied unpublished information and many references to the literature that we were not aware of. We have drawn heavily on the invaluable collection of literature contained in the *Specialist Periodical Reports of the Chemical Society on Terpenoids and Steroids*, and the present reporter, Dr. R. B. Yeats of Bishop's University, Lennoxville, Canada, generously made available the manuscript of the monoterpenoid chapter of Volume 9 before its appearance. It must be emphasized that this chapter does not attempt to give as complete coverage of the literature as the *Specialist Periodical Reports*, but it attempts to be more critical about the syntheses selected.

REFERENCES

1. W. Hoffmann, *Seifen-Oele-Fette-Wachse*, **101**, 89 (1975); **105**, 287 (1979).

2. A. Nürrenbach, *Chem. Labor Betrieb*, **28**, 171 (1977).

3. G. Ohloff, Progress in the Chemistry of Organic Natural Products, Vol. 35, Springer-Verlag, Vienna, 1978, p. 431.

4. P. Wehrli and B. Schaer, *J. Org. Chem.*, **42**, 2939 (1977).

5. P. J. R. Nederlof, M. J. Moolenaar, E. R. de Waard, and H. O. Huisman, *Tetrahedron Lett.*, 3175 (1976); *Tetrahedron*, **33**, 579 (1977). See also Refs. 64 and 65 for the use of these synthons.

6. W. G. Salmond and M. C. Sobala, *Tetrahedron Lett.*, 1695 (1977).

7. T. Sato, H. Kise, M. Seno, and T. Asahara, *Yukugaku*, **24**, 607 (1975); K. Sato, S. Inoue, and S. Morii, *Chem. Lett.*, 747 (1975). K. Sato and S. Morii, *Asahi Garasu Koffo Gijutsu Shoreikai Kenkyu Hokoku*, **29**, 199 (1976). [*Chem. Abstracts*, **89**, 197734j (1978)].

8. G. Eletti-Bianchi, F. Centini, and L. Re, *J. Org. Chem.*, **41**, 1648 (1976).

9. R. Pummerer and W. Reindel, *Ber.*, **66**, 335 (1933).

10. A. N. Pudovik and B. E. Ivanov, *Zh. Obshch. Khim.*, **26**, 2771 (1956).

11. A. A. Petrov, *Zh. Obshch. Khim.*, **13**, 481 (1943); E. J. Reist, I. G. Junga, and B. R. Baker, *J. Org. Chem.*, **25**, 1673 (1960).

12. Y. Nakatani, M. Sugiyama, and C. Honbô, *Agr. Biol. Chem.*, **39**, 2431 (1975); see also carbonylation of isoprene, J. Tsuji and H. Yasuda, *Bull. Chem. Soc. Japan*, **50**, 553 (1977).

13. H. J. E. Loewenthal, *Synth. Comm.*, **5**, 201 (1975).

14. J. Paust, W. Reif, and H. Schumacher, *Annalen*, 2194 (1976).

15. K. Takabe, T. Katagiri, and J. Tanaka, *Tetrahedron Lett.*, 4009 (1972); *Bull. Chem. Soc. Japan*, **45**, 2662 (1972); **46**, 218, 222 (1973); A. Murata, S. Tsuchiya, A. Konno, J. Tanaka, and K. Takabe, Ger. Offen. 2,542,798 (Appl. Sept. 26, 74); K. Takabe, A. Agata, T. Katagiri, and J. Tanaka, *Synthesis*, 307 (1977).

16. L. I. Zakharkin and S. A. Babich, *Izv. Akad. Nauk SSSR, Ser. Khim.*, 2099 (1976).

17. H. Yagi, E. Tanaka, H. Ishiwatari, M. Hidai, and Y. Uchida, *Synthesis*, 334 (1977).

18. W. Hoffmann, F. J. Müller, and K. von Fraunberg, Ger. Offen., 2,154,370.

19. Y. Inoue, S. Sekina, Y. Sasaki and H. Hashimoto, *J. Syn. Org. Chem. Japan*, 36, 328 (1978).

20. K. V. Laats, T. A. Kaal, I. A. Kal'ya, I. B. Kudryatsev, E. A. Muks, M. A. Tali, S. E. Teng, and A. Yu. Erm, *Zh. Org. Khim.*, 10, 159 (1974); K. V. Laats and E. A. Muks, *Zh. Org. Khim.*, 10, 162 (1974).

21. K. V. Laats, S. E. Teng, and T. O. Savich, *Zh. Org. Khim.*, 10, 164 (1974).

22. F. Clouet and J. Brossas, *Makromol. Chem.*, 180, 875 (1979).

23. J. P. Neilan, R. M. Laine, N. Cortese, and R. F. Hevk, *J. Org. Chem.*, 41, 3455 (1976).

24. I. Mochida, S. Yuasa, and T. Seiyama, *J. Catal.*, 41, 101 (1976).

25. R. Baker, A. Onions, R. J. Popplestone, and T. N. Smith, *J.C.S. Perkin II*, 1133 (1975).

26. Y. Okuda, T. Hiyama, and H. Nozaki, *Tetrahedron Lett.*, 3829 (1977).

27. A. Hoppmann and P. Weyerstahl, *Tetrahedron*, 34, 1723 (1978).

28. G. Linstrumelle, R. Lorne, and H. P. Dang, *Tetrahedron Lett.*, 4069 (1978).

29. H. Baltes, E. Steckhan, and H. J. Schäfer, *Chem. Ber.*, 111, 1294 (1978).

30. W. C. Meuly and P. S. Gradeff, U. S. Pat., 3,668,255 (June 6, 1972).

31. R. S. DeSimone, U. S. Pat., 3,976,700 (Aug. 24, 1976).

32. U. T. Bhalerao and H. Rapoport, *J. Am. Chem. Soc.*, 93, 4835 (1971).

33. W. G. Taylor, *J. Org. Chem.*, 44, 1020 (1979).

34. S. Terashima, M. Hayashi, C. C. Tseng, and K. Koga, *Tetrahedron Lett.*, 1763 (1978).

35. M. Baumann, W. Hoffmann, and H. Pommer, *Annalen*, 1626 (1976).

36. H. Westmijze, H. Kleijn, J. Meijer, and P. Vermeer, *Tetrahedron Lett.*, 869 (1977).

37. G. Rice and J. F. Pollock, U. S. Pat., 3,714,283.

38. O. P. Vig, A. K. Vig, and S. D. Kumar, *Ind. J. Chem.*, 13, 1244 (1975).

39. O. P. Vig, M. S. Bhatia, A. S. Dhindsa, and O. P. Chugh, *Ind. J. Chem.*, 11, 104 (1973); O. P. Vig, B. Ram, U. Rani, and J. Kaur, *J. Ind. Chem. Soc.*, 50, 329 (1973).

40. O. P. Vig, S. D. Sharma, M. L. Sharma, and K. C. Gupta, *Ind. J. Chem.*, B, 15, 25 (1977).

41. K. H. Schulte-Elte and M. Gadola, *Helv. Chim. Acta*, 54, 1095 (1971).

42. A. D. Dembitskii, M. I. Goryaev, R. A. Yurina, A. E. Lyuts, and S. M. Vasilyuk, *Izv. Akad. Nauk Kaz. SSR, Ser. Khim.*, 28, 45 (1978).

43. A. F. Thomas and B. Willhalm, *Tetrahedron Lett.*, 3775 (1964).

44. A. O. Chong and K. B. Sharpless, *J. Org. Chem.*, 42, 1587 (1977).

45. A. Yasuda, S. Tanaka, H. Yamamoto, and H. Nozaki, *Bull. Chem. Soc. Japan*, 52, 1752 (1979).

46. D. J. Faulkner, *Tetrahedron*, 33, 1421 (1977).

47. A. Debal, T. Cuvigny, and M. Larchevêque, *Synthesis*, 391 (1976).

48. Z. Cohen, E. Keinan, Y. Mazur, and T. H. Varkony, *J. Org. Chem.*, 40, 2141 (1975).

49. A. L. J. Beckwith, C. L. Bodkin, and T. Duong, *Chem. Lett.*, 425 (1977).

50. W. Rojahn and E. Klein, *Dragoco Rep. (Ger. ed.)*, 24, 150 (1977).

51. Y. Kuwahara, S. Ishii, and H. Fukami, *Experientia*, **31**, 1115 (1975).

52. S. Arctander, Perfume and Flavor Chemicals, Montclair, N. J., 1969, No. 2327.

53. V. Rautenstrauch, *Helv. Chim. Acta*, **56**, 2492 (1973). Various catalysts, including copper chromite, and hydrogen also effect reduction of the hydroxylamine.

54. K. Takabe, T. Katagiri, and J. Tanaka, *Tetrahedron Lett.*, 3005 (1975); A. Murata, S. Tsuchiya, H. Susuki, and H. Ikeda, Japan Kokai 78, 144,514 [*Chem. Abstracts*, **90**, 168793 (1979)].

55. K. Takabe, T. Katagiri, and J. Tanaka, *Chem. Lett.*, 1031 (1975).

56. A. Yasuda, H. Yamamoto, and H. Nozaki, *Bull. Chem. Soc. Japan*, **52**, 1752 (1979).

57. K. Takabe, T. Katagiri, and J. Tanaka, *Chem. Lett.*, 1025 (1977).

58. K. Suga, S. Watanabe, and K. Hijikata, *Chem. Ind.*, 33 (1971).

59. K. Dunne and F. J. McQuillin, *J. Chem. Soc. (C)*, 2196 (1970).

60. M. Takahashi, H. Suzuki, Y. Moro-Oka, and T. Ikawa, *Chem. Lett.*, 53 (1979).

61. J. Martel and C. Huynh, *Bull. Soc. Chim. France*, 985 (1967).

62. D. A. Evans, G. C. Andrews, T. T. Fujimoto, and D. Wells, *Tetrahedron Lett.*, 1385; 1389 (1973).

63. M. Julia and D. Arnould, *Bull. Soc. Chim. France*, 743 (1973).

64. G. Cardillo, M. Contento, M. Panunzio, and A. Umani-Ronchi, *Chem. Ind.*, 873 (1977).

65. M. Julia and D. Uguen, *Bull. Soc. Chim. France*, 513 (1976).

66. M. Julia, D. Uguen, and A. Callipolitis, *Bull. Soc. Chim. France,* 519 (1976). Recently, the coupling between sulfone and Grignard reagent in the presence of a copper complex is described for preparing olefins. No hydrogenolysis step is then required; M. Julia, A. Righini, and J.-N. Verpeaux, *Tetrahedron Lett.*, 2393 (1979).

67. G. Cardillo, M. Contento, and S. Sandri, *Tetrahedron Lett.*, 2215 (1974).

68. B. S. Pitzele, J. S. Baran, and D. H. Steinman, *Tetrahedron*, **32**, 1347 (1976).

69. J. A. Katzenellenbogen and A. L. Crumrine, *J. Am. Chem. Soc.*, **98**, 4925 (1976).

70. F. Derguini-Bouméchal, R. Lorne, and G. Linstrumelle, *Tetrahedron Lett.*, 1181 (1977).

71. B. S. Pitzele, J. S. Baran, and D. H. Steinman, *J. Org. Chem.*, **40**, 269 (1975).

72. S. Kobayashi and T. Mukaiyama, *Chem. Lett.*, 705 (1974).

73. S. N. Huckin and L. Weiler, *Can. J. Chem.*, **52**, 2127 (1974) and literature quoted therein.

74. C. P. Casey and D. F. Marten, *Synth. Comm.*, **3**, 321 (1973).

75. C. P. Casey and D. F. Marten, *Tetrahedron Lett.*, 925 (1974).

76. S. Akutagawa and S. Otsuka, *J. Am. Chem. Soc.*, **97**, 6870 (1975).

77. B. A. Patel, L.-C. Kao, N. A. Cortese, J. V. Minkiewicz, and R. F. Heck, *J. Org. Chem.*, **44**, 918 (1979).

78. M. Seno, H. Kise, and T. Sato, Japan Kokai 76 56,406 (Appl. Nov. 8, 1974).

79. B. V. Burger, C. F. Garbers, and F. Scott, *J. S. Afr. Chem. Inst.*, **29**, 143 (1976).

80. A. Alexakis, A. Commercon, J. Villiéras, and J. F. Normant, *Tetrahedron Lett.*, 2313 (1976).

81. A. Alexakis, J. F. Normant, and J. Villiéras, *J. Organomet. Chem.*, **96**, 471 (1975); A. Alexakis, Thesis, Université P. et M. Curie, Paris 1975.

82. K. Itoh, M. Fukui, and Y. Kurachi, *J.C.S. Chem. Comm.*, 500 (1977).

83. K. Sato, S. Inoue, S. Ota, and Y. Fujita, *J. Org. Chem.*, **37**, 462 (1972).

84. J. A. Katzenellenbogen and T. Utawanit, *J. Am. Chem. Soc.*, **96**, 6153 (1974).

85. N. Okukado and E. Negishi, *Tetrahedron Lett.*, 2357 (1978).

86. A. Saito, K. Ogura, and S. Seto, *Chem. Lett.*, 1013 (1975).

87. M. Julia, C. Perez, and L. Saussine, *J. Chem. Res.*, (S) 268; (M) 3401 (1978).

88. M. Julia and L. Saussine, *J. Chem. Res.*, *(S)*, 269; *(M)*, 3420 (1978).

89. K. Tsuzaki, H. Hashimoto, H. Shirahama, and T. Matsumoto, *Chem. Lett.*, 1469 (1977).

90. H. D. Durst and E. Leete, *J. Labelled Comp.*, **7**, 52 (1971).

91. M. G. Peter, W.-D. Woggon, C. Schlatter, and H. Schmid, *Helv.*, **60**, 844, 1262 (1977).

92. S. J. Rajan and J. Wemple, *J. Labelled Comp.*, **11**, 467 (1975).

93. Y. Bessière, H. Savary, and M. Schlosser, *Helv. Chim. Acta*, **60**, 1739 (1977).

94. C. D. Poulter, J. C. Argyle, and E. A. Marsh, *J. Am. Chem. Soc.*, **99**, 957 (1977).

95. B. D. Mookherjee and R. W. Trenkle, *J. Agric. Food Chem.*, **21**, 298 (1973).

96. H. Kjøsen and S. Liaaen-Jensen, *Acta Chem. Scand.*, **27**, 2495 (1973).

97. R. Kaiser and D. Lamparsky, *Tetrahedron Lett.*, 665 (1977).

98. T. Suga, T. Hirata, Y. Hirano, and T. Ito, *Chem. Lett.*, 1245 (1976).

99. D. Behr, I. Wahlberg, T. Nishida, and C. R. Enzell, *Acta Chem. Scand., B*, **32**, 228 (1978).

100. R. Tschesche, F. Ciper, and E. Breitmeier, *Chem. Ber.*, **110**, 3111 (1977).

101. P. Schreier, F. Drawert, and A. Junker, *Z. Lebensm. Unters. Forsch.*, **155**, 98 (1974); *J. Agric. Food Chem.*, **24**, 331 (1976).

102. C. Bayonove, H. Richard, and R. Cordonnier, *Compt. Rend. Acad. Sci. (C)*, **283**, 549 (1976).

103. R. B. Bates, D. W. Gosselink, and J. A. Kaczynski, *Tetrahedron Lett.*, 199 (1967).

104. S. R. Wilson, K. M. Jernberg, and D. T. Mao, *J. Org. Chem.*, **41**, 3209 (1976).

105. O. P. Vig, S. D. Sharma, S. S. Rani, and S. S. Bari, *Ind. J. Chem., B*, **14**, 562 (1976).

106. O. P. Vig, J. Chander, and B. Ram, *J. Ind. Chem. Soc.*, **49**, 793 (1972).

107. O. P. Vig, B. Ram, U. Rani, and J. Kaur, *J. Ind. Chem. Soc.*, **51**, 616 (1974).

108. E. Guittet and S. Julia, *Synth. Comm.*, **9**, 317 (1979).

109. J.-P. Morizur, G. Bidan, and J. Kossanyi, *Tetrahedron Lett.*, 4167 (1975); G. Bidan, J. Kossanyi, V. Meyer, and J.-P. Morizur, *Tetrahedron*, **33**, 2193 (1977); J. Kossanyi, J. Perales, A. Laachach, I. Kawenoki, and J.-P. Morizur, *Synthesis*, 279 (1979).

110. F. Bohlmann and H. Kapteyn, *Tetrahedron Lett.*, 2065 (1973).

111. F. Bohlmann and H.-J. Bax, *Chem. Ber.*, **107**, 1773 (1974).

112. F. Bohlmann and C. Zdero, *Phytochemistry*, **16**, 780 (1977).

113. C. M. Harring and J. P. Vité, *Naturwissenschaften*, **62**, 488 (1975). For a review of ipsenois and related terpenoids (tagetones, etc.), see Y.-R. Naves, *Riv. Ital. Essenze, Profumi, Piante off., Aromat. Syndets, Saponi, Cosmet., Aerosois*, **60**, 553 (1978). For a review of the *Ips confusus* pheromone see M. C. Birch, *Am. Sci.*, 409 (1978).

114. J. P. Vité, R. Hedden, and K. Mori, *Naturwissenschaften*, **63**, 43 (1976).

115. K. Mori, *Tetrahedron Lett.*, 1609 (1976); *Tetrahedron*, **32**, 1101 (1976). A further full paper describes the preparation of (−)-(*R*)-ipsdienol and the natural (*S*)-isomer:

K. Mori, T. Takigawa, and T. Matsuo, *Tetrahedron*, **35**, 933 (1979).

116. R. M. Silverstein, J. O. Rodin, D. L. Wood, and L. E. Browne, *Tetrahedron*, **22**, 1929 (1966).

117. M. von Schantz, K.-G. Widén, and R. Hiltunen, *Acta Chem. Scand.*, **27**, 551 (1973).

118. R. H. Fish, L. E. Browne, D. L. Wood, and L. B. Hendry, *Tetrahedron Lett.*, 1465 (1979).

119. J. C. Clinet and G. Linstrumelle, *Nouv. J. Chim.*, **1**, 373 (1977).

120. K. Kondo, S. Dobashi, and M. Matsumoto, *Chem. Lett.*, 1077 (1976).

121. R. G. Riley, R. M. Silverstein, J. A. Katzenellenboge, and R. S. Lenox, *J. Org. Chem.*, **39**, 1957 (1974).

122. A. Hosomi, M. Saito, and H. Sakurai, *Tetrahedron Lett.*, 429 (1979).

123. C. F. Garbers and F. Scott, *Tetrahedron Lett.*, 1625 (1976).

124. J. A. Katzenellenbogen and R. S. Lenox, *J. Org. Chem.*, **38**, 326 (1973).

125. S. R. Wilson and L. R. Phillips, *Tetrahedron Lett.*, 3047 (1975).

126. S. R. Wilson, L. R. Phillips, and K. J. Natalie, Jr., *J. Am. Chem. Soc.*, **101**, 3340 (1979).

127. J. Haslouin and F. Rouessac, *Bull. Soc. Chim. France*, 1242 (1977).

128. L. Skattebøl, *J. Org. Chem.*, **31**, 2789 (1966).

129. S. Karlsen, P. Frøyen, and L. Skattebøl, *Acta Chem. Scand., B*, **30**, 664 (1976).

130. W. J. Bailey and C. R. Pfeifer, *J. Org. Chem.*, **20**, 1337 (1955).

131. M. Bertrand and J. Viala, *Tetrahedron Lett.*, 2575 (1978).

132. M. Mousseron and M. Vedal, *Bull. Soc. Chim. France*, 598 (1960) (despite the errors in drawing the formulas, this really does concern the ring opening of myrcene epoxide).

133. K. B. Sharpless and R. F. Lauer, *J. Am. Chem. Soc.*, **95**, 2697 (1973).

134. K. Kondo and M. Matsumoto, *Tetrahedron Lett.*, 391 (1976).

135. K. Mori, *Agric. Biol. Chem.*, **38**, 2045 (1974).

136. Y. Masaki, K. Hashimoto, K. Sakuma, and K. Kaji, *J.C.S. Chem. Comm.*, 855 (1979).

137. K. Mori, *Tetrahedron Lett.*, 2187 (1975).

138. G. Ohloff and W. Giersch, *Helv. Chim. Acta*, **60**, 1496 (1977).

139. A. G. Armour, G. Büchi, A. Eschenmoser, and A. Storni, *Helv. Chim. Acta*, **42**, 2233 (1959).

140. W. T. Ford and M. Newcomb, *J. Am. Chem. Soc.*, **96**, 309 (1974).

141. P. Chabardes and Y. Querou, Ger. Pat. Appl. 1,811,517 (Nov. 28, 1968).

142. N. C. Hindley and D. A. Andrews, Ger. Offen. 2,353,145 (Oct. 23, 1973); Belg. Pat., 783,055 (May 5, 1972).

143. M. B. Erman, I. S. Aul'chenko, L. A. Kheifits, V. G. Dulova, Ju. N. Novikov, and M. E. Vol'pin, *Tetrahedron Lett.*, 2981 (1976); *Zh. Org. Khim.*, **12**, 921 (1976).

144. H. Pauling, D. A. Andrews, and N. C. Hindley, *Helv. Chim. Acta*, **59**, 1233 (1976).

145. P. Chabardes, E. Kuntz, and J. Varagnat, *Tetrahedron*, **33**, 1775 (1977).

146. N. Götz and R. Fischer, Ger. Offen. 2,157,035 (May 24, 1973); 2,249,372 (Oct. 9, 1972); 2,249,398 (Oct. 9, 1972).

147. Y. Ichikawa and M. Yamamoto, Jap. Pat., 133,312 (1974).

148. Y. Ichikawa and M. Yamamoto, Jap. Pat. 75,014; 75,015 (1976).

149. W. Hoffmann, *Chem. Ztg.*, 23 (1973); L. M. Polinski, J. Der Huang and J. Dorsky, Ger. Offen. 2,338,291 (July 27, 1973).

150. T. Nakai, T. Mimura, and A. Ari-Izumi, *Tetrahedron Lett.*, 2425 (1977).

151. M. Tanaka and G. Hata, *Chem. Ind.*, 202 (1977).

152. S. Torii, K. Uneyama, and M. Isihara, *J. Org. Chem.*, 39, 3645 (1974).

153. F. Bohlmann and D. Körnig, *Chem. Ber.*, 107, 1780 (1974).

154. E. V. Lassak and I. A. Southwell, *Austral. J. Chem.*, 27, 2703 (1974).

155. B. Lefebvre, J.-P. Le Roux, J. Kossanyi, and J.-J. Basselier, *Compt. Rend. Acad. Sci. (C)*, 227, 1049 (1973).

156. B. A. McAndrew and G. Riezebos, *J.C.S. Perkin I*, 367 (1972).

157. R. Couffignal and J.-L. Moreau, *Tetrahedron Lett.*, 3713 (1978).

158. E. Guittet and S. Julia, *Tetrahedron Lett.*, 1155 (1978).

159. B. Cazes and S. Julia, *Synth. Comm.*, 113 (1977).

160. D. J. J. de Villiers, C. F. Garbers, and R. N. Laurie, *Phytochemistry*, 10, 1359 (1971).

161. D. R. Adams, S. P. Bhatnagar, R. H. Cookson, and R. M. Tuddenham, *Tetrahedron Lett.*, 3197 (1974); *J.C.S. Perkin I*, 1741 (1975).

162. E. P. Blanchard, Jr., *Chem. Ind.*, 294 (1958).

163. Oee Sook Park, Y. Grillasca, G. A. Garcia, and L. A. Malonado, *Synth. Comm.*, 345 (1977).

164. O. P. Vig, B. Ram, and B. Vig, *J. Ind. Chem. Soc.*, 50, 408 (1973).

165. T. Ioshida (Yoshida?) Mezhdunar 4th Kongr. Efirnym. Maslam, 1968 (publ. 1971), p. 123 [*Chem. Abstracts*, 79, 9750 (1973)].

166. W. W. Epstein and C. D. Poulter, *Phytochemistry*, 12, 737 (1973).

167. A. F. Thomas, *Specialist Report on Terpenoids and Steroids*, Vol. 2, K. Overton, Chemical Society, 1972, p. 14.

168. C. D. Poulter, S. G. Moesinger, and W. W. Epstein, *Tetrahedron Lett.*, 67 (1972). The mass spectrum of yomogi alcohol given in this paper is actually of artemisiatriene, the alcohol presumably having dehydrated in the source (see Ref. 182); C. D. Poulter, L. L. Marsh, J. M. Hughes, J. C. Argyle, D. M. Satterwhite, R. J. Goodfellow, and S. G. Moesinger, *J. Am. Chem. Soc.*, 99, 3816 (1977); C. D. Poulter and J. M. Hughes, *J. Am. Chem. Soc.*, 99, 3824, 3830 (1977).

169. C. D. Poulter, R. J. Goodfellow, and W. W. Epstein, *Tetrahedron Lett.*, 71 (1972).

170. Y. Chrétien-Bessière, L. Peyron, L. Bénezet, and J. Garnero, *Bull. Soc. Chim. France*, 2018 (1968).

171. D. A. Otieno, G. Pattenden, and C. R. Popplestone, *J.C.S. Perkin I*, 196 (1977).

172. T. Sasaki, S. Eguchi, M. Ohno, and T. Umemura, *J. Org. Chem.*, 38, 4095 (1973).

173. J. Shaw, T. Noble, and W. W. Epstein, *J.C.S. Chem. Comm.*, 590 (1975); see also Ref. 177.

174. T. A. Noble and W. W. Epstein, *Tetrahedron Lett.*, 3933 (1977).

175. S. K. Paknikar and J. Veeravalli, *Ind. J. Chem., B*, 18, 269 (1979); W. W. Epstein and L. A. Gaudioso, *J. Org. Chem.*, 44, 3113 (1979).

176. T. A. Noble and W. W. Epstein, *Tetrahedron Lett.*, 3931 (1977).

177. E. H. Hoeger, *Proc. Mont. Acad. Sci.*, 33, 97 (1973). This paper also reports compound 155, and predates Ref. 173.

178. H. Buttkus and R. J. Bose, *J. Am. Oil Chem. Soc.*, 54, 212 (1977).

179. W. W. Epstein, L. R. McGee, C. D. Poulter, and L. L. Marsh, *J. Chem. Eng. Data*, **21**, 500 (1976); cf. *Chem. Abstracts Index*, 1971-6.

180. D. V. Banthorpe and P. N. Christon, *Phytochemistry*, **18**, 666 (1979).

181. J. P. Scholl, R. G. Kelsey, and F. Shafizadeh, *Biochem. Syst. Ecol.*, **5**, 291 (1977).

182. A. F. Thomas and W. Pawlak, *Helv. Chim. Acta*, **54**, 1822 (1971).

183. J. Boyd, W. Epstein, and G. Fráter, *J.C.S. Chem. Comm.*, 380 (1976).

184. R. G. Gaughan and C. D. Poulter, *J. Org. Chem.*, **44**, 2441 (1979). Other lyratol esters besides the acetate have been isolated from *Chrysanthemum coronarium*; F. Bohlmann and U. Fritz, *Phytochemistry*, **18**, 1888 (1979).

185. F. Bohlmann and G. Florenz, *Chem. Ber.*, **99**, 990 (1966); F. Bohlmann and M. Grenz, *Tetrahedron Lett.*, 2413 (1969).

186. S. Yamagiwa, H. Kosugi, and H. Uda, *Bull. Chem. Soc. Japan*, **51**, 3011 (1978).

187. R. Kaiser and D. Lamparsky, *Helv. Chim. Acta*, **59**, 1797 (1976).

188. B. Corbier and P. Teisseire, *Recherches*, **19**, 289 (1974).

188a. L. Re and H. Schinz, *Helv. Chim. Acta*, **41**, 1695 (1958).

189. R. A. Benkeser, *Synthesis*, 347 (1971).

190. J.-P. Pillot, J. Dunoguès, and R. Calas, *Tetrahedron Lett.*, 1871 (1976).

191. P. Gosselin, S. Masson, and A. Thuillier, Tetrahedron Lett., 2717 (1978).

192. J. F. Ruppert and J. D. White, *J. Org. Chem.*, **41**, 550 (1976).

193. Y. Okuda, S. Hirano, T. Hiyama, and H. Nozaki, *J. Am. Chem. Soc.*, **99**, 3197 (1977).

194. K. Oshima, H. Yamamoto, and H. Nozaki, *Bull. Chem. Soc. Japan,* **48**, 1567 (1975).

195. B. M. Trost and W. G. Biddlecom, *J. Org. Chem.*, **38**, 3483 (1973).

196. V. Rautenstrauch, *Helv. Chim. Acta*, **55**, 2233 (1972).

197. D. Michelot, G. Linstrumelle, and S. Julia, *Compt. Rend. Acad. Sci. (C)*, **278**, 1523 (1974); *J.C.S. Chem.Comm.*, 10 (1974); *Synth. Comm.*, 7, 95 (1977).

198. C. Huynh and S. Julia, *Synth. Comm.*, 7, 103 (1977).

199. M. Franck-Neumann and J. J. Lohmann, *Tetrahedron Lett.*, 3729 (1978). This interesting route has been extended to give direct access to chrysanthemic esters (see below); see also M. Franck-Neumann and J. J. Lohmann, *Tetrahedron Lett.*, 2397 (1979).

200. O. P. Vig, A. S. Sethi, M. L. Sharma, and S. D. Sharma, *Ind. J. Chem., B*, **15**, 951 (1977).

201. G. Peiffer, *Compt. Rend. Acad. Sci. (C)*, **258**, 3499 (1964).

202. A. W. Burgstahler and H. W. Kroeger, *Synth. Comm.*, **3**, 211 (1973).

203. B. D. Mookherjee and R. W. Trenkle, *J. Agric. Food Chem.*, **21**, 298 (1973).

204. J. C. Belsten, A. F. Bramwell, J. W. K. Burrell, and D. M. Michalkiewicz, *Tetrahedron*, **28**, 3439 (1972).

205. R. Kaiser and D. Lamparsky, *Tetrahedron Lett.*, 665 (1977).

206. Haarman and Reimer, Belg. Pat., 615,962 [*Chem. Abstracts*, **59**, 11223 (1963)].

207. M. Julia, C. Perez, and L. Saussine, *J. Chem. Res.(S)*, 311; *(M)*, 3877 (1978). There is a good collection of literature in this paper (but not Ref. 206!).

208. K. Takabe, T. Katagiri, and J. Tanaka, *Kogyo Kagaku Zasshi*, **74**, 1162 (1971).

209. J. A. Oakleaf, M. T. Thomas, A. Wu, and V. Snieckus, *Tetrahedron Lett.*, 1645 (1978).

210. K. Takabe, H. Fujiwara, T. Katagiri, and J. Tanaka, *Synth. Comm.*, **5**, 227 (1975).

211. K. Takabe, T. Katagiri, and J. Tanaka, *Tetrahedron Lett.*, 1503 (1971).

212. M. Bertrand, G. Gil, and J. Viala, *Tetrahedron Lett.*, 1785 (1977).

213. T. Sato, H. Kise, M. Seno, and T. Asahara, *Yukugaku*, **24**, 265 (1975); Hasegawa Koryo KK, Japan Kokai 75 149,608.

214. M. Takami, Y. Omura, K. Itoi, and T. Kawaguchi, Japan Kakai 78 98,915 [*Chem. Abstracts*, **90**, 23328 (1979)].

215. C. F. Garbers, J. A. Steenkamp, and H. E. Visagie, *Tetrahedron Lett.*, 3753 (1975).

216. R. Maurin and M. Bertrand, *Bull. Soc. Chim. France*, 2356 (1972). For reactions of lavandulyl bromide described in this paper, see also K. Takabe, T. Katagiri, and J. Tanaka, *Nippon Kagaku Zasshi*, **90**, 943 (1969), and this series, Vol. 2, pp. 45-46.

217. K. v. Fraunberg, U.S. Pat., 3,997,577 (July 5, 1974).

218. Relevant references in Vol. 2, pp. 49, 50.

219. T. Aratani, Y. Yoneyoshi, and T. Nagase, *Tetrahedron Lett.*, 1707 (1975). T. Nagase, S. Nakamura, T. Aratani and Y.'Yoneyoshi, Japan Kokai 74 102,650; T. Aratani, S. Nakamura, T. Nagase and Y. Yoneyoshi, Ger. Offen. 2,407,094.

220. H. Hirai and M. Matsui, *Agric. Biol. Chem.*, **40**, 169 (1976).

221. D. Holland and D. J. Milner, *J. Chem. Res. (S)*, 317; *(M)*, 3734 (1979).

222. C. D. Poulter, O. J. Muscio, and R. J. Goodfellow, *J. Org. Chem.*, **40**, 139 (1975).

223. K. Ohkata, T. Isako, and T. Hanafusa, *Chem. Ind.*, 274 (1978).

224. P. F. Schatz, *J. Chem. Educ.*, **55**, 468 (1978).

225. Y. Inouye, Y. Sugita, and M. Ohno, *Bull. Agric. Chem. Soc. Japan*, **22**, 269 (1958); *Bull. Inst. Chem. Res. Kyoto Univ.*, **38**, 8 (1960); S. Takei, T. Sugita, and Y. Inouye, *Annalen,* **618**, 105 (1958).

226. H.-D. Scharf and J. Mattay, *Chem. Ber.*, **111**, 2206 (1978).

227. P. Baeckström, *J.C.S. Chem. Comm.*, 476 (1976); *Tetrahedron*, **34**, 3331 (1978). In the latter paper, there seems to be a discrepancy between the summary and the text, concerning the direct irradiation of the isomers of **210** (R = Me).

228. M. J. Bullivant and G. Pattenden, *J.C.S. Perkin I*, 256 (1976).

229. M. Franck-Neumann and J. J. Lohmann, *Tetrahedron Lett.*, 2075 (1979).

230. R. W. Mills, R. D. H. Murray, and R. A. Raphael, *Chem. Comm.*, 555 (1971).

231. J. Ficini and J. d'Angelo, *Tetrahedron Lett.*, 2441 (1976).

232. K. Kondo, K. Matsui, and Y. Takahatake, *Tetrahedron Lett.*, 4359 (1976).

233. A. Takeda, T. Sakai, S. Shinohara, and S. Tsuboi, *Bull. Chem. Soc. Japan,* **50**, 1133 (1977).

234. M. J. Devos and A. Krief, *Tetrahedron Lett.*, 1845 (1978).

235. S. Torii, H. Tanaka and Y. Nagai, *Bull. Chem. Soc. Japan*, **50**, 2825 (1977).

236. H. Hirai, K. Ueda, and M. Matsui, *Agric. Biol. Chem.*, **40**, 153 (1976).

237. H. Hirai and M. Matsui, *Agric. Biol. Chem.*, **40**, 161 (1976).

238. C. F. Garbers, M. S. Beukes, C. Ehlers, and M. J. McKenzie, *Tetrahedron Lett.*, 77 (1978).

239. M. J. De Vos and A. Krief, *Tetrahedron Lett.*, 1891 (1979).

240. J. Kristensen, I. Thomsen, and S. Ø. Lawesson, *Bull. Soc. Chim. Belges*, **87**, 721 (1978).

241. T. Matsuo, K. Mori, and M. Matsui, *Tetrahedron Lett.*, 1979 (1976).

242. M. J. Devos, L. Hevesi, P. Bayet, and A. Krief, *Tetrahedron Lett.*, 3911 (1976); A. Krief and L. Hevesi, Belg. Pat., 827,651, Ger. Offen. 2,615,159 [*Chem. Abstracts*, 85, 177,686 (1976)].

243. M. Sevrin, L. Hevesi, and A. Krief, *Tetrahedron Lett.*, 3915 (1976).

244. H. Lehmkuhl and K. Mehler, *Annalen*, 1841 (1978). See also B. V. Maatschappij, Dutch Pat., 1974, 02879 [*Chem. Abstracts*, 83, 27684 (1975)].

245. S. C. Welch and T. A. Valdes, *J. Org. Chem.*, 42, 2108 (1977).

246. R. Sobti and Sukh Dev. *Tetrahedron*, 30, 2927 (1974); see also other variations by Y. Gopichand, A. S. Khanra, R. B. Mitra, and K. K. Chakravarti, *Ind. J. Chem.*, 13, 433 (1975); A. S. Khanra and R. B. Mitra, *Ind. J. Chem., B*, 14, 716 (1976).

247. M. Matsui, H. Yoshioka, H. Sakamoto, Y. Yamada, and T. Kitahara, *Agric. Biol. Chem.*, 29, 784 (1965); 31, 33 (1967).

248. W. Cocker, H. St. J. Lauder, and P. V. R. Shannon, *J.C.S. Perkin I*, 194 (1974).

249. W. Cocker, H. St. J. Lauder, and P. V. R. Shannon, *J.C.S. Perkin I*, 332 (1975).

250. R. B. Mitra and A. S. Khanra, *Synth. Comm.*, 7, 245 (1977).

250a. B. J. Fitzsimmons and B. Fraser-Reid, *J. Am. Chem. Soc.*, 101, 6123 (1979).

251. P. Martin, H. Greuter, and D. Belluš, *J. Am. Chem. Soc.*, 101, 5853 (1979).

252. G. Pattenden and R. Storer, *J. Label. Comp. Radiopharm.*, 12, 551 (1976).

253. F. Bohlmann, C. Zdero, and U. Faass, *Chem. Ber.*, 106, 2904 (1973).

254. A. F. Thomas and M. Ozainne, *J.C.S. Chem. Comm.*, 746 (1973).

255. J. G. MacConnell, J. H. Borden, R. M. Silverstein, and E. Stokkink, *J. Chem. Ecol.*, 3, 549 (1978).

256. J. A. Katzenellenbogen, *Science*, 194, 139 (1976); C. A. Hendrick, *Tetrahedron*, 33, 1845 (1977).

257. Vol. 2, p. 58; the full paper for this synthesis is R. C. Gueldner, A. C. Thompson, and P. A. Hedin, *J. Org. Chem.*, 37, 1854 (1972); see also J. B. Siddall, Ger. Offen. 2,056,411 [*Chem. Abstracts*, 75, 328 (1971)].

258. This type of reaction has been reviewed by S. M. Ali, T. V. Lee, and S. M. Roberts, *Synthesis*, 155 (1977).

259. H. Kosugi, S. Sekiguchi, R. Sekita, and H. Uda, *Bull. Chem. Soc. Japan*, 49, 520 (1976).

260. R. L. Cargill and B. W. Wright, *J. Org. Chem.*, 40, 120 (1975).

261. K. Mori, *Tetrahedron*, 34, 915 (1978).

262. K. Mori, S. Tamada, and P. A. Hedin, *Naturwissenschaften*, 65, 653 (1978).

263. K. Mori and M. Sasaki, *Tetrahedron Lett.*, 1329 (1979).

264. W. A. Ayer and L. M. Browne, *Can. J. Chem.*, 52, 1352 (1974). For recent modifications involving dihydro-261, see G. Rosini, A. Salomoni, and F. Squarcia, *Synthesis*, 942 (1979).

265. E. Wenkert, N. F. Golub, and R. A. J. Smith, *J. Org. Chem.*, 38, 4068 (1973).

266. E. Wenkert, D. A. Berges, and N. F. Golub, *J. Am. Chem. Soc.*, 100, 1263 (1978); N. F. Golub, *Diss. Abs. Int. (B)*, 35, 4835 (1975).

267. J. H. Tumlinson, R. C. Gueldner, D. D. Hardee, A. C. Thompson, P. A. Hedin, and J. P. Minyard, *J. Org. Chem*, 36, 2616 (1971).

268. W. E. Billups, J. H. Cross, and C. V. Smith, *J. Am. Chem. Soc.*, 95, 3438 (1973).

269. J.-C. Grandguillot and F. Rouessac, *Compt. Rend. Acad. Sci. (C)*, **277**, 1273 (1973).

270. G. Stork and J. F. Cohen, *J. Am. Chem. Soc.*, **96**, 5270 (1974).

271. R. D. Clark, *Synth. Comm.*, **9**, 325 (1979).

272. J. H. Babler, *Tetrahedron Lett.*, 2045 (1975); U.S. Pat., 3,994,953.

273. B. M. Trost and D. E. Keeley, *J. Org. Chem.*, **40**, 2013 (1975); B. M. Trost, D. E. Keeley, H. C. Arndt, and M. J. Bogdanowicz, *J. Am. Chem. Soc.*, **99**, 3088 (1977).

274. V. Rautenstrauch, *J.C.S. Chem. Comm.*, 519 (1978).

275. P. D. Hobbs and P. D. Magnus, *J.C.S. Chem. Comm.*, 856 (1974); *J. Am. Chem. Soc.*, **98**, 4594 (1976).

276. H.-U. Warnecke, *Dragoco Rep. (Ger. ed.)*, **25(9)**, 192 (1978).

277. H. Oberman, *Dragoco Rep. (Ger. ed.)*, **25(3)**, 55 (1978).

278. E. Demole and D. Berthet, *Helv. Chim. Acta*, **55**, 1866 (1972).

279. E. Demole and D. Berthet, *Helv. Chim. Acta*, **55**, 1898 (1972).

280. T. Chuman and M. Noguchi, *Agric. Biol. Chem.*, **39**, 567 (1975).

281. E. E. van Tamelen, G. M. Milne, M. I. Suffness, M. C. Rudler-Chauvin, R. J. Anderson, and R. S. Achini, *J. Am. Chem. Soc.*, **92**, 7202 (1970).

282. J. Wolinsky and W. Barker, *J. Am. Chem. Soc.*, **82**, 636 (1960).

283. T. Sakan, S. Isoe, S. B. Hyeon, R. Katsumura, T. Maeda, J. Wolinsky, D. Dickerson, M. R. Slabaugh, and D. Nelson, *Tetrahedron Lett.*, 4097 (1965).

284. J. F. Ruppert, M. A. Avery, and J. D. White, *J.C.S. Chem. Comm.*, 978 (1976).

285. Y. Matsuki, M. Kodama, and S. Itô, *Tetrahedron Lett.*, 2901 (1979).

286. See the section on iridoids in the *Specialist Periodical Reports on Terpenoids and Steroids*, Vols. 1-9 Chemical Society, London.

287. H. Obara, H. Kimura, J. Onodera, and M. Suzuki, *Chem. Lett.*, 221 (1978); *Bull. Chem. Soc. Japan*, **51**, 3610 (1978).

288. H. Obara, S. Kumazawa, J. Onodera, and H. Kimura, *Nippon Kagaku Zasshi*, 2380 (1974).

289. H. Uda, M. Maruyama, K. Kabuki, and S. Fujise, *Nippon Kagaku Zasshi*, **85**, 279 (1964).

290. T. Imagawa, N. Murai, T. Akiyama, and M. Kawanisi, *Tetrahedron Lett.*, 1691 (1979).

291. T. Akiyama, T. Fujii, H. Ishiwari, T. Imagawa, and M. Kawanisi, *Tetrahedron Lett.*, 2165 (1978).

292. K. Takai, Y. Hotta, K. Oshima, and H. Nozaki, *Tetrahedron Lett.*, 2417 (1978).

293. Y. Yamada, H. Sanjoh, and K. Iguchi, *J.C.S. Chem. Comm.*, 997 (1976).

294. Y. Yamada, H. Sanjoh, and K. Iguchi, *Chem. Lett.*, 1405 (1978).

295. J. Wolinsky, M. R. Slabaugh, and T. Gibson, *J. Org. Chem.*, **29**, 3740 (1964); see Vol. 2, p. 76. All the (1*S*)-*trans*-iridolactones corresponding to **298** have been made by R. A. Trave, A. Marchesini, and L. Garanti, *Gazz. Chim. Ital.*, **100**, 1061 (1970).

296. F. Bellesia, U. M. Pagnoni, R. Trave, G. D. Andreatti, G. Bocelli, and P. Sgarabotto, *J.C.S. Perkin II*, 1341 (1979).

297. U. M. Pagnoni, A. Pinetti, R. Trave, and L. Garanti, *Austral. J. Chem.*, **29**, 1375 (1976).

298. G. W. K. Cavill, E. Houghton, F. J. McDonald, and P. J. Williams, *Insect Biochem.*, **6**, 483 (1976).

299. C. Beaupin, J. C. Rossi, J. Passet, R. Granger, and L. Peyron, *Riv. Ital. Essenze, Profumi, Piante Off. Aromat. Syndets, Saponi, Cosmet., Aerosols*, **60**, 93 (1978).

300. J. Meinwald, T. H. Jones, T. Eisner, and K. Hicks, *Proc. Nat. Acad. Sci.*, USA, **74**, 2189 (1977).

301. M. S. Blum, J. B. Wallace, R. M. Duffield, J. M. Brand, H. M. Fales, and E. A. Sokolski, *J. Chem. Ecol.*, **4**, 47 (1978).

302. K. Yoshihara, T. Sakai, and T. Sakan, *Chem. Lett.*, 433 (1978).

303. J. Meinwald and T. H. Jones, *J. Am. Chem. Soc.*, **100**, 1883 (1978).

304. Y. Asaka, T. Kamikawa and T. Kubota, *Tetrahedron Lett.*, 1597 (1972); *Tetrahedron*, **30**, 3257 (1974).

305. J. Ficini and J. d'Angelo, *Tetrahedron Lett.*, 687 (1976).

306. B.-W. Au-Yeung and I. Fleming, *J.C.S. Chem. Comm.*, 81 (1977).

307. J. K. Whitesell and A. M. Helbling, *J.C.S. Chem. Comm.*, 594 (1977).

308. L.-F. Tietze, *Chem. Ber.*, **107**, 2491 (1974).

309. R. S. Matthews and J. K. Whitesell, *J. Org. Chem.*, **40**, 3312 (1975).

309a. S. W. Baldwin and M. T. Crimmins, *J. Am. Chem. Soc.*, **102**, 1198 (1980).

310. J. K. Whitesell and R. S. Matthews, *J. Org. Chem.*, **43**, 1650 (1978).

311. I. Kubo, I. Miura, and K. Nakanishi, *J. Am. Chem. Soc.*, **98**, 6704 (1976).

312. J. K. Whitesell, R. S. Matthews, P. K. S. Wang, and M. Helbling, A.C.S. Meeting, March 1978 (Anaheim), Abstract ORGN no. 176. The absolute stereochemistry of the alcohol **324** has been fully established: see H. Kuritani, Y. Takaoka, and K. Shingu, *J. Org. Chem.*, **44**, 452 (1979).

313. M. Nakane, C. R. Hutchinson, D. VanEngen, and J. Clardy, *J. Am. Chem. Soc.*, **100**, 7079 (1978); see also M. R. Bendall, C. W. Ford, and D. M. Thomas, *Austral. J. Chem.*, **32**, 2085 (1979), who quote further references.

314. J. K. Whitesell, R. S. Matthews, and A. M. Helbling, *J. Org. Chem.*, **43**, 784 (1978).

315. K. Nakanishi, T. Goto, S. Itô, S. Natori, and S. Nozoe, *Natural Products Chemistry*, Vol. 1, Academic Press, New York, 1974.

316. D. H. Miles, U. Kokpol, J. Bhattacharya, T. L. Altwood, K. E. Stone, T. A. Bryson, and C. Wilson, *J. Am. Chem. Soc.*, **98**, 1569 (1976).

317. H. Inouye and T. Nishioka, *Chem. Pharm. Bull. (Japan)*, **21**, 497 (1973).

318. K. Furuichi and T. Miwa, *Tetrahedron Lett.*, 3689 (1974).

319. K. Furuichi, K. Abe, and T. Miwa, *Tetrahedron Lett.*, 3685 (1974).

320. G. Büchi, J. A. Carlson, J. E. Powell, Jr., and L. F. Tietze, *J. Am. Chem. Soc.*, **95**, 540 (1973).

321. J. J. Partridge, N. K. Chadha, and M. R. Uskokovic, *J. Am. Chem. Soc.*, **95**, 532 (1973).

322. L.-F. Tietze, *Angew. Chem. Int. Ed.*, **12**, 757 (1973); *Chem. Ber.*, **107**, 2499 (1973).

323. G. Kinast and L.-F. Tietze, *Chem Ber.*, **109**, 3626 (1976).

324. L.-F. Tietze, G. Kinast, and H. C. Uzar, *Angew. Chem. Int. Ed.*, **18**, 541 (1979).

325. L.-F. Tietze, U. Niemeyer, and P. Marx, *Tetrahedron Lett.*, 3441 (1977); L.-F. Tietze and U. Niemeyer, *Chem. Ber.*, **111**, 2423 (1978).

326. C. R. Hutchinson, K. C. Mattes, M. Nakane, J. J. Partridge, and M. R. Uskokovic, *Helv. Chim. Acta*, **61**, 1221 (1978).

327. K. C. Mattes, M. T. Hsia, C. R. Hutchinson, and S. A. Sisk, *Tetrahedron Lett.*, 3541 (1977). The full paper has modified the relative stereochemistry of the bicyclic compound from which 343 is derived: S. A. Sisk and C. R. Hutchinson, *J. Org. Chem.*, 44, 3500 (1979).

328. F. A. MacKellar, R. C. Kelley, E. E. van Tamelen, and C. Dorschel, *J. Am. Chem. Soc.*, 95, 7155 (1973).

329. R. C. Kelley and I. Schetter, *J. Am. Chem. Soc.*, 95, 7156 (1973).

330. L. Panizzi, M. L. Scarpati, and G. Oriente, *Gazz. Chim. Ital.*, 90, 1449 (1960); H. C. Beyerman, L. A. van Dijck, J. Levisalles, A. Melera, and W. L. C. Veer, *Bull. Soc. Chim. France*, 1812 (1961).

331. L.-F. Tietze and H. C. Uzar, *Angew. Chem. Int. Ed.*, 18, 539 (1979).

332. B. B. Snider, D. M. Roush, and T. A. Killinger, *J. Am. Chem. Soc.*, 101, 6023 (1979).

333. A. R. Battersby, A. R. Butnett, and P. G. Parsons, *J. Chem. Soc.(C)*, 1187 (1969) give a brief historical survey.

334. R. T. Brown and C. L. Chapple, *J.C.S. Chem. Comm.*, 886 (1973); R. T. Brown, C. L. Chapple, D. M. Duckworth, and R. Platt, *J.C.S. Perkin I*, 160 (1976).

335. E. Bourquelot and H. Hérissey, *J. Pharm. Chem.*, 28, 433 (1908).

336. G. Büchi and R. E. Manning, *Tetrahedron Lett.*, 5 (26) (1960).

337. L.-F. Tietze, *Tetrahedron Lett.*, 2535 (1976).

338. C. F. Lane, *Synthesis*, 135 (1975).

339. H. Inouye, S. Tobita, and M. Moriguchi, *Chem. Pharm. Bull.*, 24, 1406 (1976).

340. R. S. Kapil, A. Schoeb, S. P. Popli, A. R. Burnett, G. D. Knowles, and A. R. Battersby, *Chem. Comm.*, 904 (1971).

341. A. Schoeb, K. Raj, R. S. Kapil, and S. P. Popli, *J.C.S. Perkin I*, 1245 (1975).

342. R. T. Brown and J. Leonard, *Tetrahedron Lett.*, 1605 (1978); *J.C.S. Chem. Comm.*, 725 (1978).

343. J. D. Wuest, A. M. Madonik, and D. C. Gordon, *J. Org. Chem.*, 42, 2111 (1977).

344. Th. Sevenet, A. Husson, and H.-P. Husson, *Phytochemistry*, 15, 576 (1976); N. G. Bisset and A. K. Choudhury, *Phytochemistry*, 13, 265 (1974).

345. B. Frank, U. Petersen, and F. Hüper, *Angew. Chem. Int. Ed.*, 9, 891 (1970).

346. A. Ferro, thèse de doctorat, University of Neuchâtel, Switzerland, 1974; *Riv. Ital. Essenze, Profumi, Piante off. Aromat. Syndets, Saponi, Cosmet., Aerosol*, 56, 611 (1974). See also A. Ferro and Y.-R. Naves, *Helv. Chim. Acta*, 57, 1141 (1974).

347. B. Singaram and J. Verghese, *Ind. J. Chem., B*, 14, 1003 (1976).

348. C. P. Mathew and J. Verghese, *J. Ind. Chem. Soc.*, 52, 997 (1975); see also B. Singaram and J. Verghese, *Ind. J. Chem., B*, 14, 479 (1976).

349. L. M. Hirschy, B. J. Kane, and S. G. Traynor, U.S. Pat. 4,136,126 (Jan. 23, 1979); S. G. Traynor B. J. Kane, M. F. Belkouski, and L. M. Hirschy, *J. Org. Chem.*, 44, 1557 (1979).

350. D. J. Coughlin and R. G. Salomon, *J. Org. Chem.*, 44, 3784 (1979).

351. M. Miyaura, H. Tagami, M. Itoh, and A. Suzuki, *Chem. Lett.*, 1411 (1974).

352. A. J. Birch and A. J. Pearson, *Tetrahedron Lett.*, 2379 (1975).

353. A. J. Birch and B. Chauncy, quoted in A. J. Birch, *J. Agric. Food Chem.*, 22, 162 (1974).

354. D. J. Thompson, *J. Organometal. Chem.*, **108**, 381 (1976).

355. A. P. Krapcho and E. G. E. Jahngen, Jr., *J. Org. Chem.*, **39**, 1322 (1974).

356. B. A. Patel, L.-C. Kao, N. A. Cortese, J. V. Minkiewicz, and R. F. Heck, *J. Org. Chem.*, **44**, 918 (1979).

357. R. M. Carman and J. K. L. Maynard, *Austral. J. Chem.*, **32**, 217 (1979).

358. J. M. Coxon, E. Dansted, M. P. Hartshorn, and K. E. Richards, *Tetrahedron*, **25**, 3307 (1969).

359. Y. Yamamoto, H. Shimoda, J. Oda, and Y. Inouye, *Bull. Chem. Soc. Japan*, **49**, 3247 (1976).

360. K. Arata, J. O. Bledsoe, Jr., and K. Tanabe, *J. Org. Chem.*, **43**, 1660 (1978).

361. U. Lipnicka and Z. Chabudziński, *Rocz. Chem.*, **49**, 307 (1975).

362. R. L. Kenney and G. S. Fisher, *J. Gas Chromat.*, **1**, 19 (1963); R. A. Jones and T. C. Webb, *Tetrahedron*, **28**, 2877 (1972).

363. A. P. Uijttewaal, F. L. Jonkers, and A. van der Gen, *Tetrahedron Lett.*, 1439 (1975); *J. Org. Chem.*, **43**, 3306 (1978).

364. A. P. Uijttewaal, F. L. Jonkers, and A. van der Gen, *J. Org. Chem.*, **44**, 3157 (1979); A. P. Uijttewaal, Proefschrift, Leiden University, 1978, and personal communication.

365. K. Noda, T. Matsuda, and Y. Ishikura, *J. Labelled Comp.*, **10**, 309 (1974).

366. B. M. Lawrence, S. J. Terhune, and J. W. Hogg, *Phytochemistry*, **13**, 1012 (1974).

367. A. F. Thomas, unpublished work.

368. S. A. Kozhin and E. I. Sorochinskaya, *Zh. Obshch. Khim.*, **44**, 944 (1974).

369. S. A. Kozhin and E. I. Sorochinskaya, *Zh. Obshch. Khim.*, **43**, 671 (1973).

370. W. Rojahn and E. Klein, *Dragoco Rep. (Ger. ed.)*, **24**, 172 (1977).

371. W. Hoffmann and W. Himmele, Ger. Offen., 2,551,172 (Nov. 14, 1975).

372. J. Debrauwere and M. Verzele, *Bull. Soc. Chim. Belges*, **84**, 167 (1975).

373. T. Nagasawa, K. Umemoto, T. Tsuneya, and M. Shiga, *Nippon Nogei Kagaku Kaishi*, **49**, 217 (1975) [*Chem. Abstracts*, **88**, 97589 (1975)].

374. D. F. Taber and B. P. Gunn, *J. Org. Chem.*, **44**, 540 (1979).

375. L. J. Dolby and M. Debono, *J. Org. Chem.*, **29**, 2306 (1964).

376. A. L. J. Beckwith and G. Phillipou, *Tetrahedron Lett.*, 79 (1974); *Austral. J. Chem.*, **29**, 1277 (1976).

377. P. Teisseire, M. Plattier, and E. Giraudi, Swiss Pat. 565,725 (French priority October 30, 1970); *Recherches*, **19**, 205 (1974).

378. A. J. Birch and G. Subba Rao, *Austral. J. Chem.*, **22**, 2037 (1969).

379. D. J. Faulkner and L. E. Wolinsky, *J. Org. Chem.*, **40**, 389 (1975).

380. H. P. Jensen and K. B. Sharpless, *J. Org. Chem.*, **40**, 264 (1975).

381. C. W. Wilson and P. E. Shaw, *J. Org. Chem.*, **38**, 1684 (1973).

382. F. Delay and G. Ohloff, *Helv. Chim. Acta*, **62**, 2168 (1979).

383. W. Knöll and C. Tamm, *Helv. Chim. Acta*, **58**, 1162 (1975); G. Feldstein and P. J. Kocienski, *Synth. Comm.*, **7**, 27 (1977).

384. P. Sundararaman and C. Djerassi, *Tetrahedron Lett.*, 2457 (1978).

385. O. P. Vig, A. S. Chahal, A. K. Vig, and O. P. Chugh, *J. Ind. Chem. Soc.*, **52**, 442 (1975).

386. T. K. John and G. S. K. Rao, *Ind. J. Chem., B*, **14**, 805 (1976).

387. R. Dulou and Y. Chrétien-Bessière, *Bull. Soc. Chim. France*, 1362 (1959).

388. H. C. Brown and C. D. Pfaffenberger, *J. Am. Chem. Soc.*, **89**, 5475 (1967); the full paper [*Tetrahedron*, **31**, 925 (1975)] contains important corrections from the preliminary communication.

389. T. Shono, A. Ikeda, J. Hayashi, and S. Hakozaki, *J. Am. Chem. Soc.*, **97**, 4261 (1975).

390. Y. Bessière and F. Derguini-Bouméchal, *J. Chem. Res. (S)*, 304; *(M)*, 3522 (1977). For further detail, see F. Derguini-Bouméchal, thèse de docteur d'état, Université P. et M. Curie, Paris 1977.

391. K. Arata, S. Akutagawa and K. Tanabe, *Bull. Chem. Soc. Japan*, **51**, 2289 (1978).

392. H. Strickler, J. B. Davis, and G. Ohloff, *Helv. Chim. Acta*, **59**, 1328 (1976).

393. Y. Fujita, S. Fujita, and H. Yoshikawa, *Bull. Chem. Soc. Japan*, **43**, 1599 (1970).

394. J. Taskinen and L. Nykänen, *Acta Chem. Scand., B*, **29**, 757 (1975).

395. S. Escher, U. Keller, and B. Willhalm, *Helv. Chim. Acta*, **62**, 2061 (1979).

396. B. Singaram and J. Verghese, *Perfum. Flavor.*, **2**, 47 (1977).

397. B. M. Trost and A. J. Bridges, *J. Am. Chem. Soc.*, **98**, 5017 (1976).

398. I. Fleming and I. Paterson, *Synthesis*, 736 (1979).

399. O. P. Vig, A. K. Sharma, J. Chander, and B. Ram, *J. Ind. Chem. Soc.*, **49**, 159 (1972).

400. O. P. Vig, M. L. Sharma, R. C. Anand, and S. D. Sharma, *J. Ind. Chem. Soc.*, **53**, 50 (1976).

401. R. E. Ireland and P. Bey, *Org. Synth.*, **53**, 63 (1973).

402. R. Robbiani, H. Bührer, H. Mändli, D. Kovačević, A. Fraefel, and J. Seibl, *Org. Mass Spec.*, **13**, 275 (1978).

403. V. K. Honwad, E. Siskovic, and A. S. Rao, *Ind. J. Chem.*, **5**, 234 (1967).

404. L. Friedman and J. G. Miller, *Science*, **172**, 1044 (1971).

405. A. Yasuda, H. Yamamoto, and H. Nozaki, *Bull. Chem. Soc. Japan*, **52**, 1757 (1979).

406. K. Sakai and O. Oda, *Tetrahedron Lett.*, 4375 (1972).

407. W. Hoffmann and W. Reif, Ger. Offen., 2,115,130.

408. T. Yamanaka and M. Yagi, Japan Kokai 77 73841.

409. Y. Nakatani and K. Kawashima, *Synthesis*, 147 (1978).

410. E. J. Corey and D. L. Boger, *Tetrahedron Lett.*, 2461 (1978).

411. T. Kobayashi, S. Kumazawa, T. Kato, and Y. Kitahara, *Chem. Lett.*, 301 (1975); Y. Kitahara, T. Kato, and T. Kobayashi, Japan Kokai 76, 113845.

412. G. Demailly and G. Solladié, *Tetrahedron Lett.*, 1885 (1977).

413. J. Tanaka, T. Katagiri, K. Takabe, and H. Fujiwara, Japan Kokai 77 42,848 [Chem. Abstracts, **87**, 117973 (1977)]. This patent also gives other reactions of imines leading to menthylamines.

414. T.-C. Chang, I. Ichimoto, and H. Ueda, *Nippon Nogei Kagaku Kaishi*, **48**, 477 (1974).

415. Y. Besière and F. Derguini-Bouméchal, *J. Chem. Res. (S)*, 204; *(M)*, 2301(1977).

416. P. Hullot, T. Cuvigny, M. Larchevêque, and H. Normant, *Can. J. Chem.*, **54**, 1098 (1976).

417. T. Nagasawa, K. Umemoto, T. Tsuneya, and M. Shiga, *Nippon Nogei Kagaku Kaishi*, **48**, 39 (1974).

418. T. Nagasawa, K. Umemoto, T. Tsuneya, and M. Shiga, *Nippon Nogei Kagaku Kaishi*,

48, 467 (1974).

419. A. F. Thomas, B. Willhalm, and J. H. Bowie, *J. Chem. Soc. (B)*, 392(1967).

420. G. Ohloff and W. Giersch, *Helv. Chim. Acta*, 51, 1328 (1968).

421. R. S. Prasad and Sukh Dev, *Tetrahedron*, 32, 1437 (1976).

422. W. C. Still, *J. Am. Chem. Soc.*, 99, 4836 (1977).

423. J. Wiemann and Y. Dubois, *Compt. Rend. Acad. Sci.*, 253, 1109 (1961); *Bull. Soc. Chim. France*, 1813 (1962). The configuration seems to have been changed between the preliminary and the full paper.

424. T. Nagasawa, K. Umemoto, T. Tsuneya, and M. Shiga, *Agric. Biol. Chem.*, 39, 2083 (1975); K. Umemoto and T. Nagasawa, *Nogoya Gakuin Univ. Rev.*, 12, 117 (1976); T. Nagasawa, K. Umemoto, and N. Hirao, *Nippon Nogei Kagaku Kaishi*, 51, 81 (1977).

425. S. Escher, personal communication.

426. J. Read and G. Swann, *J. Chem. Soc.*, 237 (1937).

427. J. P. Bain, H. G. Hunt, E. A. Klein, and A. B. Booth, U.S. Pat., 2,972,632 (to Glidden Co., Feb. 21, 1961).

428. J. A. Marshall and J. H. Babler, *Tetrahedron Lett.*, 3861 (1970).

429. V. V. Bazyl'chik, T. N. Overchuk, and P. I. Fedorov, *Zh. Org. Khim.*, 14, 2085 (1978).

430. J. A. Marshall, C. P. Hagan, and G. A. Flynn, *J. Org. Chem.*, 40, 1162 (1975).

431. H. E. Ensley, C. A. Parnell, and E. J. Corey, *J. Org. Chem.*, 43, 1610 (1978).

432. H. E. Ensley and R. V. C. Carr, *Tetrahedron Lett.*, 513 (1977).

433. T. Shono, Y. Matsumura, K. Hibino, and S. Miyawaki, *Tetrahedron Lett.*, 1295 (1974).

434. E. J. Corey and H. E. Ensley, *J. Am. Chem. Soc.*, 97, 6907 (1975); E. J. Corey, H. E. Ensley, and J. W. Suggs, *J. Org. Chem.*, 41, 380 (1976).

435. J. Solodar, *J. Org. Chem.*, 43, 1787 (1978).

436. E. Sundt, B. Willhalm, R. Chappaz, and G. Ohloff, *Helv. Chim. Acta*, 54, 1801 (1971); French Pat., 2,033,151; D. Lamparsky and P. Schudel, *Tetrahedron Lett.*, 3323 (1971); Ger. Offen., 2,043,341.

437. K. L. J. Blommaert and E. Bartel, *J. S. Afr. Bot.*, 42, 121 (1976).

438. J. C. Leffingwell and R. E. Shackelford, *Cosmetics Perfumery*, 89, 69 (1974).

439. T. Shono, I. Nishiguchi, and M. Nitta, *Chem. Lett.*, 1319 (1976).

440. G. Moinet, J. Brocard, and J.-M. Conia, *Tetrahedron Lett.*, 4461 (1972); J. Brocard, G. Moinet, and J.-M. Conia, *Bull. Soc. Chim. France*, 1711 (1973).

441. S. Torii, T. Oie, H. Tanaka, J. D. White, and T. Furuta, *Tetrahedron Lett.*, 2471 (1973); S. Torii, Japan Kokai 74 56,955, 56, 967 [*Chem. Abstracts*, 81, 120818, 120819 (1974)].

442. I. Sakata and H. Iwamura, *Agric. Biol. Chem.*, 43, 307 (1979). See also the resolution of (±)-menthol via the acetylglucose derivatives, I. Sakata and K. Koshimizu, *Agric. Biol. Chem.*, 43, 411 (1979).

443. B. Harirchian and P. D. Magnus, *Synth. Comm.*, 7, 119 (1977).

444. J. A. Turner and W. Herz, *J. Org. Chem.*, 42, 1895, 1900 (1977).

445. E. Wenkert, M. E. Alonso, B. L. Buckwalter, and K. J. Chan, *J. Am. Chem. Soc.*, 99, 4778 (1977).

445a. T. Sato, M. Tada, and T. Takahashi, *Bull. Chem. Soc. Japan*, **52**, 3129 (1979). The method is based on H. Fritel and M. Fétizon, *J. Org. Chem.*, **23**, 481 (1958).

446. A. A. Akhrem, A. M. Moiseenkov, and F. A. Lakhvich, *Dokl. Akad. Nauk SSSR*, **193**, 1053 (1970).

447. R. B. Mitra, G. D. Josh, and A. S. Khanra, *Ind. J. Chem., B*, **16**, 741 (1978).

448. A. A. Drabkina, O. V. Efimova, and Yu. S. Tsizin, *Zh. Obshch. Khim.*, **42**, 1139 (1972).

449. Yu. S. Tsizin and A. A. Drabkina, *Zh. Obshch. Khim.*, **42**, 1852 (1972).

450. H. Kayahara, H. Ueda, K. Takeo, and C. Tatsumi, *Agric. Biol. Chem.*, **33**, 86 (1969).

451. Y. Fujita, S. Fujita, and Y. Hayama, *Nippon Nogei Kagaku Kaishi*, **44**, 428 (1970).

452. A. F. Thomas, unpublished work. See also Swiss Pat. 581,592 (Appl. June 7, 1973).

453. R. Murphy and R. Prager, *Austral. J. Chem.*, **31**, 1629 (1978).

454. A. J. Birch and K. P. Dastur, *Austral. J. Chem.*, **26**, 1363 (1973).

455. J. Coll Toledano, *Rev. Real. Acad. Cienc. Exactas, Fis. Natur. Madrid*, **64**, 523 (1970) [*Chem. Abstracts*, **74**, 57198 (1971)].

456. K. Tadasa, S. Fukazawa, M. Kunimatsu, and T. Hayashi, *Agric. Biol. Chem.*, **40**, 1069 (1976); K. Tadasa, *Agric. Biol. Chem.*, **41**, 2095 (1977).

457. J. Rama Devi and P. K. Bhattacharyya, *Indian J. Biochem. Biophys.*, **14**, 288, 359 (1977).

458. C. W. Wilson III and P. E. Shaw, *J. Agric. Food Chem.*, **25**, 221 (1977).

459. M. Nomura, Y. Fujihara, and Y. Matsubara, *Nippon Kagaku Kaishi*, 1182 (1978); 305 (1979). See also O. P. Vig, S. S. Bari, and S. S. Rana, *Ind. J. Chem., B*, **17**, 171 (1979).

460. G. Ohloff, W. Giersch, K. H. Schulte-Elte, and E. sz. Kováts, *Helv. Chim. Acta*, **52**, 1531 (1969). This paper describes the separation of the isomers about C(8) of 1-menthen-8-ol, a separation that Brown and Pfaffenberger were unable to carry out in a related case.[388]

461. P. A. Hedin, A. C. Thompson, and R. C. Gueldner, *Phytochemistry*, **14**, 2087 (1975).

462. R. Kaiser, D. Lamparsky, and P. Schudel, *J. Agric. Food Chem.*, **23**, 943 (1975).

463. O. P. Vig, S. D. Sharma, G. L. Kad, and M. L. Sharma, *Ind. J. Chem.*, **13**, 439 (1975).

464. K. M. Madhyastha and P. K. Bhattacharyya, *Ind. J. Biochem.*, **5**, 102 (1968).

465. A. F. Thomas, unpublished work.

466. O. P. Vig, S. S. Bari, S. D. Sharma, and S. S. Rana, *Ind. J. Chem., B*, **15**, 1076 (1977).

467. M. S. Malinovskii and A. G. Yudasina, *Zh. Obshch. Khim.*, **30**, 1831 (1960).

468. H. C. Brown, C. G. Rao, and M. Ravindranathan, *J. Am. Chem. Soc.*, **99**, 7663 (1977).

469. J. Luteyn, H. J. W. Spronck, and C. A. Salemink, *Rec. Trav. Chim.*, **97**, 187 (1978).

470. A. J. Lundeen, U.S. Pat., 3,338,942 (Aug. 29, 1967).

471. K. J. Divakar and A. S. Rao, *Synth. Comm.*, **6**, 423 (1976).

472. F. Bohlmann, U. Niedballa, and J. Schulz, *Chem. Ber.*, **102**, 864 (1969).

473. K. J. Divakar, B. D. Kulkarni, and A. S. Rao, *Ind. J. Chem., B*, **15**, 322 (1977).

474. F. Bohlmann and C. Zdero, *Chem. Ber.*, **109**, 791 (1976).

475. F. Bohlmann and J. Kocur, *Chem. Ber.*, **109**, 2969 (1976).

476. K. J. Divakar, B. D. Kulkarni, and A. S. Rao, *Ind. J. Chem., B*, **15**, 849 (1977).

477. J. Ficini and A. M. Touzin, *Tetrahedron Lett.*, 2093, 2097 (1972).

478. H. Takahashi and M. Ito, *Chem. Lett.*, 373 (1979).

479. T. Hori and K. B. Sharpless, *J. Org. Chem.*, **43**, 1689 (1978); H. J. Reich, S. Wollowitz, J. E. Treud, F. Chow, and D. F. Wendelborn, *J. Org. Chem.*, **43**, 1697 (1978).

480. R. M. Scarborough, Jr., A. B. Smith III, W. E. Barnett, and K. C. Nicolaou, *J. Org. Chem.*, **44**, 1742 (1979).

481. F. Bondavalli, P. Schenone, S. Lanteri, and A. Ranise, *J.C.S. Perkin I*, 430 (1977).

482. D. H. R. Barton, R. K. Haynes, G. Leclerc, P. D. Magnus, and I. D. Menzies, *J.C.S. Perkin I*, 2055 (1975).

483. R. K. Haynes, *Austral. J. Chem.*, **31**, 131 (1978).

484. G. O. Schenck, K. G. Kinkel, and H. J. Mertens, *Annalen*, **584**, 125 (1953).

485. R. D. Stolow and K. Sachdev, *Tetrahedron*, **21**, 1889 (1965), give a survey of the reaction; see also R. D. Stolow and K. Sachdev, *J. Org. Chem.*, **36**, 960 (1971).

486. A. G. Gonzalez, J. Bermejo Barrera, J. L. Bermejo Barrera, and G. M. Massanet, *An. Quím.*, **68**, 319 (1972).

487. J. F. W. Keana and P. E. Eckler, *J. Org. Chem.*, **41**, 2625 (1976).

488. T. Hayashi, S. Uedono, and C. Tatsumi, *Agric. Biol. Chem.*, **36**, 690 (1972).

489. E. V. Lassak, J. T. Pinhey, B. J. Ralph, T. Sheldon, and J. J. H. Simes, *Austral. J. Chem.*, **26**, 845 (1973).

490. K. Umemoto and T. Nagasawa, *Nippon Nogei Kagaku Kaishi,* **51**, 245 (1977).

491. F. Bohlmann and C. Zdero, *Chem. Ber.,* **109**, 2653 (1976).

492. R. Kaiser, D. Lamparsky, and P. Schudel, *J. Agric. Food Chem.*, **23**, 943 (1975).

493. A. A. J. Fluck, W. Mitchell, and H. M. Perry, *J. Sci. Food Agric.*, **12**, 290 (1961).

494. H. Shibata and S. Shimizu, *Agric. Biol. Chem.*, **38**, 1741 (1974).

495. A. K. Macbeth and W. G. P. Robertson, *J. Chem. Soc.*, 3512 (1953).

496. W. Reusch, D. Anderson, and C. Johnson, *J. Am. Chem. Soc.*, **90**, 4988 (1968).

497. C. Maignan and F. Rouessac, *Bull. Soc. Chim. France*, 550 (1976).

498. M. Ohashi, S. Inoue, and K. Sato, *Bull. Chem. Soc. Japan*, **49**, 2292 (1976).

499. G. Farges and A. Kergomard, *Bull. Soc. Chim. France*, 4476 (1969); G. B. Payne, *Tetrahedron*, **18**, 763 (1962).

500. A. Kergomard and H. Verschambre, *Tetrahedron Lett.*, 835 (1975); *Tetrahedron*, **33**, 2215 (1977).

501. A. Kergomard and H. Verschambre, *Compt. Rend. Acad. Sci. (C)*, **279**, 155 (1974).

502. M. A. Schwartz and G. C. Swanson, *J. Org. Chem.*, **44**, 953 (1979).

502a. T. Prangé, D. Babin, J.-D. Fourneron, and M. Julia, *Compt. Rend. Acad. Sci. C*, **289**, 383 (1979).

503. O. P. Vig, S. D. Sharma, S. S. Bari, and M. Lal, *Ind. J. Chem., B*, **16**, 739 (1978).

504. M. Anteunis and A. De Smet, *Synthesis*, 868 (1974).

505. R. K. Thappa, K. L. Dhar, and C. K. Atal, *Phytochemistry*, **15**, 1568 (1976).

506. C. G. Casinori, G. Grandolini, L. Radics, and C. Rossi, *Experientia*, **34**, 298 (1978).

506a. A. Baraglin, G. Grandolini, C. Rossi, and C. G. Casinovi, *Tetrahedron*, **36**, 645 (1980).

507. F. Bohlmann, C. Zdero, and A. G. R. Nair, *Phytochemistry*, **18**, 1062 (1979).

508. Y.-R. Naves, *Compt. Rend. Acad. Sci.*, **249**, 562 (1959); *Bull. Soc. Chim. France*, 1871 (1959); A. F. Thomas, *Helv. Chim. Acta*, **50**, 963 (1967).

509. G. Snatzke, A. F. Thomas, and G. Ohloff, *Helv. Chim. Acta*, **52**, 1253 (1969); M.-G. Ferretti-Alloise, A. Jacot-Guillarmod, and Y.-R. Naves, *Helv. Chim. Acta*, **53**, 551 (1970).

510. T. K. Devon and A. I. Scott, *Handbook of Naturally Occurring Compounds*, Vol. 2, *Terpenes*, Academic Press, New York, 1971. See also D. E. Cane and R. H. Levin, *J. Am. Chem. Soc.*, **98**, 1183 (1976).

511. J. Romo, A. Romo de Vivar, L. Quijano, T. Rios, and E. Diaz, *Rev. Latinoamer. Quím.*, **1**, 72 (1970); F. Bohlmann and C. Zdero, *Tetrahedron Lett.*, 2419 (1969); F. Bohlmann and A. Suwita, *Phytochemistry*, **17**, 560 (1978).

512. I. A. Southwell, *Tetrahedron Lett.*, 1885 (1975).

513. N. Lander and R. Mechoulam, *J.C.S. Perkin I*, 484 (1976).

514. L. Borowiecki, and E. Reca, *Rocz. Chem.*, **45**, 493 (1971).

515. J. Ficini, A. Eman, and A. M. Touzin, *Tetrahedron Lett.*, 679 (1976).

516. O. Wallach, *Annalen*, **357**, 72 (1907); W. N. Haworth, W. H. Perkin, and O. Wallach, *J. Chem. Soc.*, **103**, 1234 (1913); for recent spectral data see G. K. Kaimal and J. Verghese, *J. Ind. Chem. Soc.*, **48**, 759 (1971); P. M. Abraham and J. Verghese, *J. Ind. Chem. Soc.*, **52**, 175 (1975).

517. S. K. Zutschi, B. K. Bamboria, and M. M. Bokadia, *Curr. Sci.*, **44**, 571 (1975).

518. E. Stahl and R. Kunde, *Tetrahedron Lett.*, 2841 (1973).

519. F. Bohlmann, A. A. Natu, and K. Kerr, *Phytochemistry*, **18**, 489 (1979).

520. F. Bohlmann, P. K. Mahanta, A. Suwita, A. A. Natu, C. Zdero, W. Dorner, D. Ehlers, and M. Grenz, *Phytochemistry*, **16**, 1973 (1977).

521. F. Bohlmann, K.-H. Knoll, C. Zdero, P. K. Mahanta, M. Grenz, A. Suwita, D. Ehlers, N. L. Van, W.-R. Abraham, and A. A. Natu, *Phytochemistry*, **16**, 965 (1977).

522. B. C. Clark, Jr., T. C. Chafin, P. L. Lee, and G. L. K. Hunter, *J. Org. Chem.*, **43**, 519 (1978).

523. W. Kraus and G. Zartner, *Tetrahedron Lett.*, 13 (1977).

524. J. Ficini and A. M. Touzin, *Tetrahedron Lett.*, 1447 (1974).

525. N. S. Zarghami and D. E. Heinz, *Phytochemistry*, **10**, 2755 (1971).

526. R. Buchecker and C. H. Eugster, *Helv. Chim. Acta*, **56**, 1121 (1973).

527. Y. Takeda, H. Nishimura, O. Kadota, and H. Inouye, *Chem. Pharm. Bull.*, **24**, 2644 (1976).

528. I. Wahlberg, K. Karlson, D. J. Austin, N. Junker, J. Roeraade, C. R. Enzell, and W. H. Johnson, *Phytochemistry*, **16**, 1217 (1977).

529. F. Jüttner, *Z. Naturforsch. C, Biosci.*, **31C**, 491 (1976).

530. H. G. W. Leuenberger, W. Boguth, E. Widmer, and R. Zell, *Helv. Chim. Acta*, **59**, 1832 (1976).

531. M. E. Kuehne and R. E. Damon, *J. Org. Chem.*, **42**, 1825 (1977).

532. S. Torii, K. Uneyama, and M. Isihara, *Chem. Lett.*, 479 (1975).

532a. T. Kametani, K. Suzuki, H. Kurobe, and H. Nemoto, *J.C.S. Chem. Comm.*, 1128 (1979).

533. F. W. Sum and L. Weiter, *J. Am. Chem. Soc.*, **101**, 4401 (1979).

534. L. E. Wolinsky and D. J. Faulkner, *J. Org. Chem.*, **41**, 597 (1976).

535. F. Rouessac and H. Zamarlik, *Tetrahedron Lett.*, 3417 (1979).

536. V. Prelog and S. Heinbach-Juhasz, *Ber.*, **74**, 1702 (1941); see also D. S. Noyce and H. I. Weingarten, *J. Am. Chem. Soc.*, **79**, 3093 (1957).

537. T. Kato, I. Ichinose, S. Kumazawa, and Y. Kitahara, *Bioorg. Chem.*, **4**, 188 (1975).

538. J. P. Alazard, R. Setton, and H. B. Kagan, quoted in H. B. Kagan, *Pure Appl. Chem.*, **46**, 177 (1976).

539. M. Kurbanov, A. V. Semenovskii, W. A. Smit, L. V. Shmelov, and V. F. Kucherov, *Izv. Akad. Nauk SSSR, Ser. Khim.*, 2451 (1971).

540. M. Kurbanov, A. V. Semenovskii, W. A. Smit, L. V. Shmelov and V. F. Kucherov, *Tetrahedron Lett.*, 2175 (1972).

541. R. Breslow, J. T. Groves, and S. S. Olin, *Tetrahedron Lett.*, 4717 (1966).

542. A. J. L. Beckwith, G. E. Gream, and D. L. Struble, *Austral. J. Chem.*, **25**, 1081 (1972).

543. L. Colombi, A. Bosshard, H. Schinz, and C. F. Seidel, *Helv. Chim. Acta*, **34**, 265 (1951); H. B. Henbest, B. L. Shaw, and G. Woods, *J. Chem. Soc.*, 1154 (1952).

544. R. N. Gedye, A. C. Parkash, and K. Deck, *Can. J. Chem.*, **49**, 1764 (1971).

545. S. Yamada, M. Shibasaki, and S. Terashima, *Tetrahedron Lett.*, 377 (1973).

546. S. Yamada, M. Shibasaki, and S. Terashima, *Tetrahedron Lett.*, 381 (1973); see correction of absolute configuration: M. Shibasaki, S. Terashima, and S.-I. Yamada, *Chem. Pharm. Bull.*, **23**, 272, 279 (1975).

547. D. J. Bennett, G. R. Ramage, and J. L. Simonsen, *J. Chem. Soc.*, 418 (1940).

548. R. Buchecker, R. Egli, H. Regel-Wild, C. Tscharner, C. H. Eugster, G. Uhde, and G. Ohloff, *Helv. Chim. Acta*, **56**, 2548 (1973).

549. R. Buchecker and C. H. Eugster, *Helv. Chim. Acta,* **56**, 2563 (1973).

550. N. Müller and W. Hoffmann, *Synthesis*, 781 (1975); Ger. Offen. 2,432,231; R. D. Clark and C. H. Heathcock, *J. Org. Chem.*, **41**, 1396 (1976).

551. E. Hunt and B. Lythgoe, *J.C.S. Chem. Comm.*, 757 (1972).

552. F. Bohlmann, C. Zdero, and M. Grenz, *Chem. Ber.*, **108**, 2822 (1975).

553. F. Bohlmann and C. Zdero, *Chem. Ber.*, **104**, 1957 (1971).

554. F. Bohlmann and A. Suwita, *Phytochemistry*, **17**, 560 (1978).

555. G. Büchi, W. Pickenhagen, and H. Wüest, *J. Org. Chem.*, **38**, 1380 (1973).

556. R. Kaiser and D. Lamparsky, *Helv. Chim. Acta*, **61**, 373 (1978).

557. T. Mukaiyama, N. Iwasawa, T. Tsuji, and K. Narasaka, *Chem. Lett.*, 1175 (1979).

558. T. Mukaiyama, T. Tsuji, and N. Iwasawa, *Chem. Lett.*, 697 (1979).

559. R. M. Coates and L. S. Melvin, Jr., *J. Org. Chem.*, **35**, 865 (1970).

560. Y. Yamada, H. Sanjoh, and K. Iguchi, *Tetrahedron Lett.*, 1323 (1979).

561. W. Cocker, K. J. Crowley, and K. Srinivasan, *J.C.S. Perkin I*, 2485 (1973).

562. J. De Pascual Teresa, I. Sanchez Bellido, M. R. Alberdialbistegui, A. Sanfeliciano, and M. Grande Benito, *An. Quím.*, **74**, 470 (1978).

563. W. Cocker, K. J. Crowley, and K. Srinivasan, *J.C.S. Perkin I*, 159 (1978).

564. L. I. Dukhanova, M. E. Perel'son, Y. E. Skylar, and M. G. Pimenov, *Chem. Natural Comp.*, **10**, 316 (1974); this paper gave the wrong structure, corrected by S. K. Paknikar, J. Veeravalli, and J. K. Kirtany, *Experientia*, **34**, 553 (1978).

565. A. Sattar, M. Ashraf, M. K. Bhatty, and N. H. Chisti, *Phytochemistry*, **17**, 559 (1978).

566. Y. Bessière and F. Ouar, *J. Labelled Comp.*, **11**, 3 (1975).

567. F. Bohlmann and C. Zdero, *Chem. Ber.*, **102**, 2211 (1969).

568. F. Bohlmann, J. Jacob, and M. Grenz, *Chem. Ber.*, **108**, 433 (1975), and other papers.

569. S. J. Torrance and C. Steelink, *J. Org. Chem.*, **39**, 1068 (1974).

570. J. Garnero and D. Joulain, *Bull. Soc. Chim. France*, 455 (1979).

571. F. Bohlmann and G. Weickgenannt, *Chem. Ber.*, **107**, 1769 (1974).

572. Y. Chrétien-Bessière and J.-A. Retamar, *Bull. Soc. Chim. France*, 884 (1963).

573. B. A. Arbuzov, A. N. Vershchagin, N. I. Gubkina, I. M. Sadykova, and S. G. Vul'fson, *Izv. Akad. Nauk SSSR, Ser. Khim.*, 1288 (1972). *Chem. Abstracts*, **77**, 101911 (1972) is incorrect.

574. A. F. Thomas and R. Guntz-Dubini, *Helv. Chim. Acta*, **59**, 2261 (1976).

575. D. Van Engen, J. Clardy, E. Kho-Wiseman, P. Crews, M. D. Higgs, and D. J. Faulkner, *Tetrahedron Lett.*, 29 (1978).

576. J. S. Mynderse and D. J. Faulkner, *J. Am. Chem. Soc.*, **96**, 6771 (1974).

577. P. Crews and E. Kho, *J. Org. Chem.*, **40**, 2568 (1975).

578. N. Ichikawa, Y. Naya, and S. Enomoto, 18th Symposium of Terpene, Essential Oils and Aromatic Chemistry, 1972, Abstracts p. 67.

579. O. J. McConnell and W. Fenical, *J. Org. Chem.*, **43**, 4238 (1978); F. X. Woolard, R. E. Moore, D. Van Engen, and J. Clardy, *Tetrahedron Lett.*, 2367 (1978).

580. A. F. Thomas, *Helv. Chim. Acta*, **56**, 1800 (1973).

581. W. Kreiser, W. Haumesser, and A. F. Thomas, *Helv. Chim. Acta*, **57**, 164 (1974).

582. O. P. Vig, S. D. Sharma, S. S. Bari, and M. Lal, *Ind. J. Chem., B*, **14**, 932 (1976).

583. K. Yoshihara and Y. Hirose, *Bull. Chem. Soc. Japan*, **51**, 653 (1978).

584. S. W. Pelletier and N. V. Mody, *J. Org. Chem.*, **41**, 1069 (1976).

585. J. Pedro de Souza and A. M. R. Gonçalves, *J. Org. Chem.*, **43**, 2068 (1978).

586. J. H. Tumlinson, R. C. Gueldner, D. D. Hardee, A. C. Thompson, P. A. Hedin, and J. P. Minyard, *J. Org. Chem.*, **36**, 2616 (1971).

587. O. P. Vig, B. Ram, and J. Kaur, *J. Ind. Chem. Soc.*, **49**, 1181 (1972).

588. J. H. Babler and T. R. Mortell, *Tetrahedron Lett.*, 669 (1972).

589. A. I. Meyers, A. Nabeya, H. W. Adickes, I. R. Politzer, G. R. Malone, A. C. Kovelesky, R. L. Nolen, and R. C. Portnoy, *J. Org. Chem.*, **38**, 36 (1973).

590. R. H. Wollenberg and R. Peries, *Tetrahedron Lett.*, 297 (1979); for the preparation of the reagent, see R. H. Wollenberg, K. F. Albizati, and R. Pereis, *J. Am. Chem. Soc.*, **99**, 7365 (1977).

591. K. Tanaka, R. Tanikaga, and A. Kaji, *Chem. Lett.*, 917 (1976); K. Tanaka, N. Yamagishi, R. Tanikaga, and A. Kaji, *Bull. Chem. Soc. Japan*, **52**, 3619 (1979).

592. R. H. Bedoukian and J. Wolinsky, *J. Org. Chem.*, **40**, 2154 (1975).

593. K. Tanaka and Y. Matsubara, *Nippon Kagaku Kaishi*, 922 (1977).

594. W. Hoffmann and E. Müller, Ger. Offen., 2,152,193.

595. P. C. Traas, H. Boelens, and H. J. Takken, *Rec. Trav. Chim.*, **95**, 308 (1976).

596. P. C. Traas, H. Boelens, and H. J. Takken, *Synth. Comm.*, **6**, 489 (1976).

597. J. Garnero, P. Buil, D. Joulain, and R. Tabacchi, 7th Int. Cong. Ess. Oils, Kyoto, Oct. 1977, paper 113.

598. B. Corbier and P. Teisseire, *Recherches*, **19**, 253 (1974).

599. S. J. Rhoads and C. F. Brandenburg, *J. Am. Chem. Soc.*, 93, 5805 (1971); S. J. Rhoads and J. M. Watson, *J. Am. Chem. Soc.*, 93, 5815 (1971).

600. P. A. Wender and M. P. Filosa, *J. Org. Chem.*, 41, 3490 (1976).

601. H. M. R. Hoffmann, *Angew. Chem. Int. Ed.*, 12, 819 (1973).

602. R. Chidgey and H. M. R. Hoffmann, *Tetrahedron Lett.*, 2633 (1977); H. M. R. Hoffmann and R. Chidgey, *Tetrahedron Lett.*, 85 (1978).

603. N. Shimizu and Y. Tsuno, *Chem. Lett.*, 103 (1979).

604. S. Hashimoto, A. Itoh, Y. Kitagawa, H. Yamamoto, and H. Nozaki, *J. Am. Chem. Soc.*, 99, 4192 (1977).

605. T. Nozake, K. Takase, and M. Ogata, *Chem. Ind.*, 1070 (1957).

606. T. Nozoe, S. Seto, K. Kikuchi, T. Mukai, S. Matsumoto, and M. Murose, *Proc. Jap. Acad.*, 26 [7], 43 (1950) for thujaplicins; T. Nozoe, T. Mukai, and T. Asao, *Bull. Chem. Soc. Japan*, 33, 1452 (1960) for β-dolabrin.

607. J. W. Cook, R. A. Raphael, and A. I. Scott, *J. Chem. Soc.*, 695 (1951).

608. W. von E. Doering and L. H. Knox, *J. Am. Chem. Soc.*, 75, 297 (1953).

609. T. Asao, T. Machiguchi, T. Kitamura, and Y. Kitahara, *Chem. Comm.*, 89 (1970).

610. K. Tanaka and A. Yoshikoshi, *Tetrahedron*, 27, 4889 (1971).

611. R. Noyori, Y. Hayakawa, M. Funakura, H. Takaya, S. Murai, R. Kobayashi, and S. Tsutsumi, *J. Am. Chem. Soc.*, 94, 7202 (1972).

612. R. Noyori, S. Makino, T. Okita, and Y. Hayakawa, *J. Org. Chem.*, 40, 806 (1975).

613. Y. Hayakawa, M. Sakai, and R. Noyori, *Chem. Lett.*, 509 (1975).

614. H. Takaya, Y. Hayakawa, S. Makino, and R. Noyori, *J. Am. Chem. Soc.*, 100, 1778 (1978).

615. T. Saito, A. Itoh, K. Oshina, and H. Nozaki, *Tetrahedron Lett.*, 3519 (1979). This method has also been used to make α-terpinyl chloride from nerol.

616. M. Franck-Neumann, F. Brion, and D. Martina, *Tetrahedron Lett.*, 5033 (1978).

617. D. A. Evans, D. J. Hart, and P. M. Koelsch, *J. Am. Chem. Soc.*, 100, 4593 (1978).

618. T. L. Macdonald, *J. Org. Chem.*, 43, 3621 (1978).

619. N. Dennis, A. R. Katritzky, and Y. Takeuchi, *Angew. Chem. Int. Ed.*, 15, 1 (1976).

620. Y. Tamura, T. Saito, H. Kiyokawa, L. C. Chen, and H. Ishibashi, *Tetrahedron Lett.*, 4075 (1977).

621. W. F. Erman, R. S. Treptow, P. Bazukis, and E. Wenkert, *J. Am. Chem. Soc.*, 93, 657 (1971).

622. C. Capellini, A. Corbella, P. Gariboldi, and G. Jommi, *Gazz. Chim. Ital.*, 107, 171 (1977).

623. S. P. Acharya, H. C. Brown, A. Suzuki, S. Nozawa, and M. Itoh, *J. Org. Chem.*, 34, 3015 (1969).

624. D. Whittaker and D. V. Banthorpe, *Chem. Rev.*, 72, 305 (1972).

625. M. Albert-Puleo, *Econ. Bot.*, 32, 65 (1978).

626. V. Hach, R. W. Lockhart, E. C. McDonald, and D. M. Cartlidge, *Can. J. Chem.*, 49, 1762 (1971).

627. C. Alexandre and F. Rouessac, *Bull. Chem. Soc. Japan*, 45, 2241 (1972).

628. Y. Gaoni, *Tetrahedron*, 28, 5525 (1972).

629. M. Higo, H. Toda, K. Suzuki, and Y. Nishida, *Ger. Offen.*, 2,814,558 (1978).

630. S. Benayache, C. Fréjaville, R. Jullien, and M. Wanat, *Rev. Ital. Essenze, Profumi, Piante Off., Aromat., Syndets, Saponi, Cosmet., Aerosols*, **60**, 118 (1978).

631. C. W. Jefford, A. Roussel, and S. M. Evans, *Helv. Chim. Acta*, **58**, 2151 (1975).

632. S. D. Sharma and R. C. Aggarwal, *Ind. J. Chem., B*, **15**, 950 (1977).

633. J. J. C. Scheffer, A. Koedam, M. J. M. Gijbels, and A. Baerheim Svendsen, *Pharm. Weekbl.*, **111**, 1309 (1976).

634. R. H. Shapiro and M. J. Heath, *J. Am. Chem. Soc.*, **89**, 5734 (1967).

635. G. M. Nano, C. Bicchi, C. Frattini, and M. Gallino, *Planta Med.*, **30**, 211 (1976); *Chim. Ind. (Milan)*, **59**, 123 (1977).

636. M. A. Johnson and M. P. Fleming, *Can. J. Chem.*, **57**, 318 (1979). This paper also corrects the earlier reports of the NMR spectra of the bornanediols.

637. A. Matsuo, Y. Uchio, M. Nakayama, Y. Matsubara, and S. Hayashi, *Tetrahedron Lett.*, 4219 (1974); Y. Uchio, *Bull. Chem. Soc. Japan*, **51**, 2342 (1978).

638. P. Lipp, *Chem. Ber.*, **80**, 165 (1947).

639. M. Gaitonde, P. A. Vatakencherry, and Sukh Dev, *Tetrahedron Lett.*, 2007 (1964); see also S. N. Suryawanshi and U. R. Nayak, *Tetrahedron Lett.*, 269 (1979), for the use of iodine chloride.

640. H. G. Richey, Jr., T. J. Garbacik, D. L. Dull, and J. E. Grant, *J. Org. Chem.*, **29**, 3095 (1964); B. H. Jennings and G. B. Herschbach, *J. Org. Chem.*, **30**, 3902 (1965); H. G. Richey, Jr., J. E. Grant, T. J. Garbacik, and D. L. Dull, *J. Org. Chem.*, **30**, 3909 (1965). These papers, particularly the first, give a full account of the discussions about the earlier incorrect structures.

641. M. Julia, D. Mansuy, and P. Detraz, *Tetrahedron Lett.*, 2141 (1976).

642. M. S. Allen, N. Darby, P. Salisbury, and T. Money, *J.C.S. Chem. Comm.*, 358 (1977); *Tetrahedron Lett.*, 2255 (1978).

643. N. Darby, N. Lamb, and T. Money, *Can. J. Chem.*, **57**, 742 (1979).

644. M. de Botton, *Compt.Rend. Acad. Sci. (C)*, **260**, 4783 (1965).

645. G. Buchbauer, *Monatsh. Chem.*, **109**, 3 (1978).

646. J. M. Midgley, W. B. Whalley, G. Buchbauer, G. W. Hana, H. Koch, P. J. Roberts, and G. Ferguson, *J.C.S. Perkin I*, 1312 (1978).

647. G. W. Hana and H. Koch, *Chem. Ber.*, **111**, 2527 (1978).

648. J. C. Fairlie, G. L. Hodgson, and T. Money, *J.C.S. Perkin I*, 2109 (1978).

649. G. L. Lange and J.-M. Conia, *Nouv. J. Chim.*, **1**, 189 (1977).

650. S. Wolff and W. C. Agosta, *Tetrahedron Lett.*, 2845 (1979).

651. R. G. Lawton, *J. Am. Chem. Soc.*, **83**, 2399 (1961); P. D. Bartlett and S. Bank, *J. Am. Chem. Soc.*, **83**, 2591 (1961).

652. J. L. Marshall, *Tetrahedron Lett.*, 753 (1971).

653. P. H. Boyle, W. Cocker, D. H. Grayson, and P. V. R. Shannon, *Chem. Comm.*, 395 (1971); *J. Chem. Soc.(C)*, 2136 (1971).

654. G. E. Gream and D. Wege, *Tetrahedron Lett.*, 535 (1964); G. E. Gream, D. Wege, and M. Mular, *Austral. J. Chem.*, **27**, 567 (1974).

655. M. Barthélémy and Y. Bessière-Chrétien, *Bull. Soc. Chim. France*, 1703 (1974); M. Barthélémy, A. Gianfermi and Y. Bessière-Chrétien, *Helv. Chim. Acta*, **59**, 1894 (1976).

656. J. De Pascual Teresa, I. Sanchez Bellido, J. F. Santos Barrueco, and Y. A. Sanfeliciano

Martin, *An. Quím.*, 74, 950 (1978).

657.   H. Indyk and D. Whittaker, *J.C.S. Perkin II*, 313, 646 (1974); P. I. Meikle, J. R. Salmon and D. Whittaker, *J.C.S. Perkin II*, 23 (1972); P. I. Meikle and D. Whittaker, *J.C.S. Chem. Comm.*, 789 (1972), and other papers.

658.   T. Gibson and W. F. Erman, *J. Am. Chem. Soc.*, 91, 4771 (1969).

659.   N. Bosworth and P. D. Magnus, *Chem. Comm.*, 618 (1971); *J.C.S. Perkin I*, 943 (1972); 2319 (1973).

660.   Y. Bessière-Chrétien and C. Grison, *Compt. Rend. Acad. Sci. (C)*, 275, 503 (1972); C. Grison and Y. Bessière-Chrétien, *Bull. Soc. Chim. France*, 4570 (1972).

661.   R. Noyori, M. Nishizawa, F. Shimizu, Y. Hayakawa, K. Maruoka, S. Hashimoto, H. Yamamoto, and H. Nozaki, *J. Am. Chem. Soc.*, 101, 221 (1979).

662.   E. Gildermeister and F. Hoffmann, *Die Aetherischen Oele*, Vol. 4, Akademie-Verlag, Berlin, 1956, p. 571.

663.   E. Demole, C. Demole, and P. Enggist, *Helv. Chim. Acta*, 59, 737 (1976).

664.   F. Müller, *Arch. Pharm.*, 238, 372 (1900); see also *Schimmelberichte*, Oct. 1910, p. 125.

665.   S. Beckmann and H. Geiger, *Chem. Ber.*, 94, 1905 (1961).

666.   T. Saeed, P. J. Sandra, and M. J. E. Verzele, *Phytochemistry*, 17, 1433 (1978).

667.   W. E. Barnett and J. C. McKenna, *Tetrahedron Lett.*, 227 (1971).

668.   H. Pommer and A. Nürrenbach, *Pure Appl. Chem.*, 43, 527 (1975).

669.   M. Baumann and W. Hoffmann, *Annalen*, 743 (1979).

670.   S. A. Monti and S. D. Larsen, *J. Org. Chem.*, 43, 2282 (1978).

671.   Y. Castanet and F. Petit, *Tetrahedron Lett.*, 3221 (1979).

672.   J. Hochmannová, L. Novotný, and V. Herout, *Coll. Czech. Chem. Comm.*, 27, 2711 (1962).

673.   K. Vokáč, Z. Samek, V. Herout, and F. Šorm, *Tetrahedron Lett.*, 1665 (1972).

674.   P. T. Lansbury and R. M. Boden, *Tetrahedron Lett.*, 5017 (1973).

675.   W. Kreiser, L. Janitschke, and W. S. Sheldrick, *J.C.S. Chem. Comm.*, 269 (1977).

676.   X. Creary, F. Hudock, M. Keller, J. F. Kerwin, Jr., and P. J. Dinnocenzo, *J. Org. Chem.*, 42, 409 (1977).

677.   R. B. Boar, D. W. Hawkins, J. F. McGhie, and D. H. R. Barton, *J.C.S. Perkin I*, 654 (1973).

678.   W. Kreiser, L. Janitschke, W. Voss, L. Ernst, and W. S. Sheldrick, *Chem. Ber.*, 112, 397 (1979).

679.   W. Kresier and L. Janitschke, *Tetrahedron Lett.*, 601 (1978); *Chem. Ber.*, 112, 408 (1979).

680.   O. R. Rodig and R. J. Sysko, *J. Org. Chem.*, 36, 2324 (1971).

681.   J. E. Oliver, *J. Label. Comp. Radiopharm.*, 13, 349 (1977).

682.   W. L. Meyer, C. E. Capshew, J. H. Johnson, A. R. Klusener, A. P. Lobo, and R. N. McCarty, *J. Org. Chem.*, 42, 527 (1977).

683.   G. Buchbauer and H. Koch, Chem. Ber., 111, 2533 (1978).

684.   F. C. Brown, E. Casadevall, P. Metzger, and D. G. Morris, *J. Chem. Res. (S)*, 335, *(M)*, 9588 (1977); N. H. Werstiuk, R. Taillefer and S. Banerjee, *Can. J. Chem.*, 56, 1140, 1148 (1978); see also S. Wolfe, H. B. Schlegel, I. G. Csizmadia, and F. Bernardi, *Can. J. Chem.*, 53, 3365 (1975).

685. R. Muneyuki, Y. Yoshimura, and K. Tori, *Chem. Lett.*, 49 (1979).

686. J. Leroy and C. Wakselman, *J.C.S. Perkin I*, 1224 (1978).

687. M. Kaneda and Y. Itaka, *Acta Cryst., B*, **28**, 1411 (1972) and Ref. 689, and K. Yamasaki, M. Kaneda, and O. Tanaka, *Tetrahedron Lett.*, 3965 (1976).

688. M. Kaneda, Y. Itaka, and S. Shibata, *Tetrahedron*, **28**, 4309 (1972)

689. M. Miyakado, N. Ohno, H. Hirai, H. Yoshioka, and T. J. Mabry, *Phytochemistry*, **13**, 2881 (1974).

690. J. T. Pinhey and I. A. Southwell, *Austral. J. Chem.*, **24**, 1311 (1971).

691. A. Matsuo, Y. Uchio, M. Nakayama, and S. Hayashi, *Bull. Chem. Soc. Japan*, **46**, 1565 (1973); A. Matsuo, M. Nakayama, T. Nakamoto, Y. Uchio, and S. Hayashi, *Agric. Biol. Chem.*, **37**, 925 (1973).

692. J. De Pascual Teresa, A. F. Barrero, E. Caballero, and M. Medarde, *An. Quím.*, **75**, 323 (1979).

693. E. Tsankova and I. Ognyanov, *Riv. Ital. Essenze, Profumi, Piante Off., Aromat., Syndets, Saponi, Cosmet., Aerosols*, **58**, 502 (1976).

694. Y. Bessière-Chrétien and C. Grison, *Bull. Soc. Chim. France*, 3103 (1970).

695. M. T. Thomas and A. G. Fallis, *Tetrahedron Lett.*, 4687 (1973); *J. Am. Chem. Soc.*, **98**, 1227 (1976).

696. M. T. Thomas, E. G. Breitholle, and A. G. Fallis, *Synth. Comm.*, **6**, 113 (1976).

697. D. V. Banthorpe and D. Whittaker, *Chem. Rev.*, **66**, 643 (1966).

698. S. D. Larsen and S. A. Monti, *J. Am. Chem. Soc.*, **99**, 8015 (1977).

699. W. Kirmse and W. Gruber, *Chem. Ber.*, **105**, 2764 (1972).

700. W. Kirmse and G. Arend, *Chem. Ber.*, **105**, 2738, 2746 (1972); W. Kirmse and R. Siegfried, *Chem. Ber.*, **105**, 2754 (1972).

701. N. Proth, *Rev. Tech. Luxemb.*, **70**, 23 (1978).

702. G. Zweifel and C. C. Whitney, *J. Org. Chem.*, **31**, 4178 (1966).

703. Y. Bessière, E. Reca, F. Chatzopoulos-Ouar, and G. Boussac, *J. Chem. Res. (S)*, 302, *(M)*, 3501 (1977).

704. G. L. Kaiser, U.S. Pat., 3,974,102; 3,974,103; 4,000,208; P. M. Koppel and W. I. Taylor, U.S. Pat., 3,987,121.

705. K. Mori, N. Mitzumachi, and M. Matsui, *Agric. Biol. Chem.*, **40**, 1611 (1976).

706. K. Mori, *Agric. Biol. Chem.*, **40**, 415 (1976).

707. J. M. Brand, J. W. Bracke, A. J. Markovetz, D. L. Wood, and L. E. Browne, *Nature*, **254**, 136 (1975).

708. T. Ishida, Y. Asakawa, M. Okano, and T. Aratani, *Tetrahedron Lett.*, 2437 (1977).

709. V. Garsky, D. F. Koster, and R. T. Arnold, *J. Am. Chem. Soc.*, **96**, 4207 (1974).

710. Y. Bessière and H. Ennaffati, unpublished work; H. Ennaffati, thèse de docteur de IIIème cycle, Université P. et M. Curie, Paris 1974.

711. G. W. Shaffer and M. Pesaro, *J. Org. Chem.*, **39**, 2489 (1974).

712. E. V. Lassak and I. A. Southwell, *Austral. J. Chem.*, **27**, 2061 (1974).

713. J. Endo and T. Nakamura, *Yakugaku Zasshi*, **98**, 789 (1978); [*Chem. Abstracts*, **89**, 135657c (1978)].

714. P. Brunetti, F. Fringuelli, and A. Taticchi, *Gazz. Chim. Ital.*, **107**, 433 (1977).

715. W. J. Gensler and P. H. Solomon, *J. Org. Chem.*, **38**, 2806 (1973).

716. B. Carlsson, H. Erdtman, A. Frank, and W. E. Harvey, *Acta Chem. Scand.*, 6, 690 (1952); H. Erdtman, W. E. Harvey, and J. G. Topliss, *Acta Chem. Scand.*, 10, 1381 (1956); T. Norin, *Ark. Kemi*, 22, 123 (1964).

717. L. Borowiecki and W. Zacharewicz, *Rocz. Chem.*, 37, 1143 (1963); K. Gollnick, and G. Schade, *Tetrahedron*, 22, 133 (1966).

718. S. Yamada, N. Takamura, and T. Mizoguchi, *Chem. Pharm. Bull. Tokyo*, 23, 2539 (1975).

719. W. Cocker, N. W. A. Geraghty, and D. H. Grayson, *J.C.S. Perkin I*, 1370 (1978).

720. C. W. Jefford and A. F. Boschung, *Helv. Chim. Acta*, 59, 962 (1976).

721. E. Cros, I. Elphimoff-Felkin, and P. Sarda, *Compt. Rend. Acad. Sci. (C)*, 286, 261 (1978).

722. R. Kaiser and D. Lamparsky, *Helv. Chim. Acta*, 59, 1797 (1976).

723. Y.-R. Naves, *Riv. Ital. Essenze, Profumi, Pianti Off., Aromat. Syndets, Saponi, Cosmet., Aerosols*, 60, 265 (1978).

724. G. Büchi, E. sz. Kováts, P. Enggist, and G. Uhde, *J. Org. Chem.*, 33, 1227 (1968).

725. A. Takeda, K. Shinhama, and S. Tsuboi, *Bull. Chem. Soc. Japan*, 50, 1903 (1977).

726. Nguên Dinh Ly and M. Schlosser, *Helv. Chim. Acta*, 60, 2085 (1977).

727. J. Štrogl, M. Janda, and I. Stibor, *Coll. Czech. Chem. Comm.*, 35, 3478 (1970).

728. A. J. Birch and J. Slobbe, *Tetrahedron Lett.*, 2079 (1976).

729. A. F. Thomas, *Helv. Chim. Acta*, 53, 605 (1970).

730. O. P. Vig, A. K. Vig, V. K. Handa, and S. D. Sharma, *J. Ind. Chem. Soc.*, 51, 900 (1974).

731. R. C. Cookson, K. H. Schulte-Elte, and A. Hauser, Ger. Offen. 28 27 383 (June 22, 1978).

732. K. H. Schulte-Elte, Swiss Pat., 603,614 (Oct. 6, 1975). The alcohol **701** had already been described by H. Diesselnkötter and P. Kurtz, *Annalen*, 679, 26 (1964).

733. L. Silver, *Chem. Eng. Prog.*, 63, 43 (1967); F. Lorenz, *Chem. Eng. Prog., Loss Prevention*, i (1967); G. V. Fesenko, E. G. Koreshkova, and C. P. Chernysh, *Khim. Farm. Zhur.*, 9, 46 (1975). It seems that the (Z)-isomer of **704** is the most dangerous.

734. P. L. N. Rao, *J. Sci. Ind. Res. (India)*, 7B, 11 (1948).

734a. N. S. Mikhailova, K. S. Rybalko, and V. I. Sheichenko, *Khim. Prir. Soedin.*, 322 (1979).

735. T. L. Peppard and J. A. Elvidge, *Chem. Ind.*, 552 (1979).

736. K. Ina and I. Suzuki, *Nippon Nogei Kagaku Kaishi*, 45, 113 (1971) [*Chem. Abstracts*, 75, 91,228 (1971)].

737. B. J. Wilson, M. R. Boyd, T. M. Harris, and D. T. C. Yang, *Nature*, 231, 52 (1971); L. T. Burka, L. Kuhnert, B. J. Wilson, and T. M. Harris, *J. Am. Chem. Soc.*, 99, 2302 (1977).

738. R. F. Abdulla and K. H. Fuhr, *J. Org. Chem.*, 43, 4248 (1978).

739. A. Hoppmann and P. Weyerstahl, *Tetrahedron*, 34, 1723 (1978).

740. P. Gosselin, S. Masson, and A. Thuillier, *J. Org. Chem.*, 44, 2807 (1979).

741. A. Zamojski and T. Kozluk, *J. Org. Chem.*, 42, 1089 (1977).

742. T. Kitamura, Y. Kawakami, T. Imagawa, and M. Kawanisi, *Synth. Comm.*, 7, 521 (1977).

743. M. Matsumoto and K. Kondo, *J. Am. Chem. Soc.*, **40**, 2259 (1975) gives a number of alternative diene starting materials related to myrcene.

744. K. Kondo and M. Matsumoto, *Tetrahedron Lett.*, 4363 (1976).

745. J. A. Turner and W. Herz, *J. Org. Chem.*, **42**, 1900 (1977).

746. S. Katsumura, A. Ohsuka, and M. Kotake, *Heterocycles*, **10**, 87 (1978).

747. S. Takahashi, *Synth. Comm.*, **6**, 331 (1976).

748. K. Inomata, S. Aoyama, and H. Kotake, *Bull. Chem. Soc. Japan*, **51**, 930 (1978).

749. K. Inomata, M. Sumita, and H. Kotake, *Chem. Lett.*, 709 (1979).

750. S. Wakayama, S. Namba, and M. Ohno, *Bull. Chem. Soc. Japan*, **43**, 3319 (1970); *Nippon Kagaku Zasshi*, **92**, 256 (1971); Japan Kokai 72 16,300.

751. P. Naegeli and G. Weber, *Tetrahedron Lett.*, 959 (1970); A. F. Thomas and M. Ozainne, *Helv. Chim. Acta*, **57**, 2062 (1974).

752. S. Wakayama, S. Namba, K. Hosoi, and M. Ohno, *Bull. Chem. Soc. Japan*, **44**, 875 (1971); **46**, 3183 (1973); S. Wakayama and S. Namba, *Bull. Chem. Soc. Japan*, **47**, 1293 (1974).

753. O. P. Vig, R. S. Bhatt, J. Kaur, and J. C. Kapur, *J. Ind. Chem. Soc.*, **50**, 37 (1973).

754. A. F. Thomas and R. Dubini, *Helv. Chim. Acta*, **57**, 2066 (1974).

755. A. F. Thomas and R. Dubini, *Helv. Chim. Acta*, **57**, 2084 (1974).

756. R. Chidgey and H. M. R. Hoffmann, *Tetrahedron Lett.*, 1001 (1978).

757. T. Kametani, H. Nemoto, and K. Fukumoto, *Heterocycles*, **6**, 1365 (1977).

758. A. S. R. Anjaneyulu, M. Bapuji, L. R. Row, and A. Sree, *Phytochemistry*, **18**, 463 (1979).

759. J. R. Sierra, W.-D. Woggon, and H. Schmid, *Experientia*, **32**, 142 (1976).

760. M. G. Peter, W.-D. Woggon, C. Schlatter, and H. Schmid, *Helv. Chim. Acta*, **60**, 740 (1977).

761. T. Okazaki, A. Ohsuka, and M. Kotake, *Nippon Kagaku Kaishi*, 359 (1973); [*Chem. Abstracts*, **78**, 136 423 (1973)].

762. D. R. Adams and S. P. Bhatnagar, *Int. Flavours and Food Additives*, **6**, 185 (1975).

763. D. R. Adams, personal communication; W. Renold and A. F. Thomas, unpublished work.

764. Y. Matsuki, M. Kodama, and S. Itô, *Tetrahedron Lett.*, 4081 (1979).

765. M. S. R. Nair and S. T. Carey, *Tetrahedron Lett.*, 1655 (1975).

766. T. Reffstrup and P. M. Boll, *Tetrahedron Lett.*, 1903 (1976); *Acta Chem. Scand., B*, **30**, 613 (1976).

767. S. C. Taneja, K. L. Dhar, and C. K. Atal, *J. Org. Chem.*, **43**, 997 (1978).

768. J. S. Patel, H. H. Mathur, and S. C. Bhattacharyya, *Ind. J. Chem. B*, **16**, 188 (1978).

769. K. Narasaka, K. Soai, Y. Aikawa, and T. Mukaiyama, *Bull. Chem. Soc. Japan*, **49**, 779 (1976).

770. T. Shono, A. Ikeda, and Y. Kimura, *Tetrahedron Lett.*, 3599 (1971).

771. M. Mühlstädt and C. Duschek, *Z. Chem.*, **11**, 459 (1971).

772. H. Ishikawa, S. Ikeda, and T. Mukaiyama, *Chem. Lett.*, 1051 (1975).

773. J. H. P. Tyman and B. J. Willis, *Tetrahedron Lett.*, 4507 (1970).

774. K. H. Schulte-Elte, Int. Pat. Appl., WO 79/00509 (Aug. 9, 1979).

Index